NEUROMETHODS

Series Editor
Wolfgang Walz
University of Saskatchewan
Saskatoon, SK, Canada

For further volumes:
http://www.springer.com/series/7657

Neuronal Network Analysis

Concepts and Experimental Approaches

Edited by

Tommaso Fellin

Department of Neuroscience and Brain Technologies, Italian Institute of Technology, Genova, Italy

Michael Halassa

Department of Psychiatry, Massachusetts General Hospital, Boston, MA, USA

Humana Press

Editors
Tommaso Fellin, Ph.D.
Department of Neuroscience
and Brain Technologies
Italian Institute of Technology
Genova, Italy

Michael Halassa, M.D., Ph.D.
Department of Psychiatry
Massachusetts General Hospital
Boston, MA, USA

ISBN 978-1-4939-5899-3 ISBN 978-1-61779-633-3 (eBook)
DOI 10.1007/978-1-61779-633-3
Springer New York Dordrecht Heidelberg London

Printed on acid-free paper

Humana Press is part of Springer Science+Business Media (www.springer.com)

Preface to the Series

Under the guidance of its founders Alan Boulton and Glen Baker, the Neuromethods series by Humana Press has been very successful since the first volume appeared in 1985. In about 17 years, 37 volumes have been published. In 2006, Springer Science+Business Media made a renewed commitment to this series. The new program focuses on methods that are either unique to the nervous system and excitable cells or which need special consideration to be applied to the neurosciences. The program strikes a balance between recent and exciting developments like those concerning new animal models of disease, imaging, in vivo methods, and more established techniques. These include immunocytochemistry and electrophysiological technologies. New trainees in neurosciences still need a sound footing in these older methods in order to apply a critical approach to their results. The careful application of methods is probably the most important step in the process of scientific inquiry. In the past, new methodologies led the way in developing new disciplines in the biological and medical sciences. For example, physiology emerged out of anatomy in the nineteenth century by harnessing new methods based on the newly discovered phenomenon of electricity. Nowadays, the relationships between disciplines and methods are more complex. Methods are now widely shared between disciplines and research areas. New developments in electronic publishing also make it possible for scientists to download chapters or protocols selectively within a very short time of encountering them. This new approach has been taken into account in the design of individual volumes and chapters in this series.

Neuherberg, Germany *Wolfgang Walz*

Preface

The study of how population dynamics emerge from the activity of single neurons is a necessary step to understanding higher functions of the brain. In the last few years, technologies that allow for monitoring and manipulating neuronal networks at a single cell resolution have been rapidly expanding. While two-photon microscopy and in vivo electrophysiology have been improved to monitor large population of neurons in awake animals, the introduction of methods to manipulate neuronal circuits (e.g., optogenetics in animals and transcranial magnetic stimulation in humans) allows for causal testing of functional hypotheses in neuroscience. The combination of these tools has transformed neuroscience into a field where reverse engineering of brain function is possible. In this volume, we provide an overview of some of the most advanced techniques for the study of neuronal circuits, highlighting current technical limitations and providing a vision of what there is yet to come.

This volume is composed of six interdependent sections, each containing an introductory chapter that gives a broad overview of a focused topic. These introductory chapters are followed by methodological ones, which, while providing a brief conceptual introduction to the specific technology discussed, contain an extensive protocol on how the experiments are performed along with invaluable practical advice. Part I discusses optical approaches to monitoring network function, focusing largely on two-photon Ca^{2+} and voltage-sensitive dye imaging in vivo. The use of structured light for the illumination of neuronal networks with complex spatio-temporal patterns is also discussed. Part II discusses extracellular and intracellular electrophysiological approaches to studying network function in brain slice preparation, with focus on synaptic plasticity, high-resolution investigation of cellular and synaptic mechanisms of oscillations, dynamic clamp technology and fast perfusion methods to studying ligand-gated ion channels. Part III transitions to a discussion of in vivo electrophysiology with chapters on laminar recordings/current source density analysis in neocortex, patch-clamp recordings in awake head-fixed rodents, multiunit recordings in the neocortex of freely behaving animals using single wires, the use of Bayesian statistics in decoding of hippocampal replay, and high-speed videography to monitor active sensing. Part IV discusses optogenetics with an overview of recently developed molecular tools and their integrated use with electrophysiology to manipulate and monitor network function. Part V uniquely combines a number of chapters on the morpho-functional study of neural circuit makeup and connectivity. Finally, Part VI discusses approaches to noninvasively studying neuronal networks in humans using transcranial magnetic stimulation for causal manipulation along with electro- and magneto-encephalography for monitoring of population activity at the millisecond timescale.

This volume is not simply a catalogue of the latest technology on neural network studies, but rather an overview of these exciting methods along with the vision of current and future leaders of circuit neuroscience. The combination of structural and functional dissection methods, along with correlative and causal tools, provides a unique practical introduction to this exciting field. We anticipate that it will be useful not only for early career neuroscientists, but also for experienced neuroscientists, to provide a practical framework for studying the brain as an organ while maintaining an appreciation for its cellular diversity.

Genova, Italy *Tommaso Fellin*
Boston, MA, USA *Michael Halassa*

Contents

Preface to the Series . . . . *v*
Preface . . . . *vii*
Contributors . . . . *xi*

PART I OPTICAL APPROACHES

Optical Interrogation of Neural Circuits 3
Tycho M. Hoogland and Peter Saggau

In Vivo Functional Imaging of the Olfactory Bulb at Single-Cell Resolution 21
Stefan Fink, Yury Kovalchuk, Ryota Homma, Bernd Schwendele, Stephan Direnberger, Lawrence B. Cohen, Oliver Griesbeck, and Olga Garaschuk

Two-Photon Imaging of Neural Activity in Awake, Head-Restrained Mice 45
Martin Wienisch, David G. Blauvelt, Tomokazu F. Sato, and Venkatesh N. Murthy

Spatial Light Modulators for Complex Spatiotemporal Illumination of Neuronal Networks 61
Francesco Difato, Marco Dal Maschio, Riccardo Beltramo, Axel Blau, Fabio Benfenati, and Tommaso Fellin

Voltage-Sensitive Dye Imaging of Cortical Function In Vivo 83
Eugene F. Civillico

PART II ELECTROPHYSIOLOGICAL APPROACHES IN VITRO

An Introduction to In Vitro Slice Approaches for the Study of Neuronal Circuitry 103
Carmen Varela, Daniel A. Llano, and Brian B. Theyel

Extracellular Recordings of Synaptic Plasticity and Network Oscillations in Hippocampal Slices 127
Gaga Kochlamazashvili, Oleg Senkov, and Alexander Dityatev

Use of Dynamic-Clamp as a Tool to Reveal the Computational Properties of Single Neurons Embedded in Cortical Circuits 149
Alberto Bacci, Antonio Pazienti, and Michele Giugliano

Fast Perfusion Methods for the Study of Ligand-Gated Ion Channels 173
Andrea Barberis

PART III ELECTROPHYSIOLOGICAL APPROACHES IN VIVO

Electrophysiological Approaches for Studying Neuronal Circuits In Vivo 191
George Dragoi

Determination of Cortical Circuit Function Using Current Source-Density Analysis In Vivo 205
Michael J. Higley

Intracellular Whole-Cell Patch-Clamp Recordings of Cortical Neurons in Awake Head-Restrained Mice ... 219
Sylvain Crochet

Investigating Sleep Homeostasis with Extracellular Recording of Multiunit Activity from the Neocortex in Freely Behaving Rats ... 237
Vladyslav V. Vyazovskiy, Umberto Olcese, and Giulio Tononi

Analysis of Hippocampal Memory Replay Using Neural Population Decoding ... 259
Fabian Kloosterman

High-Speed Videography of Embodied Active Sensing in the Rodent Whisker System ... 283
Jason T. Ritt

PART IV OPTOGENETICS

Light-Activated Ion Pumps and Channels for Temporally Precise Optical Control of Activity in Genetically Targeted Neurons ... 305
Brian Y. Chow, Xue Han, Jacob G. Bernstein, Patrick E. Monahan, and Edward S. Boyden

Integrated Optogenetic and Electrophysiological Dissection of Local Cortical Circuits In Vivo ... 339
Jessica A. Cardin

Combining Optical Stimulation with Extracellular Electrophysiology in Behaving Mice ... 357
Joshua H. Siegle

PART V MORPHO-FUNCTIONAL APPROACHES

Morphological Approaches to the Anatomical Dissection of Neuronal Circuits ... 375
David C. Lyon

The Isotropic Fractionator: A Fast, Reliable Method to Determine Numbers of Cells in the Brain or Other Tissues ... 391
Suzana Herculano-Houzel

Morpho-Functional Mapping of Cortical Networks in Brain Slice Preparations Using Paired Electrophysiological Recordings ... 405
Gabriele Radnikow, Robert Heinz Günter, Manuel Marx, and Dirk Feldmeyer

PART VI NEURONAL NETWORK ANALYSIS IN HUMANS

Combining Transcranial Magnetic Stimulation with Electroencephalography to Study Human Cortical Excitability and Effective Connectivity ... 435
Mario Rosanova, Silvia Casarotto, Andrea Pigorini, Paola Canali, Adenauer G. Casali, and Marcello Massimini

Biophysically Principled Computational Neural Network Modeling of Magneto-/Electro-Encephalography Measured Human Brain Oscillations ... 459
Stephanie R. Jones

Erratum ... *E1*

Index ... *487*

Contributors

ALBERTO BACCI • *European Brain Research Institute, Rome, Italy*
ANDREA BARBERIS • *Department of Neuroscience and Brain Technologies, Italian Institute of Technology, Genoa, Italy*
RICCARDO BELTRAMO • *Department of Neuroscience and Brain Technologies, Italian Institute of Technology, Genoa, Italy*
FABIO BENFENATI • *Department of Neuroscience and Brain Technologies, Italian Institute of Technology, Genoa, Italy*
JACOB G. BERNSTEIN • *MIT Media Lab, Department of Biological Engineering, McGovern Institute, Cambridge, MA, USA; Department of Brain and Cognitive Sciences, Massachusetts Institute of Technology, Cambridge, MA, USA*
AXEL BLAU • *Department of Neuroscience and Brain Technologies, Italian Institute of Technology, Genoa, Italy*
DAVID G. BLAUVELT • *Molecular and Cellular Biology, Harvard University, Cambridge, MA, USA; Harvard Medical School, Boston, MA, USA*
EDWARD S. BOYDEN • *MIT Media Lab, Department of Biological Engineering, McGovern Institute, Cambridge, MA, USA; Department of Brain and Cognitive Sciences, Massachusetts Institute of Technology, Cambridge, MA, USA*
PAOLA CANALI • *Department of Clinical Sciences "Luigi Sacco", University of Milan, Milan, Italy*
JESSICA A. CARDIN • *Department of Neurobiology, Yale University School of Medicine, New Haven, CT, USA*
ADENAUER G. CASALI • *Department of Clinical Sciences "Luigi Sacco", University of Milan, Milan, Italy*
SILVIA CASAROTTO • *Department of Clinical Sciences "Luigi Sacco", University of Milan, Milan, Italy*
EUGENE F. CIVILLICO • *CNS Drug Discovery and Development, Otsuka Maryland Medicinal Laboratories, Rockville, MD, USA*
BRIAN Y. CHOW • *MIT Media Lab, Department of Biological Engineering, McGovern Institute, Cambridge, MA, USA; Department of Brain and Cognitive Sciences, Massachusetts Institute of Technology, Cambridge, MA, USA*
LAWRENCE B. COHEN • *Department of Physiology, Yale University, New Haven, CT, USA*
SYLVAIN CROCHET • *INSERM/UCBL-U628, Integrated Physiology of Brain Arousal Systems, Claude Bernard University Lyon 1, Lyon, France; Laboratory of Sensory Processing, Brain Mind Institute, École Polytechnique Fédérale de Lausanne (EPFL), Lausanne, Switzerland*
MARCO DAL MASCHIO • *Department of Neuroscience and Brain Technologies, Italian Institute of Technology, Genoa, Italy*
FRANCESCO DIFATO • *Department of Neuroscience and Brain Technologies, Italian Institute of Technology, Genoa, Italy*
STEPHAN DIRENBERGER • *Max-Planck-Institute of Neurobiology, Martinsried, Germany*

ALEXANDER DITYATEV • *Department of Neuroscience and Brain Technologies, Italian Institute of Technology, Genoa, Italy*
GEORGE DRAGOI • *The Picower Institute for Learning and Memory, Massachusetts Institute of Technology, Cambridge, MA, USA*
DIRK FELDMEYER • *Department of Psychiatry, Psychotherapy and Psychosomatics, RWTH Aachen University, Aachen, Germany; Institute for Neuroscience and Medicine, INM-2, Research Centre Jülich, Jülich, Germany*
TOMMASO FELLIN • *Department of Neuroscience and Brain Technologies, Italian Institute of Technology, Genoa, Italy*
STEFAN FINK • *Department of Physiology II, University of Tuebingen, Tuebingen, Germany*
OLGA GARASCHUK • *Department of Physiology II, University of Tuebingen, Tuebingen, Germany*
MICHELE GIUGLIANO • *Department of Biomedical Sciences, University of Antwerp, Antwerp, Belgium; Brain Mind Institute, Swiss Federal Institute of Technology, Lausanne, Switzerland; Department of Computer Science, University of Sheffield, Sheffield, UK*
OLIVER GRIESBECK • *Max-Planck-Institute of Neurobiology, Martinsried, Germany*
ROBERT HEINZ GÜNTER • *Institute for Neuroscience and Medicine, INM-2, Research Centre Jülich, Jülich, Germany*
XUE HAN • *MIT Media Lab, Department of Biological Engineering, McGovern Institute, Cambridge, MA, USA; Department of Brain and Cognitive Sciences, Massachusetts Institute of Technology, Cambridge, MA, USA*
SUZANA HERCULANO-HOUZEL • *Instituto de Ciências Biomédicas, Universidade Federal do Rio de Janeiro, Rio de Janeiro, Brazil; Instituto Nacional de Neurociência Translacional, Ministério de Ciência e Tecnologia, Brasília, Brazil*
MICHAEL J. HIGLEY • *Department of Neurobiology, Program in Cellular Neuroscience, Neurodegeneration, and Repair, Yale School of Medicine, New Haven, CT, USA*
RYOTA HOMMA • *Department of Physiology, Yale University, New Haven, CT, USA*
TYCHO M. HOOGLAND • *Netherlands Institute for Neuroscience, Amsterdam, The Netherlands*
STEPHANIE R. JONES • *Massachusetts General Hospital, Athinoula A. Martinos Center for Biomedical Imaging, Charlestown, MA, USA*
FABIAN KLOOSTERMAN • *Department of Brain and Cognitive Sciences, Picower Institute for Learning and Memory, Massachusetts Institute of Technology, Cambridge, MA, USA*
GAGA KOCHLAMAZASHVILI • *Department of Neuroscience and Brain Technologies, Italian Institute of Technology, Genoa, Italy*
YURY KOVALCHUK • *Department of Physiology II, University of Tuebingen, Tuebingen, Germany*
DANIEL A. LLANO • *Picower Institute for Learning and Memory, Massachusetts Institute of Technology, Cambridge, MA, USA*
DAVID C. LYON • *Department of Anatomy & Neurobiology, School of Medicine, University of California, Irvine, CA, USA*

MANUEL MARX • *Institute for Neuroscience and Medicine, INM-2, Research Centre Jülich, Jülich, Germany*
MARCELLO MASSIMINI • *Department of Clinical Sciences "Luigi Sacco", University of Milan, Milan, Italy*
PATRICK E. MONAHAN • *MIT Media Lab, Department of Biological Engineering, McGovern Institute, Cambridge, MA, USA; Department of Brain and Cognitive Sciences, Massachusetts Institute of Technology, Cambridge, MA, USA*
VENKATESH N. MURTHY • *Molecular and Cellular Biology, Center for Brain Science, Harvard University, Cambridge, MA, USA*
UMBERTO OLCESE • *PERCRO Laboratory, Scuola Superiore Sant'Anna, Pisa, Italy*
ANTONIO PAZIENTI • *European Brain Research Institute, Rome, Italy*
ANDREA PIGORINI • *Department of Clinical Sciences "Luigi Sacco", University of Milan, Milan, Italy*
GABRIELE RADNIKOW • *Institute for Neuroscience and Medicine, INM-2, Research Centre Jülich, Jülich, Germany*
JASON T. RITT • *Department of Biomedical Engineering, Boston University, Boston, MA, USA*
MARIO ROSANOVA • *Department of Clinical Sciences "Luigi Sacco", University of Milan, Milan, Italy*
PETER SAGGAU • *Department of Neuroscience, Baylor College of Medicine, Houston, TX, USA*
TOMOKAZU F. SATO • *Molecular and Cellular Biology, Center for Brain Science, Harvard University, Cambridge, MA, USA*
BERND SCHWENDELE • *Department of Physiology II, University of Tuebingen, Tuebingen, Germany*
OLEG SENKOV • *Department of Clinical Neurobiology, German Cancer Research Center (DKFZ), Heidelberg, Germany*
JOSHUA H. SIEGLE • *Department of Brain and Cognitive Sciences, Massachusetts Institute of Technology, Cambridge, MA, USA*
BRIAN B. THEYEL • *Picower Institute for Learning and Memory, Massachusetts Institute of Technology, Cambridge, MA, USA*
GIULIO TONONI • *Department of Psychiatry, University of Wisconsin-Madison, Madison, WI, USA*
CARMEN VARELA • *Picower Institute for Learning and Memory, Massachusetts Institute of Technology, Cambridge, MA, USA*
VLADYSLAV V. VYAZOVSKIY • *Department of Psychiatry, University of Wisconsin-Madison, Madison, WI, USA*
MARTIN WIENISCH • *Molecular and Cellular Biology, Center for Brain Science, Harvard University, Cambridge, MA, USA*

Part I

Optical Approaches

Neuromethods (2012) 67: 3–20
DOI 10.1007/7657_2011_17

Published online: 1 December 2011

Optical Interrogation of Neural Circuits

Tycho M. Hoogland and Peter Saggau

Abstract

From brain slice to freely moving mouse, optical methods are being used to probe single neuron physiology and neural circuit function. Efforts in physics, engineering, and genetics have resulted in novel techniques that permit more refined optical interrogation of brain function. The field of optogenetics has enabled neural activity to be driven by light, while optical read-out of neural activity has been improved in terms of speed, spatial resolution, and imaging depth. Genetically encoded sensors and activators of neural activity can now be expressed in cell subtypes helping us to understand neural circuits with unprecedented detail and specificity. In this chapter, we discuss some of the currently available optical methods while highlighting their relative strengths and weaknesses.

Key words: Two-photon microscopy, Acousto-optic deflector, Spatial light modulator, Calcium indicator, Voltage indicator, Genetically encoded indicator, Caged compound, Optogenetics

1. Introduction

Probing nervous system function with optical tools has become as prevalent as using electrophysiological techniques while being less invasive and providing better spatial accuracy. In the span of one decade, innovations in microscopy and optics in conjunction with chemistry and genetics have greatly expanded our ability to interrogate neuronal circuits. It has become possible to use light-activated cation channels and ion pumps expressed in subsets of neurons to selectively manipulate their function, assess their connectivity, and control animal behavior (1–3). Structural changes of individual axons and dendritic spines have been imaged in the intact brain over periods of months to investigate learning and correlated dynamics of neuron morphology (4–7). We are now able to image the microorganization of receptive fields (8) and observe behavioral- and learning-correlated network activity at a fine spatial scale spanning up to hundreds of neurons (9–12). Additionally, random-access laser scanning has enabled multisite photolysis of caged neurotransmitters at physiological rates (13–15). Likewise, neural activity can be imaged at high sampling rates from many noncontiguous sites (16–18). Novel imaging approaches, such as two-photon laser scanning microscopy with temporally focused laser pulses (19) and scanless holographic microscopy (20–22), will improve our ability to optically modulate activity in groups of neurons or at the level of subcellular compartments. Utilizing novel reporters and activators of neuronal activity (3, 23, 24), these advanced imaging techniques will not only enable patterned stimulation and imaging with high spatial accuracy, but also with millisecond precision (25). Thus, by enabling precise manipulation and monitoring of spatial and temporal information at the single neuron and systems level, optical approaches will become even more important in unraveling brain function.

In this chapter, we focus on the use of two-photon laser scanning microscopy, which has enabled the monitoring of hundreds of individual neurons in anesthetized (26), awake and behaving animals (9, 11, 27, 28) as well as neurotransmitter uncaging at the level of individual dendritic spines (29, 30) or at many discrete sites simultaneously (15). We describe various microscope schemes that have been implemented to increase the rates at which neural tissue can be sampled and illuminated, to improve spatial resolution and to allow imaging in deep brain structures, as well as in freely behaving animals. We also give an overview of the optical probes that have enabled the monitoring of neural circuitry. The ensuing chapters give case-by-case examples of the application of optics in the monitoring and manipulation of neural circuits.

2. Advanced Light Microscopy to Study Neural Circuits

For many decades, light has been used to monitor and manipulate neural circuit activity. Early monitoring mainly used wide-field microscopy, employing various detection schemes including photodiode arrays and charge-coupled device (CCD) cameras to collect photons from light-driven indicators of neuronal activity (31–34). For manipulation of activity, different types of light-driven activators have been employed. Caged ions (35) and neurotransmitters (36) can be used to study synaptic signaling and integration (15, 29, 30). Light-sensitive ion channels expressed in neurons have been used more recently to optically excite subpopulations of neurons in order to reveal their connectivity (37–39), drive learning (40) and to study their effect on animal behavior (41, 42).

2.1. Two-Photon Microscopy

To study neural circuits in action, it is essential to retain the integrity of the tissue in which they are embedded. Thus, great effort has been put into imaging the intact brain. By combining the concept of two-photon absorption (43) with laser scanning microscopy (44), it has become possible to image with micrometer spatial resolution in highly light-scattering tissue, such as the living brain. Ultrafast pulsed (mode-locked) infrared lasers are used that create sufficient photon flux at the peak of the pulse to generate two-photon excitation of a fluorophore, while maintaining sufficiently low average power. The strongly reduced scattering of infrared photons results in a small focal volume, while the nonlinear nature of two-photon absorption further reduces the fluorescence excitation volume, obviating the need for the pinhole used in confocal microscopes. This enables optical sectioning at depths several hundred microns below the brain surface. Combined with multicell bolus loading of fluorescent calcium indicators (9, 45–47) or targeting of calcium sensor proteins to neurons (21, 48) activity can be monitored from hundreds of identified neurons at once in the intact brain. In addition, the small excitation volume strongly reduces photodamage, allowing for chronic in vivo imaging. Thus, two-photon microscopy has enabled imaging of structural plasticity at the single synapse level, but also behaviorally correlated activity over the course of months (6, 7, 23, 48).

2.2. Random Access Scanning Microscopy

Conventional scanning systems, which use galvanometer-driven mirrors to deflect the laser beam, perform raster scans that permit relatively slow frame rates. Faster rates of up to maximally 2 kHz are only achieved with line scans, where two-dimensional scanning is abandoned. One problem with raster scanning methods is that owing to the nonuniform distribution of neurons and their intricate processes, many locations that lack neuronal structure are sampled. Because of the inertia of galvanometer-driven mirrors, alternative techniques to steer laser beams have been developed that support visiting noncontiguous sites, in order to increase the effective scan speed. One particularly promising technique, random access scanning (15, 16, 18, 49), makes use of acousto-optic deflectors (AODs). In these devices, an electromechanical transducer produces sound waves in the radio frequency (RF) range inside a crystal (e.g., TeO_2), generating a diffraction grating (Fig. 1a). By modulating the sound frequency, first order diffracted light gets deflected at an angle proportional to the frequency. For scanning in an *xy* plane, two AODs are placed orthogonally. The method allows for very fast control (up to tens of kHz) of laser beam position at many discrete sites. For imaging purposes, the dwell time of the laser beam needs to be sufficiently high to obtain an appropriate signal-to-noise ratio. Alternatively, the laser is positioned on multiple adjacent sites of the same cell to increase the signal quality at the expense of speed (25). Two-photon microscopes equipped with AODs have been successful in sampling from many sites (Fig. 1b) at physiological rates (15, 16, 25, 50). This beam-steering

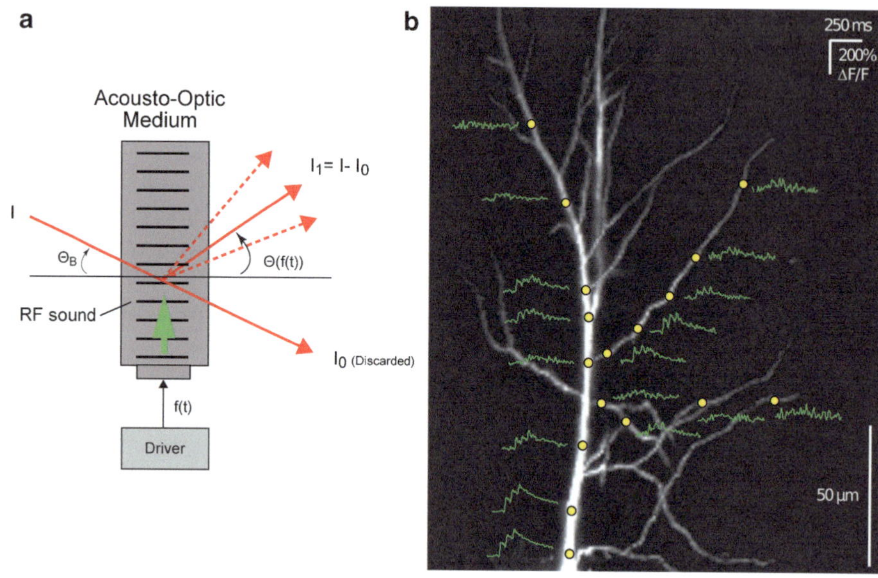

Fig. 1. (**a**) Scheme showing the function of an acousto-optic deflector (AOD). The laser beam is incident at the Bragg angle (θ_B) allowing most of the intensity *I* to be deflected into the first diffraciton order (I_1). The scan angle of I_1 $\theta(f(t))$ can be controlled by time-varying the acoustic frequency $f(t)$. For *xy* positioning, two AODs are placed orthogonally (**b**) Maximum image projection of a hippocampal pyramidal neuron imaged with a two-photon microscope with a 2D AOD laser positioning system. Calcium transients were evoked in this cell by short bursts of action potentials at 20 Hz using brief somatic current injection and sampled at 0.5 kHz.

technique has also been used for multisite photolysis of caged neurotransmitters to study synaptic integration of spatially separated inputs (13, 15). Initially, small aperture AODs limited the field-of-view, but the use of larger aperture AODs has mitigated this problem. More importantly, when using 2D AOD systems for two-photon microscopy, additional optics are required to compensate for spatial and temporal dispersion of the ultrafast laser pulses (25, 50). The reduced diffraction efficiency of AOD-based scanners is no longer a limitation with the high-power lasers available today. In fact, AOD-based laser scanning microscopes have been used in vivo (25), where more laser power is needed to image at greater depths. By using two additional AODs together with chirped and counter-propagating sound waves, random access scanners have been constructed that can focus a laser beam along the *z* axis (Fig. 2) in addition to multisite positioning in an *xy* plane, thus achieving 3D random access scanning (17). Focusing has been limited over relatively short axial ranges with high magnification objectives (~50 μm focus range with 60×, numerical aperture (NA) = 1; ~140 μm with 40×, NA = 0.8, (17, 18)). Thus, especially for in vivo applications, low magnification (e.g., 20×) with high NA could be used to increase the focusing range. The use of AODs with shorter optical path length reduces the temporal dispersion of ultrafast laser pulses at the cost of diffraction efficiency (18). In general, temporal dispersion compensation of

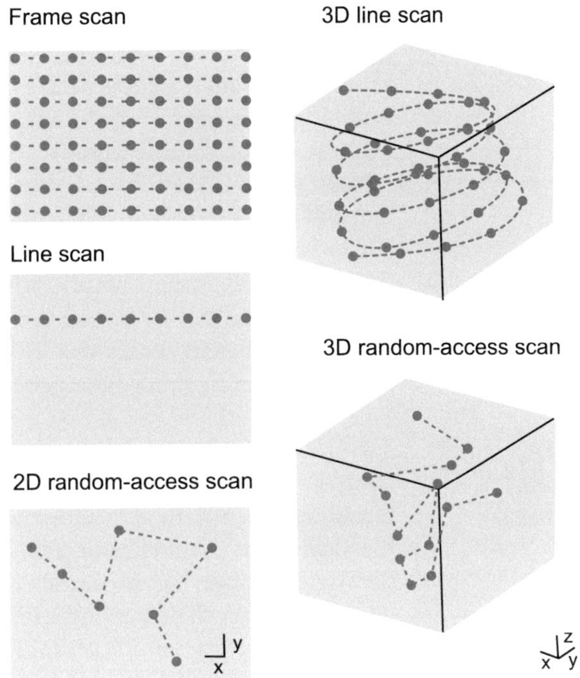

Fig. 2. Cartoon of scanning schemes that can be combined with two-photon laser scanning microscopy, including line and frame scans (using galvanometer-driven mirrors), 2D AOD scanning (using two orthogonally placed AODs in the optical path), 3D line scan (combining a fast objective piezo-electric actuator with galvanometer-driven mirrors) and 3D AOD scan (using four AODs of which two pairs are placed orthogonally in the optical path).

femtosecond pulsed lasers is straightforward (50) and, since reducing the pulse duration boosts the pulse power, it reduces the amount of average power required at the specimen.

2.3. Alternative Scanning Techniques and "Scanless" Methods

Alternative methods have been explored to increase sampling speed from volumes of tissue containing hundreds of neurons. For example, galvanometer-based scanning in conjunction with rapid axial movement of the objective under the control of a piezo-electric actuator has been used to generate 3D line scans (Fig. 2) sampling large volumes of cells (Fig. 2), although at acquisition rates (~10 Hz) comparable to those of 2D frame scans (51). In addition, multibeam two-photon microscopes, which split the laser beam into a multitude of beamlets with corresponding focal points to increase the sampling rate (52) have been used to study network dynamics in vitro (53). In contrast to the simple photodetectors (e.g., PMT) used in single-beam scanning, multibeam scanning requires imaging detectors (e.g., CCD camera). In general, this scanning technique suffers from potential multisite signal crosstalk of scattered fluorescence, and focal points have to be sufficiently separated to reduce this risk. Vector-based scanning methods have been developed in which an optimized search algorithm controls the galvanometers such that only user-specified points are imaged (54). Holographic microscopy utilizes liquid crystal on silicon (LCOS) spatial light modulators (SLMs, (20, 21, 55)), which can rapidly generate discrete and complex illumination patterns that match cellular morphology. An added advantage of using such a scanless method over one using AOD-based scanning is minimal chromatic aberration and pulse broadening caused by SLMs. Thus, neural circuits can be probed with various imaging approaches at acquisition rates required to record fast physiological signals in the brain and with sufficient spatial resolution to target or identify individual neurons, or even subcellular processes.

2.4. Controlling Neural Activity with Two-Photon Microscopy

Two-photon laser scanning microscopes can also be employed to control neural excitation or inhibition by means of neurotransmitter uncaging (15, 56), by activation of light-sensitive ion channels, such as the algal opsin channelrhodopsin-2 (57, 58), and ion pumps, such as the bacterial opsin halorhodopsin (59), and by light-induced transient disruption of a cell's membrane integrity (60). Optimization of caged neurotransmitters for photostimulation by UV and infrared lasers (61, 62) has enabled rapid patterned stimulation using AOD-based laser positioning (13, 15). Holographic microscopy has also been used to uncage neurotransmitters (54) and photoactivate ion channels (19, 22). The spatially extended illumination patterns needed for light-activated ion channels and pumps, e.g., channel- and halorhodopsins, makes holographic schemes better suited than rapid scanning with focused laser beams. With AODs many sites on a cell need to be visited severely limiting the achievable rates of stimulation as well as the number of cells that can be activated, or silenced with light-activated channels, or ion pumps. Temporal focusing of laser pulses allows to control the axial beam intensity profile independent of its lateral distribution, enabling activation of a larger number of light-sensitive ion channels and has been used in conjunction with holographic microscopy (19). The use of temporal focusing and holographic microscopy causes spatial intensity fluctuations of the illumination patterns, but these can be corrected for with phase contrast filters (22). Despite the increased use of two-photon excitation of channelrhodopsin-2 for in vivo studies, blue light sources including fiber-coupled (arrays of) light-emitting diodes will provide cheaper alternatives which can be used for photoactivation in brain slices or superficial cortical layers in vivo (42, 63, 64).

2.5. Imaging with Improved Spatial Resolution

The advantage of optical sectioning combined with a small (femtoliter, 1 μm^3) focal volume allows the use of two-photon microscopy in light-scattering tissue for imaging of small structures, such as dendritic spines (65–67), or axons (6, 68). Despite the benefit of a small excitation volume with two-photon excitation, the use of infrared excitation wavelengths will also decrease the spatial resolution according to Abbe's law ($d = \lambda/2\text{NA}$, where d = spatial resolution, NA = numerical aperture of the objective, and λ = excitation wavelength). To improve the spatial resolution of two-photon microscopy, while maintaining its advantages (e.g., reduced photobleaching and photodamage, as well as depth-resolved imaging), single photon near-infrared stimulated emission depletion (STED, (69)) of two-photon excited fluorophores can be used to achieve several-fold higher resolution beyond the traditional diffraction limit (70). This could be even further improved by using pulsed depletion light instead of continuous wave depletion, using an annular phase mask or using brighter fluorophores to permit imaging of structures in the range of tens of nanometers (70). Photoactivated localization microscopy (PALM, (71)) and stochastic optical reconstruction microscopy (STORM, (72)) are other microscope schemes that circumvent the traditional diffraction limit. PALM is a powerful technique that enables localization of fluorescently tagged proteins in cells with nanometer resolution. It uses photoactivation and subsequent bleaching of pools of individual fluorescently tagged molecules in combination with fitting of the point spread function (PSF) to the individual molecular images to determine their actual location. Application of PALM in live tissue has so far been limited to robust cell lines that can withstand repeated light exposure required to deplete an available pool of inactivated nonbleached fluorescent molecules (71, 73). STORM is essentially similar to PALM except for the type of fluorophore used. The advantage of STED is its direct applicability in live tissue (70).

2.6. Imaging at Greater Depths

In mice two-photon microscopy is suitable to image over hundreds of microns below the brain surface, but subcortical brain structures, such as the thalamus, hippocampus, striatum, or the cerebellar nuclei cannot be accessed. There are a number of ways in which two-photon microscopy could be used at greater depths, including the use of a regenerative amplifier to increase laser power and extend the imaging range from a theoretical 600–800 μm to 1 mm (74). Still, the amount of axon myelination and density of blood vessels will determine and limit the attainable depth at which can be imaged. In tandem with improvement of collection efficiency, higher laser power and dispersion compensation to minimize pulse broadening are required for the depth limit of 1 mm to be reached. A recent approach to achieve increased penetration depth adds a very different method of beam focusing to the traditional spatial focusing (51). This approach produces extremely short laser pulses only at the waist of a focused laser beam, boosting the local pulse power. The mechanism utilizes the fact that the wavelength spectrum of a short laser pulse can be spatially dispersed by a diffractive optical element, e.g., diffraction grating. Such a dispersed spectrum can be regarded as containing many beamlets of narrow spectral content with longer pulse duration and consequently lower pulse power. These low-power beamlets cannot excite significant two-photon fluorescence. Only in the focus of such a beam is the short pulse duration present and thus the high pulse power recreated that is required for useful two-photon excitation (75, 76). Still, many deep brain structures remain inaccessible to two-photon microscopy and alternatives need to be used. A crude approach is resection of the overlaying tissue. It is uncertain how the ensuing damage affects the integrity and function of the underlying brain structures. However, basic electrophysiological and behavioral responses of mice in which the hippocampus was imaged after resection of the overlaying cortex were not altered relative to controls (77). Another somewhat less invasive approach is to perform two-photon microscopy with needle-shaped gradient index (GRIN) lenses in combination with a standard objective lens (78, 79) also referred to as fluorescence micro-endoscopy (78). GRIN lenses with a diameter of several hundred microns can be inserted several millimeters into the tissue to allow for deep brain imaging. Current implementations of GRIN lens-based deep imaging suffer from limited field-of-view size using a lower NA for the central length of the lens. This is necessary to prevent focusing of the excitation light inside the high NA end piece of the lens, which can increase background fluorescence and cause self-phase modulation (79). Some progress has been made to address this issue (80). Another drawback of GRIN lenses is the poor spatial resolution (elongated PSF) as a result of optical aberrations and the low NA lenses used. The placement of a GRIN lens in series with a high NA plano-convex lens with tuned refractive indices significantly improves the PSF (81). Further miniaturization of the optics will make this a preferred approach to record from neurons and subcellular structures in deep brain structures. At present, micro-endoscopy has been miniaturized to enable imaging of deep brain structures in freely moving mice (82). Alternative approaches, such as a periscope-fiber based microscope—using single photon laser illumination—have attained imaging depths of up to 700 μm, but are also invasive due to insertion of a periscope-mirror system hundreds of microns wide (83).

2.7. Imaging in Freely Behaving Animals

It has become clear that imaging in awake, behaving mice will reveal crucial information about sensory and motor processing in the brain (77, 82, 84), since cellular response properties are strongly different in anesthetized or sleeping animals (85). Two-photon microscopy has enabled the study of networks of identified neurons and couple their response to the observed behavior in head-fixed animals that were either walking on a spherical treadmill (9, 11), or trained to perform a task while not moving (12, 86). Obviously, imaging behavior in freely moving animals would permit the study of neural circuits in a more natural setting with reduced stress and obviate the need for adapting animals to head restraint. Thus, efforts have been made to miniaturize optics to permit such a paradigm. Fluorescence excitation through optical fibers mounted to piezo-electric actuators (to generate scan patterns) have been used to image in freely behaving rats using two-photon microscopy (87). Combination of other scanning schemes that do not involve movement of the fiber has also developed (82). For example, scanning over the entrance surface of a fiber bundle has been implemented for two-photon excitation (88). A fiber optic microscope using two-photon excitation with two channel fluorescence detection has been developed to image from tens of neurons at once in freely moving rats (10). Despite several advantages, two-photon excitation microscopy is not required to image superficial signals from the mouse or rat brain. An epifluorescence-based fiberscope that did not employ scanning nor two-photon excitation demonstrated that calcium transients of many Purkinje cells can be recorded in freely moving mice (89). Another micro-endoscopic technique, employing single photon illumination and a prism inserted in the brain as part of a "periscope," was used to image calcium transients in layer V apical tufts of cortical pyramidal neurons during freely moving behavior (83). Besides freely moving rodents, imaging can be performed also in head restrained animals navigating in pseudo-realistic environments by combining a virtual reality display with a spherical treadmill (77, 90). This will benefit stability of the preparation for imaging during behavior in a more realistic environment.

3. Optical Indicators to Study Neural Circuits

Fluorescent calcium and voltage indicators are by far the most commonly used reporters of neuronal activity in imaging experiments today. Both synthetic and genetically encoded indicators have been developed. The latter can be targeted to subpopulations of neurons using viruses or by means of electroporation enabling the monitoring of activity in small networks in the intact brain. Below we give an overview of both types of indicators and their use for the optical interrogation of neural circuits.

3.1. Synthetic Calcium Indicators

Although calcium indicators report delayed and convolved representations of supra-threshold neuronal activity, they are easily targeted to cells and provide a substantially higher signal-to-noise ratio than voltage indicators. Synthetic calcium indicators (91–94) have been used for several decades to record calcium changes in individual neurons in vitro (67, 95–97) and later in vivo (98, 100). For single cell measurements, patch electrodes are used to establish whole-cell configuration and the indicator is allowed to diffuse with the internal medium of the pipette into the cell. The use of acetoxymethyl-esterified (AM) indicators which are loaded into tissue in vitro (98), or in vivo (46, 101) has enabled delivery of dyes to subcellular structures, e.g., presynaptic terminals (45) and to many neurons at once (46, 99), with limited perturbation to the intercellular milieu and concentrations that do not severely disrupt endogenous calcium buffering (102). Nonspecific esterases cleave off the AM group converting the indicator from a nonpolar to a polar molecule which is retained in the cell. Two-photon laser scanning microscopy together with multicell bolus loading (also called bulk loading) of AM-ester dyes has led to a renaissance in the study of network activity in the intact brain under anesthesia and during behavior (9, 11, 103, 104).The nonspecificity of the indicator loading has resulted in the use of the astrocytic marker sulforhodamine 101 (SR101) to distinguish astrocytes from neurons (105). However, recent studies suggest that this marker may affect neuronal excitability and should thus be used with caution (106). Analytical methods (e.g., independent component analysis, ICA) to extract neuronal and glial signals are likely to be more reliable (11, 107) and can be applied post-hoc. In addition, (commercially available) transgenic mice expressing fluorescent marker proteins, such as DsRed, and tdTomato (108) under cell-type specific promotors could be used in conjunction with bolus-loading of calcium dyes to image from specific cell groups.

3.2. Genetically Encoded Calcium Indicators

As with synthetic calcium indicators, the first generation of genetically encoded calcium indicators (GECIs), also known as fluorescent calcium indicator proteins (FCIPs), were pioneered by Roger Tsien and colleagues (109, 110). Since then, big strides have been made in improving the properties of GECIs and results in mouse brain in vivo have been encouraging with GECIs enabling detection of single action potentials with D3cpv (111), GCaMP3 (23), and YC3.60 (48). Previously, at the neuromuscular junction of Drosophila larva, detection of single action potentials with YC3.60 was not reliable (23), suggesting that the site of expression and the animal model affect the properties of the sensor protein. Despite GCaMP3 being a high dynamic range calcium sensor, with better brightness, photostability and signal-to-noise ratio than other probes, it interferes with endogenous calmodulin, a problem not encountered by the troponin-based sensor TN-XXL (112). YC3.60 has the advantage of being a Förster/fluorescence energy transfer (FRET)-based probe (where one fluorophore, the donor, decreases its fluorescence, and the other, the acceptor, increases its fluorescence during calcium binding) allowing correction of bleaching and, more importantly for imaging during behavior, motion artifacts. Despite these advantages, YC3.60 also has a calmodulin-derived calcium binding moiety that could interact with its endogenous counterpart. In D3cpv, another FRET-based dye, steric protrusions were added to the M13 domain to prevent such interactions. The kinetics of the various protein-based calcium sensors are quite different (113): the slow kinetics of D3cpv make it mostly suitable for recording neuronal signals that are sparse, for example Purkinje cell complex spikes, whereas YC3.60 has faster kinetics and is suitable for detecting neurons firing at higher frequency (48). In *Caenorhabditis elegans*, expression of YC3.60 in a subset of cells affected turning behavior, whereas GCaMP3 did not (23). To what extent this effect on behavioral phenotype extends to other animal models is unknown. Recent modifications (shortening of the linker between the calmodulin and M13 domains) of the YC2.6 calcium sensor have led to indicators with much higher affinity. These are suitable for recording not only single action potentials, but also putative subthreshold calcium increases (114). Surely, future iterations of these sensors will have improved signal-to-noise, kinetics, and reduced interference with endogenous calcium signaling. GECIs are commonly delivered using modified viruses that generate high-level expression near the site of injection, although in utero electroporation has also been used (112). Intraperitoneal injection of mannitol prior to virus injection can often facilitate spread of the virus over a larger tissue volume and reduce virus-induced cytotoxicity. Viral delivery of GECIs has made possible the long-term study of network activity in awake and behaving mice (23, 48). So far, transgenic mouse models expressing calcium-sensor proteins have not been very successful, due to low levels of expression. Initial attempts to increase expression levels were not sufficient for imaging calcium signals with (sub) cellular resolution in vivo (115). Despite these hurdles, it has now become possible to perform chronic imaging of network activity in the intact animal with cellular resolution.

3.3. Synthetic Voltage Indicators

Presently, calcium indicators only report supra-threshold activity, whereas voltage indicators have the advantage of reporting both sub-threshold and supra-threshold electrical responses with signal kinetics matching the actual changes in membrane potential. Due to their small fluorescence changes however, signal averaging is usually required, which rules out experiments that study spontaneous and single trial activity. So far, voltage indicators are thus best suited for imaging global changes in activity spanning many neurons (to allow for spatial averaging). Most experiments recording voltage signals optically have employed wide-field imaging with CCD cameras (33, 116), photodiodes (117), or photodiode arrays (118, 119) with high sampling frequencies, but poor optical sectioning and limited spatial resolution. In vitro, synthetic voltage indicators (e.g., RH414, RH795, RH1691) can be delivered by bath application, in vivo by placing a dye-infused piece of gel foam on the brain surface (33), or by injection via a pipette, containing a mixture of dye (e.g., ANNINE-6) in DMSO and pluronic, in the intact brain (120). Single cells can be targeted with synthetic dyes, such as JPW1114, or FM4-64 (121–124), although the dye takes long to diffuse because of its hydrophobicity and thus propensity to embed itself in lipid membranes. To circumvent this problem, electrodes are typically back-filled with the tip containing internal solution without any dye. The cell is patched to allow dye to fill the soma and repatched after 1–2 h, when the dye has diffused through the entire cell (121). A recent trend is the use of second harmonic generation (SHG) and two-photon microscopy to record voltage signals (122, 123), as it will enable the recording of signals over a much larger range of depths, something especially useful for in vivo imaging. In the case of SHG, background is much reduced as a result of the requirement that only correctly aligned dye molecules in the neuronal membrane generate a signal. Two-photon excitation has been used in vivo to optically record voltage signals from different cortical layers in the barrel field revealed top-down control of L1 by long-range inputs from various cortical and thalamic areas (120).

3.4. Genetically Encoded Voltage Indicators

Ideally, genetically encoded voltage indicators (GEVIs) should be targeted to subpopulations of neurons to allow better resolution of the source of the signals. The first generation of GEVIs were based on the voltage-sensing domain of the Shaker potassium channel and combined with a GFP-based fluorophore, of which FlaSh is the best known variant (125, 126). VSFP-1 is a genetically encoded voltage sensor using the potassium channel (Kv2.1) voltage-sensing domain, but has two fluorophores, enabling FRET-based measurements. The sodium channel has also been used as a template for a voltage sensor (e.g., the sodium channel protein-based activity reporting construct, SPARC, (127)). However, these first generation voltage sensors had several shortcomings: slow response times, a narrow response range to voltages, reduced fractional changes of fluorescence relative to synthetic dyes, and high background noise. One group proposed to use a hybrid approach where, farnesylated eGFP is anchored to the lipid membrane and a synthetic anion dipicrylamine (DPA) acting as an energy acceptor to generate FRET-like fluorescence changes with changes in membrane potential (128). This construct, known as hybrid voltage sensor (hVOS), has fast response times (within 500 μs) and large fractional changes of fluorescence. However, the concentrations of DPA that were necessary to generate the observed fractional changes also significantly increased cell capacitance and thereby reduced synaptically evoked action potentials. Further optimization of hVOS to work with lower concentrations of DPA has partly mitigated this problem (129). It should be noted however, that the use of DPA is strongly regulated due to its use as a calibrating agent in the detection of explosives (130). A phosphatase with a voltage sensing domain (Ci-VSP) derived from the ascidian Ciona intestinalis was fused with circularly permuted fluorescent proteins to generate second generation voltage sensors with the VSFP nomenclature (131–133). VSFP2.3 and VSFP2.4 contain two fluorophores enabling FRET imaging, whereas VSFP3.1 has faster response kinetics, but variants with single red-shifted fluorophores (mOrange2, mKate2), a potential benefit for imaging in vivo. In utero electroporation of VSFP2.3 expression plasmid results in most of the fluorescence expressed near the plasma membrane in cortical pyramidal neurons, without affecting basic synaptic physiology (24). In addition, both VSFP2.3 and a VSFP2.4 variant reported responses to individual whisker stimulation in vivo, with few ms response times. Further improvements of signal kinetics and amplitude will make GEVIs more commonly used reporters to interrogate neuronal circuits both in vitro and in vivo. If signal-to-noise ratios become sufficiently large, GEVIs may one day even replace calcium indicators as the preferred reporter of cellular and network activity in the living brain.

4. Summary

At present, various optical techniques are converging with the overall aim to improve the spatial and temporal accuracy of monitoring neural circuits and the manipulation of their activity. Two-photon excitation microscopy is widely used, enabling in vitro and vivo imaging with physiological sampling times, mostly as a result of improved imaging schemes that employ AODs for random-access scanning or SLMs for holographic microscopy. Reporters of neural activity, especially calcium indicators have improved signal-to-noise ratios (sufficient to detect single action potentials) and can be expressed either transiently (via multicell bolus loading), or in subtypes of neurons (via expression of calcium sensor proteins) to allow longitudinal studies of network function. Photolysis of caged neurotransmitters has revealed the molecular mechanisms underlying synaptic function and plasticity at the single synapse level. Algal and bacterial opsins can be expressed in neurons and selectively activated by light to drive learning and behavior. Fiberscopes have enabled imaging of deep brain structures and neural circuitry in freely moving animals. Finally, the diffraction limit has been bypassed, allowing to image structures spanning tens of nanometers (e.g., receptor complexes in synapses). With further improvements of synthetic indicators, GECIs, GEVIs, sampling time, collection efficiency, and spatial resolution, it will become possible to study the function of the same neurons at the molecular as well as the systems and behavioral level. Thus, optical tools have become indispensable in neuroscience and are likely to rival in the near future electrophysiological techniques in our attempt to understand brain function.

References

1. Cardin JA, Carlén M, Meletis K et al (2010) Targeted optogenetic stimulation and recording of neurons in vivo using cell-type-specific expression of channelrhodopsin-2. Nat Protoc 5(2):247–254
2. Tsai H, Zhang F, Adamantidis A et al (2009) Phasic firing in dopaminergic neurons is sufficient for behavioral conditioning. Science 324 (5930):1080–1084
3. Chow BY, Han X, Dobry AS et al (2010) High-performance genetically targetable optical neural silencing by light-driven proton pumps. Nature 463(7277):98–102
4. Portera-Cailliau C, Weimer RM, De Paola V et al (2005) Diverse modes of axon elaboration in the developing neocortex. PLoS Biol 3(8):e272
5. Zuo Y, Yang G, Kwon E, Gan W (2005) Long-term sensory deprivation prevents dendritic spine loss in primary somatosensory cortex. Nature 436(7048):261–265
6. De Paola V, Holtmaat A, Knott G et al (2006) Cell type-specific structural plasticity of axonal branches and boutons in the adult neocortex. Neuron 49(6):861–875
7. Holtmaat A, Wilbrecht L, Knott GW et al (2006) Experience-dependent and cell-type-specific spine growth in the neocortex. Nature 441(7096):979–983
8. Smith SL, Häusser M (2010) Parallel processing of visual space by neighboring neurons in mouse visual cortex. Nat Neurosci 13 (9):1144–1149
9. Dombeck DA, Khabbaz AN, Collman F et al (2007) Imaging large-scale neural activity with cellular resolution in awake, mobile mice. Neuron 56(1):43–57
10. Sawinski J, Wallace DJ, Greenberg DS et al (2009) Visually evoked activity in cortical cells imaged in freely moving animals. Proc Natl Acad Sci U S A 106(46):19557–19562
11. Mukamel EA, Nimmerjahn A, Schnitzer MJ (2009) Automated analysis of cellular signals from large-scale calcium imaging data. Neuron 63(6):747–760
12. Komiyama T, Sato TR, Connor DH et al (2010) Learning-related fine-scale specificity imaged in motor cortex circuits of behaving mice. Nature 464(7292):1182–1186
13. Shoham S, O'Connor DH, Sarkisov DV, Wang SS (2005) Rapid neurotransmitter uncaging in spatially defined patterns. Nat Methods 2(11):837–843
14. Gasparini S, Magee JC (2008) State-dependent dendritic computation in hippocampal CA1 pyramidal neurons. J Neurosci 26(7):2088–2100
15. Losavio BE, Iyer V, Patel S, Saggau P (2010) Acousto-optic laser scanning for multi-site photo-stimulation of single neurons in vitro. J Neural Eng 7(4):045002
16. Iyer V, Hoogland TM, Saggau P (2006) Fast functional imaging of single neurons using random-access multiphoton (RAMP) microscopy. J Neurophysiol 95(1):535–545
17. Duemani Reddy G, Kelleher K, Fink R, Saggau P (2008) Three-dimensional random access multiphoton microscopy for functional imaging of neuronal activity. Nat Neurosci 11(6):713–720
18. Kirkby PA, Srinivas Nadella KM, Silver RA (2010) A compact acousto-optic lens for 2D and 3D femtosecond based 2-photon microscopy. Opt Express 18(13):13721–13745
19. Andrasfalvy BK, Zemelman BV, Tang J et al (2010) Two-photon single-cell optogenetic control of neuronal activity by sculpted light. Proc Natl Acad Sci U S A 107(26): 11981–11986
20. Dal Maschio M, Difato F, Beltramo R et al (2010) Simultaneous two-photon imaging and photo-stimulation with structured light illumination. Opt Express 18(18):18720
21. Nikolenko V, Watson BO, Araya R et al (2008) SLM microscopy: scanless two-photon imaging and photostimulation with spatial light modulators. Front Neural Circ 25. doi:10.3389/neuro.04.005.2008
22. Papagiakoumou E, Anselmi F, Bègue A et al (2010) Scanless two-photon excitation of channelrhodopsin-2. Nat Methods 7(10): 848–854
23. Tian L, Hires SA, Mao T et al (2009) Imaging neural activity in worms, flies and mice with improved GcaMP calcium indicators. Nat Methods 6(12):875–881
24. Akemann W, Mutoh H, Perron A et al (2010) Imaging brain electric signals with genetically targeted voltage-sensitive fluorescent proteins. Nat Methods 7(8):643–649
25. Grewe BF, Langer D, Kasper H et al (2010) High-speed in vivo calcium imaging reveals neuronal network activity with near-millisecond precision. Nat Methods 7(5):399–405
26. Petzold GC, Hagiwara A, Murthy VN (2009) Serotonergic modulation of odor input to the

mammalian olfactory bulb. Nat Neurosci 12(6):784–791

27. Ferezou I, Haiss F, Gentet LJ et al (2007) Spatiotemporal dynamics of cortical sensorimotor integration in behaving mice. Neuron 56(5):907–923
28. Komiyama T, Sato TR, O'Connor DH et al (2010) Learning-related fine-scale specificity imaged in motor cortex circuits of behaving mice. Nature 464(7292):1182–1186
29. Matsuzaki M, Honkura N, Ellis-Davies GC et al (2004) Structural basis of long-term potentiation in single dendritic spines. Nature 429(6993):761–766
30. Higley MJ, Sabatini BL (2010) Competitive regulation of synaptic Ca($^{2+}$) influx by D2 dopamine and A2A adenosine receptors. Nat Neurosci 13(8):958–966
31. Grinvald A, Cohen LB, Lesher S et al (1981) Simultaneous optical monitoring of activity of many neurons in invertebrate ganglia using a 124-element photodiode array. J Neurophysiol 45(5):829–840
32. Ross WN, Werman R (1987) Mapping calcium transients in the dendrites of Purkinje cells from the guinea-pig cerebellum in vitro. J Physiol 389:319–336
33. Kleinfeld D, Delaney KR (1996) Distributed representation of vibrissa movement in the upper layers of somatosensory cortex revealed with voltage-sensitive dyes. J Comp Neurol 375(1):89–108
34. White JG, Amos WB, Fordham M (1987) An evaluation of confocal versus conventional imaging of biological structures by fluorescence light microscopy. J Cell Biol 105(1): 41–48
35. Kaplan JH, Ellis-Davies GC (1988) Photolabile chelators for the rapid photorelease of divalent cations. Proc Natl Acad Sci U S A 85(17):6571–6575
36. Callaway EM, Katz LC (1993) Photostimulation using caged glutamate reveals functional circuitry in living brain slices. Proc Natl Acad Sci U S A 90(16):7661–7665
37. Petreanu L, Huber D, Sobczyk A et al (2007) Channelrhodopsin-2-assisted circuit mapping of long-range callosal projections. Nat Neurosci 10(5):663–668
38. Wang H, Peca J, Matsuzaki M et al (2007) High-speed mapping of synaptic connectivity using photostimulation in channelrhodopsin-2 transgenic mice. Proc Natl Acad Sci U S A 104(19):8143–8148
39. Cruikshank SJ, Urabe H, Nurmikko AV, Connors BW et al (2010) Pathway-specific feedforward circuits between thalamus and neocortex revealed by selective optical stimulation of axons. Neuron 65(2):230–245
40. Huber D, Petreanu L, Ghitani N et al (2008) Sparse optical microstimulation in barrel cortex drives learned behaviour in freely moving mice. Nature 451(7174):61–64
41. Douglass AD, Kraves S, Deisseroth K et al (2008) Escape behavior elicited by single, channelrhodopsin-2-evoked spikes in zebrafish somatosensory neurons. Curr Biol 18 (15):1133–1137
42. Kravitz AV, Freeze BS, Parker PR et al (2010) Regulation of parkinsonian motor behaviours by optogenetic control of basal ganglia circuitry. Nature 466(7306):622–626
43. Göppert-Mayer M (1931) Über elementarakte mit zwei quantensprüngen. Ann Phys 401(3):273–294
44. Denk W, Strickler JH, Webb WW (1990) Two-photon laser scanning fluorescence microscopy. Science 248(4951):73–76
45. Wu LG, Saggau P (1994) Presynaptic calcium is increased during normal synaptic transmission and paired-pulse facilitation, but not in long-term potentiation in area CA1 of hippocampus. J Neurosci 14(2):645–654
46. Stosiek C, Garaschuk O, Holthoff K et al (2003) In vivo two-photon calcium imaging of neuronal networks. Proc Natl Acad Sci U S A 100(12):7319–7324
47. Sullivan MR, Nimmerjahn A, Sarkisov DV et al (2005) In vivo calcium imaging of circuit activity in cerebellar cortex. J Neurophysiol 94(2):1636–1644
48. Lütcke H, Murayama M, Hahn T et al (2010) Optical recording of neuronal activity with a genetically-encoded calcium indicator in anesthetized and freely moving mice. Front Neural Circ. doi:10.3389/fncir.2010.00009
49. Salomé R, Kremer Y, Dieudonné S et al (2006) Ultrafast random-access scanning in two-photon microscopy using acousto-optic deflectors. J Neurosci Methods 154(1–2): 161–174
50. Iyer V, Losavio BE, Saggau P (2008) Compensation of spatial and temporal dispersion for acousto-optic multiphoton laser-scanning microscopy. J Biomed Opt 8(3):460–471
51. Göbel W, Kampa BM, Helmchen F (2007) Imaging cellular network dynamics in three dimensions using fast 3D laser scanning. Nat Methods 4(1):73–79
52. Nielsen T, Fricke M, Hellweg D et al (2001) High efficiency beam splitter for multifocal multiphoton microscopy. J Microsc 201 (Pt 3):368–376

53. Tyzio R, Cossart R, Khalilov I et al (2006) Maternal oxytocin triggers a transient inhibitory switch in GABA signaling in the fetal brain during delivery. Science 314(5806): 1788–1792
54. Nikolenko V, Poskanzer KE, Yuste R (2007) Two-photon photostimulation and imaging of neural circuits. Nat Methods 4(11): 943–950
55. Golan L, Reutsky I, Farah N et al (2009) Design and characteristics of holographic neural photo-stimulation systems. J Neural Eng 6(6):066004
56. Kantevari S, Matsuzaki M, Kanemoto Y et al (2010) Two-color, two-photon uncaging of glutamate and GABA. Nat Methods 7(2): 123–125
57. Boyden ES, Zhang F, Bamberg E et al (2005) Millisecond-timescale, genetically targeted optical control of neural activity. Nat Neurosci 8(9):1263–1268
58. Rickgauer JP, Tank DW (2009) Two-photon excitation of channelrhodopsin-2 at saturation. Proc Natl Acad Sci U S A 106(35): 15025–15030
59. Han X, Boyden ES (2007) Multiple-color optical activation, silencing, and desynchronization of neural activity, with single-spike temporal resolution. PLoS One 2(3):e299
60. Hirase H, Nikolenko V, Goldberg JH et al (2002) Multiphoton stimulation of neurons. J Neurobiol 51(3):237–247
61. Matsuzaki M, Ellis-Davies GC, Nemoto T et al (2001) Dendritic spine geometry is critical for AMPA receptor expression in hippocampal CA1 pyramidal neurons. Nat Neurosci 4(11):1086–1092
62. Ellis-Davies GC, Matsuzaki M, Paukert M et al (2007) 4-Carboxymethoxy-5,7-dinitroindolinyl-Glu: an improved caged glutamate for expeditious ultraviolet and two-photon photolysis in brain slices. J Neurosci 27(25):6601–6604
63. Campagnola L, Wang H, Zylka MJ (2008) Fiber-coupled light-emitting diode for localized photostimulation of neurons expressing channelrhodopsin-2. J Neurosci Methods 169(1):27–33
64. Grossman N, Poher V, Grubb MS, Kennedy GT, Nikolic K, McGovern B et al (2010) Multi-site optical excitation using ChR2 and micro-LED array. J Neural Eng 7(1):16004
65. Yuste R, Denk W (1995) Dendritic spines as basic functional units of neuronal integration. Nature 375(6533):682–684
66. Mainen ZF, Malinow R, Svoboda K (1999) Synaptic calcium transients in single spines indicate that NMDA receptors are not saturated. Nature 399(6732):151–155
67. Sabatini BL, Svoboda K (2000) Analysis of calcium channels in single spines using optical fluctuation analysis. Nature 408(6812): 589–593
68. Tan YP, Llano I (1999) Modulation by K^+ channels of action potential-evoked intracellular Ca^{2+} concentration rises in rat cerebellar basket cell axons. J Physiol 520:165–178
69. Hell SW, Wichmann J (1994) Breaking the diffraction resolution limit by stimulated emission: stimulated-emission-depletion fluorescence microscopy. Opt Lett 19(11):780–782
70. Ding JB, Takasaki KT, Sabatini BL (2009) Supraresolution imaging in brain slices using stimulated-emission depletion two-photon laser scanning microscopy. Neuron 63 (4):429–437
71. Betzig E, Patterson GH, Sougrat R et al (2006) Imaging intracellular fluorescent proteins at nanometer resolution. Science 313 (5793):1642–1645
72. Rust MJ, Bates M, Zhuang X (2006) Sub-diffraction-limit imaging by stochastic optical reconstruction microscopy (STORM). Nat Methods 3(10):793–795
73. Manley S, Gillette JM, Patterson GH et al (2008) High-density mapping of single-molecule trajectories with photoactivated localization microscopy. Nat Methods 5(2): 155–157
74. Helmchen F, Denk W (2005) Deep tissue two-photon microscopy. Nat Methods 2(12): 932–940
75. Zhu G, van Howe J, Durst M et al (2005) Simultaneous spatial and temporal focusing of femtosecond pulses. Opt Express 13(6): 2153–2159
76. Durst ME, Zhu G, Xu C (2006) Simultaneous spatial and temporal focusing for axial scanning. Opt Express 14(25):12243
77. Dombeck DA, Harvey CD, Tian L et al (2010) Functional imaging of hippocampal place cells at cellular resolution during virtual navigation. Nat Neurosci. doi:10.1038/nn.2648
78. Jung JC, Mehta AD, Aksay E et al (2004) In vivo mammalian brain imaging using one- and two-photon fluorescence microendoscopy. J Neurophysiol 92(5):3121–3133
79. Levene MJ, Dombeck DA, Kasischke KA et al (2004) In vivo multiphoton microscopy of deep brain tissue. J Neurophysiol 91(4): 1908–1912
80. Bao H, Gu M (2009) Reduction of self-phase modulation in double-clad photonic crystal

fiber for nonlinear optical endoscopy. Opt Lett 34(2):148–150

81. Barretto RPJ, Messerschmidt B, Schnitzer MJ (2009) In vivo fluorescence imaging with high resolution microlenses. Nat Methods 6(7):511–512

82. Flusberg BA, Cocker ED, Piyawattanametha W et al (2005) Fiber-optic fluorescence imaging. Nat Methods 2(12):941–950

83. Murayama M, Pérez-Garci E, Lüscher H et al (2007) Fiberoptic system for recording dendritic calcium signals in layer 5 neocortical pyramidal cells in freely moving rats. J Neurophysiol 98(3):1791–1805

84. Dombeck DA, Graziano MS, Tank DW (2009) Functional clustering of neurons in motor cortex determined by cellular resolution imaging in awake behaving mice. J Neurosci 29(44):13751–13760

85. Vijayan S, Hale GJ, Moore CI, Brown EN, Wilson MA (2010) Activity in the barrel cortex during active behavior and sleep. J Neurophysiol 103(4):2074

86. O'Connor DH, Peron SP, Huber D et al (2010) Neural activity in barrel cortex underlying vibrissa-based object localization in mice. Neuron 67(6):1048–1061

87. Helmchen F, Fee MS, Tank DW et al (2001) A miniature head-mounted two-photon microscope. High-resolution brain imaging in freely moving animals. Neuron 31(6): 903–912

88. Göbel W, Nimmerjahn A, Helmchen F (2004) Distortion-free delivery of nanojoule femtosecond pulses from a Ti:sapphire laser through a hollow-core photonic crystal fiber. Opt Lett 29(11):1285–1287

89. Flusberg BA, Nimmerjahn A, Cocker ED et al (2008) Fluorescence microscopy in freely moving mice. Nat Methods 5(11):3–6

90. Harvey CD, Collman F, Dombeck DA et al (2009) Intracellular dynamics of hippocampal place cells during virtual navigation. Nature 461(7266):941–946

91. Bannwarth M, Correa IR, Sztretye M et al (2009) Indo-1 derivatives for local calcium sensing. ACS Chem Biol 4(3):179–190

92. Tsien RY (1980) New calcium indicators and buffers with high selectivity against magnesium and protons: design, synthesis, and properties of prototype structures. Biochemistry 19(11):2396–2404

93. Grynkiewicz G, Poenie M, Tsien RY (1985) A new generation of Ca^{2+} indicators with greatly improved fluorescence properties. J Biol Chem 260(6):3440–3450

94. Minta A, Kao JP, Tsien RY (1989) Fluorescent indicators for cytosolic calcium based on rhodamine and fluorescein chromophores. J Biol Chem 264(14):8171–8178

95. Magee JC, Johnston D (1997) A synaptically controlled, associative signal for Hebbian plasticity in hippocampal neurons. Science 275(5297):209–213

96. Brown J (1975) Rapid changes in intracellular free calcium concentration. Detection by metallochromic indicator dyes in squid giant axon. Biophys J 15(11):1155–1160

97. Graubard K, Ross WN (1985) Regional distribution of calcium influx into bursting neurons detected with arsenazo III. Proc Natl Acad Sci U S A 82(16):5565–5569

98. Helmchen F, Svoboda K, Denk W (1999) In vivo dendritic calcium dynamics in deep-layer cortical pyramidal neurons. Nat Neurosci 2(11):989–996

99. Smetters D, Majewska A, Yuste R (1999) Detecting action potentials in neuronal populations with calcium imaging. Methods 18(2):215–221

100. Waters J, Larkum M, Sakmann B et al (2003) Supralinear Ca^{2+} influx into dendritic tufts of layer 2/3 neocortical pyramidal neurons in vitro and in vivo. J Neurosci 23(24): 8558–8567

101. Garaschuk O, Milos R, Konnerth A (2006) Targeted bulk-loading of fluorescent indicators for two-photon brain imaging in vivo. Nat Protoc 1(1):380–386

102. Helmchen F, Imoto K, Sakmann B (1996) Ca^{2+} buffering and action potential-evoked Ca^{2+} signaling in dendrites of pyramidal neurons. Biophys J 70(2):1069–1081

103. Ohki K, Chung S, Ch'ng YH et al (2005) Functional imaging with cellular resolution reveals precise micro-architecture in visual cortex. Nature 433(7026):597–603

104. Ozden I, Sullivan MR, Lee HM et al (2009) Reliable coding emerges from coactivation of climbing fibers in microbands of cerebellar Purkinje neurons. J Neurosci 29(34): 10463–10473

105. Nimmerjahn A, Kirchhoff F, Kerr JN et al (2004) Sulforhodamine 101 as a specific marker of astroglia in the neocortex in vivo. Nat Methods 1(1):31–37

106. Kang J, Kang N, Yu Y et al (2010) Sulforhodamine 101 induces long-term potentiation of intrinsic excitability and synaptic efficacy in hippocampal CA1 pyramidal neurons. Neuroscience 169(4):1601–1609

107. Ozden I, Lee HM, Sullivan MR et al (2008) Identification and clustering of event patterns from in vivo multiphoton optical recordings of neuronal ensembles. J Neurophysiol 100(1): 495–503

108. Shaner NC, Campbell RE, Steinbach PA et al (2004) Improved monomeric red, orange and yellow fluorescent proteins derived from Discosoma sp. red fluorescent protein. Nat Biotechnol 22(12):1567–1572
109. Miyawaki A, Llopis J, Heim R et al (1997) Fluorescent indicators for Ca^{2+} based on green fluorescent proteins and calmodulin. Nature 388(6645):882–887
110. Miyawaki A, Griesbeck O, Heim R et al (1999) Dynamic and quantitative Ca^{2+} measurements using improved cameleons. Proc Natl Acad Sci U S A 96(5):2135–2140
111. Wallace DJ, Meyer S, Astori S et al (2008) Single-spike detection in vitro and in vivo with a genetic Ca^{2+} sensor. Nat Methods 5(9):797–804
112. Mank M, Santos AF, Direnberger S et al (2008) A genetically encoded calcium indicator for chronic in vivo two-photon imaging. Nat Methods 5(9):805–811
113. Hendel T, Mank M, Schnell B et al (2008) Fluorescence changes of genetic calcium indicators and OGB-1 correlated with neural activity and calcium in vivo and in vitro. J Neurosci 28(29):7399–7411
114. Horikawa K, Yamada Y, Matsuda T et al (2010) Spontaneous network activity visualized by ultrasensitive $Ca(^{2+})$ indicators, yellow Cameleon-Nano. Nat Methods 7(9): 729–732
115. Hasan MT, Friedrich RW, Euler T et al (2004) Functional fluorescent Ca^{2+} indicator proteins in transgenic mice under TET control. PLoS Biol 2(6):e163
116. Civillico EF, Contreras D (2006) Integration of evoked responses in supragranular cortex studied with optical recordings in vivo. J Neurophysiol 96(1):336–351
117. Kogan A, Ross W, Zecevic D et al (1995) Optical recording from cerebellar Purkinje cells using intracellularly injected voltage-sensitive dyes. Brain Res 700(1–2):235–239
118. Orbach HS, Cohen LB (1983) Optical monitoring of activity from many areas of the in vitro and in vivo salamander olfactory bulb: a new method for studying functional organization in the vertebrate central nervous system. J Neurosci 3(11):2251–2262
119. McQuiston AR, Saggau P (2003) Mu-opioid receptors facilitate the propagation of excitatory activity in rat hippocampal area CA1 by disinhibition of all anatomical layers. J Neurophysiol 90(3):1936–1948
120. Kuhn B, Denk W, Bruno RM (2008) In vivo two-photon voltage-sensitive dye imaging reveals top-down control of cortical layers 1 and 2 during wakefulness. Proc Natl Acad Sci U S A 105(21):7588–7593
121. Antić S, Zecević D (1995) Optical signals from neurons with internally applied voltage-sensitive dyes. J Neurosci 15(2):1392–1405
122. Nuriya M, Jiang J, Nemet B et al (2006) Imaging membrane potential in dendritic spines. Proc Natl Acad Sci U S A 103(3): 786–790
123. Dombeck DA, Sacconi L, Blanchard-Desce M et al (2005) Optical recording of fast neuronal membrane potential transients in acute mammalian brain slices by second-harmonic generation microscopy. J Neurophysiol 94(5): 3628–3636
124. Stuart GJ, Palmer LM (2006) Imaging membrane potential in dendrites and axons of single neurons. Pflügers Archiv: Eur j physiol 453(3):403–410
125. Siegel MS, Isacoff EY (1997) A genetically encoded optical probe of membrane voltage. Neuron 19(4):735–741
126. Guerrero G, Siegel MS, Roska B et al (2002) Tuning FlaSh: redesign of the dynamics, voltage range, and color of the genetically encoded optical sensor of membrane potential. Biophys J 83(6):3607–3618
127. Ataka K, Pieribone VA (2002) A genetically targetable fluorescent probe of channel gating with rapid kinetics. Biophys J 82(1 Pt 1): 509–516
128. Chanda B, Blunck R, Faria LC et al (2005) A hybrid approach to measuring electrical activity in genetically specified neurons. Nat Neurosci 8(11):1619–1626
129. Sjulson L, Miesenböck G (2008) Rational optimization and imaging in vivo of a genetically encoded optical voltage reporter. J Neurosci 28(21):5582–5593
130. Perret D, Marchese S, Gentili A et al (2008) LC–MS–MS determination of stabilizers and explosives residues in hand-swabs. Chromatographia 68(7–8):517–524
131. Gautam SG, Perron A, Mutoh H et al (2009) Exploration of fluorescent protein voltage probes based on circularly permuted fluorescent proteins. Front Neuroengin. doi:10.3389/neuro.16.014.2009
132. Perron A, Mutoh H, Launey T et al (2009) Red-shifted voltage-sensitive fluorescent proteins. Chem Biol 16(12):1268–1277
133. Lundby A, Akemann W, Knöpfel T (2010) Biophysical characterization of the fluorescent protein voltage probe VSFP2.3 based on the voltage-sensing domain of Ci-VSP. Eur Biophys J. doi:10.1007/s00249-010-0620-0

Neuromethods (2012) 67: 21–43
DOI 10.1007/7657_2011_1

Published online: 24 November 2011

In Vivo Functional Imaging of the Olfactory Bulb at Single-Cell Resolution

Stefan Fink, Yury Kovalchuk, Ryota Homma, Bernd Schwendele, Stephan Direnberger, Lawrence B. Cohen, Oliver Griesbeck, and Olga Garaschuk

Abstract

Functional properties of neuronal circuits can be best studied in vivo in the living mammalian brain. The use of optical methods, like two-photon calcium imaging, permits analyses of network function at single-cell resolution. This chapter provides a step-by-step description of this technique. Using mouse olfactory bulb as a model system, we compare the performance of genetically encoded calcium sensor TN-XXL and small-molecule calcium indicators; describe how to choose the right calcium indicator and how to load it into the cells of interest; discuss the use of cell type-specific markers and, finally, illustrate the application of this technique for high-resolution in vivo imaging of sensory-driven neuronal activity.

Key words: Two-photon microscopy, In vivo calcium imaging, Multicell bolus loading, Olfactory bulb, Genetically encoded calcium indicators, Small-molecule calcium indicators, TN-XXL, Fluorescence resonance energy transfer

1. Introduction

Sensory as well as other behavior-relevant signals within the mammalian brain are processed by multicellular networks. Recent evidence suggests that such networks do not comprise exclusively neurons but contain other important cellular elements as, for example, astrocytes (1–3) and microglia (4). Investigators setting out to uncover the major principles of information processing by such networks need precise and highly specific tools. Ideally, such tools should enable simultaneous monitoring of activity in a large number of identified network elements at high spatial and temporal resolution and should allow measurements in awake, behaving animals. Such experiments first became possible through the development of in vivo two-photon microscopy (5, 6). Over the last decades, in vivo two-photon microscopes were improved and miniaturized (7–9) and allow monitoring of cortical activity in awake, freely moving/behaving rodents (9–11). In addition to a microscope, however, one needs good sensor molecules capable of measuring activity in multicellular networks with single-cell or even subcellular resolution. In neurons, voltage, sodium, and calcium sensors can be used. In other cells (e.g., microglia and astrocytes), in which intracellular signaling is not immediately linked to profound changes in the cell's membrane potential and associated sodium fluxes, using a calcium sensor is more advantageous.

The use of calcium sensors for monitoring neuronal activity exploits of the fact that depolarizing electrical signals are associated with Ca^{2+} influx through numerous types of voltage-gated Ca^{2+} channels, abundantly expressed in the nervous system (12, 13). These signals are often amplified further by Ca^{2+} release from the intracellular Ca^{2+} stores (12, 13). In glial cells, Ca^{2+} signals are usually caused by an activation of various ionotropic and metabotropic receptors, Ca^{2+} release from the intracellular Ca^{2+} stores and associated Ca^{2+} entry through store-operated channels in the plasma membrane (for a review, see refs. 14–18). Thus, imaging of the intracellular calcium dynamics is a universal tool for accessing activity of complex multicellular networks.

There are two large groups of calcium indicators referred to as small-molecule calcium indicators (SMCIs, e.g., Fura-2, Oregon Green BAPTA-1, etc.) and genetically encoded calcium indicators (GECIs, (19–23)). SMCIs of high quality were first synthesized 25 years ago (24–26) and have been widely used in different areas of life sciences. These were also the first sensors used for in vivo calcium imaging of neuronal activity in the mammalian brain both at the single cell (6) and network (27) levels. SMCIs have fast kinetics (on rate: 10^8 M^{-1} s^{-1}, off rate: 100–370 s^{-1} (28, 29)), low pH and Mg^{2+} sensitivity (28), relatively high dynamic range ($F_{highCa}/F_{0Ca} = 6–40$), and two-photon excitation cross sections in the range of 2–12 GM (30, 31). So far, they are used as the "gold standard" when imaging living cells in the mammalian brain tissue.

The latest-generation GECIs include Yellow Cameleon 3.6, TN-XXL, G-CaMP3, and yellow Cameleons-Nano. These sensors have been engineered for increased signal strength and cover a range of calcium affinities for predominantly cytosolic applications, with the Cameleons-Nano more tuned for ultra high affinity to monitor resting levels of free calcium inside cells. While their maximally attainable dynamic range approaches the values for SMCIs in some cases (e.g., $F_{highCa}/F_{0Ca} \geq 11$ for G-CaMP3 (32)), other parameters, such as kinetics (on rate: 10^6 M^{-1} s^{-1}, off rate: 1–10 s^{-1} (19, 33, 34)), are generally still inferior for imaging with high temporal resolution.

In this chapter, we describe the techniques for studying in vivo function of multicellular networks by means of two-photon Ca^{2+} imaging. We use the same preparation: mouse main olfactory bulb to compare the performance of SMCIs with that of modern genetically encoded Ca^{2+} sensors (using TN-XXL as an example). The following issues are addressed: (1) how to choose a right indicator, (2) how to get it into the cells of interest, (3) how to identify different elements of multicellular networks, and (4) how to achieve good subcellular resolution. We also address the pitfalls and drawbacks of the staining techniques used and finally illustrate the application of these techniques for high-resolution in vivo imaging of sensory-driven neuronal activity.

1.1. Choosing an Indicator Dye

The following criteria have to be considered when choosing a suitable Ca^{2+} indicator: calcium affinity, kinetics, dynamic range, linearity of the response over the range of physiological Ca^{2+} concentrations, brightness, photostability, and the mode of readout (e.g., a single wavelength or a ratiometric indicator). In addition, when studying a multicellular network, the ability to obtain a cell type-specific labeling is of great interest.

Both SMCIs and GECIs comprise Ca^{2+} indicators with a broad range of Ca^{2+} affinities. Therefore, the decision about the indicator's affinity mainly depends on the range of Ca^{2+} concentrations to be measured. For detecting subtle cytosolic changes in the intracellular free Ca^{2+} concentration ($[Ca^{2+}]_i$) due to firing of one or few action potentials, high-affinity indicators are preferentially used. So far, the most widely used SMCIs for in vivo monitoring of low levels of activity are Oregon Green BAPTA-1 (OGB-1, K_d ~ 170 nM) and Fura-2 (K_d ~ 145 nM). It must be borne in mind, however, that K_d values of the indicators depend on the temperature, pH, and ionic strength of the solution and are usually higher in situ and in vivo compared to in vitro. For example, the effective K_d of Fura-2 determined in cardiomyocytes under two-photon excitation was ~180 nM (31) instead of 145 nM (see above). The newly designed family of ultrasensitive GECIs yellow Cameleon-Nano (35) includes Ca^{2+} indicators with different K_d values ranging from 15 to 140 nM. These new proteins hold big promise as reliable sensors of sparse neuronal activity. However, because of their highly nonlinear response (see Fig. 2 in ref. 35), one may need a careful selection of an indicator for a given range of activities. GCaMP3 is another GECI with a very good performance (32). Although the K_d value of GCaMP3 is higher compared to the dyes mentioned above (~660 nM), the high dynamic range of the indicator allows to detect ~60–70% of single-action potentials in vivo (32), a value which is very close to the detection probability of OGB-1 (36).

Another important decision when choosing an indicator is whether a single wavelength or a ratiometric sensor has to be used.

Ratiometric indicators show a shift in their emission or excitation spectra upon binding Ca^{2+}. However, because of the broad two-photon excitation spectra of fluorescent dyes, only the ones with emission shifts are suitable for two-photon imaging. With such a dye, changes in $[Ca^{2+}]_i$ are proportional to a ratio between the fluorescence intensities emitted at two different wavelengths. This allows correction of artifacts due to preparation movements, changes in the cell volume, bleaching, variations in illumination intensity, etc. Unfortunately, there are no good SMCI dyes with these properties. However, some good dyes (e.g., OGB-1) can become ratiometric when instead of fluorescence intensity fluorescence lifetime imaging (FLIM) is considered (see detailed description in refs. 37, 38). Among GECIs, there exist many ratiometric (e.g., FRET based) Ca^{2+} sensors. These include D3cpv, TN-XL or TN-XXL, and yellow chameleons (22, 34, 35, 39–41). Compared to single-wavelength GECIs, FRET-based indicators are less susceptible to artifacts by motion (the main source of artifacts/noise in vivo), pH, or nonlinear bleaching (33). Therefore, they seem to be indicators of choice for in vivo imaging.

Induction of a phenotype due to the presence of a sensor is another area of concern. Although it is known that some indicators (especially calmodulin-based GECIs) induce physiological as well as behavioral phenotypes (e.g., cardiomegaly induced by constitutive expression of G-CaMP2 (42), decreased turning behavior in *Caenorhabditis elegans* induced by yellow chameleon 3.6 (YC3.6), YC2.12, GCaMP1, and GCaMP2 (32)), thorough analyses of possible side effects of the indicator expression are still missing from the literature. When using SMCI, the side effects are mostly restricted to the reduction of the free intracellular Ca^{2+} concentration (Ca^{2+} buffering (43)). The side effects caused by a long-lasting overexpression of GECIs may be much more diverse as well as preparation dependent. Therefore, it is advisable to test basic physiological properties of GECI-overexpressing cells in each particular case. When studying neurons, the basic parameters to be tested include (1) resting membrane potential, (2) a threshold for action potential generation, and (3) action potential amplitude and half-width (44).

2. Materials

2.1. Labeling of Input Neurons

- Homeothermic blanket to ensure a constant body temperature; available from many providers (e.g., TSE-Systems GmbH, Bad Homburg, Germany).

- Fluorescent dye Alexa Fluor 594-dextran (MW 10,000, Invitrogen GmbH, Darmstadt, Germany).
- Triton X-100 (Sigma-Ultra, Sigma-Aldrich Chemie GmbH, Munich, Germany).
- Standard pipette solution: 150 mM NaCl, 2.5 mM KCl, and 10 mM HEPES.
- Anesthetics: For example, a mixture of ketamine (e.g., Fagron GmbH & Co KG, Barsbuettel, Germany) and xylazine (Sigma-Aldrich; 100/10 μg/g of bodyweight).
- Microloader pipette tips (Eppendorf, Westbury, NY, USA).
- Eye ointment (e.g., Bepanthen, Bayer Vital, Leverkusen, Germany).

2.2. Surgery

- Brain atlas (e.g., from Academic Press).
- Dissecting instruments (forceps, fine scissors, e.g., Fine Science Tools GmbH Heidelberg, Germany).
- Dissection microscope (e.g., Wild Heerbrugg, Gais, Switzerland).
- Scales for weighing the animals (e.g., Kern & Sohn GmbH, Balingen, Germany).
- Syringes (1 ml) and 30 G Microlance™ needles (e.g., Becton Dickinson, Madrid, Spain).
- Hand-driven micromanipulator for craniotomy (e.g., Physik Instrumente (PI) GmbH, Karlsruhe/Palmbach, Germany).
- Local anesthetic agent (e.g., xylocain® from AstraZeneca GmbH, Wedel, Germany).
- Anesthesia-monitoring system (e.g., ADInstruments GmbH, Spechbach, Germany).
- Standard external saline (mM): 125 NaCl, 4.5 KCl, 26 $NaHCO_3$, 1.25 NaH_2PO_4, 2 $CaCl_2$, 1 $MgCl_2$, 20 glucose, pH 7.4, when bubbled with 95% O_2 and 5% CO_2.
- Compressed air.
- Recording chamber with central access opening: Custom-made from a standard tissue-culture dish (Ø 35 mm, Becton Dickinson Labware Europe, Le Pont De Claix, France).
- Dental drill (e.g., Nakanishi Ultimate 500, NSK Europe GmbH, Frankfurt, Germany).
- Dental drill bits (e.g., Meisinger GmbH, Neuss, Germany) of increasingly smaller size (Ø of 0.7–0.5 mm).
- Felt polisher (e.g., from Dr. Ihde Dental, Munich, Germany).
- Cyanoacryl glue (available from many providers; e.g., UHU GmbH & Co. KG, Buehl/Baden, Germany).

2.3. Staining Neurons with SMCIs (Multicell Bolus Loading Technique)

- Glass capillaries (e.g., from Hilgenberg GmbH, Malsfeld, Germany).
- Pipette puller (e.g., PP830 from Narishige International Limited, London, UK).
- Micromanipulator (e.g., LN-Mini from Luigs & Neumann GmbH, Ratingen, Germany).
- Pressure-application device (e.g., NPI electronic GmbH, Tamm, Germany).
- Patch-clamp amplifier (e.g., EPC10 from HEKA, Lambrecht/Pfalz, Germany).
- Research grade upright biological/life science optical microscope (e.g., BX51WI or BX61WI Olympus Europa GmbH, Hamburg, Germany).
- Cold light source (e.g., Euromex microscopes, Arnheim, Holland).
- Peristaltic pump (e.g., ISMATEC SA, Glattbrugg, Switzerland).
- Filter (e.g., pore diameter of 0.45 μm, Millipore GmbH, Schwalbach/Ts., Germany).
- 20% Pluronic F-127 in DMSO (e.g., Sigma-Aldrich Chemie GmbH, Munich, Germany).
- Oregon Green 488 BAPTA-1 AM (e.g., Molecular Probes, Invitrogen GmbH, Darmstadt, Germany or TefLabs Inc, Austin, TX, USA) or other membrane-permeable calcium-indicator dyes.

2.4. In Vivo Two-Photon Calcium Imaging

- Two-photon laser-scanning microscopes are commercially available from several providers (e.g., Zeiss, Oberkochen, Germany; Leica Mikrosysteme Vertrieb GmbH, Wetzlar, Germany; Prairie Technologies Inc., Middleton, WI, USA; LaVision GmbH, Goettingen, Germany; etc.). We use the Olympus Fluoview 1000 laser-scanning microscope system (BX61WI, Olympus Europa GmbH, Hamburg, Germany) coupled to a mode-locked Ti:sapphire laser system (Mai Tai DeepSee, Newport Spectra-Physics GmbH, Darmstadt, Germany) operating at 690–1,040 nm wavelength, with a pulse width of <100 fs and repetition rates of 80 MHz. Objectives: 10× UMPlan FI 0.3 NA, 20× UMPlan FI 0.5 NA, 40× LUMPlan FLN 0.8 NA, 60× LUMPlan FLN 0.8 NA (all from Olympus Europa GmbH, Hamburg, Germany), and 40× NIR Apo 0.8 NA (Nikon GmbH, Duesseldorf, Germany).

- Optical power meter (e.g., Newport Spectra-Physics GmbH, Darmstadt, Germany).
- Low melting point agarose (Invitrogen GmbH, Darmstadt, Germany).

3. Methods

3.1. Labeling of Input Neurons

Multicellular neuronal networks can be roughly dissected into five major components: input and output neurons, local interneurons, surrounding astrocytes and microglia. In the olfactory bulb, input (olfactory receptor) neurons and their processes can be conveniently stained by applying the dextran-conjugated form of a small-molecule indicator dye into the nasal cavity (45). This protocol accommodates SMCIs, such as 10 kDa Calcium Green-1 dextran (45), as well as other fluorescent dyes, for example, 10 kDa Alexa Fluor 594-dextran (Fig. 1). The staining solution contains 3% w/v of the dye diluted in standard pipette solution and 0.063% v/v of Triton X-100 and should be freshly prepared just before the staining procedure. The animals are anesthetized by a mixture of ketamine/xylazine (100/10 μg/g of body weight) and placed onto a warming blanket while monitoring the body temperature and the breathing rate. The surgical level of anesthesia is ascertained by testing, for example, the toe pinch withdrawal reflex. With animal resting on its backside, the staining solution is applied by gently inserting the tip of the microloader 8–9 mm into the nostril and ejecting slowly 10 μl of the staining solution in two portions a few minutes apart. After the dye delivery, the animal can rest on its side, opposite to the injected nostril, until it wakes up. Four to five days post injection, the dye is usually taken up by olfactory receptor neurons and transported anterogradely along the axons to label presynaptic elements in the olfactory bulb.

3.2. Surgery

We recommend consulting a brain atlas for stereotaxic coordinates of the respective brain region of interest before starting the described procedure.

Mice are anesthetized by i.p. injection of a mixture of ketamine/xylazine (see above). Alternatively, adult mice can be anesthetized with an i.p. injection of 1.9 mg urethane per g body weight or by inhalation of isoflurane (1–1.5% (v/v) in pure O_2) depending on the intended measurements. After inducing anesthesia, the animal is placed onto a homeothermic blanket (preheated to 38°C) in front of an oxygen supply. The sensors of the anesthesia-monitoring system are affixed and the readout (e.g., the animal's temperature, blood pressure, and breathing rate) is closely monitored. As already mentioned above, we routinely apply an eye

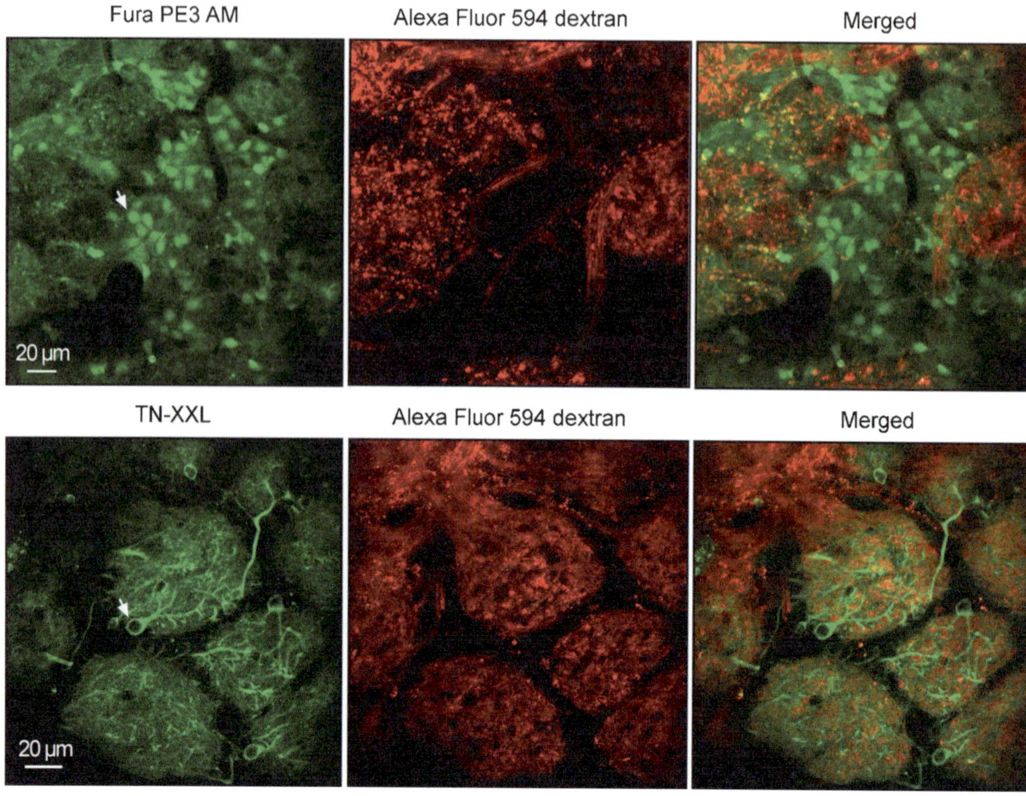

Fig. 1. Identification of cells and network elements in vivo. Juxtaglomerular cells of the mouse main olfactory bulb were either stained with Fura PE3 using the multicell bolus loading technique (MCBL, *upper panel*) or labeled with a genetically encoded calcium indicator TN-XXL by means of transgene expression (*lower panel*). The images shown here and below are maximum intensity projections of $\pm$3–5 μm, taken in 1-μm steps. The images in the *upper panel* are taken around 50 μm and those in the *lower panel* around 90 μm below the pia. Note the difference between the appearances of the cells labeled using the two staining techniques (*arrows*). While Fura PE3 labels the entire cell body, the TN-XXL labeling is clearly cytosolic (with unstained nucleus). On the other hand, dendritic processes are not visible in the Fura PE3-based stain, but can be clearly identified in preparations labeled with TN-XXL. In both panels, afferents of olfactory sensory neurons are labeled anterogradely with Alexa dextran (*middle*). The merged images on the right show presynaptic elements in *red* and the postsynaptic juxtaglomerular cells (and their processes) in *green*.

ointment to prevent the animal's eye from drying out. The surgical level of anesthesia is reached ~10 min after its induction and has to be ascertained by testing the known reflexes (e.g., the eyelid and the toe pinch withdrawal reflex). Before removing the skin over the desired brain area, we inject ~50 μl of a local anesthetic agent (e.g., 2% w/v lidocaine) subcutaneously at the location, where the skin is to be removed. Two to five minutes later, the skin can be removed with fine scissors. Any tissue debris on top of the animal's skull should be gently scraped with a surgical blade and removed completely. The surface of the bone should be dried, e.g., with a fine stream of compressed air before a custom-made recording chamber (46) can be fixed to the skull with a cyanoacryl glue. Any glue residues covering the recording area should be removed immediately. Allow the glue to dry for 10–20 min.

For thinning the skull, the recording chamber is fixed in an appropriate holder under a dissecting microscope and the skull under the opening of the recording chamber is thinned using a dental drill. Care should be taken to evenly thin the skull in order to prevent breaking of the bone during craniotomy. Especially, pressing onto the skull while drilling should be avoided. Inaccurate thinning often causes brain damage. One has to stop drilling when the brain vessels located underneath the dura become visible (the skull transparency can be improved by applying standard external saline) and to test if the bone bends when gently touched with the tweezers. When the bone is thinned, the animal is transferred into the recording setup. The sensors of the anesthesia-monitoring system are refastened. The supply of anesthetics is adjusted, if necessary, to keep physiological levels of monitored parameters (Note 1). The recording chamber is filled and perfused with a warm (37°C) standard external saline.

In general, there are two approaches allowing imaging activity of multicellular networks in vivo. The imaging can either be done through the opened or thinned skull (27). Compared to the thinned-skull preparations, opened-skull preparations provide images of a higher contrast and better depth penetration (27), but suffer more from movement artifacts and, when used for chronic recordings, may cause extensive activation of microglia and astrocytes (47). For imaging through the skull, the skull is thinned with drill bits of increasingly smaller size (Ø of 0.7–0.5 mm) to the thickness of ~10–20 μm and then the remaining bone is polished with a felt polisher (27). This is a critical step as the procedure of thinning/polishing may damage the brain tissue and should, therefore, be performed with maximal gentleness. When using genetically modified animals expressing GECIs, a craniotomy is not necessary, whereas the use of SMCIs requires at least a small craniotomy to accommodate the staining pipette. For opening the skull, we use a 30-G needle attached to a hand-driven micromanipulator instead of a freehand operation. To open the skull, mark and cut a rectangle in the bone by making four straight, slightly overlapping cuts. When the rectangle is cut, gently lift and remove the bone without touching the underlying tissue. Usually, the complete rectangle is lifted without breaking. We advice not to remove the dura. Leaving the dura in place allows protecting and stabilizing the underlying tissue (Note 2).

3.3. Cell Staining

3.3.1. Staining Neurons with SMCIs

There are several techniques allowing in vivo labeling of living cells with SMCIs. These include (1) whole-cell perfusion (6, 48), (2) in vivo electroporation (49–52), and the (3) multicell bolus loading technique (MCBL (27)). Because large multicellular networks are most conveniently stained using MCBL, this technique is described below in detail.

The acetoxymethyl (AM) ester form of an SMCI is dissolved in DMSO containing 20% Pluronic F-127 (w/v) to a final concentration of 10 mM. This solution is further diluted 1/10 to 1/100 with the standard pipette solution (composition see above) to obtain the staining solution (Note 3). The micropipette (resistance of 6–9 MΩ when filled with standard pipette solution, tip diameter ~1 μm) is pulled from a glass capillary using a pipette puller and filled with ~7 μl of staining solution. The pipette is inserted into the brain tissue using a micromanipulator. The resistance of the pipette is continuously monitored with a patch-clamp amplifier. It can transiently increase up to 15 MΩ during penetration of the dura mater. A stable increase of the pipette resistance above 20 MΩ indicates blockage of the pipette, which may render the dye ejection procedure impossible. In this case, the staining pipette should be replaced. When the pipette reaches the desired depth, the dye is pressure ejected into the brain by applying the positive pressure (70 kPa for 1 min). With the pipette tip located 150–200 μm below the cortical surface, the dye ejection stains both neurons and glia up to a depth of ~400 μm from the cortical surface (Fig. 1 (27)). When deeper cortical layers are targeted, the tip of the pipette is positioned at 650–700 μm below the pia (53). Following dye ejection, the pipette is removed and the experiment commences approximately 1 h later to allow complete deesterification of the dye.

3.3.2. Staining Neurons with GECIs

Several gene transfer techniques are available for in vivo expression of GECIs in the mammalian brain. In utero electroporation of plasmid DNA (54, 55) is a fast method that can lead to expression of a sensor in the intact mouse brain without need for long-lasting procedures to generate transgenic mice. It is possible to direct expression to the major anatomical regions of the brain, but expression is more or less scattered and not entirely reproducible. Recently, the use of this technique was extended to early postnatal age (56) and used for labeling newborn olfactory interneurons. The postnatal electroporation works well in rats and mice and provides a long-lasting (up to 1 year) staining of cells with different GFP variants, including GCaMP2.

Another efficient means to introduce foreign genes into postmitotic neurons are viral vectors (57–59). A number of different vectors are available. They differ in efficiency, preference for different cell types in the brain, insert capacity, and potential toxicity. In particular, the latest generations of lentiviral vectors and vectors based on adeno-associated virus (AAV) have proven to introduce genes into neurons with minimal adverse effects. Also this technique offers the possibility of long-term expression of the construct (for at least several months, see refs. 60, 61). Viral vectors can be used to transfect cells in many different species. However, the reproducibility of expression patterns remains a challenge. Using this technique, it is also difficult to achieve cell-type specificity of labeling because only a very limited number of cell type-specific promoters are small enough to be incorporated into the viral vectors.

Cell-type specificity can be achieved when generating a transgenic organism stably incorporating the indicator gene in its genome. Using the currently available genetic toolbox allows relatively precise targeting of a gene of interest to defined populations of cells and/or cellular subcompartments, maximal degree of reproducibility of the expression pattern from one animal to the next, and the ease of reproduction of experimental animals. However, this procedure is most labor and time-consuming and requires well-trained personnel.

3.4. In Vivo Two-Photon Calcium Imaging

Depending on the indicator used, an appropriate wavelength of the excitation light has to be determined. For two-photon excitation of common SMCIs (e.g., OGB-1, Fura-PE3, Indo-1, Calcium Green-1, and Fluo-4), an excitation light of 800 nm can be used for the initial visualization of stained neurons. As far as GECIs are concerned, 900–910-nm excitation light is used to excite GCaMPs (32, 41, 62), whereas fluorescence resonance energy transfer (FRET) indicators are excited at wavelengths between 830 and 890 nm (40, 44, 63, 64). In our laboratory, we usually excite FRET indicators at 860–890 nm. Such excitation paradigm provides sufficient basal levels of fluorescence for both CFP and YFP and thus reduces noise in ratio images (44). To separate emission of CFP and YFP, we and others use a 515-nm beam splitter (e.g., from Chroma). For FRET indicators, employing CFP/YFP 515 nm represents an isosbestic point at which indicator's fluorescence is calcium independent. In addition to the beam splitter, band-pass emission filters (e.g., 480/30 nm for CFP and 535/30 nm for YFP, Chroma) are used to separate the fluorescence of the two proteins.

To optimize the use of excitation light and, as a consequence, to increase the imaging depth within a scattering tissue like the mammalian brain, it is advisable to compensate the group-velocity dispersion (GVD) caused by the glass-containing parts of the imaging system. GVD occurs due to the fact that "reddish" part of the ultrashort (femtosecond) laser pulse travels faster through optical materials (i.e., windows, lenses, and nonlinear crystals) compared to the "bluish" one. This causes dispersive broadening of laser pulses, i.e., a decrease in the peak intensity of the excitation light. GVD can be compensated by introducing the so-called negative chirp or prechirp (i.e., by forcing the "reddish" part of the pulse to travel longer distance within the optical material). There are several devices on the market that can do this (e.g., APE GmbH Berlin Germany). We use a two-photon laser with a built-in GVD-compensation unit (Mai Tai DeepSee from Newport Spectra-Physics GmbH, Darmstadt, Germany).

3.5. Identification of Network Elements

Figure 1 illustrates two rows of images taken in vivo in the glomerular layer of the mouse olfactory bulb. In both preparations, olfactory receptor neurons were stained with 10 kDa Alexa Fluor 594-dextran as described above. Glomerular neurons were either stained with the SMCI Fura PE3 (upper row) or labeled with TN-XXL (40) by making a transgenic mouse expressing this GECI under the Thy 1.2 promoter. The left panel of the image allows comparison of the two staining techniques. The MCBL-based technique (upper row) labels virtually all cells (neurons and astrocytes) within the plane of view. The only dark spots correspond to blood vessels (27, 46, 65). In contrast, fewer cells are stained in a TN-XXL transgenic mouse (note dark areas between the glomeruli). The relative sparseness of staining as well as a high brightness of cells expressing TN-XXL allow the visualization not only of cell somata (arrows), but also of a large part of the neuron's dendritic tree. As was shown previously (see ref. 44), the use of Thy 1.2 promoter provides neuron-specific expression of GECIs leaving astrocytes unlabeled.

In a preparation stained with SMCIs using MCBL, neurons can be distinguished from astrocytes by using an astrocyte-specific marker sulforhodamine 101 (SR101, (66)). However, the SR101-based labeling of astrocytes is preparation specific. For example, SR101 fails to label astrocytes in the olfactory bulb (R. H, Y. K., L. B. C., and O. G, unpublished observations), and labels only a subset of astroglial cells in the hippocampus (67) as well as a mixed population of neurons and astrocytes in the spinal cord (68). In addition, SR101 turned out to be a highly bioactive molecule. Applied in concentrations used to label astrocytes (25–100 μM, (66)), this substance increases the excitability of neuronal tissue (Fig. 2). In the hippocampus, for example, bath application of SR101 induces a five- to sixfold increase in the amplitude of synaptically evoked population spike caused by stimulation of Schaffer collaterals and provokes synchronized repetitive firing of neurons at a frequency of 130–200 Hz (Fig. 2a,b). The SR101-evoked hyperexcitability develops rapidly (the maximal effect is reached during the first 5 min of drug application) and persists at least for 40–60 min after washout of the drug (Fig. 2a–c). Field excitatory postsynaptic currents recorded in the *stratum radiatum* of the CA1 region also undergo potentiation which persists long after the washout of SR101 (Fig. 2d, e). Thus, SR101, like other drugs inducing chemical long-term potentiation (LTP), causes long-lasting changes in neuronal excitability and the efficiency of excitatory synaptic transmission. Reducing the SR101 concentration is unlikely to solve this problem. Even when used in a concentration as low as 1 μM, SR101 still increases neuronal excitability by causing (1) a persisting reduction in the threshold for action potential firing and (2) a potentiation of excitatory synaptic transmission (69). Moreover, it induced epileptic seizures

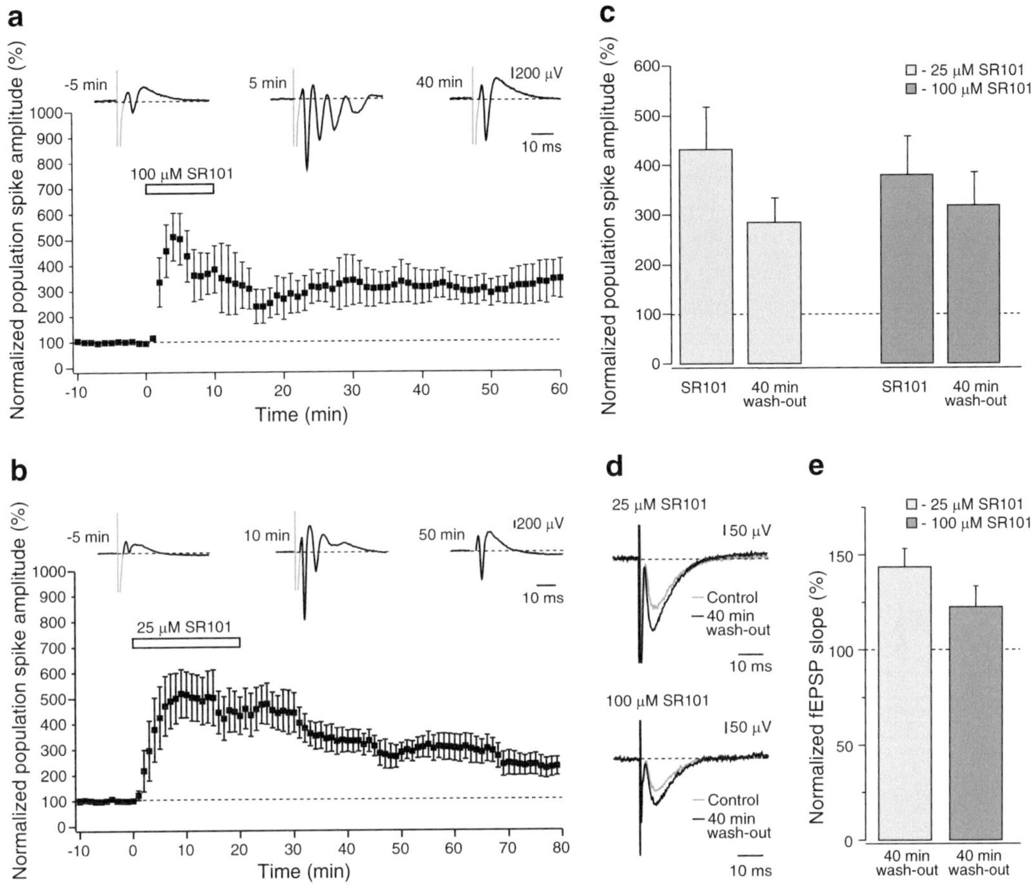

Fig. 2. Sulforodamine 101 increases neuronal excitability and induces long-term enhancement of CA3-to-CA1 synaptic transmission in hippocampal slices. (**a**, **b**) SR101-induced changes in the amplitude of the population spike recorded in the *stratum pyramidale* of the CA1 region. Data points represent mean ± S.E.M. of 5 experiments each. SR101 was bath applied during the time window marked with horizontal bars. Inserted traces (means of 12 consecutive responses) illustrate the time course of population spikes before, during, and after application of the drug. Truncated stimulation artifacts are shown in *gray*. (**c**) A bar graph illustrating an increase in the amplitude of the population spike during SR101 application (SR101) and after 40-min washout of SR101 ($n = 5$ each). (**d**) Dendritic recordings of field EPSPs illustrating SR101-induced potentiation of excitatory synaptic transmission in the *stratum radiatum*. *Upper panel*: Representative fEPSPs (averages of 12 consecutive traces), recorded before (control, *gray*) and 40 min after (*black*) application of 25 μM SR101. *Lower panel*: Similar experiment illustrating an effect of 100 μM SR101. (**e**) A bar graph summarizing the effect of SR101 on the slope of fEPSPs. When applied at 25 μM, SR101 caused an LTP of fEPSPs in all experiments ($n = 7$). 100 μM SR101 caused an LTP of fEPSPs in 3/5 experiments (illustrated) and a long-term depression of fEPSPs in two other cases. Hippocampal slices (450 μm thick) were obtained from 19- to 28-day-old C57BL/6 mice. Shaffer collaterals were stimulated at 0.2 Hz (stimulation intensity 10–30 V, pulse width 100–150 μs) with glass electrodes (tip diameter ~9 μm) filled with standard external saline. Experiments were performed at 30°C.

in vivo (69). Therefore, we would not recommend using SR101 in experiments designed to determine physiological levels of neuronal activity.

In the olfactory bulb, neurons and astrocytes can be distinguished based on the amplitude of their odor-induced responses. Although both cell types respond to odorants (70), brief (1–2-s long), applications of odorants reliably activate neurons but rarely cause glial responses. When analyzed using a set of data containing both neurons and astrocytes, odorant-evoked Ca^{2+} signals were seen in 42% of neurons (R.H, Y.K., L.B.C., and O.G, unpublished observations) and less than 30% of genetically labeled (71) astrocytes. In addition, the odorant-evoked astrocytic Ca^{2+} signals had much smaller amplitudes (<15% ΔF/F) compared to the median response amplitude (33%) in nearby neurons (R.H, Y.K., L.B.C., and O.G, unpublished observations). Fortunately, microglia, another non-neuronal cell type, are not stained in vivo using MCBL (72).

3.6. Identification of Active Cells and Cell Clusters

Fast and reliable identification of active cells/clusters is the major challenge in the experiments as the one shown in Fig. 1. To do so, one needs (1) to know the approximate size of the expected signal and (2) to have a strategy for quick identification of an active cell.

The size of the expected signal can be estimated based on in situ/in vivo calibration experiments. As shown in Fig. 3, TN-XXL is expected to give signals in the range of 10% $\Delta R/R$ in response to 4–5 and in the range of 20% $\Delta R/R$ in response to 7–10 action potentials. Note that at least in situ responses to a single-action potential can be resolved in individual trials (Fig. 3c). However, the amplitude and signal-to-noise ratio of TN-XXL-mediated Ca^{2+} signals decreased at 35°C as compared to 23°C (Fig. 3d, e) in agreement with the reduced performance of GECIs at physiological temperatures (23, 73). Thus, at 35°C, approximately 10 APs are required to reach the signal-to-noise ratio of 3. These calibration experiments suggest that the overall performance of TN-XXL is most probably not sufficient for reliable detection of in vivo activity in sparsely spiking cells (e.g., layer 2/3 cortical neurons). SMCIs seem to be much better suited for this type of experiments (36, 53). In the olfactory bulb, however, cells are frequently responding to odorants with bursts of action potentials (74–77). Under these conditions, the ability to detect a single-action potential is less of an issue. Instead, the linearity of the indicator's response becomes important. For TN-XXL, the $\Delta R/R$ signals are proportional to the number of underlying action potentials over a wide range of APs (1–40 APs, Fig. 3d (32)).

Figure 4 illustrates a technique which we use for fast detection of active cells and clusters during an experiment. To do so, we average 10–20 frames recorded (i) before and (ii) during the presentation of the sensory stimulus and subtract the two images from each other. One can refine this technique by plotting brightness of the given region of interest versus time (Fig. 4c, d) and averaging for (ii) a group of frames on the peak of the response. The ΔF image obtained in this way (Fig. 4a, b, middle panel) can then be merged with a corresponding gray-scale image (Fig. 4a, b, left panel). As illustrated on the right panel, such merged images clearly identify individual active cells/areas of the specimen.

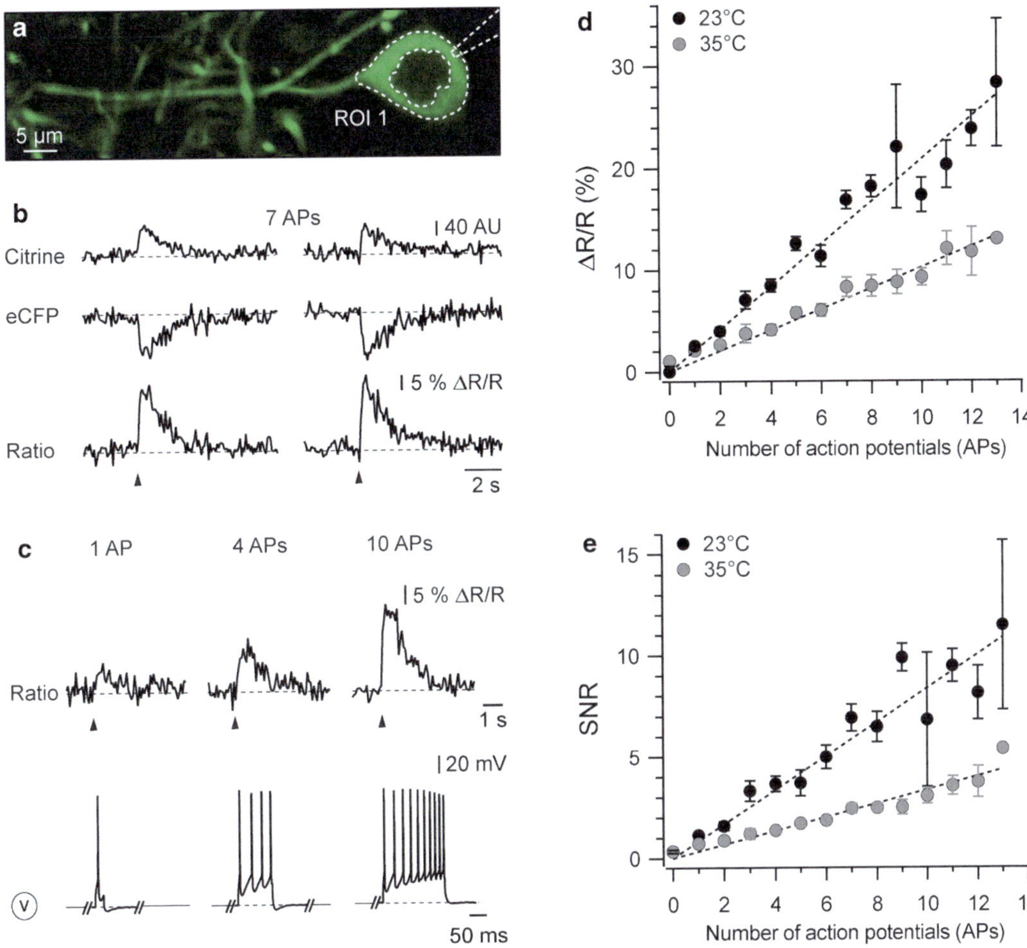

Fig. 3. Ca^{2+} sensitivity of TN-XXL in situ. (**a**) A two-photon image of a patched neuron expressing TN-XXL in a 260-μm-thick olfactory bulb slice of a 23-day-old mouse. The *dotted lines* outline the region of interest (ROI 1), where fluorescence changes in (**b**, **c**) were measured from as well as the position of the patch pipette (resistance 5 MΩ, pipette solution (in mM): 140 K gluconate, 12 KCl, 4 NaCl, 10 HEPES, 4 MgATP, 0.4 NaGTP (pH 7.3, adjusted with KOH)). (**b**) Individual raw data traces showing the change in fluorescence intensity of the donor (eCFP) and the acceptor (Citrine cp174) proteins caused by two consecutive depolarizations evoking 7 action potentials (APs) each. The *lower panel* shows the background-corrected ratio signal ($\Delta R/R$, $R =$ Citrine/eCFP). *Black arrowheads* indicate the timing of the current injection in the current clamp mode causing action potential firing. Here and in (**c**), the measurements were conducted at room temperature. (**c**) Examples of the $\Delta R/R$ signals elicited by 1, 4, and 10 APs (*upper panel*) along with corresponding voltage signals (*lower panel*) recorded from the cell shown in (**a**). (**d**, **e**) Summary graphs illustrating the dependence of $\Delta R/R$ signals (**d**) and signal-to-noise ratio (SNR) changes (**e**) on the number of underlying APs. The SNR was calculated as the ratio of average $\Delta R/R$ around the peak response to S.D. of the raw 4 s-long data trace recorded just before the stimulus onset. *Dotted lines* are linear least-square fits to the respective set of data (*black circles*: measurements conducted at 23°C, $n = 10$, P19-34; *gray circles*: measurements conducted at 35°C, $n = 8$, P32–34; Mean ± S.E.M.).

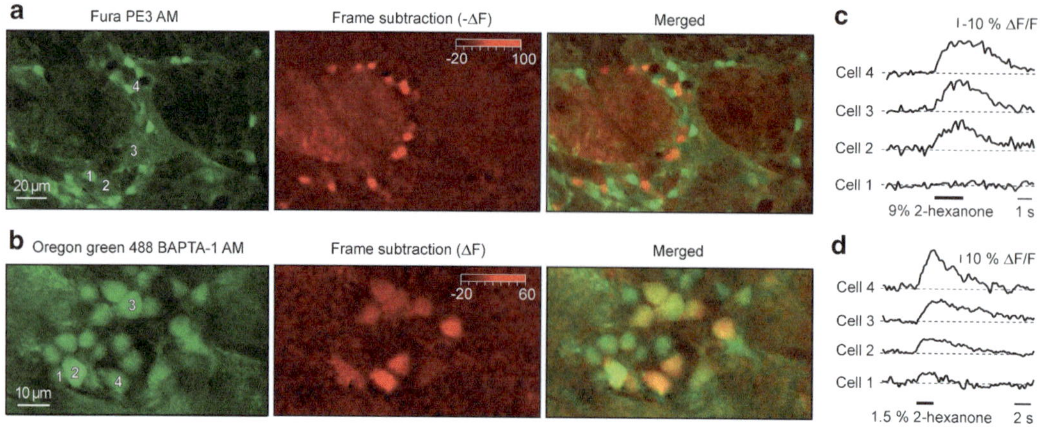

Fig. 4. Identification of active cells and cell clusters in vivo. (**a**, **b**) *Left*: Images of the glomeruli and juxtaglomerular cells in the olfactory bulb taken at lower (**a**, 80×) and higher (**b**, 160×) magnification. The cells were stained with Fura PE3 in (**a**) and with OGB-1 in (**b**) using MCBL. The *middle panels* are ΔF (frame subtraction) images obtained by subtracting an average of 10 frames recorded at the peak of odor-evoked response from a similar image recorded before the odorant presentation in (**a**) and vice versa in (**b**). The frame subtraction images were smoothed with two passes of a low-pass spatial filter. *Right panels* show merged images. Note that the active cells can be clearly identified on merged images. (**c**, **d**) Odorant-induced changes in fluorescence ($\Delta F/F$) in neurons marked with numbers in the *left panel* of (**a**) and (**b**), respectively.

This frame subtraction procedure is conveniently provided by NeuroPlex (RedShirtImaging, Decatur, GA), but can be also implemented in other imaging programs.

A similar approach can be used when imaging GECI-mediated fluorescent signals (Fig. 5). In this particular experiment, the cell shown in Fig. 5a (ROI 1) was stimulated by three consecutive presentations of the odorant 2-hexanone, applied through a custom-made flow-dilution olfactometer (78). Each of the three applications caused a decrease in the donor (eCFP) fluorescence (Fig. 5b, d) and a corresponding increase in the acceptor (Citrine) fluorescence (Fig. 5b, e), coding for an increase in $[Ca^{2+}]_i$. A Ca^{2+} transient recorded between odorant applications 2 and 3 (arrowhead in Fig. 5c) occurred spontaneously. Note that the ΔF images in D and E as well as the merged image in F not only document the Ca^{2+} signal in ROI 1, but also identify other active cells/processes, such as dendrites, in the upper left or a cell body in the lower right corner.

3.7. Imaging Odor-Evoked Calcium Signals: Examples and Applications

The experiments shown in Figs. 5 and 6 illustrate the use of TN-XXL for monitoring odor-evoked responses of juxtaglomerular neurons in the olfactory bulb. Odorant-evoked signals were not only recorded in cell somata (Figs. 5a–c and 6a, b), but also in dendrites of the neurons. Recordings of a reasonable quality were obtained even when stimulating with a low concentration of an odorant (0.3% of saturated vapor; Fig. 6b). Panel C summarizes

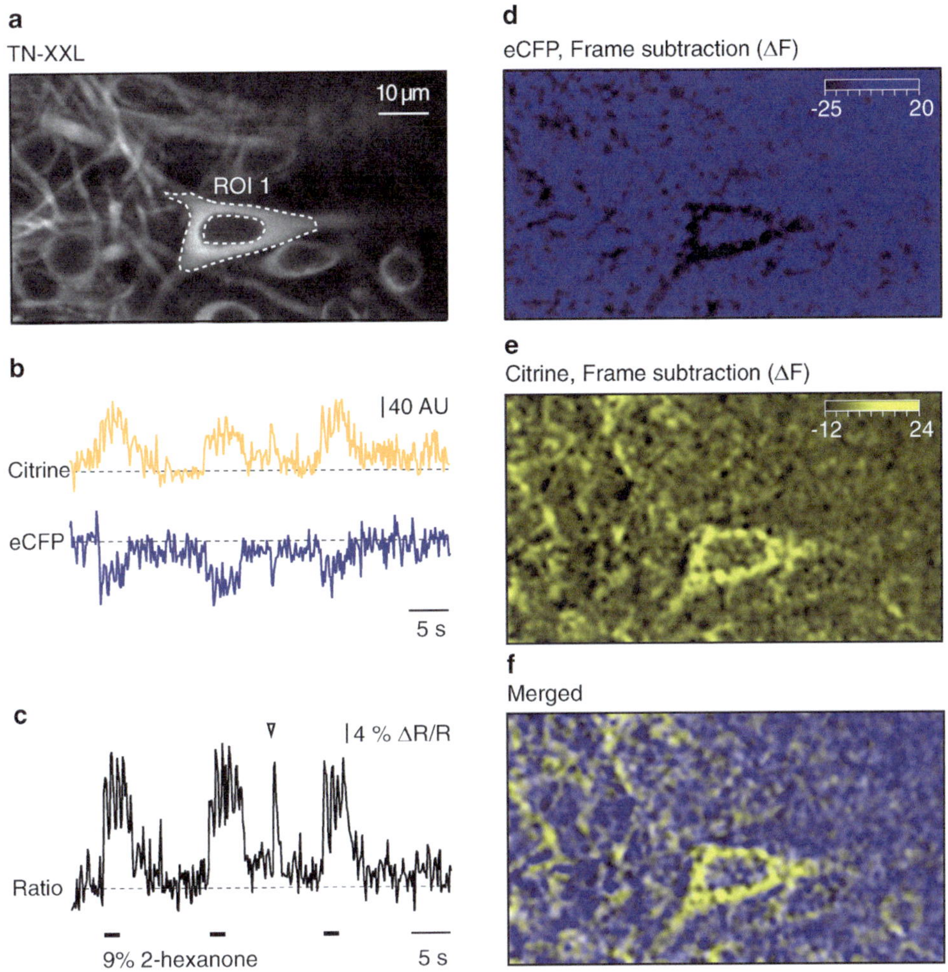

Fig. 5. In vivo identification of active cells using a FRET-based indicator TN-XXL. (**a**) An image of a large cell (ROI 1) in the glomerular layer of the olfactory bulb. (**b**) Raw data traces (recorded from ROI 1) illustrating odorant-evoked changes in fluorescence of Citrine cp174 and eCFP, respectively. (**c**) Relative changes in the corresponding Citrine/eCFP ratio. An *arrowhead* marks a spontaneously occurring Ca^{2+} transient. (**d**, **e**) ΔF (frame subtraction) images for both channels obtained by subtracting an average of 10 frames recorded before the odorant presentation from a similar image recorded at the peak of odor-evoked response. Donor channel (eCFP): *black/dark colors* mark active sites, acceptor channel (Citrine): *bright colors* mark active sites. (**f**) In the merged image, active sites appear *yellow*.

recordings conducted in five different cells stimulated with 1.5% of saturated vapor. Because TN-XXL labeling is purely cytosolic (with nucleus unstained) and dendrites of juxtaglomerular neurons are relatively thick (Figs. 5, 6), similar signal-to-noise ratios can be achieved in somatic and dendritic recordings (Fig. 6c, middle). The decay time constants of somatic and dendritic Ca^{2+} transients (1.1 ± 0.3 s, and 0.6 ± 0.1 s, respectively) were comparable to those recorded with CerTN-L15 (44) and somewhat longer than the ones obtained when using the SMCI OGB-1 (300–800 ms, (27, 53)).

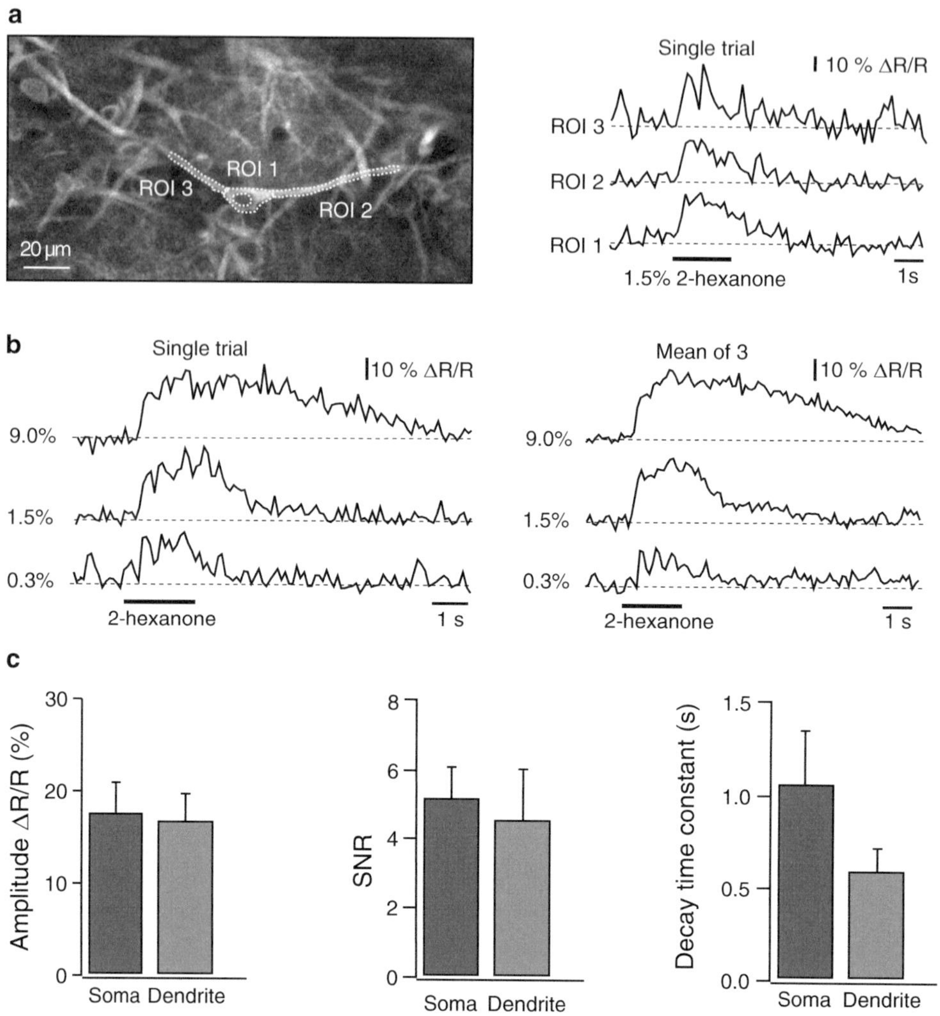

Fig. 6. Odorant-induced somatic and dendritic calcium transients in TN-XXL-labeled cells. (**a**) An in vivo image of a juxtaglomerular cell. Marked regions of interest delineate cell soma (ROI 1) and dendrites (ROI 2, 3). Note that recordings from all three regions of interest show sensory-evoked calcium transients upon stimulation with 2-hexanone. (**b**) Odorant-induced somatic Ca^{2+} transients in another cell elicited by a series of stimuli with increasing odor concentration (from 0.3 to 9% of saturated vapor). Data from single trials are shown on the left, and average of three consecutive trials on the right. Note that in this particular cell clear somatic Ca^{2+} transients were recorded even in response to a low odorant concentration (0.3% of saturated vapor). (**c**) Summary of amplitudes (*left*), signal-to-noise ratios (*middle*), and decay time constants (*right*) of somatic and dendritic sensory-evoked Ca^{2+} transients recorded in vivo using cells (n=5 each) in transgenic mice expressing TN-XXL.

Thus, the genetically encoded Ca^{2+} indicator TN-XXL is capable of capturing sensory-evoked somatic and dendritic signals in vivo. Moreover, the dye allows the recording of responses to a whole range of odorant concentrations (Fig. 6b). However, according to the calibration curves shown in Fig. 3d, e, the TN-XXL-mediated Ca^{2+} signals are caused by a burst of some 10–14 action potentials. While such a response pattern is rather typical for the olfactory bulb (74–77), it is very different from the sparse response pattern in the sensory cortices (53, 79, 80). Accordingly, sensory-evoked Ca^{2+} transients in the visual cortex were detected in 63% of OGB-1- and only in 22% of TN-XXL-labeled neurons (40). Further experiments are necessary to find out which of the modern GECIs is better suited for recording cortical neuronal activity (see also 81).

In conclusion, both SMCIs and GECIs can be used for studying the function of multicellular networks in vivo. From the future perspective, the use of GECIs is preferable because they can be introduced specifically into cellular organelles, cells, tissues, or even whole organisms of interest. Other potential advantages of GECIs lie in their ratiometric nature, ability to label long-range neuronal projections, and their applicability for repeated long-term imaging of cellular function (34, 40, 64, 82). At present, however, SMCIs still have a better dynamic range and faster binding and unbinding kinetics and, importantly, are much better tested for possible side effects. Hopefully, the next few years will see GECIs with a performance level comparable to that of SMCIs.

4. Notes

1. The quality of recordings critically depends on the good condition of the experimental animal. Therefore, continuous monitoring of the vital parameters, like respiratory and pulse rate, body temperature, arterial oxygen concentration, and blood pressure, is essential.
2. Do not remove the dura mater; removal is not required for cell staining and strongly enhances movement artifacts. The stability of recordings also critically depends on the diameter of the craniotomy. Thus, openings larger than 1 mm in diameter are often accompanied by large movement artifacts occurring at the heartbeat frequency. In this case, the stability of the recording can be improved by covering the opening with 2% w/v low melting point agarose dissolved in standard external solution.
3. Note that the pipette solution does not contain any divalent cations. This is done on purpose in order to minimize precipitation of the dye and clogging of the injection pipette. To avoid clogging, it is important to ensure that all solutions are freshly made immediately before the experiment. In addition, the staining solution can be filtered using, for example, a Millipore filter (pore diameter of 0.45 μm) if necessary.

Acknowledgments

We thank A. Weible, S. Kasperek, G. Heck, and K. Schoentag for technical assistance. This work was supported by grants from the Deutsche Forschungsgemeinschaft (SFB 596, GA 654/1-1, SFB 870), EU FP7, and the NIH NS DC005259-39.

References

1. Wang X, Lou N, Xu Q et al (2006) Astrocytic Ca^{2+} signaling evoked by sensory stimulation in vivo. Nat Neurosci 9:816–823
2. Nimmerjahn A, Mukamel EA, Schnitzer MJ (2009) Motor behavior activates Bergmann glial networks. Neuron 62:400–412
3. Halassa MM, Haydon PG (2010) Integrated brain circuits: astrocytic networks modulate neuronal activity and behavior. Annu Rev Physiol 72:335–355
4. Wake H, Moorhouse AJ, Jinno S et al (2009) Resting microglia directly monitor the functional state of synapses in vivo and determine the fate of ischemic terminals. J Neurosci 29:3974–3980
5. Denk W, Strickler JH, Webb WW (1990) Two-photon laser scanning fluorescence microscopy. Science 248:73–76
6. Svoboda K, Denk W, Kleinfeld D et al (1997) In vivo dendritic calcium dynamics in neocortical pyramidal neurons. Nature 385:161–165
7. Helmchen F, Fee MS, Tank DW et al (2001) A miniature head-mounted two-photon microscope. High-resolution brain imaging in freely moving animals. Neuron 31:903–912
8. Piyawattanametha W, Cocker ED, Burns LD et al (2009) In vivo brain imaging using a portable 2.9 g two-photon microscope based on a microelectromechanical systems scanning mirror. Opt Lett 34:2309–2311
9. Sawinski J, Wallace DJ, Greenberg DS et al (2009) Visually evoked activity in cortical cells imaged in freely moving animals. Proc Natl Acad Sci USA 106:19557–19562
10. Dombeck DA, Khabbaz AN, Collman F et al (2007) Imaging large-scale neural activity with cellular resolution in awake, mobile mice. Neuron 56:43–57
11. Dombeck DA, Harvey CD, Tian L et al (2010) Functional imaging of hippocampal place cells at cellular resolution during virtual navigation. Nat Neurosci 13:1433–1440
12. Tsien RW, Tsien RY (1990) Calcium channels, stores, and oscillations. Annu Rev Cell Biol 6:715–760
13. Berridge MJ, Lipp P, Bootman MD (2000) The versatility and universality of calcium signalling. Nat Rev Mol Cell Biol 1:11–21
14. Verkhratsky A, Orkand RK, Kettenmann H (1998) Glial calcium: homeostasis and signaling function. Physiol Rev 78:99–141
15. Farber K, Kettenmann H (2006) Functional role of calcium signals for microglial function. Glia 54:656–665
16. Pocock JM, Kettenmann H (2007) Neurotransmitter receptors on microglia. Trends Neurosci 30:527–535
17. Agulhon C, Petravicz J, McMullen AB et al (2008) What is the role of astrocyte calcium in neurophysiology? Neuron 59:932–946
18. Nedergaard M, Rodriguez JJ, Verkhratsky A (2010) Glial calcium and diseases of the nervous system. Cell Calcium 47:140–149
19. Miyawaki A, Llopis J, Heim R et al (1997) Fluorescent indicators for Ca^{2+} based on green fluorescent proteins and calmodulin. Nature 388:882–887
20. Garaschuk O, Griesbeck O, Konnerth A (2007) Troponin C-based biosensors: a new family of genetically encoded indicators for in vivo calcium imaging in the nervous system. Cell Calcium 42:351–361
21. Mank M, Griesbeck O (2008) Genetically encoded calcium indicators. Chem Rev 108:1550–1564
22. Hires SA, Tian L, Looger LL (2008) Reporting neural activity with genetically encoded calcium indicators. Brain Cell Biol 36:69–86
23. Garaschuk O, Griesbeck O (2010) Monitoring calcium levels with genetically encoded indicators. In: Verkhratsky A, Petersen OH (eds) Neuromethods 43: calcium measurement methods. Humana, New York
24. Tsien RY, Pozzan T, Rink TJ (1982) Calcium homeostasis in intact lymphocytes: cytoplasmic free calcium monitored with a new, intracellularly trapped fluorescent indicator. J Cell Biol 94:325–334
25. Grynkiewicz G, Poenie M, Tsien RY (1985) A new generation of Ca^{2+} indicators with greatly

improved fluorescence properties. J Biol Chem 260:3440–3450
26. Tsien RY (1988) Fluorescence measurement and photochemical manipulation of cytosolic free calcium. Trends Neurosci 11:419–424
27. Stosiek C, Garaschuk O, Holthoff K et al (2003) In vivo two-photon calcium imaging of neuronal networks. Proc Natl Acad Sci USA 100:7319–7324
28. Tsien RY (1980) New calcium indicators and buffers with high selectivity against magnesium and protons: design, synthesis, and properties of prototype structures. Biochemistry 19:2396–2404
29. Naraghi M (1997) T-jump study of calcium binding kinetics of calcium chelators. Cell Calcium 22:255–268
30. Xu C (2000) Two-photon cross sections of indicators. In: Yuste R, Lanni F, Konnerth A (eds) Imaging neurons: a laboratory manual. Cold Spring Harbor, New York
31. Wokosin DL, Loughrey CM, Smith GL (2004) Characterization of a range of fura dyes with two-photon excitation. Biophys J 86:1726–1738
32. Tian L, Hires SA, Mao T et al (2009) Imaging neural activity in worms, flies and mice with improved GCaMP calcium indicators. Nat Methods 6:875–881
33. Reiff DF, Ihring A, Guerrero G et al (2005) In vivo performance of genetically encoded indicators of neural activity in flies. J Neurosci 25:4766–4778
34. Mank M, Reiff DF, Heim N et al (2006) A FRET-based calcium biosensor with fast signal kinetics and high fluorescence change. Biophys J 90:1790–1796
35. Horikawa K, Yamada Y, Matsuda T et al (2010) Spontaneous network activity visualized by ultrasensitive Ca^{2+} indicators, yellow Cameleon-Nano. Nat Methods 7:729–732
36. Eichhoff G, Kovalchuk Y, Varga Z et al (2010) In vivo Ca^{2+} imaging of the living brain using multi cell bolus loading technique. In: Verkhratsky A, Petersen OH (eds) Neuromethods 43: calcium measurement methods. Humana, New York
37. Wilms CD, Schmidt H, Eilers J (2006) Quantitative two-photon Ca^{2+} imaging via fluorescence lifetime analysis. Cell Calcium 40:73–79
38. Wilms CD, Eilers J (2007) Photo-physical properties of Ca^{2+}-indicator dyes suitable for two-photon fluorescence-lifetime recordings. J Microsc 225:209–213
39. Nagai T, Yamada S, Tominaga T et al (2004) Expanded dynamic range of fluorescent indicators for Ca^{2+} by circularly permuted yellow fluorescent proteins. Proc Natl Acad Sci USA 101:10554–10559
40. Mank M, Santos AF, Direnberger S et al (2008) A genetically encoded calcium indicator for chronic in vivo two-photon imaging. Nat Methods 5:805–811
41. Mao T, O'Connor DH, Scheuss V et al (2008) Characterization and subcellular targeting of GCaMP-type genetically-encoded calcium indicators. PLoS One 3:e1796
42. Tallini YN, Ohkura M, Choi BR et al (2006) Imaging cellular signals in the heart in vivo: cardiac expression of the high-signal Ca^{2+} indicator GCaMP2. Proc Natl Acad Sci USA 103:4753–4758
43. Neher E (1995) The use of fura-2 for estimating Ca^{2+} buffers and Ca^{2+} fluxes. Neuropharmacology 34:1423–1442
44. Heim N, Garaschuk O, Friedrich MW et al (2007) Improved calcium imaging in transgenic mice expressing a troponin-C based biosensor. Nat Methods 4:127–129
45. Wachowiak M, Cohen LB (2001) Representation of odorants by receptor neuron input to the mouse olfactory bulb. Neuron 32:723–735
46. Garaschuk O, Milos RI, Konnerth A (2006) Targeted bulk-loading of fluorescent indicators for two-photon brain imaging in vivo. Nat Protoc 1:380–386
47. Xu HT, Pan F, Yang G et al (2007) Choice of cranial window type for in vivo imaging affects dendritic spine turnover in the cortex. Nat Neurosci 10:549–551
48. Jia H, Rochefort NL, Chen X et al (2010) Dendritic organization of sensory input to cortical neurons in vivo. Nature 464:1307–1312
49. Nagayama S, Zeng S, Xiong W et al (2007) In vivo simultaneous tracing and Ca^{2+} imaging of local neuronal circuits. Neuron 53:789–803
50. Nevian T, Helmchen F (2007) Calcium indicator loading of neurons using single-cell electroporation. Pflugers Arch 454:675–688
51. Kitamura K, Judkewitz B, Kano M et al (2008) Targeted patch-clamp recordings and single-cell electroporation of unlabeled neurons in vivo. Nat Methods 5:61–67
52. Judkewitz B, Rizzi M, Kitamura K et al (2009) Targeted single-cell electroporation of mammalian neurons in vivo. Nat Protoc 4:862–869
53. Kerr JN, Greenberg D, Helmchen F (2005) Imaging input and output of neocortical networks in vivo. Proc Natl Acad Sci USA 102:14063–14068

54. Tabata H, Nakajima K (2001) Efficient in utero gene transfer system to the developing mouse brain using electroporation: visualization of neuronal migration in the developing cortex. Neuroscience 103:865–872
55. Tabata H, Nakajima K (2008) Labeling embryonic mouse central nervous system cells by in utero electroporation. Dev Growth Differ 50:507–511
56. Chesler AT, Le Pichon CE, Brann JH et al (2008) Selective gene expression by postnatal electroporation during olfactory interneuron neurogenesis. PLoS One 3:e1517
57. Klein RL, Hamby ME, Gong Y et al (2002) Dose and promoter effects of adeno-associated viral vector for green fluorescent protein expression in the rat brain. Exp Neurol 176:66–74
58. Dittgen T, Nimmerjahn A, Komai S et al (2004) Lentivirus-based genetic manipulations of cortical neurons and their optical and electrophysiological monitoring in vivo. Proc Natl Acad Sci USA 101:18206–18211
59. Teschemacher AG, Wang S, Lonergan T et al (2005) Targeting specific neuronal populations using adeno- and lentiviral vectors: applications for imaging and studies of cell function. Exp Physiol 90:61–69
60. Wong LF, Goodhead L, Prat C et al (2006) Lentivirus-mediated gene transfer to the central nervous system: therapeutic and research applications. Hum Gene Ther 17:1–9
61. Coura Rdos S, Nardi NB (2007) The state of the art of adeno-associated virus-based vectors in gene therapy. Virol J 4:99
62. Diez-Garcia J, Akemann W, Knopfel T (2007) In vivo calcium imaging from genetically specified target cells in mouse cerebellum. Neuroimage 34:859–869
63. Hendel T, Mank M, Schnell B et al (2008) Fluorescence changes of genetic calcium indicators and OGB-1 correlated with neural activity and calcium in vivo and in vitro. J Neurosci 28:7399–7411
64. Wallace DJ, Zum Alten Borgloh SM, Astori S et al (2008) Single-spike detection in vitro and in vivo with a genetic Ca^{2+} sensor. Nat Methods 5:797–804
65. Garaschuk O, Milos RI, Grienberger C et al (2006) Optical monitoring of brain function in vivo: from neurons to networks. Pflugers Arch 453:385–396
66. Nimmerjahn A, Kirchhoff F, Kerr JND et al (2004) Sulforhodamine 101 as a specific marker of astroglia in the neocortex in vivo. Nat Methods 1:31–37
67. Kafitz KW, Meier SD, Stephan J et al (2008) Developmental profile and properties of sulforhodamine 101-labeled glial cells in acute brain slices of rat hippocampus. J Neurosci Methods 169:84–92
68. Johannssen HC, Helmchen F (2010) In vivo Ca^{2+} imaging of dorsal horn neuronal populations in mouse spinal cord. J Physiol 588:3397–3402
69. Kang J, Kang N, Yu Y et al (2010) Sulforhodamine 101 induces long-term potentiation of intrinsic excitability and synaptic efficacy in hippocampal CA1 pyramidal neurons. Neuroscience 169:1601–1609
70. Petzold GC, Albeanu DF, Sato TF et al (2008) Coupling of neural activity to blood flow in olfactory glomeruli is mediated by astrocytic pathways. Neuron 58:897–910
71. Hirrlinger PG, Scheller A, Braun C et al (2005) Expression of reef coral fluorescent proteins in the central nervous system of transgenic mice. Mol Cell Neurosci 30:291–303
72. Eichhoff G, Brawek B, Garaschuk O (2010) Microglial calcium signal acts as a rapid sensor of single neuron damage in vivo. Biochim Biophys Acta. doi:10.1016/j.bbamcr.2010.10.018
73. Nakai J, Ohkura M, Imoto K (2001) A high signal-to-noise Ca^{2+} probe composed of a single green fluorescent protein. Nat Biotechnol 19:137–141
74. Wellis DP, Scott JW (1990) Intracellular responses of identified rat olfactory bulb interneurons to electrical and odor stimulation. J Neurophysiol 64:932–947
75. Buonviso N, Amat C, Litaudon P et al (2003) Rhythm sequence through the olfactory bulb layers during the time window of a respiratory cycle. Eur J Neurosci 17:1811–1819
76. Cang J, Isaacson JS (2003) In vivo whole-cell recording of odor-evoked synaptic transmission in the rat olfactory bulb. J Neurosci 23:4108–4116
77. Tan J, Savigner A, Ma M et al (2010) Odor information processing by the olfactory bulb analyzed in gene-targeted mice. Neuron 65:912–926
78. Vucinic D, Cohen LB, Kosmidis EK (2006) Interglomerular center-surround inhibition shapes odorant-evoked input to the mouse olfactory bulb in vivo. J Neurophysiol 95:1881–1887
79. Tolhurst DJ, Smyth D, Thompson ID (2009) The sparseness of neuronal responses in ferret primary visual cortex. J Neurosci 29:2355–2370

80. Wolfe J, Houweling AR, Brecht M (2010) Sparse and powerful cortical spikes. Curr Opin Neurobiol 20:306–312
81. Lutcke H, Murayama M, Hahn T et al (2010) Optical recording of neuronal activity with a genetically-encoded calcium indicator in anesthetized and freely moving mice. Front Neural Circuits 4:9
82. Boldogkoi Z, Balint K, Awatramani GB et al (2009) Genetically timed, activity-sensor and rainbow transsynaptic viral tools. Nat Methods 6:127–130

Neuromethods (2012) 67: 45–60
DOI 10.1007/7657_2011_18

Published online: 1 December 2011

Two-Photon Imaging of Neural Activity in Awake, Head-Restrained Mice

Martin Wienisch, David G. Blauvelt, Tomokazu F. Sato, and Venkatesh N. Murthy

Abstract

Two-photon microscopy has become an invaluable tool for visualizing the activity of neuronal populations at cellular resolution in vivo. Imaging typically requires restraining the head of the animal underneath the objective of a dedicated optical setup and experiments are therefore often performed under anesthesia. Here, we describe a method that allows imaging in awake mice with minimal motion artifacts and without the need for extensive training of the animal. We detail the necessary surgical procedures to chronically implant a small, lightweight headplate and to create a clear window for imaging. The design of a simple apparatus capable of stably accommodating the headplate while the mouse is positioned on a wheel with spring suspension is presented. When used in combination with a multiphoton microscope, this approach greatly facilitates optical recordings in nonanesthetized animals and opens the door to many projects that can bridge the gap between neural activity and behavior.

Key words: Multiphoton imaging, In vivo, Mouse, Brain, Microscopy, Neural activity, Awake, Anesthetized, Laser scanning, Neural ensemble, Head restrained

1. Introduction

Two-photon microscopy has emerged as a powerful tool to observe the dynamics of neuronal circuits in living animals at high spatial resolution (1, 2). Most notably, multiphoton excitation of fluorescent calcium indicators has now become a widely used method that allows simultaneous functional imaging of hundreds of neurons at cellular resolution several hundred micrometers deep within the brain tissue (for example, see ref. 3). However, since movements of the brain relative to the microscope objective cause imaging artifacts, optical recordings have traditionally been performed in anesthetized, head restrained animals. While this approach is relatively simple to implement and has led to a number of exciting discoveries, it has two main disadvantages: First, any drug used for anesthesia can potentially interfere with the activity of the observed neuronal ensembles (4, 5). For example, commonly used anesthetics for rodents often enhance inhibitory synaptic transmission and/or suppress excitatory neurotransmission (6). Secondly, anesthesia renders it impossible to directly observe the neuronal activity linked to a specific behavioral task or to motor activity. Similarly, intravital population imaging has been successfully used for visualizing sensory evoked neuronal dynamics in anesthetized rodents and cats (3, 7, 8). However, active sensation of a nonanesthetized animal could result in significantly different neural coding patterns.

In this chapter, we describe a method that allows high-resolution imaging in awake and behaving mice. This is achieved by first chronically securing a small and lightweight titanium headplate onto the skull of the animal. After creating an imaging window, the headplate can then be reversibly fixed to a specific holder while the mouse is placed on top of a spring-loaded wheel serving as a treadmill. This apparatus offers a simple solution for achieving optical recordings without major motion artifacts and without the need for extensive animal training.

1.1. Multiphoton Imaging in Awake Animals

The desire for imaging neuronal activity in awake, behaving animals has led to the development of two strategies. First, one can make use of flexible optical fibers and miniature microscopes mounted on the head of a freely moving animal (9). Using such a "two-photon fiberscope," a recent study succeeded in functionally imaging neurons in freely moving rats (10). While this approach offers the most flexibility to record neural activity occurring during natural behavioral paradigms, it usually has limited image quality and is susceptible to motion artifacts. Furthermore, despite ongoing advances, head-mounted microscopes are still relatively bulky and are therefore not feasible for mice. This represents a major drawback since transgenic mice expressing fluorescent proteins or functional probes in defined cell populations will be an essential tool to decipher specific neuronal coding patterns.

The second strategy is imaging head restrained, awake animals (Fig. 2). Several useful designs that allow functional imaging in head-fixed mice and rats exist but differ with respect to their possible behavioral readouts, their ease of construction and the extent of animal training required. The simplest solution is to immobilize the head of the animal by a surgically implanted metal post or a headplate that can be stably clamped with screws. One study used this method to passively record cortical neuronal activity during different brain states in rats (5). In addition to head-fixation, Komiyama et al. (11) placed a mouse into a narrow tube and let the animal use its front paws to grip the tube edge to further control movements. In order to combine functional multiphoton imaging with behavior, they also implemented an odorant delivery system and a water reward port and trained the mouse in a lick/no-lick task. Similarly, a recent study succeeded in chronically imaging neural activity in head-fixed mice while performing a visual discrimination task (12). These studies show that it is possible to train head-fixed animals to remain still during tasks and perform simultaneous population imaging. However, considering the long durations that some in vivo experiments require, it is useful to provide additional means to help the animal tolerate the imaging setup without anxiety or discomfort. One simple possibility is to give the animal some sense of freedom of motion.

In a recent innovation, head-fixed mice were placed on a freely rotating styrofoam exercise ball (13). This configuration gives an animal the two degrees of freedom that it naturally has. Moreover, it provides the additional advantage that movements of the animal on the ball can be measured. Dombeck et al. (13, 14) used this information to detect running-correlated neural activity in both hind-limb sensory and motor cortex. Nevertheless, this technique can be suboptimal for long duration restraint. It can take multiple days to train animals to stay on the sphere comfortably as it can very freely rotate laterally, causing discomfort for the mouse. Nimmerjahn et al. (15) combined head-restrained imaging with an undamped exercise ball offering one degree of freedom. To simplify the setup even further, we omitted using a sphere altogether and replaced it with a styrofoam disk (Figs. 1a and 2b, c) on spring suspensions. It is easy to build, requires only a little training to acclimate the animal, shows minimal imaging artifacts due to motion, and if desired, still provides the option to quantify one dimensional walking/running behavior.

1.2. Present and Future Perspectives

Head-restraining allows the use of dedicated microscopes that are specialized for optical recordings in vivo. However, the true strength of the head-fixed paradigm is that its usefulness is not limited to high-speed or high-resolution imaging. By keeping the skull and brain in a fixed location, pharmacology, electrophysiology, and optical manipulation methods requiring stereotaxic coordinates are much more easily done than in fully mobile animals. If one needs to target small specific brain areas with characteristic neural activity or obtain antidromic spikes, multiple penetrations can be performed using tungsten electrodes. Although head-fixing places some limits on behavior, many movements, such as breathing, licking, eye blinks, grooming, and forward running, are still intact and are easier to monitor compared to freely moving animals. These diverse motor outputs may facilitate a much deeper understanding of behavioral effects, and the function of specific neural cell assemblies underlying them. For example, activity correlated to motor behavior, such as running and grooming (14, 15) or licking (11), has been visualized using head-fixed mice. Similarly, Andermann et al. succeeded in recording cellular activity in the visual cortex occurring during a visual orientation discrimination task (12). These studies greatly benefit from the ease with which a variety of additional behavioral paradigms can be implemented when using head restrained animals. Such paradigms include

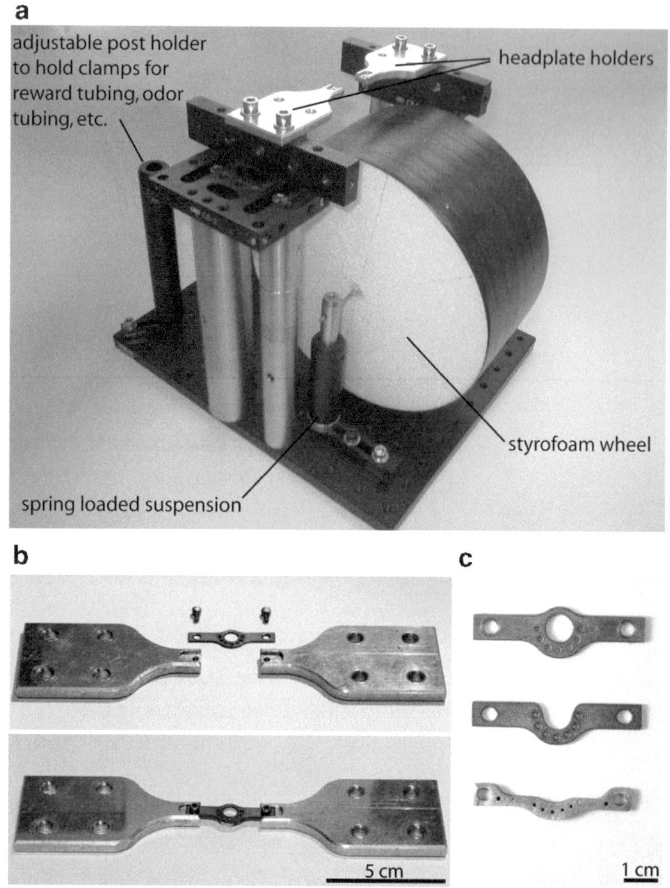

Fig. 1. Head restraining using a wheel on spring suspensions. (**a**) A simple styrofoam wheel (8 in. diameter, 4 in. width) placed on spring loaded suspensions is used as a treadmill. The wheel was adjusted so that a mouse could comfortably sit on the surface while an implanted headplate was used for head-restraining. Only short training durations are sufficient to acclimatize the mouse to the apparatus. Post holders provide modular addition of components that could be used for behavior, such as odor delivery tubes, reward ports, lick detectors, or air puff delivery. (**b**) Disassembled and assembled headplate and headplate holders. To provide flexible mounting options headplate holders feature several 1/4-20 through holes and have #4-40 tapped holes for attachment of the headplate. Also shown is an exemplar headplate with #4-40 through holes and corresponding #4-40 screws. (**c**) Design of three different headplates. In all examples smaller holes around the central rim that can be filled with acrylic adhesive provide additional stability. While the closed design (*upper panel*) offers the best stability and protection of the craniotomy, open designs (*middle* and *lower panel*) provide more accessibility to the imaging window and cover a smaller area of the skull. For details, see Note 4.2.

administrating water rewards, monitoring licking or running and using auditory or olfactory feedback cues. Another feature that may be hard to normally employ is a negative reinforcing stimulus. Punishments, such as air puffs, can now be administered easily with the

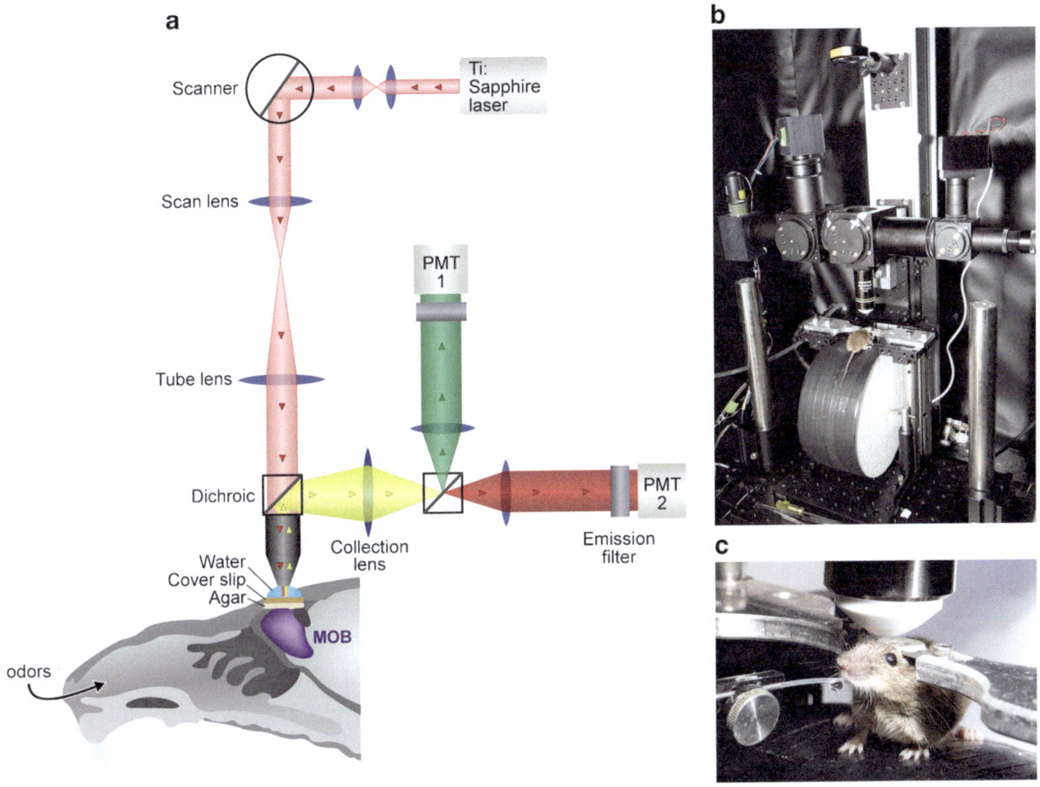

Fig. 2. Two-photon microscopy in awake, head-restrained mice using a wheel as treadmill. (**a**) Simplified layout of the optical system. *PMT* photomultiplier tube, *MOB* main olfactory bulb. A femtosecond laser source provides near-infrared pulses. The laser beam size is expanded before and after hitting the scan mirrors to fill the back pupil of the objective lens. A lens placed close to the objective efficiently collects the emitted fluorescent light after two-photon excitation. The light is then further split into its *green* and *red* spectral components and detected separately by two PMTs. An imaging window over the MOB allowed us to visualize odor-evoked neuronal activity in awake mice. LED illumination and one-photon epifluorescence light-path not shown. (**b**) Photograph showing the integration of the treadmill apparatus in the custom-made two-photon setup. Note that the size of the treadmill might make it difficult to use with a commercial microscope. (**c**) Example of a head-restrained mouse undergoing two-photon imaging. Photograph shows the relative dimensions of the animal, objective, headplate, and headplate holder. Head-fixation makes it easy to integrate additional components that are useful for setting up certain behavioral paradigms. As an example, the picture shows a reward tube used to deliver water during intertrial sessions.

animal restrained. As the behavior of the animal can be easily monitored while visualizing the activity of single neurons with multiphoton microscopy, head-fixed imaging will also be useful for experiments addressing decision making, associative learning, and sensory integration. Furthermore, since the headplate is well tolerated by the animal and can be quickly attached or removed from the holder, this system also provides an excellent foundation for long-term experiments allowing chronic imaging of the same brain region (12, 16). We have imaged neural activity in principal neurons of the olfactory bulb in awake, head-restrained mice (Fig. 3a, b).

Odor-evoked responses could be imaged at cellular resolution at depths greater than 300 μm (Fig. 3c).

A general drawback of in vivo multiphoton microscopy is the limited number of neurons that can be visualized. First, even though depth penetration is better than with conventional techniques, it is impossible to image brain tissue located more than 1 mm from the surface. Imaging deeper regions thus either requires removal of the overlaying tissue or will necessitate implantation of small periscopic devices (17). Another issue is the number of neurons that can be imaged during each frame scan in single trials. Typically, the image is built by scanning individual pixels in one specific horizontal (x–y) plane and only neurons located in this specific depth can be studied. Performing a full x–y scan in multiple planes is often not an option because it is slow compared to fast activity changes in the neural network. However, fast three-dimensional scanning approaches are being developed (18, 19) and can be easily combined with head-restrained imaging.

The apparatus for head-fixed imaging presented in this chapter also has certain limitations. Compared to head-mounted microscopes, both the number and types of behavioral paradigms are still rather limited. Other designs for head-fixed imaging utilizing spherical treadmills with three degrees of freedom will make it easier in the future to let the animal navigate in a virtual reality system (13, 20). Also, accessing brain areas that are not on the dorsal surface usually involves more invasive surgical procedures, and is not well suited for imaging in nonanesthetized animals. Even with these limitations, head-fixed methodology provides a balance between the ease of access and manipulations on awake animals and the behavioral outputs, and opens the door to many projects that can bridge the gap between neural activity and behavior.

2. Materials

2.1. Headplate Attachment Surgery

- Custom-cut titanium headplate (Fig. 1c).
- Stereotaxic surgery apparatus.
- Ketamine/xylazine cocktail (100 and 10 mg/kg, Webster).
- Vetbond (3M).
- Jet Repair Acrylic Adhesive (Lang Dental).
- Homeothermic Blanket (NP50-7053-F, Harvard Apparatus).

2.2. Craniotomy

- Custom made headplate surgery setup.
- Ketamine/xylazine cocktail.
- Glass cover slip (5 mm diameter).

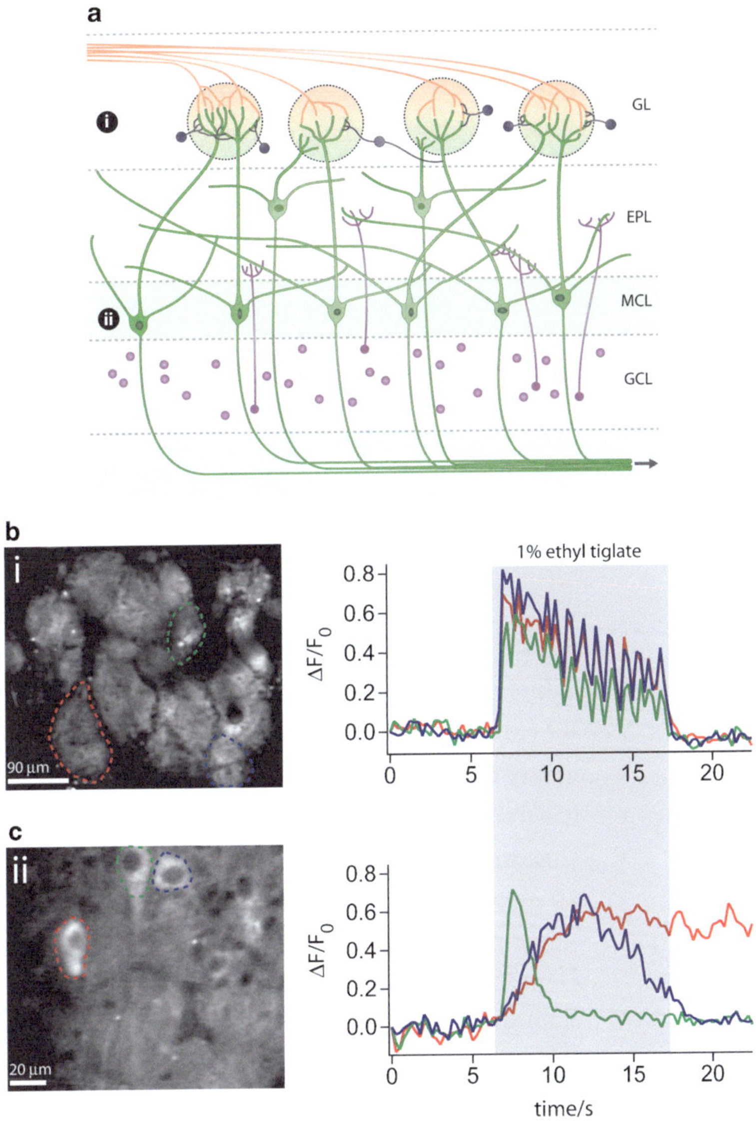

Fig. 3. Odor-evoked activity of mitral cells in different layers of the olfactory bulb. (**a**) Simplified schematics showing the layers and the neuronal circuitry of the mouse olfactory bulb. *GL* glomerular layer, *EPL* external plexiform layer, *MCL* mitral cell layer, *GCL* granule cell layer. Olfactory receptor neurons (ORNs) in the nose project their axons to the olfactory bulb and form anatomical units called glomeruli. Here, the axons from ORNs make synapses on the apical dendrites of mitral/tufted (M/T) cells which are the major excitatory output channels from the bulb. The incoming sensory information is modulated by at least two inhibitory microcircuits: Periglomerular cells in the GL and by granule cell dendrites in the EPL. Two-photon microscopy allows the study of odor-evoked activity of mitral cells that express the genetically encoded calcium indicator GCaMP2 (34) in both the glomerular layer (*i*) and the cell body layer (*ii*). For simplicity, additional cell types of the olfactory bulb are not represented. (**b**) Odor-evoked calcium dynamics in tufts of mitral cell dendrites in the glomerular layer, located about 50 µm below the dura mater. *Left*: Averaged resting fluorescence of GCaMP2 in mitral cell tufts. Scale bar is 90 µm. *Right:* Odor-evoked responses of three selected glomeruli as indicated on *left*. Fluctuations of about 2 Hz during the stimulus are caused by the sniffs of the animal. (**c**) Odor-evoked calcium dynamics in somata of mitral cells in the MCL, located about 250 µm below the dura mater. *Left*: Averaged resting fluorescence of GCaMP2 in somata of mitral cells. *Right:* Odor-evoked responses of three selected mitral cell bodies as indicated on *left*. Neither in (**b**) nor (**c**) algorithms to correct for motion artifacts were necessary.

- Jet Repair Acrylic Adhesive.
- artificial cerebrospinal fluid (aCSF) solution (135 mM NaCl, 5.4 mM KCl, 5 mM HEPES, and 1.8 mM $CaCl_2$, pH 7.4).
- Buprenorphine (0.5 mg/kg).

2.3. In Vivo Imaging

2.3.1. Custom-Made Head-Fixed Apparatus (Fig. 1a)

- 8 in. diameter × 4 in. tall smooth foam disk (Plasteel Corp., Inkster, MI) coated with tape or plastic film as treadmill.
- Opto-mechanics for assembly of head mounting structure, including table clamps (CL2, Thorlabs Inc., Newton, NJ), pillar post spacers (RS03 or similar to adjust wheel height), mounting posts (P8, RS6 and RS2), post holders with spring-loaded thumbscrews (PH3-ST) with corresponding posts (TR4) for holding the disk, as well as construction bases and rails (CB-2 and CR-1, Newport Corp., Irvine, CA) for attachment of the headplate holder.
- Breadboard of at least 8 in. × 8 in. (Thorlabs Inc., Newton, NJ) as general base.
- Metal rod of <3 mm as axle for the wheel (McMaster-Carr, Robbinsville, NJ).
- Steel compression spring (soft) <1/2 in. diameter, e.g., 1986 K9 (McMaster-Carr, Robbinsville, NJ) as suspension.
- Epoxy adhesive.
- Custom-machined headplate holders (Fig. 1b).
- Custom-machined headplates (Fig. 1c).
- 10 ml luer lock syringe.
- Blunt end needle (e.g., 23 G, 0.5 in., Warner Instruments, Hamden, CT).
- Tubing (e.g., PE-50, 0.58 ID, Warner Instruments, Hamden, CT).
- 10% sucrose solution.

2.3.2. Two-Photon Laser-Scanning Microscope

We used a custom-built two-photon microscope (7) based on the design of the Kleinfeld lab (21) for in vivo imaging (Fig. 2a). The key components of our system are:

- Mai Tai HP Ultrafast Ti:Sapphire laser (Spectra-Physics, Santa Clara, CA).
- Pockels Cell Attenuator Model 302RM/350-50-BK (Conoptics Inc., Danbury, CT).
- High sensitivity photomultiplier tubes (R3896 or H7422-40, Hamamatsu Photonics, Hamamatsu City, Japan).
- Galvanometer Scanner (model 6210, Cambridge Technology, Lexington, MA).

- 20 × 0.95NA water immersion objective, IR coated (XLUMPLFL/IR Olympus, Center Valley, PA).
- Laser beamsplitter (680dcxr, Chroma, Bellows Falls, VT).
- Collection optics, including dichroic mirror (FF555-Di02, Semrock, Rochester, NJ), emission filters (FF01-525/50, FF01-607/70) and laser blocking filter (FF01-750/SP).
- Input/output board (PCI-6115, National Instruments, Austin, TX).
- LabVIEW based, custom-written software code for hardware control and data acquisition.
- CCD camera (Guppy-F046C, Allied Vision Technologies, Stadtroda, Germany) and super bright blue LED for epifluorescence (22).
- Translation and z-stages (462-XY-M and VP25-XA, Newport Corp., Irvine, CA) to move sample and objective, respectively.

3. Methods

3.1. Headplate Attachment Surgery

1. Anesthetize the animal by injecting intraperitoneally the appropriate amount of the ketamine/xylazine cocktail.
2. When you have ensured that the animal is fully anesthetized (see Note 4.1.1), shave off the fur around the head of the animal and cover the eyes with veterinary eye ointment or Vaseline.
3. Attach the animal to a stereotaxic surgery apparatus and ensure that the animal is kept warm with a homeothermic heating blanket set to 37°C. Clean off the top of the skin with ethanol, iodine, or other disinfectant.
4. Make a single midline anterior–posterior incision through the skin using a scalpel or scissors.
5. Separate the skin and keep the skull exposed with skin clips or other preferred method.
6. Scrape away the thin membrane that covers the skull and thoroughly dry the skull, ensuring that all bleeding has stopped.
7. Using a hand-held dental drill, drill very slightly into the skull around the circumference of the desired craniotomy site (see Note 4.1.2).
8. Cover the entire skull and some of the surrounding tissue with a thin layer of Vetbond.
9. Place the headplate on the skull, ensuring that the craniotomy site is not covered.

10. Mix the powder and liquid parts of the Jet Repair Acrylic Adhesive set and apply it around the headplate. Do not cover the site of the craniotomy.
11. After the adhesive is dry, remove the animal from the stereotaxis and allow it to recover (see Note 4.1.3).

3.2. Craniotomy

1. Anesthetize the animal by injecting intraperitoneally the appropriate amount of the ketamine/xylazine cocktail.
2. When you have ensured that the animal is fully anesthetized (see Note 4.1.1), cover the eyes with veterinary eye ointment or Vaseline.
3. Attach the animal with its headplate to the custom made headplate surgery setup and maintain the animal's body heat with the heating blanket.
4. Clean out any debris from the area around the craniotomy and disinfect the surface.
5. Find the site of the craniotomy, which was previously marked by drilling slightly into the skull (see Step 3.1.7). Begin drilling around the circumference of the craniotomy site (see Note 4.1.4).
6. Once you finish drilling (see Note 4.1.5), carefully lift off the skull piece with a scalpel, ensuring that you do not damage the brain.
7. Cover the brain with aCSF and make sure that the surface is always moist. Sponge up any blood and wait till the bleeding completely stops (see Note 4.1.6).
8. Heat up agar (1.2% in aCSF) until it completely liquefies. Draw up a small amount into a pipette and wait for it to cool—its temperature should be tested on the back of the hand or wrist.
9. Once the agar is cool enough, drop a small amount onto the craniotomy site and quickly place a 5 mm glass cover slip on top. Push down onto the cover slip with forceps, forcing out the excess agar, and wait for the agar to harden. Once the agar has hardened remove any excess agar with a scalpel or light vacuum.
10. Mix the Jet Repair Acrylic powder and liquid, and apply the mixture around the edges of the cover slip, ensuring that it does not cover the craniotomy site.
11. Remove the animal from the surgery setup. It can either be imaged on the same day or allowed to recover before imaging later.

3.3. Head-Restraining Apparatus

Headplates were made of titanium, 0.032 in. thick (Fig. 1c). They consisted of a large central hole, 5 mm in diameter, through which craniotomies were performed, and a through hole for #4-40 screws at each end of the wings jutting out from the headplates.

Depending on the specific experiment, we have also used headplates of slightly different design (see Note 4.2.2). Custom-machined headplate holders with 1/4 in. through holes were secured to sturdy posts on an aluminum or steel breadboard with 1/4 in.-20 screws (Fig. 1b). These headplate holders had tapped holes for a #4-40 screw and were spaced apart so that the headplate through holes would match the tapped holes of the headplate holders. The center of a styrofoam wheel (8 in. diameter, 4 in. wide Smoothfoam, Plasteel Corp) was pierced with a thin metal rod acting as an axle. This axle was fixed to the wheel with epoxy and inserted into the holes of two metal rods (1/2 in. diameter rods, Thorlabs Inc.) in a tube (1/2 in. post holders, Thorlabs Inc.); the metal rods were damped by a soft spring. The wheel was positioned so that an animal, when implanted with the headplate, would be comfortably placed on the freely rotating wheel (Figs. 1a and 2c).

3.4. Animal Training

After 48 h for recovery, the animal was acclimated to being placed on the wheel for short periods of time daily for 1 week. The animals were water restricted when not on the wheel, and given free access to sugar water while on the wheel. We typically use thin PE tubing as reward delivery tube (Fig. 2c), connected to a syringe with a blunt needle and filled with sucrose solution. The animal was trained gradually to increasing amounts of restraint and duration on the wheel. For the first day or two, the animal was not head-fixed, but merely held by the tail and was allowed to explore the wheel. Afterward, longer and longer durations of head-fixing were performed. After acclimation, the last part of the surgery was performed (see section on "Craniotomy," above).

All animal procedures were performed according to guidelines of the US National Institutes of Health and were approved by the Institutional Animal Care and Use Committee at Harvard University.

4. Notes

4.1. Surgical Preparation

1. After injecting the animal with the anesthesia cocktail, it takes several minutes before it is fully anesthetized. Once it stops moving, its breathing will start to slow down and become more regular. Once this occurs, you can test for the level of anesthesia by firmly pinching its foot. If the animal squirms, it is not fully anesthetized and may need more of the cocktail. However, ensure that you wait at least 5–10 min before giving it more anesthesia. Sometimes, the animal just needs more time for the anesthesia to take full effect, and giving a booster

prematurely can result in death by overdose. Large and obese animals often need more time to become fully anesthetized because the adipose tissue can absorb the anesthesia; often, these animals are difficult to anesthetize. Using isoflurane for inhalation anesthesia is a very good alternative if a precision vaporizer is available. The depth of the anesthesia achieved via this method can be more easily adjusted and recovery of the animal is significantly faster.

2. Do not drill too deeply. This step is not to expose the brain or remove the skull piece. It is simply to mark the site of the craniotomy for later reference. Once the headplate is attached to the skull, many of the natural reference points on the skull will be lost.
3. While the animal is still anesthetized, allow it to recover in a warm environment without overheating it. Placing the cage on a heating blanket is a good option. Once the animal wakes up, take out the blanket and allow it to recover for at least 1–2 days before reanesthetizing it. If indicated, administer the analgesic, buprenorphine, intraperitoneally.
4. The goal when drilling through the skull is to thin the skull in a circle until it is so thin that the skull piece can be lifted off, exposing the brain. However, it is very important not to damage the dura matter that covers the surface of the brain. While drilling technique should be catered to the researcher's individual preferences, some ways to drill with more control include supporting the drill between the index and middle finger of the nondrilling hand and drilling with wide horizontal motions, like a painter. Mastering the drilling technique takes some practice.
5. It is also very difficult to know when to stop drilling, and knowing that the skull is thin enough takes some experience. However, there are a few things to look for while drilling. First, know how thick the skull is on average over the craniotomy site. Often, the skull thickness can change across a single craniotomy. For instance, if the site is very close to the lateral part of the skull, the lateral part of the site will often be thicker since the angle of the skull changes as it curves around the head. Second, look for the skull to become more transparent as it gets thinner. As the skull gets very thin, surface blood vessels become visible and CSF will sometimes seep through the skull. Finally, gently press the middle of the skull piece with forceps or a scalpel. If it is sufficiently thin, the skull will give a little.
6. Depending on the quality of the drilling, the amount of bleeding will vary. Most of the time, at least a small amount of microbleeding will occur. These usually stop on their own

in a few minutes, and the researcher simply needs to sponge up the blood. For more severe cases of bleeding, Gelfoam (Pfizer) mixed with aCSF can be used to stop the bleeding. If that does not work, it is likely that the brain is too damaged to be imaged, and the animal should be euthanized.

7. Instead of putting the cover slip on top of the agar, the craniotomy can also be filled with a Kwik-Sil plug on a cover slip (13), the cover slip might be laid directly on the dura mater and glued to the skull (23) or an imaging window can be created using a thinned skull preparation (24).

4.2. Head-Restraining Apparatus

1. If the headplate is not fixed properly, it might become loose and even detach from the skull. There are several strategies to assure a strong bond between the metal plate and the skull. First, it is pivotal to make sure that both skull and metal plate are dry before gluing, e.g., using cotton swabs or blowing air. It might also help to carefully drill a number of shallow indentations into the skull to roughen the surface before gluing. Second, our headplates typically contain a number of additional small holes within their central rim (Fig. 1c). When gluing the headplate to the skull of the animal these holes can be filled with acrylic adhesive and significantly improve the stability of the attachment. Finally, it is essential to train the animal on the wheel before attaching the headplate to the headplate holder. If the mouse is calm and stress-free, only minimal forces will be applied to the headplate-holder connection.
2. Another potential problem might be caused by the size of the headplate. While a relatively big headplate with a central hole is generally easier to attach and offers the most stability, it also has certain disadvantages. For example, if the brain region of interest is located toward the edge of the skull it might be difficult to place the central hole over the area of interest for the craniotomy and still have sufficient space reserved for gluing. Additionally, if the headplate has to be placed onto an anterior part of the skull, it might cover the eyes of the mouse and cause distress. These problems can become even more relevant when working with younger mice. Three slight variations of the basic headplate design offer possible advantages for a given experimental setup and make the head-fixed-imaging approach more flexible. The first design simply employs a narrowed rim on one side of the hole, making it easier to glue to anterior or posterior edges of the skull (Fig. 1c, top). The second design omits the rim on one side altogether and is particularly useful when access to the craniotomy is of importance, e.g., in order to bring in pipettes or electrodes to the brain surface through the narrow space

between the objective and the sample (Fig. 1c, middle). A third design offers minimal size and weight and can be used if even more space or accessibility is needed (Fig. 1c, bottom). This headplate is similar to a bar with only slight curvature and is easily fixed anywhere on the skull independent of the site of craniotomy. However, smaller headplates are more difficult to glue to the skull in a horizontal fashion and might be more prone to detachment. Also more care has to be taken to stably secure the cover slip covering the craniotomy since the "open" headplate design offers less protection.

4.3. Imaging System

While our imaging system is completely custom-built following the design of the Kleinfeld lab (21), a number of alternative designs have been published. These are either alternate custom-designed instruments (25) or are hybrid systems, using modified commercial laser scanning microscopes (26–29). Recently, commercial vendors also offer dedicated multiphoton imaging systems with improved performance, such as Leica Microsystems, Olympus, Prairie Technologies, and Zeiss. However, custom-made systems offer far more flexibility, can achieve higher sensitivity as well as speed and exhibit much more space and accessibility at a fraction of the costs of a commercial microscope. Notably, fitting the head-restrained apparatus described here into a commercial system built around a standard microscope stand can be a major challenge (Fig. 2b). Lastly, freely available control and acquisition software codes for two-photon laser scanning microscopy, such as ScanImage (30) or MPScope (31) facilitate the implementation of self-built systems and further expand their versatility.

4.4. Brain Motion During Imaging

1. The method presented in this chapter has the capability of producing raw imaging data that can be used and analyzed in its unprocessed form. Nevertheless, the occurrence of some degree of artifacts caused by motion of the animal is inevitable. These are best prevented by spending more time training and acclimating the animal before imaging. We found it useful to give the water-deprived animal a small drop of water in between individual imaging trials during one session to keep it calm and move less to prevent possible motion artifacts (Fig. 2c). It is also possible to behaviorally condition the animal to be relatively still during the time of image acquisition by using "rewarding" or "punishing" cues, if applicable.
2. Remaining motion artifacts within the imaging plane (*XY* direction) can be corrected offline, e.g., using a Hidden Markov Model for movements occurring during scanning of individual images (13), image alignment based on the Lucas–-Kanade algorithm (32), or algorithms for 2D translation image registration for displacements in the image sequence

(33). Movements of the sample orthogonal to the imaging plane (*Z* direction) are much harder to correct offline and usually necessitate recording of a new trial. If many motion artifacts occur in the *Z* direction, scanning small volumes instead of planes might be helpful (12). However, this will reduce imaging speed and/or resolution.

Acknowledgments

This work was supported, in part, by a Human Frontier Science Foundation fellowship (M.W.), and by Harvard University (V.N.M.).

References

1. Denk W, Strickler JH, Webb WW (1990) Two-photon laser scanning fluorescence microscopy. Science 248:73–76
2. Grewe BF, Helmchen F (2009) Optical probing of neuronal ensemble activity. Curr Opin Neurobiol 19:520–529
3. Ohki K, Chung S, Ch'ng YH et al (2005) Functional imaging with cellular resolution reveals precise micro-architecture in visual cortex. Nature 433:597–603
4. Rinberg D, Koulakov A, Gelperin A (2006) Sparse odor coding in awake behaving mice. J Neurosci 26:8857–8865
5. Greenberg DS, Houweling AR, Kerr JN (2008) Population imaging of ongoing neuronal activity in the visual cortex of awake rats. Nat Neurosci 11:749–751
6. Hara K, Harris RA (2002) The anesthetic mechanism of urethane: the effects on neurotransmitter-gated ion channels. Anesth Analg 94:313–318, table of contents
7. Petzold GC, Albeanu DF, Sato TF et al (2008) Coupling of neural activity to blood flow in olfactory glomeruli is mediated by astrocytic pathways. Neuron 58:897–910
8. Stosiek C, Garaschuk O, Holthoff K et al (2003) In vivo two-photon calcium imaging of neuronal networks. Proc Natl Acad Sci U S A 100:7319–7324
9. Helmchen F, Fee MS, Tank DW et al (2001) A miniature head-mounted two-photon microscope. high-resolution brain imaging in freely moving animals. Neuron 31:903–912
10. Sawinski J, Wallace DJ, Greenberg DS et al (2009) Visually evoked activity in cortical cells imaged in freely moving animals. Proc Natl Acad Sci U S A 106:19557–19562
11. Komiyama T, Sato TR, O'Connor DH et al (2010) Learning-related fine-scale specificity imaged in motor cortex circuits of behaving mice. Nature 464:1182–1186
12. Andermann ML, Kerlin AM, Reid RC (2010) Chronic cellular imaging of mouse visual cortex during operant behavior and passive viewing. Front Cell Neurosci 4:3
13. Dombeck DA, Khabbaz AN, Collman F et al (2007) Imaging large-scale neural activity with cellular resolution in awake, mobile mice. Neuron 56:43–57
14. Dombeck DA, Graziano MS, Tank DW (2009) Functional clustering of neurons in motor cortex determined by cellular resolution imaging in awake behaving mice. J Neurosci 29:13751–13760
15. Nimmerjahn A, Mukamel EA, Schnitzer MJ (2009) Motor behavior activates Bergmann glial networks. Neuron 62:400–412
16. Holtmaat A, Bonhoeffer T, Chow DK et al (2009) Long-term, high-resolution imaging in the mouse neocortex through a chronic cranial window. Nat Protoc 4:1128–1144
17. Murayama M, Larkum ME (2009) In vivo dendritic calcium imaging with a fiberoptic periscope system. Nat Protoc 4:1551–1559
18. Gobel W, Kampa BM, Helmchen F (2007) Imaging cellular network dynamics in three dimensions using fast 3D laser scanning. Nat Methods 4:73–79
19. Duemani Reddy G, Kelleher K, Fink R et al (2008) Three-dimensional random access

multiphoton microscopy for functional imaging of neuronal activity. Nat Neurosci 11:713–720

20. Dombeck DA, Harvey CD, Tian L et al (2010) Functional imaging of hippocampal place cells at cellular resolution during virtual navigation. Nat Neurosci 13(11):1433–1440
21. Tsai PS, Nishimura N, Yoder EJ et al (2002) Principles, design and construction of a two photon scanning microscope for in vitro and in vivo studies. In: Frostig R (ed) Methods for in vivo optical imaging. CRC, Boca Raton, FL, pp 113–171
22. Albeanu DF, Soucy E, Sato TF et al (2008) LED arrays as cost effective and efficient light sources for widefield microscopy. PLoS One 3: e2146
23. Mostany R, Portera-Cailliau C (2008) A craniotomy surgery procedure for chronic brain imaging. J Vis Exp (12), e680, DOI: 10.3791/680
24. Yang G, Pan F, Parkhurst CN et al (2010) Thinned-skull cranial window technique for long-term imaging of the cortex in live mice. Nat Protoc 5:201–208
25. Mainen ZF, Maletic-Savatic M, Shi SH et al (1999) Two-photon imaging in living brain slices. Methods 18:231–239, 181
26. Muller M, Schmidt J, Mironov SL et al (2003) Construction and performance of a custom-built two-photon laser scanning system. J Phys D Appl Phys 36:1747–1757
27. Nguyen QT, Callamaras N, Hsieh C et al (2001) Construction of a two-photon microscope for video-rate Ca(2+) imaging. Cell Calcium 30:383–393
28. Nikolenko V, Nemet B, Yuste R (2003) A two-photon and second-harmonic microscope. Methods 30:3–15
29. Majewska A, Yiu G, Yuste R (2000) A custom-made two-photon microscope and deconvolution system. Pflugers Arch 441:398–408
30. Pologruto TA, Sabatini BL, Svoboda K (2003) ScanImage: flexible software for operating laser scanning microscopes. Biomed Eng Online 2:13
31. Nguyen QT, Tsai PS, Kleinfeld D (2006) MPScope: a versatile software suite for multiphoton microscopy. J Neurosci Methods 156:351–359
32. Greenberg DS, Kerr JN (2009) Automated correction of fast motion artifacts for two-photon imaging of awake animals. J Neurosci Methods 176:1–15
33. Guizar-Sicairos M, Thurman ST, Fienup JR (2008) Efficient subpixel image registration algorithms. Opt Lett 33:156–158
34. Diez-Garcia J, Matsushita S, Mutoh H et al (2005) Activation of cerebellar parallel fibers monitored in transgenic mice expressing a fluorescent Ca2+ indicator protein. Eur J Neurosci 22:627–635

Neuromethods (2012) 67: 61–81
DOI 10.1007/7657_2011_3

Published online: 9 December 2011

Spatial Light Modulators for Complex Spatiotemporal Illumination of Neuronal Networks

Francesco Difato, Marco Dal Maschio, Riccardo Beltramo, Axel Blau, Fabio Benfenati, and Tommaso Fellin

Abstract

The introduction of fluorescent probes and light-sensitive molecules and the recent development of optogenetics are tremendously contributing to our understanding of neuronal circuit function. In parallel with the development of these optical tools, new technologies for the illumination of neural tissue with complex spatiotemporal patterns have been introduced. Here, we describe a method for generating spatially modulated illumination by using liquid crystal on silicon spatial light modulators (LCOS-SLMs). The theoretical background and the description of working principles of LCOS-SLMs are presented together with a detailed experimental procedure to install LCOS-SLMs on conventional two-photon laser scanning microscopes and perform experiments on neuronal cells. In combination with the development of light-sensitive proteins with cell-specific and subcellularly localized expression, this technical approach has the potential to open new horizons for the optical investigation of neuronal circuits.

Key words: Digital holography, Structured light, Two-photon microscopy, Imaging, Photostimulation, Uncaging

1. Introduction

Due to its low invasiveness and the ability to monitor large numbers of cells while maintaining single-cell resolution, optical microscopy currently represents a fundamental tool for the investigation of the central nervous system. In particular, the combination of optical microscopy with the development of fluorescent indicators, caged compounds, and genetically engineered light-sensitive proteins (1–7) is bringing new and previously unachievable insights into neuronal network function in vitro as well as in the intact brain. Currently, the two most common configurations of optical microscopy used in neuroscience laboratories are wide-field illumination and laser scanning imaging. In wide-field microscopy, the whole field of view is simultaneously illuminated, allowing fast image acquisition or fast

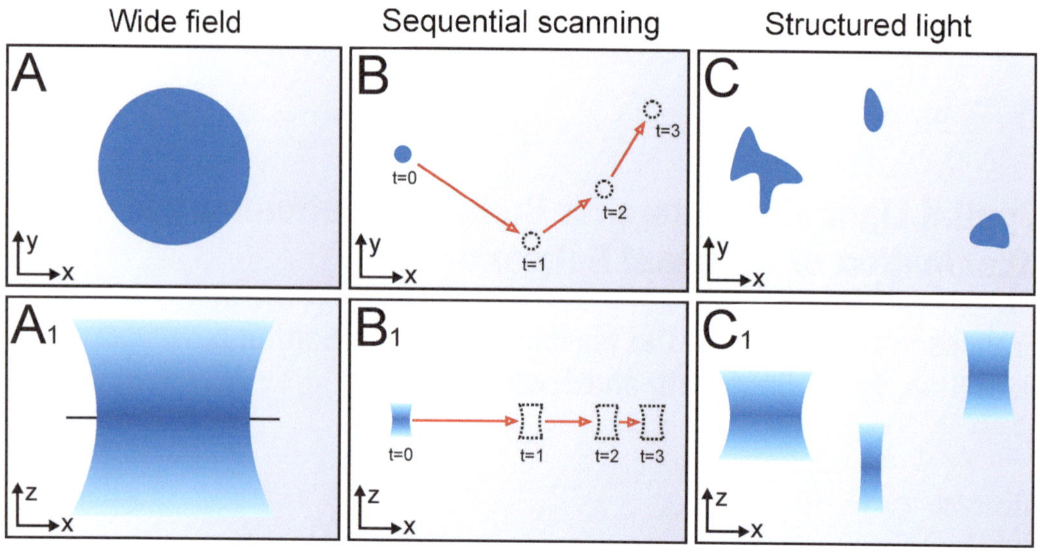

Fig. 1. Advantages of structured light illumination. (**a**, **b**) While wide-field microscopy (a) results in the illumination of the whole field of view (*blue disk*) preventing the projection of complex spatial patterns, in scanning imaging (b) the image is formed by sequentially illuminating the field of view with a diffraction-limited spot (*blue disk*, $t = 0$) which is steered across the sample (*open circles*, $t = 1, 2, 3$) according to the desired trajectory (*red arrows*). (**c**) Structured light illumination offers the possibility to shape the light in the *x*–*y* plane to any desired pattern, thus simultaneously illuminating complex structures at the sample plane. (**a_1–c_1**) The *x*–*z* profiles under the different experimental conditions are shown. Note that with structured light complex three-dimensional patterns can be obtained and the compression of the *z* dimension of different *x*–*y* shapes can be achieved with temporal focusing (41, 49) (see also Sect. 6).

repetitive stimulation, but preventing the application of complex spatial light patterns (Fig. 1a). Differently, in laser scanning microscopy, a diffraction-limited laser spot is sequentially deflected in the field of view, allowing the selective illumination of portions of the sample that depend on the scanning trajectory (Fig. 1b). This configuration leads to an increase of the spatial but to a significant loss in the time resolution of the optical system. Both approaches, thus, have intrinsic limitations with respect to the degree of complexity with which spatiotemporal patterns of light can be projected onto the biological sample. Illumination with structured light represents an innovative alternative to overcome these limitations. In this experimental approach, the laser wave front is engineered (or shaped) to simultaneously and selectively illuminate only specific regions of interest in a given field of view (Fig. 1c). This technique offers flexibility in the pattern of illumination that cannot be achieved with more traditional optical approaches and gives the opportunity of imaging/stimulating simultaneously multiple portions of a given cell or different cells within a neuronal network.

Over the course of years, various techniques have been developed to sculpt the light wave front (8–10), including phase modulation of the laser beam. Initially applied in astronomy to develop adaptive systems to correct optical aberration induced by atmospheric turbulence (11, 12) or to filter out the effects of specimen-induced aberrations (13), light phase modulation has been more recently applied to the investigation of the nervous system (14). Among different devices that generate phase modulation (15), liquid crystals on silicon spatial light modulators (LCOS-SLMs) are increasingly recognized as preferred tools. These devices have been largely used for optical tweezers applications (16–21) and more recently rediscovered as active tools for distortion minimization in two-photon in vivo microscopy (22, 23) or to perform imaging or photostimulation of neuronal circuits with complex spatiotemporal patterns (14, 24, 25). In this chapter, we present the theory of operation of LCOS-SLMs together with a detailed experimental protocol for their integration into commercially available scanning microscopes for functional imaging/uncaging experiments on neuronal networks with sculpted light.

1.1. Light Wave Front as the Superposition of Multiple Spherical Components

The working principle of wave front engineering with LCOS-SLMs is based on the Huygens principle, which states that an arbitrary wave front can be considered as the envelope of spherical waves emitted by point sources. In this view, the calculation of the complex field distribution $U(x,y)$ at a certain distance z in the propagation direction is obtained from the integration of a set of spherical waves generated by a collection of points in the source plane $\Sigma(\xi,\zeta)$ at $z = 0$ as given by the Huygens–Fresnel diffraction formula (26):

$$U(x, y) = \frac{1}{j\lambda} \iint_{\Sigma} U(\xi, \zeta) \frac{e^{jkr}}{r} \cos\beta \, ds,$$

where $U(\xi, \zeta)e^{jkr}/r$ is the diverging spherical wave originating in the point (ξ,ζ), separated by a distance r from the point (x,y) in the observation plane, β the angle formed by the vector r and the z direction, j the imaginary unit, k the wave number, and λ the light wavelength (Fig. 2). The previous diffraction formula, rearranged by considering $\cos\beta$ as the z/r ratio and taking into account the expression for r derived by its Taylor series truncated at the quadratic term, in the far-field approximation, results in the following Fraunhofer diffraction integral:

$$U(x, y) = \frac{e^{jkz}}{j\lambda z} e^{j\frac{k}{2z}(x^2+y^2)} \iint_{-\infty}^{+\infty} U(\xi, \zeta) e^{-j\frac{2\pi}{\lambda z}(x\xi+y\zeta)} d\xi \, d\zeta.$$

Besides the first quadratic phase factor, the last equation describes $U(x,y)$ as the two-dimensional Fourier transform of the initial complex field distribution $U(\xi,\zeta)$ and demonstrates that the amplitude and phase of the complex field at coordinates (x, y) in the observation plane are determined by the input Fourier components at frequencies $(x/\lambda z, y/\lambda z)$ in the source plane (26). As a

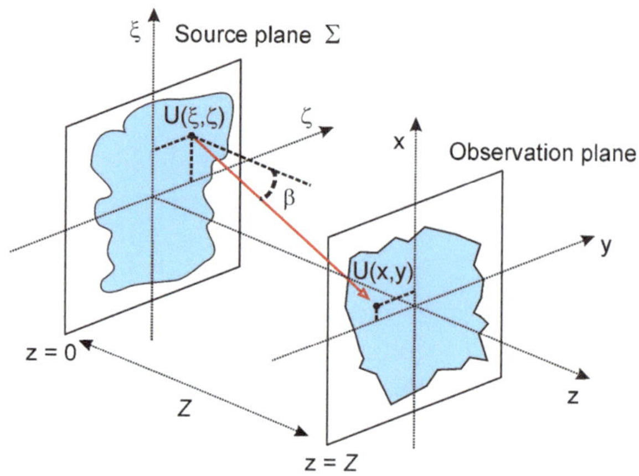

Fig. 2. Light propagation between Fourier planes. The complex field distribution $U(x,y)$ at a certain distance Z in the direction of propagation results from the integration of spherical waves generated by a collection of points in the source plane at $z = 0$.

consequence, a certain diffraction pattern observed in the far field is determined by the propagation of a specific wave front generated by a proper map of spatial frequency components. Considering a condition of uniform illumination, this map of spatial frequencies is completely described by a distribution of phase delays within the illuminated area at the source plane.

1.2. Shaping the Light Wave Front by Phase Modulation with LCOS-SLMs

In light of the theory described in the previous section, it is clear that an effective strategy to engineer a light wave front is by modulating the phase of its spherical components (Fig. 3a). To achieve this aim, current technology has focused on certain kinds of birefringent materials in which a change in molecular orientation results in a change in the effective refraction index. This is the case for nematic liquid crystals which are commonly used as components of LCOS-SLMs devices (Fig. 3b). Indeed, nematic liquid crystals have typically a rod-like molecular structure with one unique symmetry axis of anisotropy, called optic axis or director. This leads to the existence of two different refractive indices for different polarizations, an extraordinary refractive index (n_e) for light polarized perpendicularly to the optic axis, and an ordinary refractive index (n_o) for light polarized in parallel to the optic axis (Fig. 3c). LCOS-SLMs are composed of a matrix of active cells, each containing nematic liquid crystals the orientation of which can be individually controlled through the application of a voltage difference (Fig. 3b). Considering a cell where the liquid crystal molecules are all aligned with one another, when an electric field is applied across the liquid crystal layer along the light propagation direction (z), a progressive point-by-point reorientation of the

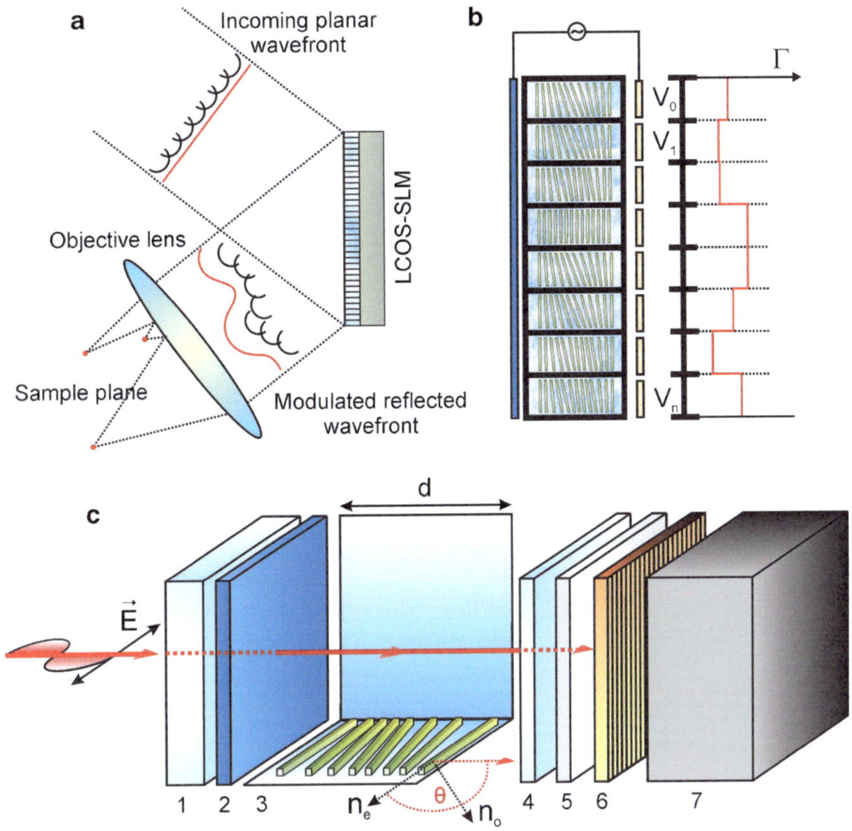

Fig. 3. LCOS-SLM working principle. (**a**) Schematic view of the effect of an LCOS-SLM in shaping an incident planar laser wave front (*red straight line*). By modulating the phase of the spherical components (*black semicircles*), the LCOS-SLM modifies the wave front of the reflected light beam (*curved red line*). This change in phase that is introduced in the Fourier space by the LCOS-SLM results in the generation of structured light illumination at the sample plane (behind the objective lens). (**b**) Zoom in showing the structure of the LCOS-SLM as a matrix of active cells. By controlling the liquid crystal orientation through voltage, each cell (or pixel) can differentially modulate the phase of the light (Γ) impinging upon it. (**c**) The structure of a single cell is shown at an expanded scale. *Legend*: (1) protective cover glass, (2) transparent indium thin oxide (ITO) electrode layer, (3) liquid crystal layer, (4) backplane alignment layer, (5) planar dielectric mirror, (6) electrode pixel matrix of aluminum pads, (7) complementary metal oxide semiconductor (CMOS) driving circuitry.

molecules occurs and a change in the effective refraction index n_{eff} profile is induced according to the following equation:

$$n_{eff}(\theta(z)) = \frac{n_{\mathrm{o}} n_{\mathrm{e}}}{\sqrt{n_{\mathrm{o}}^2 \sin^2 \vartheta(z) + n_{\mathrm{e}}^2 \cos^2 \vartheta(z)}},$$

where ϑ is the angle between the director and the direction of light propagation. Because the degree of reorientation depends on the z position within the liquid crystal layer, the total phase delay Γ generated by a cell of thickness d results from the following integration (27, 28):

$$\Gamma = \frac{2\pi}{\lambda} \int_{-d/2}^{+d/2} (n_{eff}(\vartheta(z)) - n_{\mathrm{o}}) \mathrm{d}z.$$

It is, thus, clear that the response of the director distribution, $\vartheta(z)$, to the application of external fields and the time required for liquid crystal reorientation represent the most important features determining the performance of a liquid crystal cell, both in terms of the reliability and precision of the phase modulation and in terms of the refresh rate.

An analytical expression for the time and voltage dependencies of the phase delay $\Gamma(t, V)$ is based on approximated solutions of the Oseen–Frank (29) and Ericksen–Leslie set of equations (30). It is derived by minimizing the total free energy of the system, including the elastic and viscous components. For a specific liquid crystal compound, the solution of these equations shows that (1) a bias potential greater than the Frèedericksz threshold has to be applied across the liquid crystal cell in order to induce molecule reorientation (30). (2) The cell modulation capability in terms of π fractions, also called phase stroke, depends on the ratio between the cell thickness d and the wavelength λ, and is proportional to the material birefringence $n_e - n_o$. (3) Nematic liquid crystal molecules have typical settling times that are proportional to the square of the cell thickness d and to the viscosity coefficient η while they are inversely proportional to the splay elastic constant (29). All these parameters must be considered in the design of liquid crystal-based SLMs and contribute to define the temporal and optical performance (i.e., the maximum refresh rate and the spectral range) of the device.

1.3. Structure of Commercial LCOS-SLMs

The core of reflective LCOS-SLMs (Fig. 3c) is a matrix of active pixels, each generally composed of (1) a protective cover glass with an antireflection broadband or specific coating; (2) a transparent indium thin oxide (ITO) layer serving as ground electrode and being rubbed or coated by a vapor-deposited SiO2 alignment layer to provide the proper director orientation and pretilting angle for liquid crystal molecules; (3) the liquid crystal layer; (4) a backplane alignment layer; (5) a planar dielectric mirror which enhances the light utilization efficiency and the fill factor; (6) a layer with aluminum electrode pads forming the pixel matrix; and (7) the complementary metal oxide semiconductor (CMOS) driving and addressing circuitry realized with very large-scale integration (VLSI) technology. Depending on the reciprocal orientation of the two alignment layers, the direction of the external electric field, and the polarization state of the incoming light, LCOS-SLMs operate in different modulation modes. In particular, for the configuration with parallel alignment layers, pure phase ($\alpha = 0°$), or a combination of phase and amplitude ($0 < \alpha < 90°$), modulation can be achieved by altering the angle α between the polarization of the incoming light and the liquid crystal optical axis (see Sect. 1). Commercial two-dimensional liquid crystal matrices with layouts ranging from 256×256 to 1900×1000 and pixel sizes ranging between 8 × 8 and 40 × 40 μm^2 have an effective active area of the order of several mm^2 and maximum spatial resolutions in the range of 20–33 line pairs/mm (9). LCOS-SLMs are currently available with optical windows covering wavelengths ranging from UV to IR and are subjected to a calibration process to linearly map the 8-bit levels to the designed phase stroke for each specific wavelength.

1.4. Structured Light Illumination for the Optical Investigation of Neuronal Function

Initially applied to aberration correction optics and for designing optical tweezers (see Sect. 1), LCOS-SLM technology has been recently applied to the study of the central nervous system. Single-photon holographic uncaging of caged 4-methoxy-7-nitroindolinyl-caged-L-glutamate (MNI-glutamate) has been performed on cerebellar and hippocampal neurons in brain slice preparation in combination with intracellular recordings to measure photoactivation-induced currents mediated by the α-amino-3-hydroxyl-5-methyl-4-isoxazole-propionate (AMPA) receptor. These initial studies demonstrated that the shaping of the excitation light combined with classical electrophysiology recordings represents an extremely useful tool for the functional mapping of ion channel distribution throughout different subcellular compartments (14). In a similar experimental approach, but using two- rather than single-photon excitation, localized photostimulation of dendritic spines with holographic illumination has demonstrated the potentials of this technique for studying neuronal input summation properties and suprathreshold activation of pyramidal neurons in brain slices (25). Besides electrophysiological recordings, holographic uncaging experiments have been performed in combination with Ca^{2+} imaging to detect and characterize the activation of AMPA receptors in glial cells in hippocampal slices. In this study, an image quality enhancement method was introduced to generate holographic illumination based on reliable volumetric rendering of the cell soma distribution, which was derived from wide-field excitation stacks (24). Two-photon holographic illumination has not only been used for photoactivation or photo-uncaging experiments, but also for fast scanless fluorescence imaging both in vitro and in situ (25, 31). This particular application of SLM-based microscopy allows the simultaneous imaging of multiple regions of interest (as for example different neurons or different portions of a neuron) in a given field of view at high acquisition frequency (tens to hundreds of Hz).

Recent technical advances in this field of research include a portable holographic microscope (32) and a dual microscope, which combines two independently tunable lasers with an SLM-based holographic module and a conventional scan head. This latter system combines the versatility of scanless holographic excitation for imaging or photostimulation together with the reliability of standard galvo-steered spot uncaging and high-resolution multiphoton scanning imaging (31). The dual microscope can be used in two principal configurations: holographic imaging combined with galvo-steered uncaging and conventional scanning imaging combined with structured light uncaging. The potentials of this system for the study of brain function have been demonstrated with MNI-glutamate uncaging and Ca^{2+} imaging experiments on neuronal cultures (31).

1.5. Present and Future Perspectives

Structured light illumination obtained with LCOS-SLM technology offers many advantages when compared to common imaging and photostimulating approaches. Traditional point scanning systems based on galvo mirror devices or innovative beam steering configurations with acousto-optical deflectors are highly reliable and unparalleled in their switching rates. Nonetheless, they are limited by being single-spot illumination techniques which do not allow the simultaneous illumination of the sample at multiple locations or with arbitrary two/three-dimensional patterns (Fig. 1, see also Sect. 2). These considerations become particularly relevant in applications, where the low spatial density and the fast temporal relaxation dynamics of light-excited molecules are main factors determining the efficacy of the optical stimulation (33). For example, structured light illumination obtained with temporal focusing (to achieve thin, disk-like stimulation volumes with lateral diameters two orders of magnitude larger than diffraction-limited spots, see Sect. 6) has been used for efficient two-photon excitation of Channelrhodopsin 2 (34). From this point of view, the flexible beam shaping capabilities provided by holographic microscopy based on LCOS-SLMs represent a very promising technical approach to be used in combination with optogenetics and, more in general, with photoactivable proteins which may require complex patterns of stimulation. Indeed, a recent study (35) shows effective two-photon stimulation of channelrhodopsin with structured light illumination with modulation techniques derived from the theory of the phase contrast (36, 37) highlighting the importance of this approach for the activation of sparse neuronal assemblies with complex spatiotemporal patterns (33).

For imaging applications, holographic excitation systems can be used to illuminate arbitrary regions of interest within the field of view while simultaneously detecting the emitted fluorescence with a fast CCD camera. The major advantage of this parallel approach is its high acquisition rate, which is only limited by the signal-to-noise (S/N) ratio of the excited fluorescence and the camera frame rate. It, thus, outperforms acquisition system based on the combination of galvo mirror scanning and photomultiplier tube detection. For example, if a field of view is divided into a 512 × 512 pixel array and scanned with a 4.4-μs dwell time per pixel, a frame rate lower than one frame per second (fps) is obtained. In contrast, using a scanless holographic configuration on both brain slices and cultured neuron preparations, spontaneous Ca^{2+} signals recorded with acquisition rates up to 70 fps have been reported (25, 31).

The considerations discussed above together with the prospects that future generation of LCOS-SLMs will have even higher refresh rates and larger spectral windows suggest that single- and multiphoton structured light illumination offers unique advantages compared to more traditional optical approaches. In combination with genetically encoded and cell-specific indicators/actuators, the complexity of the light stimuli and the flexibility with which they can be applied to the sample with this technique promise to bring new and fundamental insights into our understanding of brain circuits.

2. Materials

- Opto-mechanics for mounting freestanding optics and cage assembly, including bases (e.g., BA1, Thorlabs Inc., Newton, NJ), post holders (PH1/M–PH6/M), posts (TR20/M–TR300/M), right-angled clamps (RA90/M), flipping mounts (FM90), kinematic mounts (KM100/KCB1), cage assembly rods (ER8-ER05), square cage plates (CP02/M).
- High reflective ($R > 99\%$) NIR dielectric mirrors (BB1-E03) (Thorlabs Inc., Newton, NJ).
- Mounted achromatic IR lenses (AC254-100-B-ML, AC254-300-B-ML, AC254-100-B-ML, AC254-030-B-ML) (Thorlabs Inc., Newton, NJ).
- Half waveplate (lambda/2 B. Halle Nachfl GmbH).
- Research grade upright biological/life science optical microscope (Olympus BX61W, Milan, Italy) with the following components: bright-field illumination assembly in transmission (halogen bulb); fluorescence illumination assembly (mercury lamp); rotating filter wheel for filter cubes, Z objective motorization; objectives 20× XLUMPLFL20XW 0.95 NA, 40× LUMPLFL40XW 0.8 NA, 60× LUMFL, 1.1 NA.
- Two-photon system Ultima IV from Prairie Technologies (Madison, WI) with the following main components: Chameleon Ultra II Ti:Sapphire source (Coherent, Milan, Italy); Pockels cell modulator (350–80 LA-BK, Conoptics, Danbury, CT); scanhead equipped with short pass dichroic mirror (FF670-SDi01, Semrock, Rochester, NJ), IR blocking filter (ET750sp-2p8), and emission filters 530/50 nm and 590/40 nm (Chroma, Fuerstenfeldbruck, Germany).
- Orca R2 CCD camera (Hamamatsu, Milan, Italy).

- Reflective X10468-07 LCOS spatial light modulator (Hamamatsu, Milan, Italy).
- CARPE Autocorrelator with external detector (APE GmbH, Berlin, Germany).
- Chameleon PreComp precompressor unit (Coherent, Milan, Italy).
- Optical power meter (PM100 with S121B sensor).
- LabVIEW (National Instruments, Austin, TX)-based application for CCD camera control and acquisition.
- LabVIEW (National Instruments, Austin, TX)-based application for spatial light modulator configuration.
- Calcium indicator Fluo-4AM (Invitrogen, Milan, Italy).
- MNI-glutamate (Tocris, Bristol, UK).

3. Methods

3.1. Combining an LCOS-SLM with a Commercial Two-Photon Laser Scanning Microscope

The LCOS-SLM has to be positioned in such way within the optical pathway of the microscope that the phase map generated by the LCOS-SLM is projected onto the back focal plane of the objective (Fig. 4). This optical design results in the projection, at the sample plane, of the Fourier transform distribution of the phase map generated by the LCOS-SLM. Given that, in commercial systems, the scan head galvo mirrors are usually optically conjugated to the back focal plane of the objective (via a 4f telescope composed of the tube lens/scan lens, Fig. 4) (38), LCOS-SLM conjugation with the pupil objective can be achieved by simply conjugating the LCOS-SLM with the galvo mirror plane (via a 4f telescope, Fig. 4). In 4f arrangement, the two lenses of the telescope are placed at distance equal to the sum of the two respective focal lengths. The image at the back focal plane of the first lens (L_3) is thereby relayed to the front focal plane of the second lens (L_4), with a magnification factor of f_4/f_3 (f_3 and f_4 respective focal lengths of L_3 and L_4). This optical design ensures the projection of the wave front generated by the LCOS-SLM onto the back focal plane of the objective without vignetting, thus avoiding the loss of high-frequency components located at the image periphery (39). Moreover, this strategy represents a convenient solution as scan heads are usually compact and difficult to modify with new optical components. The distance, d, between the LCOS-SLM and the galvo mirrors is thus:

$$d = 2f_3 + 2f_4.$$

The focal lengths of L_3 and L_4 are chosen to obtain the proper magnification of the beam diameter to match the dimension of the

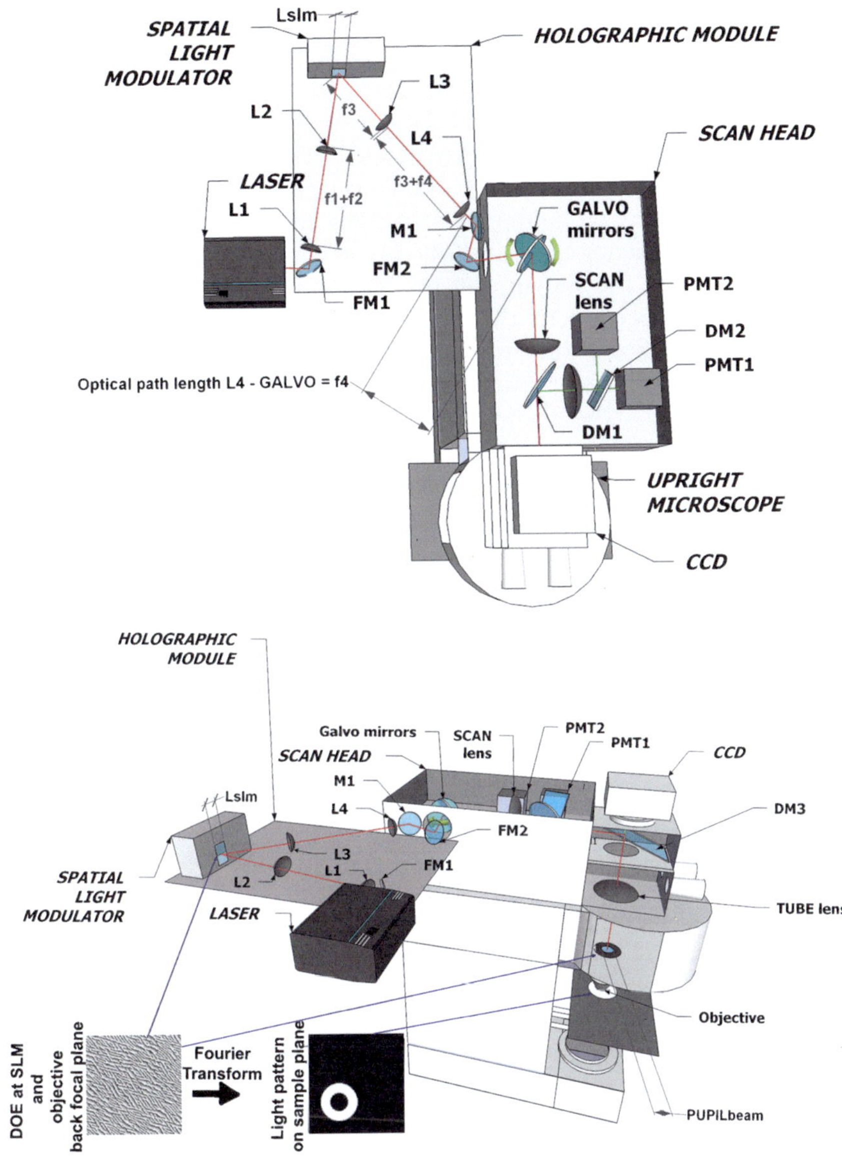

Fig. 4. Layout of the optical setup. *Legend*: FM_1, FM_2, flipping mirrors; M_1, turning mirror; L_1, L_2, L_3, L_4, plano-convex lenses; DM_1, 660-nm long-pass dichroic mirror; DM_2, 575-nm long-pass dichroic mirror; DM_3, 660-nm short-pass dichroic mirror; PMT_1, photomultiplier #1; PMT_2, photomultiplier #2; CCD, CCD camera.

pupil of the microscope objective. Given the magnification of the telescope composed of the scan and tube lens of the microscope (M_{MICR}), M_{T2} can be expressed as:

$$M_{\mathrm{T2}} = \frac{f_4}{f_3} = \left(\frac{\emptyset_{\mathrm{PUPILbeam}}}{\emptyset_{\mathrm{SLMbeam}}}\right) / M_{\mathrm{MICR}}$$

$$M_{\mathrm{MICR}} = \frac{f_{\mathrm{tubelens}}}{f_{\mathrm{scanlens}}}$$

with $\emptyset_{\mathrm{PUPILbeam}}$ being the diameter of the laser beam at the back aperture of the objective, $\emptyset_{\mathrm{SLMbeam}}$ the diameter of the laser beam at the LCOS-SLM plane, and f_{tubelens} and f_{scanlens} the focal lengths of the tube and scan lenses, respectively, which are fixed characteristics of the scanning microscope. In some experimental configurations, a slight underfilling of the back aperture of the microscope objective may be preferred. This prevents the clipping out of the high-frequency content of the phase map and improves the light efficiency of the system, but results in a decrease in the effective numerical aperture of the objective (40). This might be useful for some photostimulation applications in which light efficiency is more important than spot size (31).

The diameter $\emptyset_{\mathrm{SLMbeam}}$ of the beam impinging on the LCOS-SLM surface is determined by the lateral dimensions of the SLM surface (L_{SLM}, see also Fig. 4). It is recommended to maximize $\emptyset_{\mathrm{SLMbeam}}$ in order to illuminate the maximum number of pixels of the LCOS-SLM chip, thereby increasing the high-frequency content of the projected diffractive optical element (DOE). With the help of a second telescope (lenses L_1 and L_2 in Fig. 4) between the laser and the LCOS-SLM, the diameter $\emptyset_{\mathrm{LASER}}$ of the laser beam exiting the laser source is enlarged to $\emptyset_{\mathrm{SLMbeam}}$. The magnification factor (M_{T1}) of this telescope is:

$$M_{\mathrm{T1}} = \frac{f_2}{f_1} = \frac{\emptyset_{\mathrm{SLMbeam}}}{\emptyset_{\mathrm{LASER}}} = \frac{L_{\mathrm{SLM}}}{\emptyset_{\mathrm{LASER}}}.$$

The use of a cage assembly (Thorlabs Inc., Newton, NJ) to mount the telescopes is recommended as it simplifies their alignment. The correct alignment of the telescope is obtained by matching the laser spot position, in a plane far from the telescope, with and without the telescope lenses in the optical path.

3.2. Optimizing the Design of the Holographic Pathway

After defining the optical conjugation of the LCOS-SLM with the galvo mirrors of the scan head, a series of parameters must be considered to optimize the properties of the holographic path (see also Sects. 1 and 3). First, in order to achieve a compact and space-saving design, the LCOS-SLM position should be set close to the entrance port of the scan head. Second, the diffraction efficiency of the LCOS-SLM depends on the angle of incidence with which light impinges on it. To maximize the diffraction efficiency, the angle of incidence between the laser light and the direction orthogonal to the LCOS-SLM plane should be kept to a minimum (as small as the dimensions of the optical components allow). Nonetheless, it should be considered that the angle of incidence also determines the distance at which the incident and reflected beam can be separated, thus determining the overall physical size of the holographic path. Smaller angles mean larger distances at which the two beams can be separated. An incidence angle of about

10° represents a good experimental compromise since it does not significantly affect the diffraction efficiency while allowing the separation of the two beams at a reasonable distance from the LCOS-SLM plane (e.g.: with a $Ø_{SLMbeam}$ of 10 mm, the two beams can be separated at about 10 cm from the LCOS-SLM). Third, it is recommendable to mount the LCOS-SLM onto two orthogonally positioned linear translators. In this way, the LCOS-SLM can be moved in a plane perpendicular to the direction of the laser propagation offering the possibility to center the position of the SLM chip on the laser beam without changing the angle of incidence.

3.3. Alignment of the Optical System

When redirecting the light going to the scan head onto the LCOS-SLM (mirror FM_1, Fig. 4), it is important to ensure that the laser beam is parallel to the optical table and centered in the optical axis of the light pathway. A convenient solution is to put the first (FM_1 in Fig. 4) and last (FM_2) mirrors of the holographic path on flip mounts to have the possibility to switch back and forth from the standard configuration of the scanning microscope to the holographic configuration. Moreover, the addition of a mirror (M_1) close to FM_2 facilitates the alignment of the beam in the scan head.

Once the holographic module is coupled to the scan head (see previous Sects. 1 and 2), it is necessary to fine-tune the alignment of the laser beam to the optical axis of the microscope. This aim can be achieved in eight steps as follows: (1) Ensure Köhler illumination with any kind of specimen, then remove the microscope objective and the sample, and set the galvo mirrors in their center position. (2) Close the field diaphragm of the halogen bulb to have a reference light spot at the objective housing port and center the laser beam on it (use mirrors M_1 and FM_2 to perform this task; FM_2 controls the laser position with respect to the reference spot while M_1 is used to center the light beam with respect to the entrance port of the scan head). (3) Repeat the same procedure with a microscope objective mounted on the microscope and a reflecting mirror at the sample plane while observing the shape of the reflected laser spot with a CCD camera mounted on the camera port. (4) Move FM_2 and M_1 sequentially to generate an undistorted laser spot at the center of field of view. Control the inclination of the laser beam with respect to the optical axis of the microscope with the same two mirrors (M_1 and FM_2) by varying the focus of the objective (the laser spot changes in size but it should remain symmetric and in the same position in the images acquired with the CCD). (5) Switch on the SLM control unit and project a DOE to generate a grid of several spots evenly distributed on the sample plane. (6) Move the LCOS-SLM with the two orthogonal translators to center the projected DOE on the laser beam and obtain undistorted spots of similar intensity on the CCD. (7) In the center of the field of view, a bright spot, which represents the light component undiffracted by the LCOS-SLM (also called the "zero order"), is now visible. To remove the zero order component, different strategies can be used, including a reflecting grating to separate the first ($m = 1,-1$) from the zero order and the use of simple optical functions on the LCOS-SLM to direct only the diffracted portion of light to the scan head (41). Probably, one of the easiest methods to remove the zero order is by spatial filtering (25, 31). This can be done, for example, by placing a small piece of aluminum foil mounted on a glass coverslip at the Fourier plane of the first lens (L_3, Fig. 4) of the secondary telescope. This optical plane is conjugated to the sample plane and thus the projected hologram is visible, and the zero order can be filtered out by aligning the aluminum foil with the spot corresponding to the zero order component. (8) Adjust the distance of the lenses in one of the two telescopes to fine-tune the collimation of the beam. Put a fluorescent sample on the microscope stage (e.g., immunostained cells), flip down mirrors FM_1 and FM_2, and acquire a reference image with the scanning system and the PMTs. Flip mirrors FM_1 and FM_2 up and leave the LCOS-SLM switched off (the LCOS-SLM works as a reflecting mirror when turned off). Because in this configuration the laser is entering the scan head as a simple Gaussian beam, the scanning microscope works under normal conditions (raster scanning by galvo mirrors and images acquired with PMTs). During continuous imaging, fine-tune the distance between L_1 and L_2 to obtain the same focus position on the sample, as in the reference image (42).

3.4. Projecting Complex Spatial Light Patterns at the Sample Plane

At this point, it is possible to switch from a holographic scanless microscope to a laser scanning microscope by simply switching on/off the LCOS-SLM control unit without changing/removing any optical component. If the LCOS-SLM is switched on, it is possible to vary the field of view of the laser scanning system by changing the projected DOE on the LCOS-SLM. This procedure moves the position of the holographic spot at the sample plane, which is then raster scanned by the galvo mirrors. In this experimental configuration, phase-only modulation of the laser beam at the back focal plane of the microscope objective is obtained without altering the path of the beam within the scan head. In contrast, for scanless holographic imaging, it is necessary to direct emission light into the camera port. This is achieved by placing a shortwave-pass dichroic mirror (DM_3, Fig. 4) to direct excitation light to the objective and the emission light to the CCD. A set of shortwave-pass emission filters to reject excitation light, and appropriate emission filters to select the emission bandwidth must then be placed in front of the CCD camera. It is now possible to project several holographic patterns at the sample plane (see Sect. 4 for details on the generation of the DOE) and acquire fluorescence signals at a frame rate limited only by the camera sensitivity.

3.5. Structured Light Illumination for Functional Imaging and Uncaging Experiments on Neuronal Networks

The light shaping properties achieved with holographic illumination allow the design of functional experiments that overcomes some of the major limitations of traditional wide-field and laser scanning approaches (see Sect. 1). Here, we first describe a protocol for holographic imaging of fluorescence signals on neuronal cultures at high acquisition rates and then a protocol for uncaging experiments with structured light illumination. We specifically focus on the procedures to optimize the optical setup while details on the preparation of neuronal cultures can be found elsewhere (43).

3.5.1. Imaging Experiment

In scanless imaging experiments, the LCOS-SLM can be used to tailor the laser light either into extended regions of interest or into a two-dimensional distribution of points according to the structure of the biological sample. When combined with the use of fluorescent reporter molecules, as for example the Ca^{2+} indicators Fluo-4 or Oregon Green BAPTA, this approach allows the simultaneous monitoring of specific neuronal subpopulations or subcellular compartments of a given cell at high acquisition frequencies. A high-resolution image of the biological sample is first obtained with the two-photon laser scanning system and PMTs; based on this, various regions of interest in the field of view are identified (Fig. 5a, see also Sect. 5). The experimental protocol then requires a brief calibration step, which depends on the selected working wavelength, in order to establish a precise correspondence between the holographic addressable area and the PMT field of view. This can be achieved by projecting a phase mask corresponding, at the sample plane, to a reference gridded pattern while maintaining the galvo mirror in a reference position. Once the calibration procedure is performed, a simple software code can be used to extract the binary masks representing the desired illumination patterns from the images acquired in scanning mode. These patterns are then used to calculate

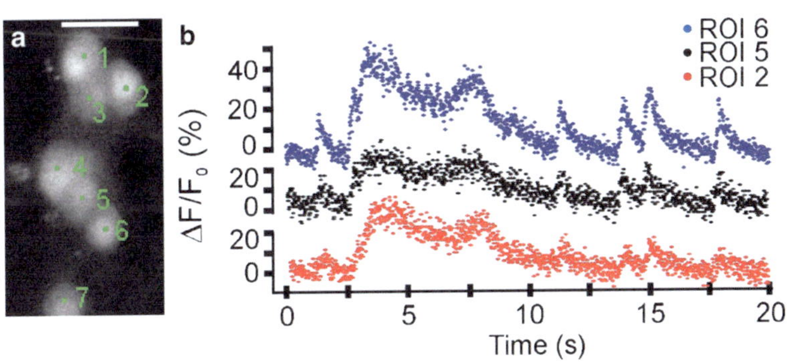

Fig. 5. Fast holographic Ca^{2+} imaging on neuronal networks. (**a**) Fluorescence image showing Fluo-4 loaded neurons in culture. Based on this image, multiple regions of interest corresponding to different cells (*green dots* numbered 1–7) are identified and simultaneously imaged with structured light ($\lambda_{imaging} = 830$ nm). *Scale bar* 15 μm. (**b**) Time course of the fluorescence signals showing spontaneous Ca^{2+} oscillations. Acquisition rate: 71 fps. Modified from (31).

the phase map to be sent to the LCOS-SLM control unit. Signals emitted from the different excited regions can then be collected in parallel by a fast CCD. Particular care has to be taken during high frame rate acquisition (>50 Hz) to find the best compromise between a good signal-to-noise ratio of the fluorescence signal variations and the power density that is continuously being delivered to the sample. Indeed, while the photodamage threshold limits the maximum power density per illuminated voxel at the sample plane, the signal of the emitted fluorescent light strongly depends upon the excitation power. As a reference, using two-photon excitation light at 830 nm, power density values of >10 mW per voxel have been reported to induce no photodamage during long-lasting Ca^{2+} recordings in brain slice preparation at frame rates around 60 fps (25). For similar acquisition rates, a power value of <4 mW per voxel has been used to measure Fluo-4 signals in neuronal cultures (Fig. 5b) (31).

3.5.2. Uncaging Experiment

The capability of shaping light into arbitrary three-dimensional illumination patterns with LCOS-SLM finds its natural application in photostimulation experiments, where light excitation is used to trigger conformational changes in synthetic molecules leading to the release of an effector molecule as, for example, in the case of MNI-glutamate uncaging. This approach can also be applied to light-gated proteins, as those of the opsin family, which can be expressed in neurons to control their excitability (35). The fine control of the three-dimensional and temporal profiles of the two-photon illumination that can be achieved with LCOS-SLM is fundamental to obtain reliable and effective stimulation while preserving the health of the preparation. Besides the specific working wavelengths, which depend upon the particular molecule that is photostimulated, the experimental protocol for photostimulation is similar to the one described for imaging experiments in terms of hardware configuration, calibration steps, and generation of phase maps. In photostimulation experiments though, a fine control of the illumination time interval and power density has to be performed. In the case of two-photon laser sources, this is achieved by introducing an electro-optic modulation unit, such as a Pockel cell, along the optical path (25, 31). The most important factor to be considered is the energy density required for photostimulation: at 720 nm (the optimal two-photon wavelength for MNI-glutamate uncaging), an energy density of about 10 $\mu J/\mu m^2$ has been reported to be sufficient for the holographic illumination of cell body areas in neuronal cultures (Fig. 6) (31). At similar wavelengths, the recording of excitatory postsynaptic currents (EPSCs) in brain slices after spine stimulation required pulses with energy densities close to100 $\mu J/\mu m^2$ (25). The electrophysiological recording of neuronal activity is undoubtedly the most sensitive and accurate method to track the photostimulus-induced effects with high temporal

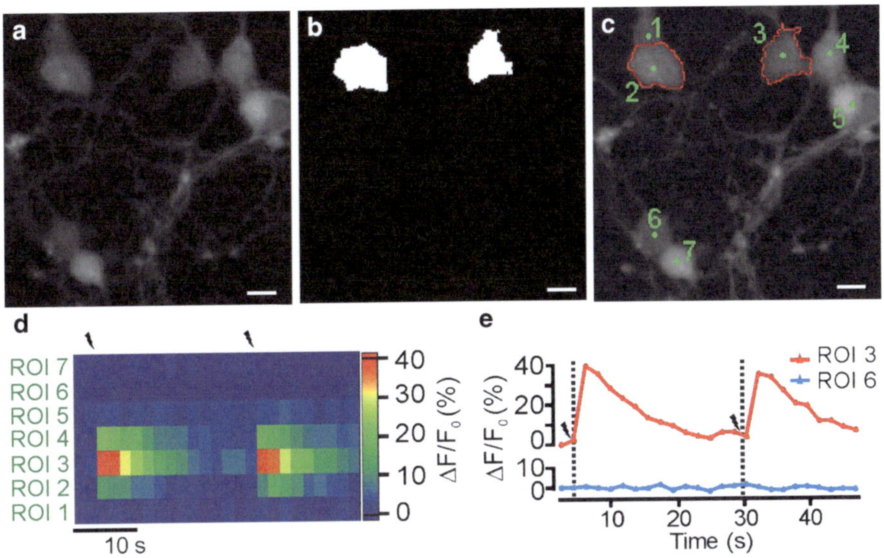

Fig. 6. Spatially defined activation of neurons with two-photon holographic uncaging of MNI-glutamate. (**a–c**) A fluorescence image of the field of view (**a**) is taken and used to generate an image mask (**b**) to configure the LCOS-SLM to illuminate only the desired regions of interest [*white regions* in (**b**) and areas delimited by a *red line* in (**c**)]. (**c**) Also shows the seven regions of interest in which Fluo-4 signals are monitored with conventional laser scanning microscopy at 0.54 Hz. (**d**, **e**) Time course of the Fluo-4 fluorescence. The *arrows* indicate the time of delivery of the photostimulus events. Modified from (31).

resolution (14, 25). Nonetheless, optical approaches for detecting the light-induced activity changes represent a valid alternative (24, 31).

4. Notes

4.1. Light Polarization

The LCOS-SLM is sensitive to the polarization of the incident light (see Sect. 3). It acts as a phase-only modulator for light that is linearly polarized in the direction corresponding to the liquid crystal orientation. Therefore, it is convenient to place a half-wave plate (RAC 5.2.10 achromatic $\lambda/2$ retarder—B. Halle Nachfl GmbH) between the laser source and the LCOS-SLM to linearly polarize the laser beam to match the LCOS-SLM optimal orientation, corresponding to the orientation of the liquid crystal director.

4.2. Generation of Three-Dimensional Patterns of Illumination

Another advantage of integrating a holographic apparatus into a scanning system is the ability to easily correct the z position of the excitation spot via software without the need of moving the objective while preserving the properties of the point spread function within a limited z range. This is obtained by imposing a phase map that produces a change in the collimation properties of the light beam (44). Thus, complex three-dimensional patterns of illumination can be generated.

4.3. Group Dispersion Velocity

An important aspect when using holographic illumination with multiphoton pulsed laser sources is the group velocity dispersion (GVD) introduced by the LCOS-SLM device. By using a CARPE autocorrelator (APEBerlin, Berlin, Germany), Dal Maschio et al. recently reported the broadening of femtosecond pulses generated by an LCOS-SLM-based holographic module (31). In the 760–980-nm range, the average GVD introduced by the holographic module with 20× objective (0.95 NA) is approximately 20,000 fs^2 at the sample plane, a value that can be completely compensated with commercial pulse compensators.

4.4. Configuration of the Diffractive Optical Element

A key aspect for the efficient use of LCOS-SLMs is the control and configuration of the liquid crystal active matrix. A phase map, also called DOE, has to be generated by a specific software. It is based on the desired illumination pattern at the sample plane and on the laser beam properties at the LCOS-SLM plane. Among the different algorithms available for holographic projection (45), the intuitive "Gratings and Lenses" approach and the versatile Gerchberg–Saxton algorithm are illustrated here. The Grating and Lenses model relies on the superposition of the phase characteristics of two basic optical components: gratings which produce lateral shifts and lenses which produce axial shifts. The capability of the gratings to steer a beam relies on the fact that the same wave front modification induced by a linear phase profile, because of the 2π periodicity of a wave, can be obtained by decomposing it in a saw-tooth phase profile, where every period resembles the original $0–2\pi$ phase modulation (phase folding). Similarly, in the case of the axial shift, the change of the wave front convergence position obtained by a lens is achieved by means of the projection of a circular phase map characterized by a periodic pattern with radial symmetry. The Grating and Lenses approach can also be used to generate multiple beams that can be independently controlled in a three-dimensional volume, but does not allow the projection of more complex and extended patterns at the sample plane (46). A more robust approach is the Gerchberg–Saxton algorithm (47), which is a Fourier transform-based method, that iteratively converges toward the phase distribution required at the LCOS-SLM plane to produce the desired intensity distribution at the sample plane. The algorithm is first initialized with a complex field with random phase and constant amplitude. By taking the Fourier transform, a complex field at the image plane is then calculated. At this stage, the image field is modified preserving the phase information but substituting the calculated amplitude with the target amplitude distribution. In the following step, the resulting field is back propagated to the hologram plane by means of an Inverse Fourier Transform. The field resulting at the hologram plane is modified keeping the phase information but replacing the amplitude distribution with a constant distribution. After few iterations, the argument of the field at the hologram plane converges toward the phase map required to produce the target intensity at the sample plane. This algorithm can be used to generate multiple spots or even arbitrary two-dimensional distributions at the focal plane and can be integrated with the Grating and Lenses algorithm in order to achieve a multifocal projection of complex three-dimensional patterns.

4.5. Dimensions of the Holographic Field

At the sample plane, holographic illumination can be achieved generally in a subregion of the field of view (holographic field). The dimensions of the holographic field are determined by the maximum angle of deflection that is introduced by the LCOS-SLM. Theory sets this limit equal to:

$$u_{\max} = \frac{\lambda}{2M_{T2}d_{\text{pitch}}} f_0,$$

where $u_{\max}$ is the maximum displacement of the laser spot on the sample plane, λ the wavelength of the laser source, d_{pitch} the pixel pitch of the LCOS-SLM chip, and f_o the focal length of objective (48). Diffraction efficiency of the device sets an experimental limit for $u_{\max}$.

4.6. Temporal Focusing

With the implementation of temporal focusing, the spectral components of a short laser pulse are geometrically dispersed by means of a grating to enhance the axial localization of the two-photon excitation (34, 41, 49). In this experimental configuration, a short pulse is obtained at the image plane of a lens projecting system while its shape is temporally stretched in the out-of-focus regions along the propagation direction (below and above the focal plane), leading to an effective two-photon absorption mostly limited to the focal plane. This optical design helps in preserving the z confinement of a bidimensional pattern with slow phase variations, but can lead to effective power reduction (49).

Acknowledgments

We thank Gian Michele Ratto for critical reading of the manuscript. This work was supported by grants from MIUR PRIN program to F. Benfenati, Telethon-Italy (GGP09134 to F. Benfenati and GGP10138 to T. Fellin), and by the San Paolo "Programma in Neuroscienze" grant to F. Benfenati and T. Fellin.

References

1. Zhang J, Campbell RE, Ting AY et al (2002) Creating new fluorescent probes for cell biology. Nat Rev Mol Cell Biol 3:906–918
2. Giepmans BN, Adams SR, Ellisman MH et al (2006) The fluorescent toolbox for assessing protein location and function. Science 312:217–224
3. Kramer RH, Fortin DL, Trauner D (2009) New photochemical tools for controlling neuronal activity. Curr Opin Neurobiol 19:544–552
4. Gorostiza P, Isacoff E (2007) Optical switches and triggers for the manipulation of ion channels and pores. Mol Biosyst 3:686–704
5. Airan RD, Hu ES, Vijaykumar R et al (2007) Integration of light-controlled neuronal firing and fast circuit imaging. Curr Opin Neurobiol 17:587–592
6. Zhang F, Aravanis AM, Adamantidis A et al (2007) Circuit-breakers: optical technologies for probing neural signals and systems. Nat Rev Neurosci 8:577–581
7. Knopfel T, Lin MZ, Levskaya A et al (2010) Toward the second generation of optogenetic tools. J Neurosci 30:14998–15004
8. McManamon PF, Watson EA (2009) A review of phased array steering for narrow-band electrooptical systems. Proc IEEE 97:1078–1096
9. Savage N (2009) Digital spatial light modulators. Nat Photonics 3:170–172
10. Maurer C, Jesacher S, Bernet M et al (2011) What spatial light modulators can do for optical microscopy. Laser Photonics Rev 5:81–101
11. Tyson RK (1991) Principles of adaptive optics. Academic Press, London
12. Hardy JW (1998) Adaptive optics for astronomical telescopes. Oxford University Press, Oxford
13. Neil MA, Juskaitis R, Booth MJ et al (2000) Adaptive aberration correction in a two-photon microscope. J Microsc 200 (Pt 2):105–108
14. Lutz C, Otis TS, DeSars V et al (2008) Holographic photolysis of caged neurotransmitters. Nat Methods 5:821–827
15. Booth MJ (2007) Adaptive optics in microscopy. Philos Transact A Math Phys Eng Sci 365:2829–2843
16. Eriksen R, Daria V, Gluckstad J (2002) Fully dynamic multiple-beam optical tweezers. Opt Express 10:597–602
17. Melville H, Milne G, Spalding G et al (2003) Optical trapping of three-dimensional structures using dynamic holograms. Opt Express 11:3562–3567
18. van der Horst A, Forde NR (2008) Calibration of dynamic holographic optical tweezers for force measurements on biomaterials. Opt Express 16:20987–21003
19. Bowman R, Gibson G, Padgett M (2010) Particle tracking stereomicroscopy in optical tweezers: control of trap shape. Opt Express 18:11785–11790
20. Cojoc D, Difato F, Ferrari E et al (2007) Properties of the force exerted by filopodia and lamellipodia and the involvement of cytoskeletal components. PLoS One 2:e1072
21. Mejean CO, Schaefer AW, Millman EA et al (2009) Multiplexed force measurements on live cells with holographic optical tweezers. Opt Express 17:6209–6217
22. Ji N, Milkie DE, Betzig E (2010) Adaptive optics via pupil segmentation for high-resolution imaging in biological tissues. Nat Methods 7:141–147
23. Heintzman R (2010) Correcting distorted optics: back to the basic. Nat Photonics 7:108–110
24. Zahid M, Velez-Fort M, Papagiakoumou E et al (2010) Holographic photolysis for multiple cell stimulation in mouse hippocampal slices. PLoS One 5:e9431
25. Nikolenko V, Watson BO, Araya R et al (2008) SLM microscopy: scanless two-photon imaging and photostimulation with spatial light modulators. Front Neural Circuits 2:5–19
26. Goodman JW (2005) Introduction to Fourier optics. Roberts & Company, Greenwood Village, CO
27. Khoo IA (2007) Liquid crystals. Wiley, Hoboken, NJ
28. Vicari L (2003) Optical applications of liquid crystals. Institute of Physics Publishing Ltd, London
29. Efron U (1994) Spatial light modulator technology: material, devices and applications. Marcel Dekker Inc., New York, NY
30. Dayton D, Browne S, Gonglewski J et al (2001) Characterization and control of a multielement dual-frequency liquid-crystal device for high-speed adaptive optical wave-front correction. Appl Opt 40:2345–2355

31. Dal Maschio M, Difato F, Beltramo R et al (2010) Simultaneous two-photon imaging and photo-stimulation with structured light illumination. Opt Express 18:18720–18731
32. Nikolenko V, Peterka DS, Yuste R (2010) A portable laser photostimulation and imaging microscope. J Neural Eng 7:045001
33. Peron S, Svoboda K (2011) From cudgel to scalpel: toward precise neural control with optogenetics. Nat Methods 8:30–34
34. Andrasfalvy BK, Zemelman BV, Tang J et al (2010) Two-photon single-cell optogenetic control of neuronal activity by sculpted light. Proc Natl Acad Sci USA 107:11981–11986
35. Papagiakoumou E, Anselmi F, Begue A et al (2010) Scanless two-photon excitation of channelrhodopsin-2. Nat Methods 7:848–854
36. Palima D, Alonzo CA, Rodrigo PJ et al (2007) Generalized phase contrast matched to Gaussian illumination. Opt Express 15:11971–11977
37. Gluckstad J, Palima D (2010) Generalized phase contrast. Springer in association with Canopus Academic Publishing Limited, Dordrecht, NE
38. Lee WM, Reece PJ, Marchington RF et al (2007) Construction and calibration of an optical trap on a fluorescence optical microscope. Nat Protoc 2:3226–3238
39. Martin-Badosa E, Montes-Usategui M, Carnicer A et al (2007) Design strategies for optimizing holographic optical tweezers setups. J Opt A Pure Appl Opt 9:S267–S277
40. Helmchen F, Denk W (2005) Deep tissue two-photon microscopy. Nat Methods 2:932–940
41. Papagiakoumou E, de Sars V, Oron D et al (2008) Patterned two-photon illumination by spatiotemporal shaping of ultrashort pulses. Opt Express 16:22039–22047
42. Shaevitz JW, Fletcher DA (2007) Enhanced three-dimensional deconvolution microscopy using a measured depth-varying point-spread function. J Opt Soc Am A Opt Image Sci Vis 24:2622–2627
43. Doering LC (2010) Protocols for neural cell culture. Springer, Heidelberg
44. Daria VR, Stricker C, Bowman R et al (2009) Arbitrary multisite two-photon excitation in four dimensions. Appl Phys Lett 95:093701-1–093701-3
45. Liesener J, Reicherter M, Haist T et al (2000) Multi-functional optical tweezers using computer-generated holograms. Opt Commun 185:77–82
46. Leach J, Wulff K, Sinclair G et al (2006) Interactive approach to optical tweezers control. Appl Opt 45:897–903
47. Zalevsky Z, Mendlovic D, Dorsch RG (1996) Gerchberg–Saxton algorithm applied in the fractional Fourier or the Fresnel domain. Opt Lett 21:842–844
48. Golan L, Reutsky I, Farah N et al (2009) Design and characteristics of holographic neural photo-stimulation systems. J Neural Eng 6:066004
49. Papagiakoumou E, de Sars V, Emiliani V et al (2009) Temporal focusing with spatially modulated excitation. Opt Express 17:5391–5401

Neuromethods (2012) 67: 83–99
DOI 10.1007/7657_2011_2

Published online: 1 December 2011

Voltage-Sensitive Dye Imaging of Cortical Function In Vivo

Eugene F. Civillico

Abstract

Voltage-sensitive dye imaging (VSDI) is the optical interrogation of transmembrane voltage using an exogenous membrane-bound probe. When bulk-applied to intact tissue, voltage-sensitive dyes allow the measurement at high temporal and spatial resolution of subthreshold population membrane potential dynamics in regions of interest of arbitrary size and shape. This is particularly useful in vivo for mapping of input patterns to surface brain structures in order to study neuronal integration at the population level. This chapter provides an introduction to the use of VSDI in vivo. The first half consists of a discussion of the mechanism of action of optical voltage probes and the resulting implications for the interpretation of VSDI data, and offers a perspective on useful future applications and improvements. The second half includes a detailed protocol for VSDI in the barrel cortex of the anesthetized mouse, followed by further notes on practical details of the method.

Key words: VSDI, Voltage-sensitive dye, Optical recording, Physiology

1. Introduction: Theory and Application

Studies of the optical changes observed in an axon during the propagation of an action potential (1) led to the idea that the optical properties of excitable tissue could be used to monitor transmembrane voltage. Although such measurements are possible without the addition of exogenous molecular probes, signal-to-noise ratio is greatly enhanced by their use; accordingly, the first voltage-sensitive dye recordings were made with merocyanine 540 in the squid giant axon (2). Probe design has evolved since that time, with signal-to-noise ratio continually increasing, and deleterious effects on tissue continually decreasing (3–10). In vitro, voltage-sensitive dye imaging (VSDI) experiments have illuminated the thalamocortical and intracortical progression of synaptic activity (11, 12), the frequency-dependence of intracortical propagation (13), and the functional relationships between hippocampal and cortical projections (14).

1.1. Mechanism

Interpretation of VSDI data is enhanced by a correct biophysical understanding of the mechanism of action of a voltage probe. Potentiometric dyes are linear indicators of membrane potential over physiological ranges (7, 15, 16). When applied topically in vivo, commonly used dyes stain the supragranular cortical layers most intensely (17, 18) and are taken up preferentially by dendrites and cell bodies. Because layers 2/3 are primarily neuropil, the signal source in vivo is considered to be mostly from dendrites (3, 13, 19). Some contribution also comes also from glial cells (8, 20).

A basic requirement for transducing transmembrane voltage to fluorescence is that a fluorophore localize to the cell membrane. In contrast to other fluorescent probes, VSDs are not cell-filling dyes. For this reason and because they stain cell types indiscriminately, when bulk-applied in vivo or in brain slices, voltage-sensitive dyes do not produce resolvable signals from single cells.[1] Therefore, although the resulting signals often share some of the dynamics of single-cell membrane potential deflections, they ought not generally to be interpreted as such. Rather, these signals are best understood as a population measurement. In this regard, VSDI recordings are not unlike local field potential (LFP) recordings. In both cases, on the one hand, the decoding of the population signal into single neuron records is an intractable inverse problem, but on the other hand, a high-temporal resolution kinetic readout of neuronal activity is obtained on a millisecond time scale matching the time scale of the elemental units of perception and behavior. Appropriate quantitative measures for VSDI and LFP data are critical to achieving reproducibility, which is a key step toward maturity and acceptance for an experimental technique. In this regard, new data reduction and visualization tools, such as those emerging from the new discipline of systems biology, may be enormously helpful in the near future.

As alluded to above, the VSD signal originating from bulk-stained tissue cannot be straightforwardly mapped to the activity of individual constituent cells. Studies in which VSDI was performed simultaneously with intracellular recording (13, 18, 29, 30) provided a keystone for interpretation of the VSD signal; however, a fundamental limitation stems from the fact that, in the absence of perfect neuronal (and glial) synchrony and isopotentiality over the membrane seen by a detector, the contribution of an individual cell to the VSD signal from that detector is an "inverse problem" without a unique solution.

[1]This is not the case when voltage probes are confined to the membrane of a single cell by intracellular application, e.g., with a microelectrode. Several investigators have made excellent use of this laborious technique to study integration in single cells (21–24, 26–28). Single neurons grown in culture may also be resolved following staining by bath application (25).

The VSD signal from a given CCD detector or photodiode is the averaged fluorescent signal from all stained membrane in a three-dimensional volume of tissue. The x and y dimensions of the volume are determined by the detector size and the magnification of the optics, and the z dimension by the focal depth of the objective lens used and the depth of staining. In the coronal or sagittal slice preparation, the entire cortical depth is generally accessible for imaging. In vivo, the situation is more complicated. Factors affecting the cortical depth of the collected light in vivo include staining time, hydrophobicity of the dye, and individual variation between animals in the translucency of the cortex, which could be exacerbated by age or strain differences.

The subthreshold response of the different parts of any single cell will be spread out over many detectors, and mixed within an individual pixel or photodiode with the responses of many other cells. Different dendrites and the axon will contribute to different pixels; the cell body may dominate the pixel in which it lies, or it may be "blocked" by a stained glial cell. The action potentials from any single cell will travel rapidly through minute subvoxel volumes, and will not generally be detectable with current technology. Signals from other types of cell membranes, including glial cells, blood vessels, and even the intracellular organelles of the neurons themselves, will contribute to varying degrees.

1.2. Mapping and Beyond

Despite its extensive use in vitro, VSDI in vivo is still a new method. As of this writing, there have been approximately 1300 VSDI studies indexed in PubMed since merocyanine 540 was first used to track action potentials in a leech central neuron (6). These include approximately 250 studies of the cerebral cortex, of which only 50 were in vivo. For all the reasons outlined in the previous section, like the local field potential, VSDI in vivo is most useful for illuminating the population subthreshold dynamics of a neural system. When applied as described in the protocol to follow (Sects. 2–4), surface brain structures such as the neocortical or cerebellar cortices are the optimal targets (Fig. 1), although a hippocampal or thalamic preparation created by aspiration of overlying neocortex followed by staining of the exposed surface is not out of the realm of possibility.

Mapping is the first and most obvious use for VSDI in vivo. Information is encoded in sensory cortices in distributed representations with varying scales of temporal and spatial correlation. Recording of the aggregate membrane potential of neural tissue allows the observation of activity from different areas of a sensory or projection map, on a time scale that permits the resolving of population synaptic events. Valuable insight into the rules of information representation can be obtained by recording the responses to

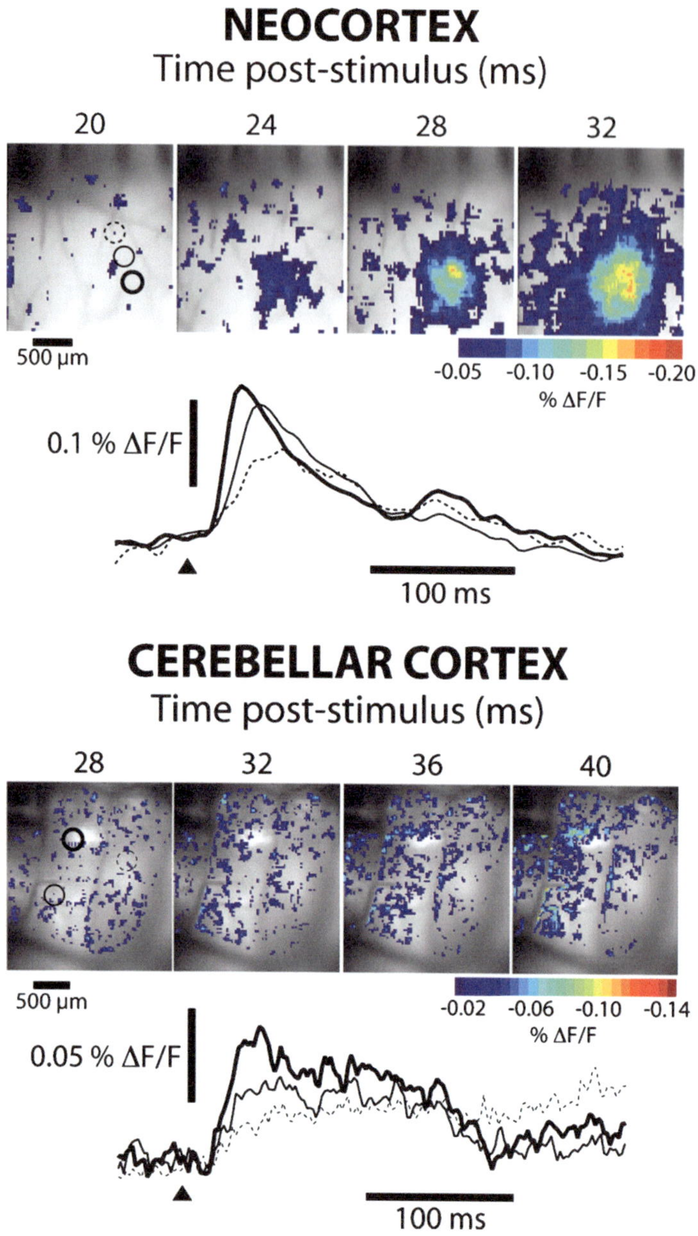

Fig. 1. VSDI reveals contrasting spatiotemporal patterns in the population subthreshold response to similar stimuli in primary sensory (*top*) and cerebellar cortex (*bottom*). Stimulus in each case was a 100 ms ramp-and-hold whisker deflection. Traces below the images are the average signal from the areas marked by *circles* in the leftmost image of each set.

simple stimuli and the interactions between such responses at high temporal and spatial resolution over large areas. VSDI has been used extensively to map spatiotemporal activity patterns in the supragranular neocortical layers, which are accessible by epifluorescent imaging from the surface of the intact brain. In this way, the technique has contributed to the understanding of orientation columns in visual cortex (31) and isofrequency strips in auditory cortex (32). Recently, the technique has been vigorously applied to the study of the rodent barrel cortex in vivo (17, 18, 30, 33–38). When interpreting mapping studies it is important to keep in mind the technique's bias toward synchronous subthreshold activity. In some sense, the VSDI experimenter is actually observing the activity of the neuronal population *projecting to* the area being imaged.

While combined optical and intracellular studies (13, 18, 29, 30) served as a validation of the VSD method because the intracellular recordings tracked closely the VSD signal from their immediate neighborhood, such studies also raise interesting questions. While the VSDI signal from intact cortex shows that many neuronal processes from many different cells receive synchronous subthreshold input, it tells the investigator little about spike output. In combination with spiking measurements, however, VSDI could make a critical contribution to this fundamental question: what is discarded between the widespread and synchronous subthreshold signal and the sparse and asynchronous spike records in a local cortical circuit? With spike-triggered averaging methods analogous to those applied successfully to LFP recordings, it may be possible in the future to use VSDI to generate maps of the postsynaptic impact of a single cell's sodium spikes in vivo. This would be a fundamental finding in neuronal integration.

1.3. Why VSDI?

Voltage-sensitive dyes confer numerous advantages over electrode-based methods. Single-neuron recordings are inherently limited in scope, since electrodes must be targeted to particular columnar locations, often blindly. Multielectrode arrays overcome this limitation, but any individual array is constrained to a predetermined layout, which may not be in register with the functional regions of interest, especially in highly variable in vivo preparations. VSDI, by contrast, allows the measurement of membrane potential dynamics, at fine temporal and spatial scale, of functionally significant regions of arbitrary size and shape. This is particularly useful when the size of the functionally relevant area is not known in advance.

VSDI is sometimes compared unfavorably with imaging of the cytosolic calcium concentration, a method that has exploded in popularity in recent years. It is sometimes asserted, for example, that optical voltage probes produce a "small signal" relative to calcium indicators. This is true, but irrelevant, because the two classes of indicator have different goals and uses. VSDI in intact tissue provides a high-resolution, high-speed view of the evolving subthreshold population input to a neuronal network. Such data is critical to understanding neuronal integration. Calcium is quite promiscuous in its involvement in cell processes; more importantly, somatic calcium lacks a clearly defined relationship to either the input or the output of a neuron. While work continues on algorithms to translate somatic calcium transients into spike trains (39), such algorithms will likely vary by cell type according to the density and types of calcium channels expressed. A more direct readout of membrane potential is critical to the study of neuronal integration.

The major disadvantage of VSDI in vivo has been alluded to above: with current technology, the method does not provide activity readouts from discrete neurons. Nevertheless, the brain is not a printed circuit board, and it may be beneficial to take a different view of excitable tissue. While connections between neurons are often conveniently visualized as a network of lines through empty space, linking nodes, triangle-shaped pyramids, and spherical interneurons, without glia or blood vessels, the biological reality is less convenient. Much progress has been made in physiology by dividing the brain into cells, which are obviously important metabolic and processing units; however, subcellular processes have influence on one another due to physical proximity which may or may not be related to the location of the cell body from which they originate. This influence may be mediated by a number of "nonclassical" methods of intercellular communication present in the mammalian brain, including dendrodendritic synapses (40), gap junctions between networks of interneurons (41, 42), and ephaptic interactions (43–47). While single unit recordings have shown that action potential firing in cortex is sparse and asynchronous, VSD recordings report a synchrony of nonspiking dynamics that cannot be observed in recordings of single cells but is quite clear from another population subthreshold signal, the LFP. There is much to be learned from understanding this coordinated subthreshold activity. While VSDI cannot report the spiking of individual cells or resolve components corresponding to different cellular processes, its limitations result in a signal from the brain as it actually is: a tightly packed matrix of excitable membranes. Thus, while one reductionist approach to brain function seeks an intimate understanding of the behaviors of single cells, there is a case to be made for a complementary reductionist approach that seeks an intimate understanding of the behavior of small volumes of membrane.

1.4. Future Improvements

Notwithstanding the advantages detailed above, the power of VSDI to provide a spatially and temporally precise measure of neuronal population dynamics is significantly enhanced by combination with other methods which will more precisely localize the signal or otherwise facilitate its interpretation. Genetic targeting of dye molecules, or of enzymes that activate them, to particular subpopulations of cells (48, 49, 50) would remove some of the ambiguity associated with the source of the signal. A particularly valuable achievement would be the development of probes that do not stain glia or blood vessels. The use of multisite electrodes to identify large numbers of single neurons within a volume (51) could be combined with VSD imaging to allow recording of subthreshold input and spiking output from the same tissue (33, 52). This could allow the characterization of recurring input–output relations, network "motifs" of processing, in cortex and elsewhere. The use of multisite probes with vertically oriented recording sites, together with better modeling of the extracellular space (53) could remove some of the ambiguity associated with the depth of staining and provide further insight into the relationship between membrane potential and extracellular current flow throughout the cortical depth. Finally, the use of nonlinear microscopy in combination with VSDs will allow the precise localization of optical measurements of membrane potential in three dimensions at ever-increasing spatial and temporal resolution (25, 54–57).

2. Materials

2.1. Surgery

- Large (e.g., model 1404) or small-animal (e.g., model 1900) stereotaxic frame
  (see Sect. 4)
  David Kopf Instruments
- Model 926 mouse adapter with Model 921 ear cups
  Both ear cup bars and the nose piece were adjusted with custom machined inserts to position the head about 10 cm above the plane of the stereotaxic frame rails
  David Kopf Instruments
- Dental drill
- Micro steel burr drill bit
  George Tiemann & Co
  #160-1293, 0.24 mm
- Dumostar #55 forceps
  Fine Science Tools
  11252–40
- Double-end excavator
  Sullivan-Schein
  100–7153

2.2. Voltage-Sensitive Dye Application

- RH795
  Invitrogen
  R-649
- RH1691/1692
  Optical Imaging

- Gelfoam absorbable gelatin sponge
  09-0396-05
  6510-00-064-4858
  Pharmacia & Upjohn (Pfizer)

2.3. Optics

- 12-V halogen lamp
  Olympus
- Upright microscope
  Olympus BX-50
- Filters

  Chroma Technologies
  Excitation: 540 ± 20 nm
  Emission: long-pass at 600 nm
- Objective lens
  Olympus XLFluor 4×/340
  NA 0.28
- Camera adapter
  Olympus U-TV0.35xC
  The combination of the 4× objective and 0.35× camera adapter resulted in an imageable area of 1 by 2-mm and a pixel area of 22 by 22 μm (484 μm^2) for the MiCam01 CCD chip.
- CCD camera
  MiCam series: MiCam01, MiCam02, MiCam Ultima
  SciMedia
- Stable power supply for excitation light
  Kepco ATE 25-20M
  500 W

2.4. Software

- Acquisition controlled by BrainVision software (SciMedia).
- Post hoc data analysis in IGOR Pro (Wavemetrics) using custom routines available from the author as part of the IGOR-IMP package.

3. Methods

NOTE: This protocol was developed for the purpose of fast imaging of response dynamics in the barrel cortex of an anesthetized mouse; however, it should be a reasonable starting point, together with the scientific literature, for developing protocols for other experimental models.

3.1. Surgery

1. Anesthetize mouse. A suitable anesthetic is ketamine-xylazine at a dosage of 100 mg/kg i.p., 20 mg/kg i.p. respectively,

administered intraperitoneally. Administration methods and doses for other options, for example urethane, halothane, or isofluorane, are available in the scientific literature. Correct dosing and administration will result in visible motor impairment within seconds, and deep anesthesia within 20 min.

NOTE 1: If the preparation will be anesthetized for the duration of the experiment, prepare suitable means of maintaining the anesthesia. This may include heart rate or CO_2 monitoring, EEG monitoring, etc. In our preparations, we administer supplemental anesthesia (25 mg/kg, 5 mg/kg) as necessary to maintain cortical slow oscillations, as assessed with EEG/LFP, and weak or absent foot withdrawal reflex.

Follow the animal care guidelines of your institution.

NOTE 2: If the anesthesia will be allowed to wane to produce an alert head-fixed preparation, consult detailed protocols available elsewhere (36, 58, 59) for instructions. The remainder of this protocol presumes anesthesia.

2. Mount the mouse in a stereotaxic frame. The head shape and ear position of mice will not accommodate standard ear bars, so use serrated or smooth "ear cups." Position the head of the animal under the objective lens of an upright fluorescence microscope. If the animal is in a stereotaxic frame sized for rodents, it may be possible to simply move the entire frame and preparation to a suitable location for microscopy. Our system is modified from a rig used for electrophysiological experiments in anesthetized cats. An Olympus BX50 is mounted on rails and can slide forward when the preparation is complete. Custom-machined L-shaped ear-cup bars position the mouse's head above the plane of the stereotaxic frame, within the focal range of the microscope.
3. Remove the skin from above the entire skull using fine surgical scissors. If necessary, trim hair around the edges of the incision. Push or wipe away any remaining fluid from the skull. It may be advantageous to gently polish the entire exposed skull with a dental excavator. This will make for a clearer view and cleaner drilling during the craniotomy.
4. Moisten a Q-tip with ethanol and use it to dry the exposed skull. After the ethanol has completely evaporated, the surface should appear uniform, white in color, not glistening or shiny, with red blood vessels apparent throughout.
5. Use a dental drill with a microsteel burr bit (see Sect. 2.1) to make a craniotomy over the barrel cortex. The larger the craniotomy, the easier it will be to orient the images with respect to the barrel field. A larger craniotomy will also allow for flexibility in the case of bleeding during the durotomy (see step 6). A craniotomy which extends 2 mm in the

anterior–posterior direction starting from bregma, and from 2 to 4 mm from the midline will be sufficient in most animals to expose most of the posteromedial barrel subfield (PMBSF). The craniotomy should be no larger than necessary for the question at hand; larger craniotomies are more prone to mechanical artifacts.

NOTE 3: Other cortical areas may be studied by adjusting the craniotomy position, including M1, S2, and parts of V1. To obtain accurate measurements of activated areas it is necessary to have the ability to adjust the angle of either the brain surface (via the stereotaxic frame) or the microscope. With care and a sufficiently large detector it is possible to study interarea or even interhemispheric dynamics (60). The cerebellar cortex is also accessible (Fig. 1; (61)).

6. Remove the meninges from the brain surface. Overlaying the exposed cortex is an elastic translucent membrane which can be difficult to see but which will become more obvious as it begins to dry out a few minutes after the craniotomy. This membrane should be removed. *This is the most difficult part of the preparation. The slightest nick to the tissue under the meninges will cause bleeding which will make good recordings next to impossible.* We have found it useful to do this with a dedicated pair of #55 forceps. The membrane can be gently lifted in one place until it tears, and then drawn to the side. Flaps will form at the edges of the incision, which can be gripped with the forceps and pulled away. A useful tool for retraction of this membrane may be constructed by bending the tip of a fine gauge syringe tip into a hook, which can then be easily manipulated with a small hemostat. Once this membrane is removed it is essential to drip saline over the exposed brain every few minutes. When this procedure has been done successfully you will find that the brain surface appears more translucent.

   NOTE 4: A recent report (62) indicates that dye may be applied and the brain imaged through this membrane. This would certainly have advantages, in that it eliminates what in our experience is the most common cause of an unusable preparation; however, given that the membrane will stain, it seems to us that it will contribute to the background without contributing to the functional neural signal, and so the trade-off must be carefully considered.

7. If desired, extracellular electrodes for single-unit or field potential recordings may be inserted at this point, taking care to keep them at the edge of the area to be imaged. The electrophysiological recordings may be optimized during the staining period to follow. An advantage of inserting electrodes is that they may serve as a scaffold to which the bubble of dye solution

will attach by surface tension, helping to keep the staining solution in place and the tissue hydrated.

3.2. Staining

8. Prepare the dye solution. We have been successful with a 1 mg/mL solution of RH795 (Invitrogen) in 0.9% saline. RH795 will dissolve in saline with sonication and gentle warming. If necessary a small amount of Pluronic F-127 may be added. The dye should be kept in the dark when not in use. We have not experimented with freezing aliquots, though this may be possible.
9. Following (17), soak a 1 mm^3 piece of Gelfoam (Upjohn Pharmacia) in the warmed dye solution and use the forceps to place it on the cortex. Add additional dye a few drops at a time every few minutes for 60–90 min. After this time, carefully remove the gelfoam with forceps, and rinse the brain surface with saline to remove unbound dye. While the brain is being stained, if electrodes were inserted in step 7, hand stimulation of the whiskers with audio feedback from the cortical LFPs can be used to determine the approximate location of the electrodes and stained area within the PMBSF.

4. Notes

4.1. Stereotaxic Frame

While it may seem to be overkill for a rodent, the use of a large-animal stereotaxic frame such as the Kopf model 1404 had a number of practical advantages. A great deal of space was available around the mouse's head for tactile, visual, or auditory stimulation. In addition, "roll" rotation of the frame allowed the barrel cortex to be rotated into the horizontal plane, facilitating imaging. In order to use a large frame with a small animal, we made use of a small platform on an adjustable riser to support the mouse's body.

4.2. Imaging Setup Details

Light source: 12-V halogen lamp powered by a stabilized power supply.

Filters: For RH795, excitation light bandpass-filtered to pass 540 $\pm$ 20 nm. Light emitted from the preparation long-pass filtered at 600 nm.

Detector: Photodiode array or high-sensitivity fast CCD camera. We have been successful with the Micam01 (BrainVision, Japan). This instrument has a detector array of 96 $\times$ 64 pixels (87 $\times$ 60 imageable) and can achieve frame rates upward of 1,000 Hz, though in practice we found a frame rate of 250 Hz (4 ms/frame) to give the best balance of speed and signal.

Optical recording was controlled by the MiCam software. For orientation at the start of the experiment, some autoaveraged recordings of 10–20 trials were used. Data to be used for analysis were collected as batches of single trials, with no online blank subtraction or online averaging. Data was written to a binary format described in the MiCam documentation.

4.3. Fractional Fluorescence

Fractional fluorescence, or $\Delta F/F$, has become the standard way of reporting VSD data, although care must be taken with this measure in vivo due to the wide variability of the baseline stain according to nonneuronal features of the preparation (for example, the increased baseline noise from pixels imaging blood vessels). To facilitate interpretation, the in vivo researcher must control for staining inhomogeneity by visual inspection of the preparation prior to recordings, since the correction for uneven staining by calculation of fractional fluorescence will not be sufficient to correct the large differences that may obtain in vivo.

4.4. Differentiation of Neuronal and Blood-Flow Related Signals

Care must be taken to ensure that recorded VSD signals are not compromised by hemodynamic responses. Data should be analyzed no more than 300–400 ms after a stimulus or reference time point. This time is short enough to exclude the blood-flow/Hb absorbance changes seen in cortical tissue in response to sensory stimulation, which only begin after several hundred milliseconds. For example, in ref. 63, blood-flow related signals were imaged from 1.5 to 2.5 s after the stimulus, whereas LFP and MUA were integrated over 0–300 ms after the stimulus. In ref. 64, their Fig. 3b, the thin trace representing the "early signal" or what would later be called the "initial dip" corresponding to the increase in deoxyhemoglobin by oxygen delivery to neurons takes almost 1 s to start. The same time course is also seen in ref. 65. Newer probes such as RH1691 (10), with a shifted absorbance spectrum, are not sensitive to hemoglobin changes.

In a background image of the intact cerebral cortex stained with a voltage-sensitive dye, large blood vessels are immediately obvious, and smaller blood vessels can be seen upon closer scrutiny. To the extent that they obscure the neuronal signal, these vessels can be corrected for and worked around in the analysis of the VSD data (see, for example, (34)); however, there are of course blood vessels too small to be seen in the background image, which will have unknown effects on the signal. The inclusion of these structures in the pixel decreases the percentage of neuronal membrane in the pixel, and therefore the amplitude of the fraction change from baseline fluorescence for a given neuronal response.

4.5. Data Analysis Strategy

As is the case with many newer systems-oriented techniques, a majority of the effort to obtain a physiological result from VSD imaging consists of post hoc data analysis and reduction. The following workflow is suggested:

Identify preparations of sufficient quality to reward detailed analysis.

Within each preparation, identify trials free of large-scale mechanical artifacts.

Within each trial, identify stained pixels (i.e., exclude blood vessels) using a brightness threshold.

Optionally screen stained pixels for low baseline noise by measuring the standard deviation over 100 ms or more of unstimulated recording.

If applicable, combine pixels into larger regions of interest (ROIs). In barrel cortex, the barrel columns are ideal for this. A protocol for projecting barrel column outlines onto the VSD images post hoc is described in Sect. 4.8. Simply dividing the cortex into squares of equal size would also be useful. Performing analysis on signals from ROIs larger than 1 pixel lowers the effective spatial resolution of the data, but increases the signal-to-noise ratio. The research question determines the optimal trade-off here.

As a complement to binning pixels to produce time-varying optical signals, the dimensionality of VSD data sets may be reduced in other ways, such as by computing parameters only possible with imaging, for example, the area giving a signal above a certain threshold as a function of time (34), the path taken by the activated region's "center of mass", or the shape of the signal amplitude distribution as a function of time.

4.6. Software

We used custom routines written in Igor Pro (Wavemetrics) for postprocessing, visualization, and quantification of optical recordings. Many of these routines have been assembled into the IGOR-IMP package, which may be obtained by contacting the author.

4.7. Electrical Stimulation

The use of physiologically realistic sensory stimuli, i.e., controlled movements of the whiskers or presentation of audible tones, is obviously fundamental to understanding computational function. The details of such methods are best described in modality-specific literature. Here we will confine ourselves to the description of electrical stimulation. The circuit properties of cortical networks may also be probed using direct epicortical electrical stimulation. Such stimulation has a long history of use to explore network dynamics, both in vivo and in vitro. Experiments in which electrical stimuli are delivered to one location while recordings are made from another have revealed functional projections between areas, as well as frequency- and modulation-dependent properties of thalamocortical and corticocortical (13) connections. Delivery of a brief (100 μs) electrical stimulus elicits a barrage of action potentials from cells and axons located in the vicinity of the electrode (the presynaptic or direct response). Subsequently, activity developing over tens or hundreds of milliseconds following the stimulus in the recorded area (the postsynaptic response) is indicative of the strength and nature of the connection from the area activated directly. When optical recording methods are used, a large amount of data is generated by such a stimulus, since the response may be observed at thousands of discrete locations. The optical response is thought to be generated mainly by the dendrites postsynaptic to the cells excited directly (13, 19).

Suitable electrical stimuli include single 100 μs pulses of 0.1–0.3 mA intensity delivered through the recording electrodes in a bipolar configuration. In a healthy stained cortex, the response to an electrical stimulus spreads across all visible barrel cortex. This property may be exploited to assess the health of the preparation during an experiment.

4.8. Barrel Binning

At the conclusion of an imaging experiment, make two fiducial marks by twice advancing a single tungsten electrode into the cortex. Capture reference images containing the location of the signals to be studied, the fiducial marks, and the electrodes if used, which will also leave fiducial marks.

Perfuse the mouse transcardially with 4% paraformaldehyde in 0.1 M sodium phosphate buffer (PBS).

Postfix the brain overnight in the same fixative. Dissect the cortex and flatten it by pressing gently between two clean microscope slides submerged in PBS.

Cut 100 μm thick tangential sections on a vibratome. To reveal the barrels, treat tissue with 3,3′-diaminobenzidine (DBA, Sigma D-5905) and Cytochrome C from horse heart (Sigma C-2506) as described in ref. 66 with the following modifications:

Wash sections in 0.1 M PBS (3 × 10 min) at room temperature and incubate at room temperature for 15 min in a mixture of 0.1 M PBS with 10% methanol (Fisher Scientific BP1105-1) and 1% hydrogen peroxide (Sigma H-1009).

Wash again in PBS (3 × 10 min) and shake in the dark overnight at 4°C in 0.1 M PBS containing 4 g sucrose, 50 mg DBA (Sigma), and 30 mg of cytochrome oxidase per 100 mL of phosphate buffer (PB).

The following day, wash tissue in PBS, mount on subbed glass slides, dehydrate, and coverslip.

The fiducial marks made by the electrodes may be used to align the barrel outlines from histology with the fractional fluorescence images. This allows binning of pixels into signals corresponding to the average activity within barrel-columns while also increasing the signal-to-noise ratio.

References

1. Cohen LB, Keynes RD, Hille B (1968) Light scattering and birefringence changes during nerve activity. Nature 218:438–441
2. Davila HV, Salzberg BM, Cohen LB et al (1973) A large change in axon fluorescence that provides a promising method for measuring membrane potential. Nat New Biol 241:159–160
3. Grinvald A, Lieke EE, Frostig RD et al (1994) Cortical point-spread function and long-range lateral interactions revealed by real-time optical imaging of macaque monkey primary visual cortex. J Neurosci 14:2545–2568
4. Kuhn B, Fromherz P, Denk W (2004) High sensitivity of Stark-shift voltage-sensing dyes by one- or two-photon excitation near the red spectral edge. Biophys J 87:631–639
5. Loew LM, Cohen LB, Salzberg BM et al (1985) Charge-shift probes of membrane potential. Characterization of aminostyrylpyridinium dyes on the squid giant axon. Biophys J 47:71–77
6. Salzberg BM, Davila HV, Cohen LB (1973) Optical recording of impulses in individual neurones of an invertebrate central nervous system. Nature 246:508–509
7. Salzberg BM, Obaid AL, Senseman DM et al (1983) Optical recording of action potentials from vertebrate nerve terminals using potentiometric probes provides evidence for sodium and calcium components. Nature 306:36–40
8. Salzberg BM (1989) Optical recording of voltage changes in nerve terminals and in fine neuronal processes. Annu Rev Physiol 51:507–526
9. Salzberg BM, Obaid AL, Bezanilla F (1993) Microsecond response of a voltage-sensitive merocyanine dye: fast voltage-clamp measurements on squid giant axon. Jpn J Physiol 43 (Suppl 1):S37–S41
10. Shoham D, Glaser DE, Arieli A et al (1999) Imaging cortical dynamics at high spatial and temporal resolution with novel blue voltage-sensitive dyes. Neuron 24:791–802
11. Laaris N, Carlson GC, Keller A (2000) Thalamic-evoked synaptic interactions in barrel cortex revealed by optical imaging. J Neurosci 20:1529–1537
12. Laaris N, Keller A (2002) Functional independence of layer IV barrels. J Neurophysiol 87:1028–1034
13. Contreras D, Llinas R (2001) Voltage-sensitive dye imaging of neocortical spatiotemporal dynamics to afferent activation frequency. J Neurosci 21:9403–9413
14. Ang CW, Carlson GC, Coulter DA (2005) Hippocampal CA1 circuitry dynamically gates direct cortical inputs preferentially at theta frequencies. J Neurosci 25:9567–9580
15. Cohen LB, Salzberg BM, Grinvald A (1978) Optical methods for monitoring neuron activity. Annu Rev Neurosci 1:171–182
16. Cohen LB, Salzberg BM (1978) Optical measurement of membrane potential. Rev Physiol Biochem Pharmacol 83:35–88
17. Kleinfeld D, Delaney KR (1996) Distributed representation of vibrissa movement in the upper layers of somatosensory cortex revealed with voltage-sensitive dyes. J Comp Neurol 375:89–108
18. Petersen CC, Grinvald A, Sakmann B (2003) Spatiotemporal dynamics of sensory responses in layer 2/3 of rat barrel cortex measured in vivo by voltage-sensitive dye imaging combined with whole-cell voltage recordings and neuron reconstructions. J Neurosci 23:1298–1309
19. Yuste R, Tank DW, Kleinfeld D (1997) Functional study of the rat cortical microcircuitry with voltage-sensitive dye imaging of neocortical slices. Cereb Cortex 7:546–558
20. Konnerth A, Obaid AL, Salzberg BM (1987) Optical recording of electrical activity from parallel fibres and other cell types in skate cerebellar slices in vitro. J Physiol 393:681–702
21. Antic S, Major G, Zecevic D (1999) Fast optical recordings of membrane potential changes from dendrites of pyramidal neurons. J Neurophysiol 82:1615–1621
22. Antić S, Major G, Chen WR et al (1997) Fast voltage-sensitive dye recording of membrane potential changes at multiple sites on an individual nerve cell in the rat cortical slice. Biol Bull 193:261
23. Grinvald A, Salzberg BM, Lev-Ram V et al (1987) Optical recording of synaptic potentials from processes of single neurons using intracellular potentiometric dyes. Biophys J 51:643–651
24. Zecević D, Antić S (1998) Fast optical measurement of membrane potential changes at multiple sites on an individual nerve cell. Histochem J 30:197–216
25. Dombeck DA, Sacconi L, Blanchard-Desce M et al (2005) Optical recording of fast neuronal membrane potential transients in acute mammalian brain slices by second-harmonic generation microscopy. J Neurophysiol 94:3628–3636

26. Zhou WL, Yan P, Wuskell JP et al (2007) Intracellular long-wavelength voltage-sensitive dyes for studying the dynamics of action potentials in axons and thin dendrites. J Neurosci Methods 164:225–239
27. Palmer LM, Stuart GJ (2009) Membrane potential changes in dendritic spines during action potentials and synaptic input. J Neurosci 29:6897–6903
28. Palmer LM, Clark BA, Gründemann J et al (2010) Initiation of simple and complex spikes in cerebellar Purkinje cells. J Physiol 588:1709–1717
29. Petersen CC, Sakmann B (2001) Functionally independent columns of rat somatosensory barrel cortex revealed with voltage-sensitive dye imaging. J Neurosci 21:8435–8446
30. Petersen CC, Hahn TT, Mehta M et al (2003) Interaction of sensory responses with spontaneous depolarization in layer 2/3 barrel cortex. Proc Natl Acad Sci USA 100: 13638–13643
31. Nelson DA, Katz LC (1995) Emergence of functional circuits in ferret visual cortex visualized by optical imaging. Neuron 15:23–34
32. Song WJ, Kawaguchi H, Totoki S et al (2006) Cortical intrinsic circuits can support activity propagation through an isofrequency strip of the guinea pig primary auditory cortex. Cereb Cortex 16:718–729
33. Berger T, Borgdorff A, Crochet S et al (2007) Combined voltage and calcium epifluorescence imaging in vitro and in vivo reveals subthreshold and suprathreshold dynamics of mouse barrel cortex. J Neurophysiol 97:3751–3762
34. Civillico EF, Contreras D (2005) Comparison of responses to electrical stimulation and whisker deflection using two different voltage-sensitive dyes in mouse barrel cortex in vivo. J Membr Biol 208:171–182
35. Civillico EF, Contreras D (2006) Integration of evoked responses in supragranular cortex studied with optical recordings in vivo. J Neurophysiol 96:336–351
36. Crochet S, Petersen CC (2006) Correlating whisker behavior with membrane potential in barrel cortex of awake mice. Nat Neurosci 9:608–610
37. Ferezou I, Bolea S, Petersen CC (2006) Visualizing the cortical representation of whisker touch: voltage-sensitive dye imaging in freely moving mice. Neuron 50:617–629
38. Ferezou I, Haiss F, Gentet LJ et al (2007) Spatiotemporal dynamics of cortical sensorimotor integration in behaving mice. Neuron 56:907–923
39. Yaksi E, Friedrich RW (2006) Reconstruction of firing rate changes across neuronal populations by temporally deconvolved Ca2+ imaging. Nat Methods 3:377–383
40. Rall W, Shepherd GM, Reese TS et al (1966) Dendrodendritic synaptic pathway for inhibition in the olfactory bulb. Exp Neurol 14:44–56
41. Fukuda T, Kosaka T, Singer W et al (2006) Gap junctions among dendrites of cortical GABAergic neurons establish a dense and widespread intercolumnar network. J Neurosci 26:3434–3443
42. Traub RD, Whittington MA, Stanford IM et al (1996) A mechanism for generation of long-range synchronous fast oscillations in the cortex. Nature 383:621–624
43. Anastassiou CA, Perin R, Markram H et al (2011) Ephaptic coupling of cortical neurons. Nat Neurosci 14:217–223
44. Blinder KJ, Pumplin DW, Paul DL et al (2003) Intercellular interactions in the mammalian olfactory nerve. J Comp Neurol 466:230–239
45. Bokil H, Laaris N, Blinder K et al (2001) Ephaptic interactions in the mammalian olfactory system. J Neurosci 21:RC173
46. Ren J, Momose-Sato Y, Sato K et al (2006) Rhythmic neuronal discharge in the medulla and spinal cord of fetal rats in the absence of synaptic transmission. J Neurophysiol 95:527–534
47. Vreugdenhil M, Bracci E, Jefferys JG (2005) Layer-specific pyramidal cell oscillations evoked by tetanic stimulation in the rat hippocampal area CA1 in vitro and in vivo. J Physiol 562:149–164
48. Hinner MJ, Hubener G, Fromherz P (2004) Enzyme-induced staining of biomembranes with voltage-sensitive fluorescent dyes. J Phys Chem B 108:2445–2453
49. Ng DN, Fromherz P (2011) Genetic targeting of a voltage-sensitive dye by enzymatic activation of phosphonooxymethyl-ammonium derivative. ACS Chem Biol 6:444–451
50. Akemann W, Mutoh H, Perron A, Rossier J, Knöpfel T (2010) Imaging brain electric signals with genetically targeted voltage-sensitive fluorescent proteins. Nat Methods 7:643–649
51. Buzsáki G (2004) Large-scale recording of neuronal ensembles. Nat Neurosci 7:446–451
52. Kerr JN, Greenberg D, Helmchen F (2005) Imaging input and output of neocortical networks in vivo. Proc Natl Acad Sci USA 102:14063–14068
53. Pettersen KH, Devor A, Ulbert I et al (2006) Current-source density estimation based on

inversion of electrostatic forward solution: effects of finite extent of neuronal activity and conductivity discontinuities. J Neurosci Methods 154:116–133

54. Dombeck DA, Blanchard-Desce M, Webb WW (2004) Optical recording of action potentials with second-harmonic generation microscopy. J Neurosci 24:999
55. Fisher JAN, Salzberg BM, Yodh AG (2005) Near infrared two-photon excitation cross-sections of voltage-sensitive dyes. J Neurosci Methods 148:94–102
56. Kuhn B, Denk W, Bruno RM (2008) In vivo two-photon voltage-sensitive dye imaging reveals top-down control of cortical layers 1 and 2 during wakefulness. Proc Natl Acad Sci USA 105:7588–7593
57. Fisher JAN, Barchi JR, Welle CG, Kim GH, Kosterin P, Obaid AL, Yodh AG, Contreras D, Salzberg BM (2008) Two-photon excitation of potentiometric probes enables optical recording of action potentials from mammalian nerve terminals in situ. J Neurophysiol 99:1545–1553
58. Petersen CCH (2011) Voltage-sensitive dye imaging of cortical spatiotemporal dynamics in awake behaving mice. In Helmchen F and Konnerth A (eds) Imaging in Neuroscience: A Laboratory Manual. CSHL Press, New York
59. Dombeck DA, Khabbaz AN, Collman F et al (2007) Imaging large-scale neural activity with cellular resolution in awake, mobile mice. Neuron 56:43–57
60. Mohajerani MH, McVea DA, Fingas M et al (2010) Mirrored bilateral slow-wave cortical activity within local circuits revealed by fast bihemispheric voltage-sensitive dye imaging in anesthetized and awake mice. J Neurosci 30:3745–3751
61. Cohen D, Yarom Y (1998) Patches of synchronized activity in the cerebellar cortex evoked by mossy-fiber stimulation: questioning the role of parallel fibers. Proc Natl Acad Sci USA 95:15032–15036
62. Lippert MT, Takagaki K, Xu W et al (2007) Methods for voltage-sensitive dye imaging of rat cortical activity with high signal-to-noise ratio. J Neurophysiol 98:502–512
63. Devor A, Dunn AK, Andermann ML et al (2003) Coupling of total hemoglobin concentration, oxygenation, and neural activity in rat somatosensory cortex. Neuron 39:353–359
64. Frostig RD, Lieke EE, Ts'o DY et al (1990) Cortical functional architecture and local coupling between neuronal activity and the microcirculation revealed by in vivo high-resolution optical imaging of intrinsic signals. Proc Natl Acad Sci USA 87:6082–6086
65. Kim DS, Duong TQ, Kim SG (2000) High-resolution mapping of iso-orientation columns by fMRI. Nat Neurosci 3:164–169
66. Wong-Riley M (1979) Changes in the visual system of monocularly sutured or enucleated cats demonstrable with cytochrome oxidase histochemistry. Brain Res 171:11–28

Part II

Electrophysiological Approaches In Vitro

Neuromethods (2012) 67: 103–125
DOI 10.1007/7657_2011_19

Published online: 1 December 2011

An Introduction to In Vitro Slice Approaches for the Study of Neuronal Circuitry

Carmen Varela, Daniel A. Llano, and Brian B. Theyel

Abstract

The acute slice preparation can be a powerful tool to study brain networks that would otherwise be difficult to manipulate at the synaptic and cellular levels. In the first part of this chapter, we discuss the specific challenges of preparing brain slices to study neural networks, and we review solutions to overcome problems that can be faced during slice preparation and maintenance. In addition, we describe slice preparations that preserve the connectivity between multiple brain areas, such as hippocampal and thalamocortical slices.

In the second part, we introduce several techniques that can be used to stimulate specific cells or networks in acute slices. We begin by reviewing methods for electrical stimulation, glutamate-based stimulation, and optogenetic stimulation. An additional procedure that combines the use of laser photostimulation with flavoprotein autofluorescence is also presented. We discuss advantages and disadvantages of these methods for neural network investigation in the acute slice preparation.

Key words: Slice preparation, Slice maintenance, Electrical stimulation, Laser photostimulation, Optogenetic stimulation, Flavoprotein autofluorescence

1. Introduction

Although brain slices were used since the 1930s to study brain metabolism (1), it was the extensive work of Henry McIlwain's group in the 1950s and 1960s that introduced the slice preparation as a reliable method for the study of brain physiology (2–4). Their work established an acute preparation with relatively well-preserved metabolism that offered a powerful environment for controlled manipulations. A critical breakthrough in the development of the technique was the observation of resting membrane potentials similar to those measured in in vivo preparations (5, 6). A further advance came with the first demonstration of simultaneous stimulation and recording, which opened the door to the study of neuronal circuitry in vitro (7).

More recently, a number of methods have been developed that maintain viable brain slices in culture for months (8, 9).

Such long-term organotypic cultures are beneficial in the study of the long-term effect of chemical compounds, ischemia, and the study of neurodegeneration and development. However, the acute slice preparation is more commonly used to study neuronal networks, and we focus on this method for the present review (for a review on organotypic slice cultures, see 10, for a protocol reference 11, and for applications 12).

2. Preparation of Brain Slices to Study Neuronal Circuitry

Several detailed descriptions of the instrumentation, as well as protocols for the preparation of brain slices, exist in the literature (13–19). We focus on the steps of slice preparation that are critical for neural network study.

2.1. Selection of Animal Model

The selection of an animal model can critically affect the outcome of slice physiology experiments. Rats and mice are commonly used since their small size permits the retention of a greater proportion of network connections using the typical slice thickness (300–400 μm). Cats, ferrets, and guinea pigs are other common choices, and relatively few primate slice experiments have been done (20–22). Young rodents (~1–2-weeks old) have often been used in the preparation of brain slices. However, older animals may be required for certain neuronal network studies. For example, it has been shown that spindle activity in the dorsal thalamus (which relies on the connectivity with the thalamic reticular nucleus) develops only after postnatal day 22 in ferrets (23). In addition, certain firing patterns, such as the repetitive bursting of some neocortical cells, are virtually absent in ferrets less than 3 months old (24) and in mice less than 1 month old (25). The synaptic response to repetitive activation of certain neocortical synapses in the rat has also been shown to change from depression to facilitation during the first postnatal month (26).

Brain removal for slice preparation in older animals, and, in general, in animals larger than young rodents (e.g., young cats), offers several challenges (18). The thicker skull slows down the removal of the brain, increasing the time between death and placement of the brain into artificial cerebrospinal fluid (aCSF) and, therefore, increasing the risk of hypoxic-ischemic injury. A second problem becomes apparent when the slices are under the microscope. Visibility through the slice is largely reduced due to the increased myelination density compared to juvenile animals. This makes it difficult to visualize cells and obtain whole-cell patch clamp recordings from healthy cells (commonly found in the depth of the slice). However, a number of steps can be taken during slice preparation to prevent hypoxia and improve cell visibility (see sections below).

2.2. Slice Preparation

The success or failure of a slice electrophysiology experiment is often determined at the brain removal stage, as it involves a brief period of ischemia which, if protracted, can render a tissue slice useless. Oxygen is commonly supplied to brain slices in the form of a gas mixture (95% O_2, 5% CO_2) bubbled into the aCSF, but irreversible hypoxic-ischemic damage can occur between the time of decapitation and the placement of slices into the oxygenated aCSF, as well as in the deeper portions of the slice if the superfusion rate and/or oxygenation level is not sufficient. Several strategies can prevent damage caused by lack of oxygen and the related toxicity. For example, when using larger species such as cats, the animal can be kept alive under deep anesthesia while the skull is being removed (16). With rodents, after deep anesthesia, the animal can be transcardially perfused with chilled, oxygenated aCSF (18). This is intended to cool down the brain, thus reducing metabolic demand as well as the release of glutamate and associated excitotoxicity. The perfusion also serves to clear blood from the vasculature, which can also help with cell visualization in the slice (18).

Another common method to limit tissue damage is to modify the aCSF used for sectioning and incubation of the slices (Table 1). Brain hypoxia is accompanied by release of glutamate that can cause neuronal death through several mechanisms (27). First, the entrance of sodium during cell depolarization mediated by glutamate may trigger multiple events potentially leading to cell death, such as the influx of chloride and water (causing cell swelling), a sodium-induced increase in cytosolic calcium, or a decrease in intracellular pH (28, 29). These risks can be mitigated by replacing sodium chloride with sucrose in the aCSF (18, 30). Another mechanism of glutamate toxicity depends on the influx of calcium into the cell through NMDA receptors (31), which leads to the activation of a number of degrading enzymes (27). In this respect, the use of ionotropic glutamate receptor blockers, such as ketamine and kynurenic acid (32–35), or general blockers of synaptic transmission (36) can help preserve the viability of cells in the slice.

Paradoxically, a hyperoxic environment, which can be toxic, can also develop in portions of the slice (37) due to the surplus of oxygen bubbled into the aCSF (95% O_2). Oxygen derivatives can interact with unsaturated fatty acids in cell membranes (lipid peroxidation) and affect synaptic transmission and plasticity (38, 39). Oxygen derivatives can also inhibit glutamate uptake transporters, thereby contributing to excitotoxicity (38). Thus, including antioxidants in the aCSF could be particularly important when preparing brain slices for neuronal circuitry analysis. A number of compounds can be added to the aCSF during slice preparation and incubation to prevent cellular damage by oxidative stress (Table 1). For example, ascorbate is an important antioxidant that is not synthesized in the brain, and is quickly lost during slice preparation (37, 40–43), and including ascorbate in the aCSF has

Table 1
aCSF modifications suggested to prevent hypoxic-ischemic and/or oxidative damage in the slice preparation

Substance	General effect	Potential problems	Use and references
Receptor and channel blockers (often glutamate receptor blockers)	Prevent excitotoxicity	Some blockers may decrease ATP levels (e.g., thiopental (152)) or be difficult to wash out (e.g., phenytoin (34))	*Ketamine*: 1 mM during slice preparation (34, 36), 25 μM in recording chamber (32) *D-APV*: 10–100 μM in recording chamber (32) *APH*: 200 μM during preparation (36) *Kynurenate*: 10 mM during preparation (153), 5–10 μM in recording chamber (32) *Barbiturates*: Thiopental: 400 μM in recording chamber (154), 600 μM in recording chamber (152, 155); pentobarbital: 10–100 μM in recording chamber (156) *Lidocaine*: 10–100 μM during slice preparation (34) *Tetrodotoxin*: 200 μg/l in recording chamber (157) *Phenytoin*: 100–200 mg/kg i.p. before decapitation, and then 10 μM during incubation (34)
Calcium: magnesium ratio	General blockade of synaptic transmission (prevents excitotoxicity). Preferred method when channel or receptor blockade is an issue (36)	Ratio should be carefully chosen to prevent effects on cell excitability (34, 158)	0 calcium: 10 mM magnesium during slice preparation (32, 34, 36)
Sucrose instead of sodium chloride	Reduces swelling and lysis	May enhance GABAergic transmission and affect network plasticity (e.g., LTP (159)); sucrose concentration should be chosen to preserve aCSF osmolarity (18)	252 mM during slice preparation (30)

(continued)

Table 1 (continued)

Substance	General effect	Potential problems	Use and references
Choline chloride instead of sodium chloride	Reduces swelling and lysis	None found	110 mM during slice preparation (160–162)
Ascorbate	Prevents oxidative damage, therefore also reducing excitotoxicity, swelling, and lysis	None found	0.4 mM preserves the in vivo level in slices kept at room temperature; at higher temperatures, higher concentrations are required (37) 0.4–1 mM during preparation and recording (18) 1.3 mM during preparation only (161)
Lactate	Helps support metabolism during recovery from hypoxia (163)	Still controversial as a substitute to glucose (see 164, 165)	3–6 mM during preparation and recording (163)
Pyruvate	Supports metabolism, prevents oxidative damage and excitotoxicity	It may be neuroprotective only in certain conditions (166)	2 mM (18) 5 mM during preparation and recording (98)

been found to prevent the formation of peroxides and decrease the swelling of brain slices (37, 41, 43).

On the other hand, Moyer and Brown (18) chose to modify the preparation technique, rather than the composition of the aCSF, to improve slice quality. They reported that keeping constant osmolarity across all aCSF solutions used throughout the experiment, as well as preincubating the slices in 36°C aCSF for 30 min, substantially increased the viability of the tissue prepared from aging or aged rats.

2.3. Blocking the Brain to Study Long-Range Connections

The correct blocking and positioning of the brain with respect to the cutting blade are critical in determining the success with which the circuit of interest is included in the slice. With short-range connections, the pre- and postsynaptic components of the circuit can often be included in slices sectioned in one of the anatomical planes (coronal, parasagittal, horizontal) at the standard thicknesses (~300–400 μm). Procedures for obtaining a variety of slice preparations that require sectioning in atypical planes are available in the literature (see references in subsequent paragraphs). A few approaches are reviewed in brief below.

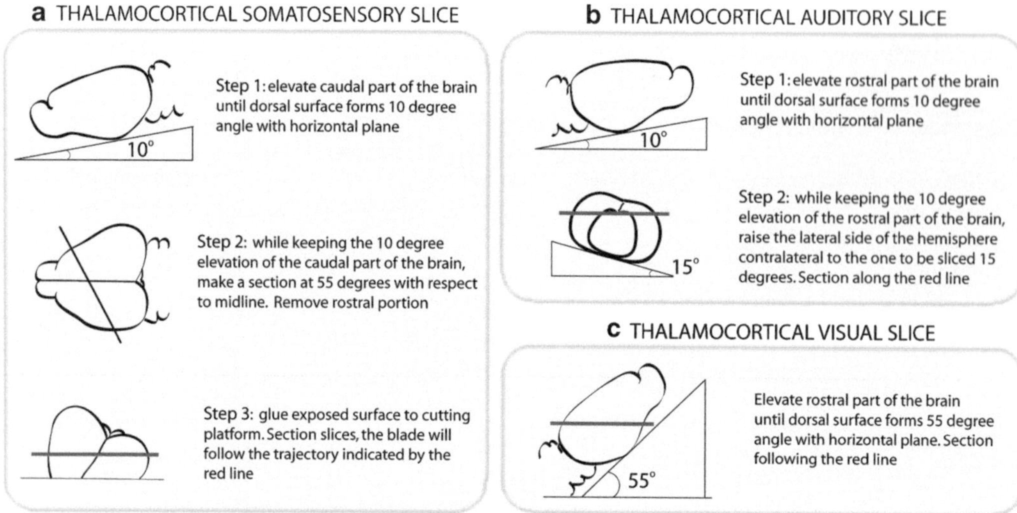

Fig. 1. Schematic representation of the position and blocking of a rodent brain to obtain sensory thalamocortical slices. *Red line* indicates the path that the blade follows when sectioning. (**a**) Somatosensory slice. (**b**) Auditory slice. (**c**) Visual slice. For initial description of these slice preparations, see references 52, 53, 55, respectively.

One of the most extensively used slices has been the hippocampal transverse slice, in which the sections are prepared by cutting close to perpendicular to the longitudinal hippocampal axis (i.e., parallel to the lamellae) to preserve the connections between all subregions of the hippocampal formation, from dentate gyrus to CA1 (14, 44–46). Due to the curvature of the hippocampus, a number of sectioning techniques have been developed. Cutting in a horizontal plane preserves the intrahippocampal connectivity in slices from intermediate to ventral hippocampus. A slight deviation from the horizontal plane allows the inclusion of parahippocampal regions in the hippocampus-entorhinal cortex slice (47–51). Dorsal hippocampus is best studied with sagittal or coronal sections. Alternately, the hippocampus can be dissected free before sectioning to produce slices from dorsal through ventral hippocampus (excepting the extreme septal and temporal poles).

Thalamocortical preparations (Fig. 1) have also become popular more recently, and a variety of methods have been developed to preserve connectivity in this complex network. The first approach, developed by Agmon and Connors (52), included the somatosensory thalamus and both feed-forward and feedback connections with somatosensory cortex. This was followed by slices including the auditory thalamocortical (and corticothalamic) pathway (53, 54), and slices including the visual thalamocortical pathway (55). In addition to sensory thalamocortical slices, preserved connectivity has also been demonstrated in slices between the midline thalamus (including the mediodorsal nucleus among others) and the anterior cingulate cortex (56). The potential of the thalamocortical slices goes beyond the study of the thalamocortical synapse because thalamocortical and corticothalamic fibers travel across the thalamic reticular nucleus in their way to the cortex or the dorsal thalamus. Thus, some of these slices are also useful for studies of the network including the thalamic reticular nucleus (57, 58). Likewise, the thalamocortical preparation may also be used to study the intracortical spread of thalamic stimulation (59) as well as the specific and nonspecific thalamocortical projections (60, 61). Furthermore, thalamocortical preparations may include currently unexplored connections; such is the case of the auditory thalamocortical slice, which may also include thalamostriatal connections (62).

Within the neocortex itself, coronal, parasagittal, or modifications of these planes have been used to study cortical networks within and across functional regions. For example, coronal slices (63) or slightly off-coronal slices (64) include reciprocal connections between primary (V1) and secondary (lateromedial) visual cortex. In the somatosensory cortex, the plane of sectioning can be modified in order to preserve connectivity across particular barrels. For example, Finnerty et al. (65) cut slices at 45° with respect to the sagittal plane in order to obtain, in the same section, barrels receiving input from each of the whisker rows on the rat snout. Additional examples of slices that preserve both pre- and postsynaptic components include the mamillothalamic projection to the anterior dorsal thalamic nucleus (66) and the inferior colliculus projection to the medial geniculate body (67).

However, including both pre- and postsynaptic components of a circuit in their entirety is not always possible, and one alternative has been to target the axons of presynaptic components of the circuit under study while recording the postsynaptic cells. Examples of this strategy can be found in cerebellar slice (68), visual corticothalamic and retinogeniculate pathways (69), hippocampo-accumbens pathway (70), cortico-cortical between hemispheres (71, 72), and corticothalamic projections to the anterior dorsal nucleus (66).

2.4. Maintenance During Recording/Stimulation

The final step to study neural networks in the slice is to ensure that the slices maintain physiological properties during the experiment. Even when all the steps are taken to increase the likelihood of obtaining slices with normal physiology, it should be remembered that this is, nevertheless, an artificial preparation. Indeed, there have been reports of differences in the levels of metabolites, ion homeostasis, and respiration rates in the slice compared to the in vivo condition (73). It has been suggested that most of these changes can be explained by three observations (73): (1) mechanical cell damage during sectioning affects at least 50 μm on each side of the slice (because metabolic parameter estimations are commonly calculated per total slice volume or weight and a fraction of the slice volume contains dead cells, metabolic parameter estimations in the slice likely underestimate the in vivo values);

(2) hypoxic-ischemic damage during preparation; and (3) overall low spontaneous activity due to diminished neuromodulatory input, which has the secondary effect of reducing metabolic demand. Thus, taking measures to prevent damage from hypoxia (as described in previous sections) and recording from deeper parts of the slice will ensure that conditions more closely mimic the in vivo preparation. In addition, metabolic parameters (e.g., ATP levels) are the lowest right after preparation, and they recover to steady state in 2–4 h (14) with most of the recovery occurring in the first hour after slice preparation. Accordingly, incubation for at least 1 h before recording helps to ensure that the investigators are conducting their experiments at steady state. Once slices are in the recording chamber, superfusion rates can affect the oxygen level inside the tissue. Hypoxic conditions have been found at 150 μm of depth with superfusion rates of 3 ml/min or less (74). Therefore, superfusion rates of 3–6 ml/min in chambers with a total volume of 0.5 ml are recommended to provide appropriate oxygenation (75). Because the mechanical damage and level of hypoxia may vary from day to day causing variability in the data across recording sessions, care should be taken to standardize slice preparation across experiments.

2.5. Reliability of Cellular and Network Properties

Despite the substantial differences in the experimental conditions of neurons in the slice preparation compared to those found in vivo, several studies have suggested that many of the neuronal firing properties described in the slice are similar to those found in vivo provided that (1) the solution bathing the slice resembles in vivo CSF; (2) the temperature is kept within 5–10°C of the physiological temperature; and (3) the animal is old enough for the expected firing properties to have developed (24, 25, 76–81). Indeed, several types of characteristic firing patterns that have been described in neocortical cells (e.g., fast spiking, intrinsic bursting, etc.) have been found in slices as well (in vitro: 82–84; in vivo: 85). In addition, other specific firing properties, such as bistable states in thalamic neurons, or spontaneous irregular firing in striatal cholinergic cells have been observed both in vivo and in the slice (thalamus in vitro: 86–89; thalamus in vivo: 86, 90; striatum in vitro: 91; striatum in vivo: 92).

Network properties in the slice also appear to, in many cases, reproduce results found in vivo. For example, network rhythmic activity in the slice preparation has been reported in a variety of brain regions, including slow-wave oscillations in cortical slices (93), theta rhythm in hippocampal slices (94), and spindle oscillations in thalamic slices (95). Significantly, postsynaptic potentials recorded in vitro also resemble those recorded in vivo (45) and components of local field potentials elicited by pathway stimulation in slices are similar to those reported in vivo (14, 45, 96).

Maintenance of normal network physiology relies on maintenance of normal synaptic transmission. In addition to the parameters discussed in previous sections for obtaining slices with normal metabolic and cellular properties, attention should be paid to factors that may affect neurotransmission. For example, normal hippocampal network activity can be affected by oxygen levels; it has been shown that hippocampal sharp-wave ripples and fast oscillations are substantially reduced when perfusion rates are too low to provide normal oxygen levels (74). The concentration of the most common energy source used in aCSF, glucose, can also affect some connections (97). Further, the choice of energy source was found by Holmgren et al. (98) to dramatically affect the postsynaptic effects of GABA by altering E_{GABA} and the resting potential, leading the authors to propose the use of aCSF containing more physiological combinations of energy sources (such as pyruvate and ketones in addition to glucose).

Likewise, concerns have been raised regarding the scarcity of synaptic activity and the lack of physiological levels of modulatory neurotransmitters, which are not commonly used in aCSF (99, 100). The lack of synaptic activity can be problematic when studying synaptic plasticity because it is well-known that synaptic strength depends on previous spiking history (101). Furthermore, ongoing spiking activity contributes to gain control (102) and the reliability (103) of input–output relationships. Low spontaneous activity may be related to a number of factors, such as the selective elimination of excitatory inputs during sectioning. A possible solution to this problem would be to mimic the excitatory and inhibitory potentials a cell normally receives in vivo with electrical current injection (102, 103). Low oxygen levels caused by the use of low superfusion rates for the aCSF may also contribute to decreased activity. This is suggested by the recent finding of increased frequency of spontaneous EPSPs and IPSPs when the aCSF is perfused at high rates (75). The lack of physiological levels of modulatory neurotransmitters could contribute to the reduced synaptic activity observed in brain slices; modulatory neurotransmitters affect resting membrane potentials, as well as intrinsic response properties (104–106) and synaptic behavior (107). A variety of modulatory neurotransmitters (such as taurine, D-serine, or the previously discussed ascorbate) are present in the normal CSF, but these neurotransmitters are not routinely included in the artificial CSF used with brain slices. It has been suggested that the aCSF used for recording could be made more similar to the normal CSF by including small concentrations of those neuromodulators, as observed in vivo (75). Investigation of network properties in the slice typically requires the introduction of controlled input to the system. Below, we introduce three forms of stimulation that can be used in the slice: electrical stimulation, glutamate-based stimulation, and optogenetic stimulation.

3. Forms of Stimulation of Neural Activity in the Slice

3.1. Electrical Stimulation

Electrical stimulation offers the capacity to stimulate neurons with a virtually limitless variety of stimulation protocols using tools that can be as simple as a broken-tip glass electrode. The major liability of this technique is the lack of precision in the actual elements being stimulated. Despite this, electrical stimulation has found wide use due to its ease of application and flexibility.

The easiest and lowest-cost method to stimulate brain slices electrically is to use a broken-tip (tip size 2–10 μm) glass microelectrode filled with aCSF, with a reference electrode placed in the fluid portion of the tissue chamber. A point of caution with this approach is that the stimulation current path runs from the stimulation point to reference electrode, potentially stimulating neural structures in this pathway. Metal electrodes are typically made of tungsten and often constructed in a side-by-side or concentric bipolar configuration, which limits the spread of stimulation current. Bipolar electrodes permit lower stimulation strengths to be used (108), are less likely to cause electrolytic damage to neural tissue, and have been shown to produce lower stimulus artifacts than monopolar stimulation (109). In addition, bipolar electrodes allow the investigator to precisely control the orientation of stimulation, which may be important for activation of axons (108). However, we have found concentric bipolar electrodes to be too bulky to stimulate small structures in the slice. The primary disadvantage to such electrodes is the potential additional time and/or expense in their fabrication, compared to glass electrodes. Other approaches involve the use of arrays of electrodes, which allow arbitrary spatiotemporal patterns of stimuli to be applied to the slice (110–112).

Several studies have examined the sensitivity of different neuronal substructures to extracellular current injection. Nowak and Bullier concluded that electrical stimulation in the slice preparation activated axons, rather than cell bodies (113). They further demonstrated that inactivation of initial axonal segments via depolarization block had very little influence over stimulation threshold, suggesting that axonal branches, rather than initial segments, are responsible for triggering action potentials caused by extracellular current injection (114). It is also worth noting that manipulation of stimulation parameters or configurations can selectively activate different elements within a peripheral nerve; this was initially shown in vivo and recent work has suggested that these approaches may be used in the slice (115). A detailed description of these techniques is beyond the scope of this chapter, but the reader is referred to 116–120 for further reading.

Major issues to consider when developing an electrical stimulation paradigm are the efficacy of stimulation and damage to neural tissue. Most protocols call for trains of short-duration stimuli (e.g., <1 ms) since the efficacy of stimulation is a decreasing exponential function of stimulus duration (121), and long-duration stimuli do not permit the clearance of toxic metabolites that are generated during electrical stimulation. Most investigators use stimulation rates <50 pulses per second, as higher rates have been associated with depolarization block (122). Notably, depolarization block has been seen at 50 pulses per second with pulse durations of 100 μs, which yields a duty cycle of 0.5%. This suggests that the fundamental limiting factor is not related to persistent depolarization provided at high pulse rates. The actual waveforms of the individual pulses may also influence experimental outcome. For example, monophasic square-wave stimuli are commonly used because they offer high efficacy of axonal stimulation (121). In contrast, biphasic pulses are less likely to cause tissue damage, but have a lower efficacy of axonal stimulation. More complicated waveforms have been described for specific purposes, such as exponential waveforms to limit neuronal damage or quasi-trapezoidal/multistage waveforms for selective stimulation of subpopulations of neurons (121, 123, 124).

3.2. Glutamate-Based Stimulation

One liability of electrical stimulation in the slice is that electrical stimulation can trigger anti- and orthodromic activity in axons, dendrites, and cell bodies. This is often a hurdle that must be overcome by a series of controls to ensure that there is no stimulation of axons "passing through" the vicinity of stimulation. Glutamate-based stimulation avoids this problem since axons are not excitable by glutamate (125–127).

Glutamate-based stimulation is typically achieved via direct microapplication or laser-based photostimulation. Glutamate may be applied directly and locally via either microiontophoresis or pressure injection using a micropipette. Pressure injection involves the action of a high-speed valve to eject small volumes (generally, in the single-digit nanoliter range) of glutamate onto groups of neurons (121, 122). L-glutamate (1–2 mM) can be dissolved in aCSF and loaded into an injection pipette (1–5 μm tip diameter) and injection pressures of approximately 1–10 psi are typically used. The actual volume of glutamate ejected is proportional to the third power of the tip diameter. Therefore, if injection volume is important to a particular experimental question, each pipette should be individually calibrated (using a calibrated ocular to measure drop diameter) or a pipette fabrication approach that produces highly consistent tip size should be used. Microiontophoresis of glutamate is possible since glutamate carries a net negative charge at pH 8. Iontophoresis is generally done with pipettes (tip diameter 1–2 μm) with L-glutamate concentrations of approximately 1 mM (123, 124). Higher concentrations (20–200 mM) may be required when smaller tip sizes are used (125, 126). Negative current pulses are used to iontophorese glutamate while positive current is used as a holding current between stimuli. One potential disadvantage to iontophoresis is the direct influence of tonic holding or injection currents on nearby neurons, and the relatively high glutamate concentrations necessary for this technique, which may produce receptor desensitization. While both approaches solve the fibers of passage problem, both techniques can be too cumbersome (compared to laser-based stimulation techniques; see below) for stimulation across an array of sites, and are therefore unsuitable for systematically mapping synaptic inputs. The main advantage of microapplication of glutamate via either iontophoresis or pressure injection is the relatively low cost and ease of setting up such a system.

One approach to enable multifocal stimulation with glutamate is to use scanning laser technology to focally activate glutamate. Toward this end, "caged" glutamate (L-glutamic acid α-(4,5-dimethoxy-2-nitrobenzyl) ester) was synthesized with a molecular cage that renders the glutamate molecule unable to bind to receptors (127). Exposure to UV light (355 nm) removes the cage enabling the glutamate to bind. Newer nitroindoline derivatives are more stable than the earlier generation molecules, retain rapid kinetics of photorelease, and are more commonly used for photostimulation (128). This technique yields high spatial resolution in two dimensions, with suprathreshold excitation being limited to an ~50–100-μm cylinder through the tissue sample (129). The speed of mapping, coupled with the relative ease of implementation with free software developed by Pologruto et al. (130), has allowed laser photostimulation to be extensively used to map synaptic inputs to specific areas in brain slices, both locally and across relatively long distances (131, 132).

One initial concern with laser-based photostimulation is that photostimulation of one group of neurons may cause a synaptically coupled neuron to spike ("synaptic driving"), which would confound the interpretation of presumed monosynaptic input maps generated using laser photostimulation. This issue was addressed by Katz and Dalva (133), who found that while recording extracellularly from cortical neurons spikes could only be elicited while stimulating over the cell body or proximal dendrites. Similar controls have been done by other investigators (129, 132), suggesting that synaptic driving in the cortical slice is unlikely.

One drawback to "traditional" single-photon-based photostimulation is the spread of activation caused by uncaging of glutamate molecules in the fluid along the light path and light scatter caused by the brain tissue outside of the plane of focus. This effectively limits the resolution of the technique such that it is unable to selectively stimulate individual neuronal substructures, such as dendritic spines. An approach to dealing with this problem is to liberate caged glutamate using two-photon stimulation, where a high degree of spatial resolution can be achieved since only small volumes of tissue have the photon density to liberate glutamate. Two-photon uncaging is typically done with forms of caged glutamate with a large two-photon cross section (which is a measure of the probability of two-photon absorption), such as MNI-glutamate, and may be coupled with two-photon calcium imaging. As such, this technique is perfectly suited for visualization of calcium ion flux resulting from the activation of a single dendritic spine (134–136).

One note of caution regarding caged compounds is their potential to interact with receptors in the absence of photostimulation. For example, nitroindolinyl-caged GABA and nitroindolinyl-caged glutamate have been shown to interact with GABA receptors (128, 137), nitroindolinyl-caged glycine has been shown to interact with glycine receptors (128), and α-carboxy-2-nitrobenzyl-caged glutamate has been shown to interact with NMDA receptors (138). In the case of nitroindolinyl-caged glutamate, many investigators (including the authors) have used this compound without observation of seizure activity and with the ability to observe potent $GABA_A$ receptor activation (131). However, a new version of caged glutamate based on ruthenium chemistry (RuBi-glutamate) has been described which has a lower potential to cause GABA blockade. The high quantum efficiency of RuBi-glutamate allows lower concentrations of caged glutamate to be used, causing less GABA blockade (137).

3.3. Optogenetic Stimulation

An emerging technology for neuronal stimulation in the slice is the incorporation of genetically encoded light-sensitive channels into target neurons (for more details, see Part IV ("optogenetics") in this volume). The most commonly incorporated channel, channelrhodopsin2, is an algae-derived cation channel that opens in the presence of blue light (peak conductance at 470 nm). The reversal potential of the channelrhodopsin2 channel is approximately 0 mV; therefore, light-induced conductance in these channels depolarizes neurons. Channelrhodopsin constructs are typically combined with a fluorescent marker protein, such as yellow-fluorescent protein, to allow visualization of transfected neurons during experimentation. These constructs are generally delivered to target neurons via germline encoding, transfection via viral vector, or electroporation. The reader is referred to 138 and to the chapter by Cardin in the present volume for further description of approaches to channelrhodopsin2 gene delivery.

The main utility of "optogenetic" approaches is to achieve stimulation in a select population of neurons, targeted either genetically as belonging to a particular functional cell class (e.g., parvalbumin-containing interneurons (139)) by projection target (e.g., cortico-cortical neurons (140)) or other characteristics, such as cortical layer of origin (72). Once transfected, expression of channelrhodopsin2 occurs throughout the cell, including dendrites and axon. The latter feature is particularly appealing for the study of long-range connections in the slice. When using electrical stimulation, the experimenter could be exciting an unknown, but potentially large, population of axons in addition to those of interest for the study. An improved approach would be to selectively transfect the projections of interest with channelrhodopsin2, cut slices of any desired orientation, and then selectively stimulate the axons using light of the appropriate wavelength. This strategy was successfully used to study somatosensory thalamocortical projections in the mouse (141).

Several technical issues are important to consider when comparing the utility of optogenetic approaches to more traditional approaches, such as electrical stimulation and glutamate uncaging. One issue is temporal resolution. Although the onset kinetics of channelrhodopsin2 are relatively fast and spike timing has high precision, there is a slow decay of activation lasting on the order of tens of milliseconds (142, 143). In addition, channelrhodopsin2 shows substantial desensitization after initial activation (143, 144). These factors limit the frequency with which action potentials can be driven in presynaptic axons. It should be noted that several new variations of the channelrhodopsin molecule have been engineered to have faster kinetics, and may be able to achieve more consistent high-frequency stimulation (145). An additional consideration is the compatibility of optogenetic approaches with other recording or imaging technologies. It is possible that direct illumination of a glass recording electrode may create an electrical artifact coinciding with the light pulse. Cardin et al. (146) described strategies to minimize these artifacts via use of a shorter wire or by coating the electrode with a nonreflective opaque substance (146) (for more details, see chapters by Siegle and Cardin in this volume). Coupling optogenetic stimulation with activation imaging is potentially complicated by the overlapping excitation spectra of traditional activation indicators and channelrhodopsin2. Such overlap would not permit selective stimulation and imaging to be separated in time (e.g., to stimulate with a pulse, then image over time) since the excitation light used for imaging would continue to cause stimulation. Newer red-shifted variants of channelrhodopsin (145) and/or imaging fluorophores (147) permit the flexible combination of optogenetic stimulation with fluorescence-based imaging (147, 148).

3.4. Laser Photostimulation Combined with Flavoprotein Autofluorescence as a Tool to Assess Connectivity in the Slice

A common problem faced by the slice physiologist is how to determine the degree of retained neural connectivity in a slice once it has been cut. For example, thalamocortical slices have been described for the study of the mouse somatosensory, auditory, and visual systems (52, 54, 55). Typically, only one slice (at best) per mouse has retained connectivity between thalamus and cortex. A rapid, noninvasive and sensitive method has been developed to assess directional connectivity in the slice using laser photostimulation coupled to flavoprotein autofluorescence (LPFA). This approach allows the user to rapidly assess which slices are usable for the study of long-range connections. In addition, LPFA is a powerful tool that one can use to rapidly determine topographical maps of neuronal connectivity and to facilitate the development of novel slice preparations.

In LPFA, slices are bathed in caged glutamate, and a UV laser (355 nm) is used to focally uncage glutamate and orthodromically stimulate small groups of neurons in the slice. Flavoprotein autofluorescence (FA) imaging captures light produced in the projection field of the stimulated neurons. FA has recently been adapted for in vitro use in combination with a variety of stimulation methods (62, 66, 149). This technique relies on the green fluorescence (~520 nm) of endogenous mitochondrial flavin moieties, flavin mononucleotide (FMN), and flavin adenine dinucleotide (FAD) when they are exposed to blue light (~460 nm), and thus does not require dye loading. This avoids time-consuming equilibration with voltage- or calcium-sensitive dyes, which may alter the health of the slice and may produce heterogeneous uptake (150, 151). Under illumination at 460 nm, FMN and FAD increase their fluorescence by as much as 15–25% in response to neuronal stimulation, giving FA imaging one of the highest signal-to-noise ratios of the imaging methods employed to investigate network-level slice activity. Additionally, FA signal changes are sensitive to subthreshold postsynaptic activity and have a spatial resolution of approximately 100–200 μm (62).

Although FA can be coupled to electrical stimulation or local application of glutamate via pressure injection, we have found that photostimulation of glutamate combined with FA imaging offers the most rapid and noninvasive means to investigate slice connectivity. This combination has been used for rapid topographic mapping of connectivity within a slice, novel slice development (66), and connectivity assessment on a slice-by-slice basis (61). LPFA can be particularly useful when multiple structures are synaptically connected in a single slice, and can be used to identify connected loci for subsequent experimentation (Fig. 2). In consideration of practicality, adding FA imaging capability to an epifluorescence-equipped in vitro electrophysiology rig is straightforward. Beyond the appropriate filter set, the only additional materials are a moderate sensitivity camera and associated image acquisition software. The major drawback to this technique is the slow time course of activation, which is measured on the order of seconds. This potentially limits the ability to use FA to map the precise temporal evolution of a neuronal response, but is not a practical limitation when used for long-range spatial mapping. Thus, FA imaging offers a practical, cost-effective approach to rapidly assessing connectivity in vitro.

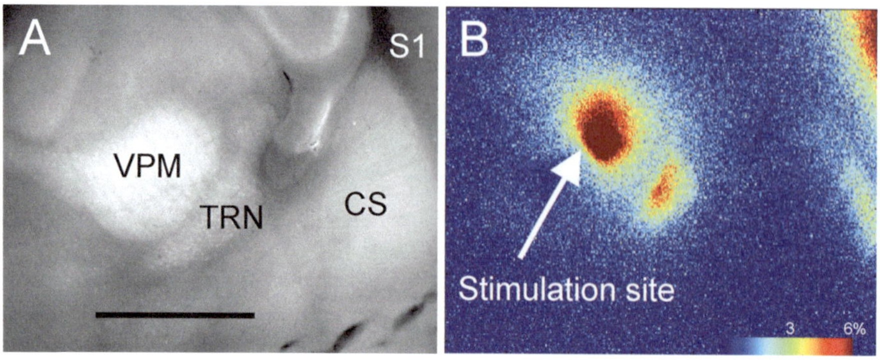

Fig. 2. Illustration of the use of LPFA to identify multiple connected areas in the mouse brain slice. (**a**) Bright-field image of the living slice at 5× magnification. *VPM* ventral posterior medial nucleus of the thalamus; *TRN* thalamic reticular nucleus; *CS* corpus striatum; *S1* primary somatosensory field. (**b**) $\delta f/f$ image of flavoprotein autofluorescence after laser photostimulation of caged glutamate in the VPM. Activation loci are seen in the TRN, CS, and the ventral portion of S1BF. Image is derived from a single trial consisting of a train of 20 pulses, 10-ms pulse duration, interpulse interval = 40 ms, pulse amplitude = 67 mW. Scale bar = 1 mm. For methodological details, see 62.

4. Summary

In summary, a wide spectrum of techniques is available for in vitro manipulation of neuronal circuitry. The remainder of this section focuses on implementation of some of the newest methodologies available to study neuronal networks in the slice preparation.

Acknowledgments

We thank Jennie Z. Young and Iraklis Petrof for helpful comments on this chapter.

References

1. Elliot K, Wolfe L (1962) Brain tissue respiration and glycolysis. In: Thomas C (ed) Neurochemistry, 3rd edn. Springfield, Illinois
2. McIlwain H, Cheshire J (1950) Metabolic maintenance of the inorganic and creatine phosphates of brain tissue in vitro. Biochem J 47:xviii
3. Buchel L, McIlwain H (1950) Narcotics and the inorganic and creatine phosphates of mammalian brain. Br J Pharmacol Chemother 5:465–473
4. McIlwain H, Buchel L, Cheshire J (1951) The inorganic phosphate and phosphocreatine of Brain especially during metabolism in vitro. Biochem J 48:12–20
5. Li CL, McIlwain H (1957) Maintenance of resting membrane potentials in slices of mammalian cerebral cortex and other tissues in vitro. J Physiol 139:178–190

6. Hillman HH, McIlwain H (1961) Membrane potentials in mammalian cerebral tissues in vitro: dependence on ionic environment. J Physiol 157:263–278
7. Yamamoto C, McIlwain H (1966) Potentials evoked in vitro in preparations from the mammalian brain. Nature 210:1055–1056
8. Gahwiler BH (1981) Organotypic monolayer cultures of nervous tissue. J Neurosci Methods 4:329–342
9. Rambani K, Vukasinovic J, Glezer A et al (2009) Culturing thick brain slices: an interstitial 3D microperfusion system for enhanced viability. J Neurosci Methods 180:243–254
10. Gahwiler BH, Capogna M, Debanne D et al (1997) Organotypic slice cultures: a technique has come of age. Trends Neurosci 20:471–477
11. Gahwiler BH, Thompson SM, Muller D (1999) Preparation and maintenance of organotypic slice cultures of CNS tissue. Curr Protoc Neurosci 6:6.11.1–6.11.11
12. Noraberg J, Poulsen FR, Blaabjerg M et al (2005) Organotypic hippocampal slice cultures for studies of brain damage, neuroprotection and neurorepair. Curr Drug Targets CNS Neurol Disord 4:435–452
13. Hatton GI, Doran AD, Salm AK et al (1980) Brain slice preparation: hypothalamus. Brain Res Bull 5:405–414
14. Alger B, Dhanjal S, Dingledine R et al (1984) Brain slice methods. In: Dingledine R (ed) Brain slices, 1st edn. Plenum Press, New York
15. Reid KH, Edmonds HL Jr, Schurr A et al (1988) Pitfalls in the use of brain slices. Prog Neurobiol 31:1–18
16. Edwards FA, Konnerth A, Sakmann B et al (1989) A thin slice preparation for patch clamp recordings from neurones of the mammalian central nervous system. Pflugers Arch 414:600–612
17. Finkel A, Bookman R (1997) The electrophysiology setup. Curr Protoc Neurosci 6:6.1.1–6.1.6
18. Moyer J, Brown T (2002) Patch-clamp techniques applied to brain slices. In: Walz W et al (eds) Patch-clamp analysis: advanced techniques, Ith edn. Humana Press, Totowa, New Jersey
19. Debanne D, Boudkkazi S, Campanac E et al (2008) Paired-recordings from synaptically coupled cortical and hippocampal neurons in acute and cultured brain slices. Nat Protoc 3:1559–1568
20. Ramcharan EJ, Gnadt JW, Sherman SM (2000) Burst and tonic firing in thalamic cells of unanesthetized, behaving monkeys. Vis Neurosci 17:55–62
21. Briggs F, Callaway EM (2001) Layer-specific input to distinct cell types in layer 6 of monkey primary visual cortex. J Neurosci 21:3600–3608
22. Zarrinpar A, Callaway EM (2006) Local connections to specific types of layer 6 neurons in the rat visual cortex. J Neurophysiol 95:1751–1761
23. McCormick DA, Trent F, Ramoa AS (1995) Postnatal development of synchronized network oscillations in the ferret dorsal lateral geniculate and perigeniculate nuclei. J Neurosci 15:5739–5752
24. Brumberg JC, Nowak LG, McCormick DA (2000) Ionic mechanisms underlying repetitive high-frequency burst firing in supragranular cortical neurons. J Neurosci 20:4829–4843
25. Llano DA, Sherman SM (2009) Differences in intrinsic properties and local network connectivity of identified layer 5 and layer 6 adult mouse auditory corticothalamic neurons support a dual corticothalamic projection hypothesis. Cereb Cortex 19:2810–2826
26. Reyes A, Sakmann B (1999) Developmental switch in the short-term modification of unitary EPSPs evoked in layer 2/3 and layer 5 pyramidal neurons of rat neocortex. J Neurosci 19:3827–3835
27. Choi DW (1994) Calcium and excitotoxic neuronal injury. Ann N Y Acad Sci 747:162–171
28. Rothman SM (1985) The neurotoxicity of excitatory amino acids is produced by passive chloride influx. J Neurosci 5:1483–1489
29. Syntichaki P, Tavernarakis N (2003) The biochemistry of neuronal necrosis: rogue biology? Nat Rev Neurosci 4:672–684
30. Aghajanian GK, Rasmussen K (1989) Intracellular studies in the facial nucleus illustrating a simple new method for obtaining viable motoneurons in adult rat brain slices. Synapse 3:331–338
31. Lehmann A, Jacobson I (1990) Ion dependence and receptor mediation of glutamate toxicity in the immature rat hippocampal slice. Eur J Neurosci 2:620–628
32. Clark GD, Rothman SM (1987) Blockade of excitatory amino acid receptors protects anoxic hippocampal slices. Neuroscience 21:665–671
33. Rothman SM, Thurston JH, Hauhart RE et al (1987) Ketamine protects hippocampal neurons from anoxia in vitro. Neuroscience 21:673–678
34. Aitken PG, Breese GR, Dudek FF et al (1995) Preparative methods for brain slices: a discussion. J Neurosci Methods 59:139–149
35. Strasser U, Lobner D, Behrens MM et al (1998) Antagonists for group I mGluRs

attenuate excitotoxic neuronal death in cortical cultures. Eur J Neurosci 10:2848–2855

36. Feig S, Lipton P (1990) N-methyl-D-aspartate receptor activation and Ca^{2+} account for poor pyramidal cell structure in hippocampal slices. J Neurochem 55:473–483
37. Rice ME (1999) Use of ascorbate in the preparation and maintenance of brain slices. Methods 18:144–149
38. Mattson MP (1998) Modification of ion homeostasis by lipid peroxidation: roles in neuronal degeneration and adaptive plasticity. Trends Neurosci 21:53–57
39. Pedersen JZ, Bernardi G, Centonze D et al (1998) Hypoglycemia, hypoxia, and ischemia in a corticostriatal slice preparation: electrophysiologic changes and ascorbyl radical formation. J Cereb Blood Flow Metab 18:868–875
40. Bell JL, McIlwain H, Thomas J (1956) The composition of isolated cerebral tissues; ascorbic acid and cozymase. Biochem J 64:332–335
41. Kovachich GB, Mishra OP (1983) The effect of ascorbic acid on malonaldehyde formation, K^+, Na^+ and water content of brain slices. Exp Brain Res 50:62–68
42. Rice ME, Perez-Pinzon MA, Lee EJ (1994) Ascorbic acid, but not glutathione, is taken up by brain slices and preserves cell morphology. J Neurophysiol 71:1591–1596
43. Brahma B, Forman RE, Stewart EE et al (2000) Ascorbate inhibits edema in brain slices. J Neurochem 74:1263–1270
44. Skrede KK, Westgaard RH (1971) The transverse hippocampal slice: a well-defined cortical structure maintained in vitro. Brain Res 35:589–593
45. Lynch G, Schubert P (1980) The use of in vitro brain slices for multidisciplinary studies of synaptic function. Annu Rev Neurosci 3:1–22
46. Sayer RJ, Redman SJ, Andersen P (1989) Amplitude fluctuations in small EPSPs recorded from CA1 pyramidal cells in the guinea pig hippocampal slice. J Neurosci 9:840–850
47. Walther H, Lambert JD, Jones RS et al (1986) Epileptiform activity in combined slices of the hippocampus, subiculum and entorhinal cortex during perfusion with low magnesium medium. Neurosci Lett 69:156–161
48. Jones RS, Heinemann U (1988) Synaptic and intrinsic responses of medical entorhinal cortical cells in normal and magnesium-free medium in vitro. J Neurophysiol 59:1476–1496
49. Dreier JP, Heinemann U (1990) Late low magnesium-induced epileptiform activity in rat entorhinal cortex slices becomes insensitive to the anticonvulsant valproic acid. Neurosci Lett 119:68–70
50. Boulton CL, von Haebler D, Heinemann U (1992) Tracing of axonal connections by rhodamine-dextran-amine in the rat hippocampal-entorhinal cortex slice preparation. Hippocampus 2:99–106
51. Empson RM, Heinemann U (1995) The perforant path projection to hippocampal area CA1 in the rat hippocampal-entorhinal cortex combined slice. J Physiol 484(Pt 3):707–720
52. Agmon A, Connors BW (1991) Thalamocortical responses of mouse somatosensory (barrel) cortex in vitro. Neuroscience 41:365–379
53. Metherate R, Cruikshank SJ (1999) Thalamocortical inputs trigger a propagating envelope of gamma-band activity in auditory cortex in vitro. Exp Brain Res 126:160–174
54. Cruikshank SJ, Rose HJ, Metherate R (2002) Auditory thalamocortical synaptic transmission in vitro. J Neurophysiol 87:361–384
55. MacLean JN, Fenstermaker V, Watson BO et al (2006) A visual thalamocortical slice. Nat Methods 3:129–134
56. Lee CM, Chang WC, Chang KB et al (2007) Synaptic organization and input-specific short-term plasticity in anterior cingulate cortical neurons with intact thalamic inputs. Eur J Neurosci 25:2847–2861
57. Zhang L, Jones EG (2004) Corticothalamic inhibition in the thalamic reticular nucleus. J Neurophysiol 91:759–766
58. Lam YW, Sherman SM (2010) Functional organization of the somatosensory cortical layer 6 feedback to the thalamus. Cereb Cortex 20(1):13–24
59. Laaris N, Carlson GC, Keller A (2000) Thalamic-evoked synaptic interactions in barrel cortex revealed by optical imaging. J Neurosci 20:1529–1537
60. Llinás RR, Leznik E, Urbano FJ (2002) Temporal binding via cortical coincidence detection of specific and nonspecific thalamocortical inputs: a voltage-dependent dye-imaging study in mouse brain slices. Proc Natl Acad Sci U S A 99:449–454
61. Theyel BB, Lee CC, Sherman SM (2010) Specific and nonspecific thalamocortical connectivity in the auditory and somatosensory thalamocortical slices. Neuroreport 21:861–864
62. Llano DA, Theyel BB, Mallik AK (2009) Rapid and sensitive mapping of long-range connections in vitro using flavoprotein

autofluorescence imaging combined with laser photostimulation. J Neurophysiol 101:3325–3340

63. Shao Z, Burkhalter A (1996) Different balance of excitation and inhibition in forward and feedback circuits of rat visual cortex. J Neurosci 16:7353–7365
64. Dong H, Shao Z, Nerbonne JM et al (2004) Differential depression of inhibitory synaptic responses in feedforward and feedback circuits between different areas of mouse visual cortex. J Comp Neurol 475:361–373
65. Finnerty GT, Roberts LS, Connors BW (1999) Sensory experience modifies the short-term dynamics of neocortical synapses. Nature 400:367–371
66. Petrof I, Sherman SM (2009) Synaptic properties of the mammillary and cortical afferents to the anterodorsal thalamic nucleus in the mouse. J Neurosci 29:7815–7819
67. Lee CC, Sherman SM (2010) Topography and physiology of ascending streams in the auditory tectothalamic pathway. Proc Natl Acad Sci U S A 107:372–377
68. Garthwaite J, Batchelor AM (1996) A biplanar slice preparation for studying cerebellar synaptic transmission. J Neurosci Methods 64:189–197
69. Turner JP, Salt TE (1998) Characterization of sensory and corticothalamic excitatory inputs to rat thalamocortical neurones in vitro. J Physiol 510(Pt 3):829–843
70. Matthews RT, Coker O, Winder DG (2004) A novel mouse brain slice preparation of the hippocampo-accumbens pathway. J Neurosci Methods 137:49–60
71. Vogt BA, Gorman AL (1982) Responses of cortical neurons to stimulation of corpus callosum in vitro. J Neurophysiol 48:1257–1273
72. Petreanu L, Huber D, Sobczyk A et al (2007) Channelrhodopsin-2-assisted circuit mapping of long-range callosal projections. Nat Neurosci 10:663–668
73. Lipton P, Whittingham T (1984) Energy metabolism and brain slice function. In: Dingledine R (ed) Brain slices, 1st edn. Plenum Press, New York
74. Hajos N, Ellender TJ, Zemankovics R et al (2009) Maintaining network activity in submerged hippocampal slices: importance of oxygen supply. Eur J Neurosci 29:319–327
75. Hajos N, Mody I (2009) Establishing a physiological environment for visualized in vitro brain slice recordings by increasing oxygen supply and modifying aCSF content. J Neurosci Methods 183:107–113
76. Schwartzkroin P (1977) Further characteristics of CA1 neurons recorded intracellularly in the hippocampal in vitro slice preparation. Brain Res 128:53–68
77. Schwartzkroin P, Altschuler R (1977) Development of kitten hippocampal neurons. Brain Res 134:429–444
78. Thompson SM, Masukawa LM, Prince DA (1985) Temperature dependence of intrinsic membrane properties and synaptic potentials in hippocampal CA1 neurons in vitro. J Neurosci 5:817–824
79. McCormick DA, Prince DA (1987) Postnatal development of electrophysiological properties of rat cerebral cortical pyramidal neurones. J Physiol 393:743–762
80. Cepeda C, Walsh JP, Buchwald NA et al (1991) Neurophysiological maturation of cat caudate neurons: evidence from in vitro studies. Synapse 7:278–290
81. Ramoa AS, McCormick DA (1994) Developmental changes in electrophysiological properties of LGNd neurons during reorganization of retinogeniculate connections. J Neurosci 14:2089–2097
82. Connors BW, Gutnick MJ, Prince DA (1982) Electrophysiological properties of neocortical neurons in vitro. J Neurophysiol 48:1302–1320
83. McCormick DA, Connors BW, Lighthall JW et al (1985) Comparative electrophysiology of pyramidal and sparsely spiny stellate neurons of the neocortex. J Neurophysiol 54:782–806
84. Yang CR, Seamans JK, Gorelova N (1996) Electrophysiological and morphological properties of layers V-VI principal pyramidal cells in rat prefrontal cortex in vitro. J Neurosci 16:1904–1921
85. Nowak LG, Azouz R, Sánchez-Vives MV et al (2003) Electrophysiological classes of cat primary visual cortical neurons in vivo as revealed by quantitative analyses. J Neurophysiol 89:1541–1566
86. Crunelli V, Hughes SW (2010) The slow (<1 Hz) rhythm of non-REM sleep: a dialogue between three cardinal oscillators. Nat Neurosci 13(1):9–17
87. Smith GD, Cox CL, Sherman SM et al (2000) Fourier analysis of sinusoidally driven thalamocortical relay neurons and a minimal integrate-and-fire-or-burst model. J Neurophysiol 83:588–610
88. Li J, Bickford ME, Guido W (2003) Distinct firing properties of higher order thalamic relay neurons. J Neurophysiol 90:291–299

89. Landisman CE, Connors BW (2007) VPM and PoM nuclei of the rat somatosensory thalamus: intrinsic neuronal properties and corticothalamic feedback. Cereb Cortex 17:2853–2865
90. Mukherjee P, Kaplan E (1995) Dynamics of neurons in the cat lateral geniculate nucleus: in vivo electrophysiology and computational modeling. J Neurophysiol 74:1222–1243
91. Bennett BD, Wilson CJ (1999) Spontaneous activity of neostriatal cholinergic interneurons in vitro. J Neurosci 19:5586–5596
92. Wilson CJ, Chang HT, Kitai ST (1990) Firing patterns and synaptic potentials of identified giant aspiny interneurons in the rat neostriatum. J Neurosci 10:508–519
93. Sánchez-Vives MV, McCormick DA (2000) Cellular and network mechanisms of rhythmic recurrent activity in neocortex. Nat Neurosci 3:1027–1034
94. Konopacki J, Golebiewski H, Eckersdorf B et al (2000) In vitro recorded theta-like activity in the limbic cortex: comparison with spontaneous theta and epileptiform discharges. Acta Neurobiol Exp (Wars) 60:67–85
95. von Krosigk M, Bal T, McCormick DA (1993) Cellular mechanisms of a synchronized oscillation in the thalamus. Science 261:361–364
96. Shaw C, Teyler TJ (1982) The neural circuitry of the neocortex examined in the in vitro brain slice preparation. Brain Res 243:35–47
97. Cox DW, Bachelard HS (1982) Attenuation of evoked field potentials from dentate granule cells by low glucose, pyruvate + malate, and sodium fluoride. Brain Res 239:527–534
98. Holmgren CD, Mukhtarov M, Malkov AE et al (2010) Energy substrate availability as a determinant of neuronal resting potential, GABA signaling and spontaneous network activity in the neonatal cortex in vitro. J Neurochem 112:900–912
99. Tsumoto T (1992) Long-term potentiation and long-term depression in the neocortex. Prog Neurobiol 39:209–228
100. Steriade M (2001) Impact of network activities on neuronal properties in corticothalamic systems. J Neurophysiol 86:1–39
101. Davis GW (2006) Homeostatic control of neural activity: from phenomenology to molecular design. Annu Rev Neurosci 29:307–323
102. Chance FS, Abbott LF, Reyes AD (2002) Gain modulation from background synaptic input. Neuron 35:773–782
103. Mainen ZF, Sejnowski TJ (1995) Reliability of spike timing in neocortical neurons. Science 268:1503–1506
104. Madison DV, Nicoll RA (1982) Noradrenaline blocks accommodation of pyramidal cell discharge in the hippocampus. Nature 299:636–638
105. Madison DV, Nicoll RA (1984) Control of the repetitive discharge of rat CA 1 pyramidal neurones in vitro. J Physiol 354:319–331
106. Goaillard JM, Vincent P (2002) Serotonin suppresses the slow afterhyperpolarization in rat intralaminar and midline thalamic neurones by activating 5-HT(7) receptors. J Physiol 541:453–465
107. Marder E, Thirumalai V (2002) Cellular, synaptic and network effects of neuromodulation. Neural Netw 15:479–493
108. Ranck JJB (1975) Which elements are excited in electrical stimulation of mammalian central nervous system: a review. Brain Res 98 (3):417–440
109. Neagu B, Strominger NL, Carpenter DO (2005) Use of bipolar parallel electrodes for well-controlled microstimulation in a mouse hippocampal brain slice. J Neurosci Methods 144(2):153–163
110. Heck D (1995) Investigating dynamic aspects of brain function in slice preparations: spatiotemporal stimulus patterns generated with an easy-to-build multi-electrode array. J Neurosci Methods 58(1–2):81–87
111. Heuschkel MO, Fejtl M, Raggenbass M et al (2002) A three-dimensional multi-electrode array for multi-site stimulation and recording in acute brain slices. J Neurosci Methods 114 (2):135–148
112. Tass PA, Silchenko AN, Hauptmann C et al (2009) Long-lasting desynchronization in rat hippocampal slice induced by coordinated reset stimulation. Phys Rev E 80(1):011902
113. Nowak LG, Bullier J (1998) Axons, but not cell bodies, are activated by electrical stimulation in cortical gray matter I. Evidence from chronaxie measurements. Exp Brain Res 118 (4):477–488
114. Nowak LG, Bullier J (1998) Axons, but not cell bodies, are activated by electrical stimulation in cortical gray matter II. Evidence from selective inactivation of cell bodies and axon initial segments. Exp Brain Res 118 (4):489–500
115. FitzGerald JJ, Lacour SP, McMahon SB et al (2009) Microchannel electrodes for recording and stimulation: in vitro evaluation. IEEE Trans Biomed Eng 56(5):1524–1534

116. Fang ZP, Mortimer JT (1991) A method to effect physiological recruitment order in electrically activated muscle. IEEE Trans Biomed Eng 38(2):175–179
117. Fang Z-P, Mortimer J (1991) Alternate excitation of large and small axons with different stimulation waveforms: an application to muscle activation. Med Biol Eng Comput 29 (5):543–547
118. Grill WM, Mortimer JT (1993) Selective activation of distant nerve fibers. In: Engineering in Medicine and Biology Society. Proceedings of the 15th Annual International Conference of the IEEE
119. Grill WM, Mortimer JT (1995) Stimulus waveforms for selective neural stimulation. IEEE Eng Med Biol 14(4):375–385
120. Grill WM, Mortimer JT (1997) Inversion of the current-distance relationship by transient depolarization. IEEE Trans Biomed Eng 44 (1):1–9
121. McCaman RE, McKenna DG, Ono JK (1977) A pressure system for intracellular and extracellular ejections of picoliter volumes. Brain Res 136:141–147
122. Sakai M, Swartz BE, Woody CD (1979) Controlled micro release of pharmacological agents: measurements of volume ejected in vitro through fine tipped glass microelectrodes by pressure. Neuropharmacology 18:209–213
123. Cormier RJ, Mauk MD, Kelly PT (1993) Glutamate iontophoresis induces long-term potentiation in the absence of evoked presynaptic activity. Neuron 10:907–919
124. Schwindt PC, Crill WE (1997) Local and propagated dendritic action potentials evoked by glutamate iontophoresis on rat neocortical pyramidal neurons. J Neurophysiol 77:2466–2483
125. Cash S, Yuste R (1999) Linear summation of excitatory inputs by CA1 pyramidal neurons. Neuron 22:383–394
126. Milojkovic BA, Radojicic MS, Goldman-Rakic PS et al (2004) Burst generation in rat pyramidal neurones by regenerative potentials elicited in a restricted part of the basilar dendritic tree. J Physiol 558:193–211
127. Wilcox M, Viola RW, Johnson KW et al (1990) Synthesis of photolabile precursors of amino acid neurotransmitters. J Org Chem 55(5):1585–1589
128. Canepari M, Nelson L, Papageorgiou G et al (2001) Photochemical and pharmacological evaluation of 7-nitroindolinyl-and 4-methoxy-7-nitroindolinyl-amino acids as novel, fast caged neurotransmitters. J Neurosci Methods 112(1):29–42
129. Dantzker JL, Callaway EM (2000) Laminar sources of synaptic input to cortical inhibitory interneurons and pyramidal neurons. Nat Neurosci 3(7):701
130. Pologruto T, Sabatini B, Svoboda K (2003) ScanImage: flexible software for operating laser scanning microscopes. Biomed Eng Online 2(1):13
131. Llano DA, Theyel BB, Mallik AK et al (2009) Rapid and sensitive mapping of long-range connections in vitro using flavoprotein autofluorescence imaging combined with laser photostimulation. J Neurophysiol 101 (6):3325–3340
132. Shepherd GMG, Pologruto TA, Svoboda K (2003) Circuit analysis of experience-dependent plasticity in the developing rat barrel cortex. Neuron 38(2):277–289
133. Katz L, Dalva M (1994) Scanning laser photostimulation: a new approach for analyzing brain circuits. J Neurosci Methods 54 (2):205–218
134. Carter AG, Sabatini BL (2004) State-dependent calcium signaling in dendritic spines of striatal medium spiny neurons. Neuron 44 (3):483–493
135. Matsuzaki M, Ellis-Davies GC, Nemoto T et al (2001) Dendritic spine geometry is critical for AMPA receptor expression in hippocampal CA1 pyramidal neurons. Nat Neurosci 4(11):1086–1092
136. Sobczyk A, Scheuss V, Svoboda K (2005) NMDA receptor subunit-dependent (Ca^{2+}) signaling in individual hippocampal dendritic spines. J Neurosci 25(26):6037–6046
137. Fino E, Araya R, Peterka DS et al (2009) RuBi-glutamate: two-photon and visible-light photoactivation of neurons and dendritic spines. Front Neural Circuits 3:1–9
138. Maier W, Corrie JE, Papageorgiou G et al (2005) Comparative analysis of inhibitory effects of caged ligands for the NMDA receptor. J Neurosci Methods 142:1–9
139. Sohal VS, Zhang F, Yizhar O et al (2009) Parvalbumin neurons and gamma rhythms enhance cortical circuit performance. Nature 459(7247):698–702
140. Lima SQ, Hromádka T, Znamenskiy P et al (2009) PINP: a new method of tagging neuronal populations for identification during in vivo electrophysiological recording. PLoS One 4(7):e6099
141. Cruikshank SJ, Urae H, Nurmikko AV et al (2010) Pathway-specific feedforward circuits

between thalamus and neocortex revealed by selective optical stimulation of axons. Neuron 65(2):230–245

142. Bamann C, Kirsch T, Nagel G et al (2008) Spectral characteristics of the photocycle of channelrhodopsin-2 and its implication for channel function. J Mol Biol 375(3):686–694
143. Schoenenberger P, Gerosa D, Oertner TG (2009) Temporal control of immediate early gene induction by light. PLoS One 4(12): e8185
144. Nagel G, Szellas T, Huhn W et al (2003) Channelrhodopsin-2, a directly light-gated cation-selective membrane channel. Proc Natl Acad Sci U S A 100(24):13940–13945
145. Lin JY (2011) A user's guide to channelrhodopsin variants: features, limitations and future developments. Exp Physiol 96(1):19–25
146. Cardin JA, Carlén M, Meletis K et al (2010) Targeted optogenetic stimulation and recording of neurons in vivo using cell-type-specific expression of channelrhodopsin-2. Nat Protoc 5(2):247–254
147. Airan RD, Hu ES, Vijaykumar R et al (2007) Integration of light-controlled neuronal firing and fast circuit imaging. Curr Opin Neurobiol 17(5):587–592
148. Gradinaru V, Zhang F, Ramakrishnan C et al (2010) Molecular and cellular approaches for diversifying and extending optogenetics. Cell 141:154–165
149. Shibuki K, Hishida R, Murakami H et al (2003) Dynamic imaging of somatosensory cortical activity in the rat visualized by flavoprotein autofluorescence. J Physiol 549(3):919–927
150. Hopt A, Neher E (2001) Highly nonlinear photodamage in two-photon fluorescence microscopy. Biophys J 80(4):2029–2036
151. Obaid AL, Loew LM, Wuskell JP et al (2004) Novel naphthylstyryl-pyridinium potentiometric dyes offer advantages for neural network analysis. J Neurosci Methods 134(2):179–190
152. Kass IS, Abramowicz AE, Cottrell JE et al (1992) The barbiturate thiopental reduces ATP levels during anoxia but improves electrophysiological recovery and ionic homeostasis in the rat hippocampal slice. Neuroscience 49(3):537–543
153. Christie BR, Eliot LS, Ito K et al (1995) Different Ca^{2+} channels in soma and dendrites of hippocampal pyramidal neurons mediate spike-induced Ca^{2+} influx. J Neurophysiol 73(6):2553–2557
154. Sasaki R, Hirota K, Roth SH et al (2005) Anoxic depolarization of rat hippocampal slices is prevented by thiopental but not by propofol or isoflurane. Br J Anaesth 94(4):486–491
155. Wang T, Raley-Susman KM, Wang J et al (1999) Thiopental attenuates hypoxic changes of electrophysiology, biochemistry, and morphology in rat hippocampal slice CA1 pyramidal cells. Stroke 30(11): 2400–2407
156. Shibata S, Kagami-ishi Y, Ueki S et al (1992) Neuroprotective effect of WEB 1881 FU (nebracetam) on an ischemia-induced deficit of glucose uptake in rat hippocampal and cerebral cortical slices and CA1 field potential in hippocampal slices. Jpn J Pharmacol 58 (3):243–250
157. Boening JA, Kass IS, Cottrell JE et al (1989) The effect of blocking sodium influx on anoxic damage in the rat hippocampal slice. Neuroscience 33(2):263–268
158. Piccolino M, Pignatelli A (1996) Calcium-independent synaptic transmission: artifact or fact? Trends Neurosci 19(4):120–125
159. Kuenzi FM, Fitzjohn SM, Morton RA et al (2000) Reduced long-term potentiation in hippocampal slices prepared using sucrose-based artificial cerebrospinal fluid. J Neurosci Methods 100(1–2):117–122
160. Magee JC, Avery RB, Christie BR et al (1996) Dihydropyridine-sensitive, voltage-gated Ca^{2+} channels contribute to the resting intracellular Ca^{2+} concentration of hippocampal CA1 pyramidal neurons. J Neurophysiol 76 (5):3460–3470
161. Hoffman DA, Johnston D (1998) Downregulation of transient K^{+} channels in dendrites of hippocampal CA1 pyramidal neurons by activation of PKA and PKC. J Neurosci 18 (10):3521–3528
162. Mainen ZF, Maletic-Savatic M, Shi SH et al (1999) Two-photon imaging in living brain slices. Methods 18(2):231–239, 181
163. Schurr A, Payne RS, Miller JJ et al (1997) Brain lactate is an obligatory aerobic energy substrate for functional recovery after hypoxia: further in vitro validation. J Neurochem 69(1):423–426
164. Yamane K, Yokono K, Okada Y (2000) Anaerobic glycolysis is crucial for the maintenance of neural activity in guinea pig hippocampal slices. J Neurosci Methods 103(2):163–171
165. Cater HL, Chandratheva A, Benhan CD et al (2003) Lactate and glucose as energy substrates during, and after, oxygen deprivation in rat hippocampal acute and cultured slices. J Neurochem 87(6):1381–1390
166. Matthews CC, Zielke HR, Parks DA et al (2003) Glutamate-pyruvate transaminase protects against glutamate toxicity in hippocampal slices. Brain Res 978(1–2):59–64

167. Merrill DR, Bikson M, Jefferys JGR (2005) Electrical stimulation of excitable tissue: design of efficacious and safe protocols. J Neurosci Methods 141(2):171–198
168. Jensen AL, Durand DM (2009) High frequency stimulation can block axonal conduction. Exp Neurol 220(1):57–70
169. Fang ZP, Mortimer JT (1991) Selective activation of small motor axons by quasitrapezoidal current pulses. IEEE Trans Biomed Eng 38(2):168–174
170. Millar J, Barnett TG (1997) The Zeta pulse: a new stimulus waveform for use in electrical stimulation of the nervous system. J Neurosci Methods 77(1):1–8
171. Callaway EM, Katz L (1993) Photostimulation using caged glutamate reveals functional circuitry in living brain slices. Proc Natl Acad Sci U S A 90(16):7661–7665
172. Christian EP, Dudek FE (1988) Characteristics of local excitatory circuits studied with glutamate microapplication in the CA3 area of rat hippocampal slices. J Neurophysiol 59 (1):90–109
173. Goodchild AK, Dampney RA, Bandler R (1982) A method for evoking physiological responses by stimulation of cell bodies, but not axons of passage, within localized regions of the central nervous system. J Neurosci Methods 6(4):351–363
174. Zhang F, Gradinaru V, Adamantidis AR et al (2010) Optogenetic interrogation of neural circuits: technology for probing mammalian brain structures. Nat Protoc 5(3):439–456

Neuromethods (2012) 67: 127–147
DOI 10.1007/7657_2011_4

Published online: 8 December 2011

Extracellular Recordings of Synaptic Plasticity and Network Oscillations in Hippocampal Slices

Gaga Kochlamazashvili, Oleg Senkov, and Alexander Dityatev

Abstract

Activity-dependent strengthening and weakening of synaptic weights, manifested as long-term potentiation (LTP) and depression (LTD), are two major mechanisms that are thought to be involved in creating memory traces in the brain. Oscillations of neuronal activity, especially in the θ (4–12 Hz), γ (30–100 Hz), and "ripple" (130–200 Hz) frequency bands, are also fundamental phenomena that are believed to contribute to learning and memory. However, the interplay between oscillations and plasticity is still not understood. These brain phenomena are rarely considered together when synaptic plasticity is studied. In this chapter, we summarize the existing knowledge in the field, describe protocols that can be used to induce LTP in seven major excitatory synaptic pathways in hippocampal slices, and introduce a procedure to investigate synaptic plasticity and induce high-frequency oscillations under one experimental paradigm.

Key words: LTP, LTD, Synaptic plasticity, Learning, Theta, Gamma, Oscillation

1. Introduction

The phenomenon of LTP of synaptic transmission has been first discovered in vivo in anesthetized rabbits. In 1966, Terje Lomo first registered the potentiation of field excitatory postsynaptic potentials (fEPSPs) in granular cells in the hippocampal dentate gyrus (DG) after the application of repetitive trains of stimuli to the main input of the DG—perforant path (PP). Then, with Tim Bliss, he further investigated this phenomenon in the beginning of 1970s (1). Since then, research regarding LTP, as a leading cellular and network model of learning and memory, made a gigantic leap in deciphering the molecular and synaptic mechanisms of the encoding/decoding of memory traces (2). In vivo studies of LTP and the opposite phenomenon (i.e., LTD) widely paralleled the majority of in vitro experiments and brought ample new information about experience- and activity-dependent synaptic plasticity in freely behaving animals. Experiments using anesthetized animals (especially rats) also extended our knowledge about how the brain perceives, operates and learns in paradigms such as associative eyelid conditioning, stimulus discrimination in the barrel cortex, ocular dominance and auditory plasticity (3). During 40 years of LTP research, synaptic plasticity was found in nearly all input/output extrahippocampal, cortical, thalamic, and cerebellar excitatory synapses, including the amygdala (4–6), cerebellum (7, 8), and the prefrontal (9), entorhinal (10), visual (11), and somatosensory (12) cortices.

The phenomenon of brain oscillations is also well and long known in neuroscience and clinical studies since the discovery in the late twentieth of the last century Berger's electroencephalographic (EEG) waves registered from the surface of a human scalp (13, 14). From that time, brain oscillations recorded with the use of different invasive and noninvasive techniques and paradigms had been widely explored, and a strong link between brain waves and perception, cognition, learning and memory in humans and animals has been established (15). However, little is known about how these two ultimate brain phenomena interact during encoding, decoding, storage and retrieval of memory traces in normal and pathological conditions. The major goal of this chapter is to overview the studies, in which these two phenomena were addressed together and to outline one of options how to methodologically merge in one experimental scope investigation of synaptic plasticity and network oscillations at different synapses of mouse hippocampal formation in vitro.

1.1. Oscillations In Vivo

The mammalian brain constantly generates a variety of oscillations that differ in amplitude and frequency (from 0.05 to 500 Hz), depending on the brain state and ongoing behavior (16–18). Brain oscillations in the θ (4–12 Hz) and γ (30–100 Hz) frequency ranges are most prominent in the areas with the highest cognitive and computational loads (e.g., in the hippocampus and cortex), suggesting their potential roles in perception, learning and memory. The θ rhythm can be readily observed in all layers and regions of the excitatory trisynaptic circuitry of the hippocampus, but it is most regular in frequency and largest in amplitude in the *stratum lacunosum-moleculare (str. lac-mol.)* of the CA1 region. A θ rhythm that is smaller (by factor of 5–10) in power can also be recorded in vivo in many cortical and subcortical structures (e.g., the subicular, entorhinal, perirhinal, cingulate, and prefrontal cortices and the amygdala). In rodents, there are two types of hippocampal θ rhythms: high θ (8–12 Hz), which depends on ongoing motor activity and the engagement of the brain while exploring the environment, and low cholinergic-dependent θ (5–8 Hz), which is triggered by medial septum activation and can be fully erased by atropine or scopolamine. The highest-amplitude θ rhythm occurs during exploratory behavior and REM sleep. The firing of many cells in the hippocampus, especially principal pyramidal neurons, is phase-locked to the θ rhythm (19), which means that they mostly fire in a certain phase of the rhythm. In particular, so-called "place cells", which encode space and create a navigational map for the animal are locked into the θ rhythm. The phase of firing of place cells relative to the θ rhythm changes systematically when the animal approaches a specific location (20, 21).

An important role for the hippocampal θ rhythm in synaptic plasticity is suggested by several observations: (1) The most prominent θ rhythm occurs while exploring a new environment and generating a spatial map of this environment. (2) Spatial learning is impaired when the θ rhythm is disrupted (e.g., via a lesion in the medial septal nucleus) (22). (3) The excitation of hippocampal principal cells is phase-coupled to the θ waves. (4) Trains of stimuli that are delivered to mimic the θ pattern induce LTP more readily than stimulation at other frequencies.

Several recent studies strengthen a link between oscillations and plasticity. High-frequency stimulation (HFS, 200 Hz, 15 stimuli, 10 bursts, 10-s interburst interval, with an intensity of 40% from the maximal response) of the medial PP induces an increase in the θ power and facilitates γ oscillations in the DG. These changes occurred in rats that could develop LTP or short-term potentiation (STP) but not in a group of animals that failed to show potentiation. The increased γ and θ rhythms in the DG of freely moving rats can persist up to 24 h after tetanization (23, 24). The delivery of HFS (3 bursts, 5 pulses of 400 Hz in each burst) to the CA3–CA1 synapses of freely moving rats was also effective for inducing long-lasting (1 h) LTP (if tetanization of CA3 axons was applied at the θ peaks recorded extracellularly in the CA1 stratum radiatum) or LTD (if it was applied at the θ troughs). These experiments clearly demonstrated that LTD can be induced in the hippocampus by a physiological pattern of stimulation and without using artificial long trains of low-frequency stimulation (LFS) (25).

The same effect was observed in urethane-anesthetized rats when the θ rhythm was generated by sensory stimulation (26). Sensory stimulation can induce θ-like oscillations in anesthetized animals. For instance, a tail-pinch can be used to trigger the θ rhythm. The application of brief bursts of stimuli (5 pulses, 200 Hz) during different phases of a tail-pinch-evoked θ rhythm can mimic naturally occurring neuronal activity much more closely than either standard HFS or LFS. A single burst of 5 pulses at 200 Hz delivered to a θ peak can evoke 20% potentiation of CA1 synapses that is stable for 60 min. Notably, LTP can be gradually increased up to 50–60% from baseline if three bursts are applied. Using a similar paradigm, LTP can be evoked in other synapses (e.g., PP-DG granule cell layer synapses), if the tetanus is delivered to a positive peak of θ in urethane-anesthetized rats (27, 28). Such bidirectional synaptic modifications induced by the pairing of θ waves and HFS can persist up to 48 h in freely moving rats (29).

In another interesting work, the firing patterns of CA1 place cells were analyzed during REM sleep (30), which is known to be vital for memory consolidation. When rats experienced a familiar environment, the place cells tended to fire in troughs of θ waves during the subsequent REM sleep, which is consistent with the induction of LTD. However, after the rats were exposed to a novel maze, place cell firing shifted to the peaks of the θ oscillations, which is consistent with the induction of LTP. Such a transition in the θ phase from a novel to an explored context is built up gradually over several days, as rats became familiar with the novel environment. It was concluded that, perhaps, the remote memory traces that are encoded in a familiar place are erased during subsequent REM sleep, whereas the traces of recent experiences that are acquired while facing a novel environment are strengthened.

1.2. Oscillations In Vitro

Oscillating network activity can also be induced and manipulated pharmacologically in brain slices. For example, robust θ oscillations can be induced by perfusion with cholinergic agonists, such as carbachol. A series of studies has shown that only one burst (4 pulses, 100 Hz) of Schaffer collateral/commissural pathway stimulation administered during the peak of carbachol-induced oscillations resulted in the LTP of CA1 fEPSPs, and one burst presented at the trough depotentiated these synapses (31, 32).

Cholinergic activation can induce higher-frequency oscillations in addition to the θ rhythm. For example, temporally stable (up to 1 h) and very coherent network activity at γ frequencies (about 30–35 Hz) can be induced at the CA3 *stratum pyramidale* and *radiatum (str. pyr., str. rad.)* by the bath application of carbachol to rat hippocampal slices (33). The γ oscillations in these two layers are phase-reversed and alternate periodically, similar to those reported in vivo (34). During γ oscillations, pyramidal cells fire at the negative peak of the oscillations, which is followed by the discharge of interneurons that form a recurrent loop with pyramidal cells. This gives rise to the precise timing of firing of pyramidal cell assemblies at γ frequencies. Fast-spiking perisomatic inhibitory interneurons are the most active cell type during γ oscillations that have a strong phase coupling to the γ rhythm, whereas the majority of interneurons that target the dendritic regions of pyramidal cells do not exhibit significant phase-coupled firing.

In the CA1 and DG regions of the hippocampus, large-amplitude β (14–30 Hz) and γ oscillations can be elicited by the application of glutamate or by the tetanic stimulation of afferents (35, 36). These and other studies indicate that the role of oscillations in synaptic plasticity in vitro can be addressed by several different paradigms. First, a pharmacological approach can be utilized; synchronous oscillations can be induced via application of different activators of neuronal activity, such as agonists of cholinergic muscarinic, glutamate metabotropic or kainate receptors, or via changing the ionic composition of the extracellular solution (e.g., by decreasing Ca^{2+} and increasing K^{+} concentrations). Second, electrical stimulation (tetanization) of presynaptic terminals with different stimulation protocols can be used to evoke short and locked to stimulus-onset oscillations. Such tetanization can be applied to different neuronal pathways within one brain structure or even to distinct structures (e.g., hippocampus and amygdala). Thus, the role of long-range synchronization in synaptic plasticity can be studied. To facilitate the induction of oscillations, a combination of the aforementioned methods can be used.

It is well known that in contrast to in vivo recordings, spontaneous oscillations are rarely detected in hippocampal slices under basal conditions. However, prominent hippocampal atropine-resistant θ rhythm (3–10 Hz) is readily detectable in the whole hippocampus in vitro recording (37), where majority of its intrinsic pathways are preserved. In this elegant study, authors demonstrated existence of multiple self-sustained cholinergic-, medial septum-, and CA3-independent theta oscillators located along the longitudinal axis of CA1 area of the hippocampus. Their θ oscillations could be blocked by antagonists of $GABA_A$ or AMPA/kainate receptors, but not with muscarinic acetylcholine receptor antagonists. By contrast, carbachol-induced θ activity in slices does not require involvement of inhibition through $GABA_A$ receptors, and CA3 neurons take part in θ generation. Thus, multiple network and molecular mechanisms are responsible for generating θ oscillations in slice and whole hippocampus preparations.

1.3. Synaptic Plasticity and Oscillations

Although the majority of studies in the synaptic plasticity field were conducted using an artificial forced synchronized activation of afferents, a few available publications suggest an important link between plasticity and naturally occurring or induced spontaneous oscillations. Oscillations at the θ and γ frequencies may provide temporal windows in which distinct neuronal populations can fire synchronously, allowing temporally functional assemblies to be formed to encode a common feature of the environment or a task (38). Such temporal segregation via oscillations, in turn, favors synaptic plasticity, perception, learning, and memory. A recent study conducted by Moser's group strengthens this idea of routing flow of information by oscillations (39) and supports the view that two different frequencies of γ oscillations, slow (25–50 Hz) and fast (65–140 Hz), are responsible for different functions in the hippocampus. Fast γ synchronizes CA1 and medial entorhinal cortex during encoding spatial information, whereas slow CA1 γ is locked to CA3 and supports retrieval of stored memory. These two γ oscillations predominantly occur at different phases of θ cycle and mostly on different θ cycles in CA1. Moreover, fast γ power is maximal at the θ phase when place cells are most likely to fire, and slow γ occurred primarily at the θ phase when CA1 place cells fire with low probability, due to CA3 feedforward inhibition of CA1 area. These results provide a mechanism for temporal segregation of information transmitted to CA1 from the entorhinal cortex and CA3.

Synaptic weights can be modulated by the θ rhythm and higher-frequency oscillations by the "peak-trough" rule. For example, in the rat visual cortex, the synapses of the pyramidal cells in layer II exhibited LTP when the EPSPs coincided with the peaks of the oscillations in the β and γ frequency bands (20 and 40 Hz, respectively), but they exhibited LTD if the EPSPs coincided with the troughs. These results suggest that, when neuronal networks engage in high-frequency oscillations, synaptic plasticity remains exquisitely sensitive to the timing of the discharges. The precise synchronization of discharges serves as a signature of relatedness during distributed processing. Thus, the temporal relations among the discharges of interconnected neurons are important variables in synaptic plasticity. Synapses undergo LTP when the EPSP precedes the postsynaptic spike by within a few tens of milliseconds, and they undergo LTD when the EPSP follows the spike, exhibiting so-called "spike time-dependent plasticity" (40).

1.4. Future Directions for Research

In following methodological sections, we first describe protocols for the induction of LTP by theta-burst stimulation (TBS) or HFS in several subtypes of hippocampal synapses, which have distinct mechanisms of synaptic plasticity. For example, LTP in CA3-CA1 synapses depends on signaling via NMDA-Rs, whereas LTP is NMDA-R independent in mossy fiber-CA3 synapses. To understand the role of a particular molecule in synaptic plasticity, it is tempting to investigate how manipulating the expression of that molecule may affect plasticity in distinct types of synapses (41) or to use various paradigms for the induction of synaptic plasticity, which involve distinct induction mechanisms (42). Owing to specific roles played by different hippocampal synapses in cognitive functions (3), it is attractive to analyze all of the major excitatory synaptic connections in the hippocampal circuitry in combination with a behavioral analysis of cognition. This approach can help to merge understanding of synaptic and cognitive functions.

Second, we introduce an HFS protocol (20 pulses at 100 Hz at a high stimulation intensity) that induces high-frequency oscillations in the pyramidal layer, which are followed by the LTP of CA3–CA1 synapses. This protocol mimics a commonly found situation in which a short sensory stimulus is followed by induced oscillations. Importantly, when 20 pulses are applied at 100 Hz but with a stimulation intensity that does not induce oscillations, no LTP is induced. Furthermore, mice that are deficient in anion exchanger 3, exhibit more prominent oscillations and increased LTP following these oscillations, as compared to wild-types (Kochlamazashvili and Dityatev, unpublished data). These data suggest that a pairing of synaptic activity with neuronal oscillations, even with the considerable delay of a 100 ms, could contribute to the enforcement of the stimulated synapses. Possible mechanisms may involve the back-propagation of the spikes that are generated during these oscillations to the synapses that are located on dendrites of pyramidal cells. These spikes can activate postsynaptic voltage-gated L-type Ca^{2+} channels or relieve the Mg^{2+} block of NMDA-Rs bound to glutamate that was released during HFS. Further work is needed to elucidate these mechanisms. Furthermore, by adding another electrode to stimulate independent afferent pathways during induced oscillations, this paradigm could be extended to investigate the mechanisms of oscillation- and spike time-dependent plasticity.

2. Materials

2.1. Materials for Slice Preparation

1. A vessel connected by tubing to a regulated CO_2 supply (for mouse sedation).
2. A vibratome (e.g., VT 1200S, Leica).
3. Razor blades (a new blade is used for each slice preparation).
4. A mounted-line gas mixture (95% O_2/5% CO_2) with a regulator.
5. A diffuser (aquarium air stone, with tubing of an appropriate diameter to connect the regulator and the air stone).
6. Agar blocks and solution (agarose from Sigma Aldrich).
7. A water bath (e.g., TW 12, Julabo).
8. A plastic chamber for incubating the brain slices.
9. Rapid glue (Vetbond).
10. A small weighing spatula, scalpel, forceps, large scissors, small sharp-nosed dissecting scissors and a soft small paintbrush for transferring the slices (FST).

2.2. Materials for fEPSP Recording in Hippocampal Slices

1. Three micromanipulators (Scientifica).
2. An upright microscope (BX61WI Olympus).
3. An amplifier (EPC10 HEKA).
4. A stimulus isolator unit (A360 World Precision Instruments).
5. A workstation (Dell).
6. A recording chamber (Warner Instruments).

7. A perfusion system, consisting of a 500-ml plastic container with an aquarium stone for babbling the artificial cerebrospinal fluid (ACSF) and polyethylene tubing for circulating the ACSF under the control of a peristaltic pump (Gilson).
8. A mounting-line heater to control the temperature of the ACSF (Ala Scientific Instruments).
9. A homemade platform for the slices, consisting of a nylon mesh glued onto a metal half-circle to perfuse the slices with ACSF from both sides more effectively.
10. Small pieces of nylon mesh (7 × 10 mm, Millipore) with a hole (2 mm in diameter) in the middle.
11. Two small metal loads to fix the mesh onto the recording platform.
12. Glass pipettes for extracellular recordings (Hilgenberg).
13. A glass microelectrode puller (Sutter Instruments).

3. Methods

Here, we describe the methods that are used for the extracellular recordings of synaptic plasticity and oscillations in vitro. These include the preparation of the hippocampal slices, the recording and analysis of fEPSPs, the induction of short- and long-term synaptic plasticity in different synaptic connections in the hippocampus and the induction of oscillatory activity in the CA1 region using a short tetanic stimulation.

Importantly, the hippocampal slices remain viable for several hours, and stable recordings can be carried out at visually controlled recording sites. Bath application of the determined concentrations of the pharmacological agents, which can be easily washed in and out, is an exceptionally useful advantage of using in vitro methods for pharmacological studies. The drugs can also be applied using visually guided micropipettes to discrete neuronal regions that consist of well-defined hippocampal layers and subareas. In such preparations, the cell bodies of the pyramidal neurons lie in a single densely packed layer that is clearly apparent. This layer is divided into several distinct regions, including CA1, CA2, and CA3. The dendrites of the pyramidal cells in the CA1 region form a thick band (*str. rad.*), where they receive synapses from the axons of the pyramidal cells located in the CA3 region. The CA3–CA1 synapse is the most widely studied synapse in the hippocampus due to its simple identification. Therefore, here we concentrate on protocols to study this synapse and only briefly describe the protocols for recording LTP in the other major excitatory hippocampal synapses.

3.1. Methods for Slice Preparation

Hippocampal slices for synaptic plasticity experiments can be prepared from juvenile, adult or old mice. We could reliably record LTP from 10-day- to 2-year-old mice (see Notes, Sect. 1) A mouse is placed in a closed transparent plastic vessel, into which CO_2 is slowly released via regulated mounted-line tubing until the mouse is sedated. Next, the mouse is quickly decapitated using large scissors, and the head is placed in a Petri dish filled with ice-cold oxygenated (with a mixture of 95% O_2 and 5% CO_2) dissection artificial cerebrospinal fluid (dACSF), containing the following substances (in mM): sucrose 250, $NaHCO_3$ 24, glucose 25, KCl 2.5, NaH_2PO_4 1.25, $CaCl_2$ 2, $MgSO_4$ 1.5, pH 7.35–7.4. The mouse head is fixed using forceps inserted into the eye-sockets; the skin is then cut from the skull and removed using small sharp-nose scissors. The bones are cut up to the parietal bones using small scissors, with one sharp end inserted underneath the skull from the spinal cord side. With small bent forceps, the occipital, interparietal, parietal, and frontal bones are broken and removed from the brain surface. The brain is then quickly removed using a small spatula and placed into another Petri dish with the same ice-cold oxygenated dACSF. It is important to minimize time that is required for this step (to 1–1.5 min) to avoid the ischemic death of the brain tissue. The cerebellum is then dissected away, and the brain is cut along the midline into two equal hemispheres using a sharp scalpel. One hemisphere is immediately glued (using a rapid glue) onto the object-mounting platform with a cutting surface (for recordings in the CA1 field) adjacent to a 1-cm^3 cubic 20%-agar block, which is glued to the mounting platform prior to brain dissection and is used as a support for the brain tissue during slicing. The brain hemisphere is covered with 10% warm (kept in a water bath) liquid agar (to embed the brain tissue and fill the gap between it and the agar block) and placed in a vibratome buffer tray filled with ice-cold, continuously oxygenated dACSF. The other hemisphere is placed into the same buffer tray to be sliced later. For CA3 recordings, the ventral surface of each hemisphere is cut at the angle of 20°, and this surface is glued to the platform (43). For recordings in other synapses, either sagittal or horizontal slicing is used, as indicated below. The cerebral hemisphere that is attached to the platform is then cut into slices using the vibratome (see Notes, Sect. 2). The agar, which is used to cover the brain tissue, protects the brain from deformation during cutting and from the unwanted release of the brain hemisphere from the platform. The slices are separated from their agar surroundings using a small soft paintbrush and transferred to the slice incubation chamber using a small spatula. The slices are maintained in ACSF (120 mM NaCl instead of 250 mM sucrose in dACSF) for 1.5–2 h before their use in the incubation chamber (see Notes, Sect. 3). The ACSF should be previously bubbled for at least 30 min at room temperature before slice transfer.

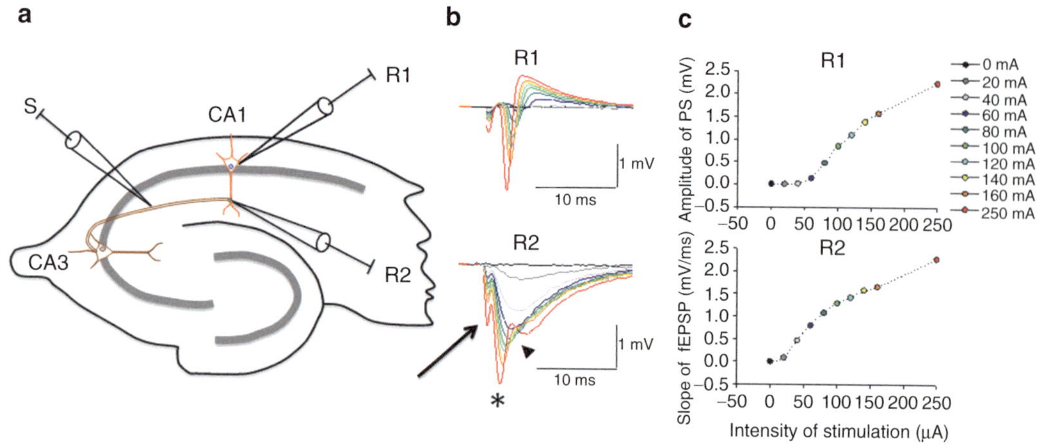

Fig. 1. Design of extracellular recordings in the CA3–CA1 connections. (**a**) Schematic illustration of a hippocampal slice and the positioning of electrodes for the recording of fEPSPs and population spikes (PSs) in the CA1 region. S, stimulating electrode; R1, recording electrode in the *stratum pyramidale*; R2, recording electrode in the *stratum radiatum*. (**b**) Traces representing responses in R1 and R2 evoked by stimulation with increasing stimulus strengths (from 0 to 160 μA with 20-μA steps and a stimulation intensity of 250 μA). The *upper panel* shows the appearance of PSs with increasing stimulation intensities at electrode R1. The *lower panel* shows recordings of the fEPSPs (a peak indicated by *asterisk*) from *stratum radiatum*. The early component preceding fEPSP is a presynaptic volley (indicated by *arrow*). At 160 μA (but not at lower intensities), the "supramaximal" fEPSP is evoked in the *striatum radiatum*, which is superimposed with a PS (indicated by *arrowhead*). A single supramaximal stimulation at 250 μA was given to show the responses at the stimulation strength that was most effective for inducing oscillations (see Fig. 4). At 250 μA, both the fEPSP and PS are clearly distinguishable. (**c**) Stimulus intensity-response curves for the PS amplitude (*upper panel*) and the slope of the fEPSPs (*lower curve*). The color code represents measurements at different stimulation intensities.

3.2. Methods for fEPSP Recording In Vitro

For fEPSP recording in hippocampal slices, we used the procedures described below. A hippocampal slice is transferred into the recording chamber (see Notes, Sect. 4) using a paintbrush and spatula. Good-quality slices can be recognized under a light microscope at 4× magnification by clearly visible pyramidal and granule cell layers (which are lighter than the surrounding tissue). The recording and stimulating glass pipettes are prepared in advance with previously determined shapes and filled with ACSF. The electrodes for the extracellular recording of the fEPSPs usually have resistances of 2 M Ω, whereas the stimulation electrode has a resistance less of than 1 M Ω. We used one stimulation electrode (S) and two recording electrodes (R1 and R2, Fig. 1). R1 is used for the recording of population spikes (PSs) in the *str. pyr.*, and R2 is used for recording fESPSs in the *str. rad.* (Fig. 1). All three electrodes are gently placed onto the surface of the slice in a configuration such that the recording electrodes are about 200–250 μm apart from the stimulating electrode, which is located in the *str. rad.* (Fig. 1a). It is important to first place the stimulating electrode slightly away from the pyramidal cell layer and then slowly deepen it within the tissue. Recording electrode R2 is placed in the *str. rad.* at about the same depth and distance from the pyramidal cell layer (Fig. 1a). Recording electrode R1 is located in the pyramidal cell layer at about the same distance from the stimulating electrode (Fig. 1a). Stimulation (50–70 μA, 0.2 ms) is applied to CA3 axons every 20 s, and the recording and stimulating electrodes are slowly advanced into the slice until the maximal fEPSP amplitude is obtained.

After the baseline recording is stable for at least 10 min, a stimulus–response curve is generated to analyze the efficacy of basal synaptic transmission. Following progressively increasing stimulation strengths, the synaptic responses rise until a PS appears in the decaying phase of the fEPSP (a "supramaximal" response). The experiment continues when this response satisfies several criteria: (1) The amplitude of the supramaximal fEPSP should not be less than 1.5 mV. (2) The amplitude of the presynaptic fiber volley should be at least 2.5-fold less than the amplitude of the fEPSPs. (3) The submaximal fEPSPs should not be contaminated by other components. In Fig. 1, we show two recording sites and the fEPSPs that were recorded at the *str. pyr.* and *str. rad.* (Fig. 1a).

3.3. Induction of LTP In Vitro

LTP is a persistent increase in synaptic responses that is commonly induced by HFS or TBS being applied to presynaptic fibers. For LTP recordings, the following experimental design is used:

1. We recorded a stable baseline and determined the stimulation intensity at which a supramaximal response is elicited (Fig. 1).
2. With a stimulation intensity generating fEPSPs with 50% of the slope of the supramaximal fEPSPs, we recorded a baseline every 20 s for at least 10 min. During the baseline recordings, several aspects should be taken into consideration (see Notes, Sect. 4).
3. After a stable baseline of the appropriate duration is recorded, LTP is induced (Fig. 2). An efficient protocol is applying four trains of TBS every 20 s. Each TBS consists of 8 bursts delivered at 5 Hz, and each burst consists of 4 pulses at 100 Hz. However, different numbers of TBS and HFS episodes and pharmacological manipulations are used to induce different forms of LTP in various hippocampal synapses (Table 1). These protocols share similar features with synaptic associative memory mechanisms, such as rapid induction, synapse specificity, associative interactions, persistence, and a dependence on correlated synaptic activity.
4. TBS induces a drastic potentiation of the fEPSP magnitude. The initial phase of potentiation lasts about 2 min and is called short-term potentiation (STP). It is followed by early LTP (here called LTP for simplicity) that reaches its maximum within about 10–20 min after TBS (Fig. 2b) and may persist for several hours. To measure LTP, we use the average slope of the fEPSPs that are recorded 50–60 min after TBS and express it as a

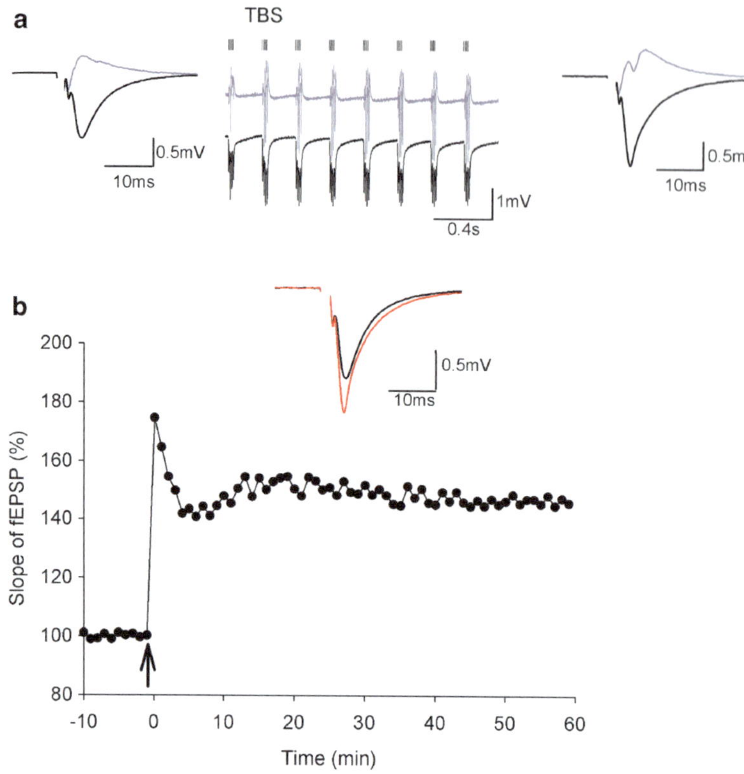

Fig. 2. LTP recordings in the CA3–CA1 connection in the hippocampus. (**a**) Representative traces of fEPSPs recorded in *stratum pyramidale* (*gray*) and *stratum radiatum* (*black*) are shown on the *left panel*). The average of 10 min of baseline (30 sweeps) is presented. Representative traces of TBS applied to induce LTP are shown in the *middle panel*). Four such TBSs were applied every 20 s. (Only the first TBS is presented in the graph). Recordings in the *stratum pyramidale* (*upper gray*) reveal the appearance of pop spikes during TBS, which were not detectable in the *stratum radiatum*. Recordings from the *stratum radiatum* during TBS reveal the facilitation of the fEPSPs. Responses averaged 50–60 min after LTP induction are presented in the *right panel*. The fEPSPs and PSs show significant increases; the appearance of the PS in the *stratum pyramidale* and an increase in the amplitude of the fEPSPs in the *stratum radiatum*. (**b**) A profile of TBS-induced LTP. (*Upper*) An overlay of the representative fEPSPs before (*black trace*) and after (*red trace*) the induction of LTP, as shown individually in *panel* (**a**). (*Lower*) Measurements of the slope of the fEPSP before, during and after LTP induction. Stimulation was delivered every 20 s. Significant STP (1–2 min) and LTP (50–60 min) of the fEPSPs was detected after TBS (indicated by *arrow*). The mean slope of the fEPSPs recorded 0–10 min before TBS is defined as 100%.

percentage (where 10 min of baseline recording is 100%). It is important to control the amplitude of the presynaptic volleys (a small, short-lasting component that precedes the fEPSP) after LTP induction. This is a good measure of the number of stimulated axons, and its stability indicates that the same number of presynaptic fibers is recruited before and after LTP induction.

Table 1
The protocols used for the induction of LTP in different excitatory hippocampal synapses

Axonal projections	Postsynaptic area	Stimulation intensity	Pattern of stimulation	Pharmacology
PP	DG	100%	5×HFS	Picrotoxin
	CA1, CA3	100%	2×HFS	Picrotoxin
MF	CA3	Sufficient to elicit 40- to 60-μV responses	2×HFS	AP5
CA3	CA1	50%	2×HFS	No
	CA3	50%	2×HFS	No

PP perforant path; *MF* mossy fibers; *DG* dentate gyrus; *CA1*, *CA3* areas of the hippocampus (41)

3.4. Induction of LTP by HFS in Different Synaptic Pathways in the Hippocampus

As summarized in the Table 1 and Fig. 3, LTP can be reliably induced in seven major excitatory connections in hippocampal slices using several trains of HFS. Each train includes 100 pulses delivered at 100 Hz. The intertrain interval is 20 s, and the pulse duration is 0.1–0.2 ms.

1. To stimulate the associational/commissural projections and record the CA3–CA3 fEPSPs, one can place the stimulating electrode either in the *str. rad.* of the CA3 field and thus activate the associational fibers orthodromically, or one can place the electrode in the *str. rad.* of the CA1 field and activate the associational projections antidromically. We suggest using the latter approach to avoid the direct stimulation of the local CA3 interneurons. Because the CA3–CA3 synapses are located in the *str. rad.* of the CA3 field, the recording electrode is placed in this area. LTP is induced by two trains of HFS, with the stimulation intensity providing the fEPSPs with a magnitude of 50% of the supramaximal response.
2. To stimulate the Schaffer collateral/commisural projections and record the CA3–CA1 fEPSPs, the stimulating and recording electrodes are placed in the *str. rad.* of the CA1 region. For LTP induction, two trains of HFS are applied, with the stimulation intensity providing the fEPSPs with a magnitude of 50% of the supramaximal response (44, 45).
3. To record the direct perforant path-CA1 responses, we place the stimulating and recording electrodes in the *str. lac-mol.* of the CA1 field. The stimulating electrode is placed close to the subiculum. Because there is a high probability of activating the perforant path fibers that project to the DG (which may cause

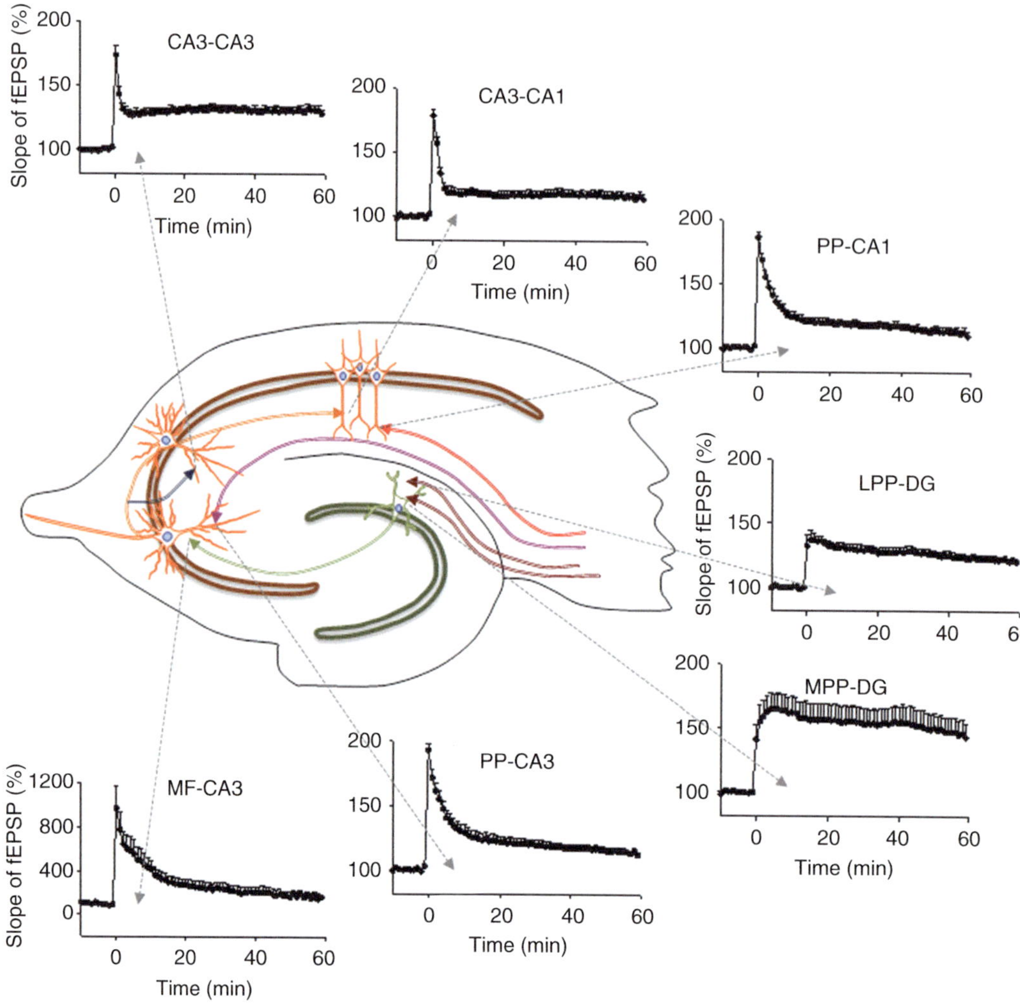

Fig. 3. A schematic illustration of the seven major excitatory synaptic connections and forms of LTP in the hippocampus. The synaptic connections from CA3 to CA3, from CA3 to CA1, from the dentate gyrus to CA3 and from the perforant path to all these regions are shown in a color-coded manner. The short- and long-term potentiation profiles are shown that were recorded in seven major excitatory synapses in the hippocampus of wild-type mice with a C57BL6 background (for details, see ref. (41)). The data represent mean + SEM. The mean slope of the fEPSPs recorded 0–10 min before HFS is taken as 100%. For all recordings, the number of slices per LTP trace was on average 8 and the number of animals was 4. Reproduced with modifications and permission from ref. (41).

the polysynaptic activation of CA1 neurons), we cut the slices using a glass pipette through the DG and along the border between the CA1 and CA3 regions (46). LTP is induced by two trains of HFS delivered with a stimulation intensity that elicits supramaximal responses in disinhibited slices. For disinhibition, 100 μM picrotoxin is applied to the slices 10 min before and during HFS.

4. To record the lateral perforant path (LPP)-DG fEPSPs, the stimulating and recording electrodes are placed in the outer

molecular layer of the DG. The responses are identified by the application of paired-pulse stimulation with 50-ms interstimulus intervals, and only the fEPSPs with paired-pulse facilitation are analyzed. LTP is induced by five trains of HFS using a supramaximal stimulation intensity in disinhibited slices (47). Disinhibition is required to induce LTP in mature perforant path synapses (48) and is achieved by the application of 100 μM picrotoxin 10 min before and during HFS.

5. To record the medial perforant path (MPP)-DG fEPSPs, the stimulating and recording electrodes are placed in the inner molecular layer, and only the responses with paired-pulse depression are recorded further. LTP is induced as in the LPP-DG connections.

6. To record the direct perforant path-CA3 responses, a stimulating electrode is placed in the *str. lac-mol.* of the CA1 field near the subiculum, and a recording electrode is placed in the *str. lac-mol.* of the CA3 field. To avoid the disynaptic activation of the CA3 pyramidal neurons, the slices are cut with a glass electrode through the DG (49). LTP is induced by two trains of HFS delivered with a supramaximal stimulation intensity. The slices are disinhibited with 100 μM picrotoxin.

7. To record the mossy fiber-CA3 fEPSPs, a stimulating electrode is placed in the DG near the internal side of granule cell layer, and a recording electrode is placed in the *stratum lucidum* (*str. luc.*) of the CA3 field. To avoid polysynaptic responses, a low stimulation intensity is used to elicit 40- to 60-μV fEPSPs. The mossy fiber responses are identified by their short duration and a frequency-dependent facilitation of >200% that is elicited in response to a 0.33-Hz stimulation (50). Two trains of HFS are used for the induction of LTP (44). LTP at this synapse is known to be NMDA receptor-independent; therefore, 50 μM AP5 is applied 15 min before and during HFS. To confirm that the recorded field EPSPs are evoked by the stimulation of the mossy fibers and are not contaminated by the associational/commissural pathway, a metabotropic glutamate receptor agonist (2 μM DCGIV) is applied at the end of each experiment, and only the responses that are inhibited by >80% are assumed to be elicited by mossy fiber stimulation (51).

3.5. Induction of Oscillations by High-Frequency Stimulation

The recordings are performed in a submerged chamber at 30°C. Schaffer collateral commissural fibers are stimulated in the *str. rad.* with a bipolar tungsten stimulation electrode (WPI) or a glass pipette. After stable baseline recordings, the threshold of oscillation generation and their amplitude and duration are analyzed. Short trains of HFS are applied every 5 min with 50, 100, 150, 200, 250, 300, 350 and 400 μA stimulation intensities. Short HFS contains 20 pulses that

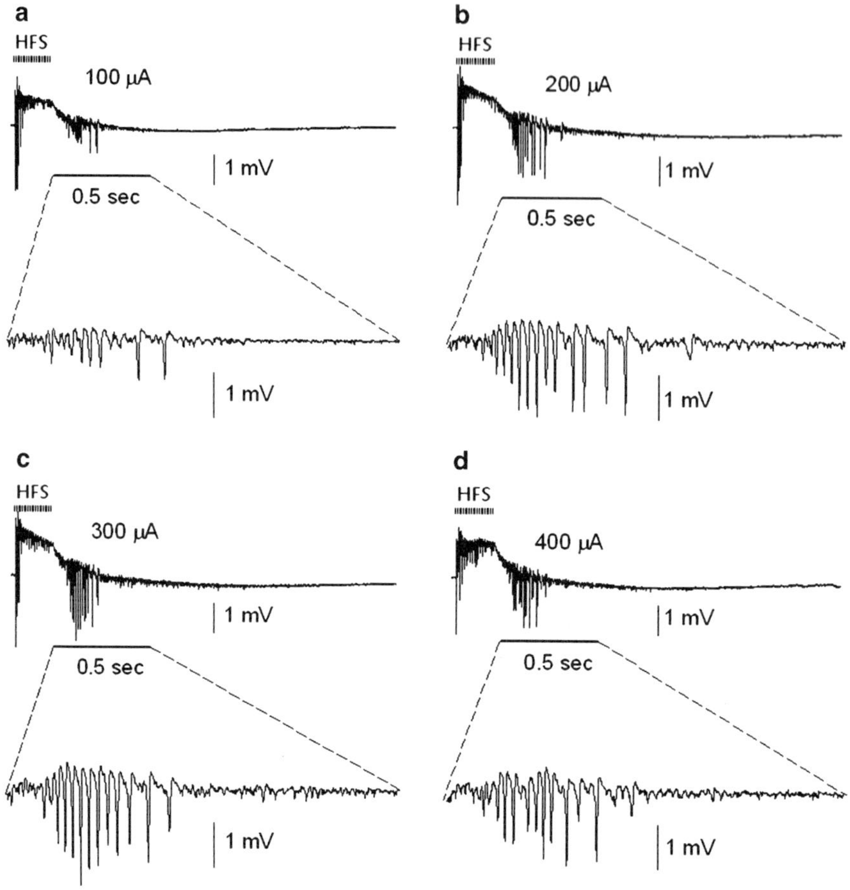

Fig. 4. HFS-induced high-frequency oscillations of neuronal activity. Representative traces of HFS-induced oscillatory activity recorded at the *stratum pyramidale* with different stimulation intensities (100, 200, 300, and 400 μA in **a**, **b**, **c**, and **d**, respectively). Each HFS contained 20 pulses at 100 Hz and was delivered in a 5-min interval. During HFS (*upper*), PSs are inhibited. Within about 0.2–0.3 s after HFS, induced oscillatory activity is generated. (Traces with a higher resolution after the subtraction of slow potentials are shown in the *lower panels*.) Oscillations are most prominent at stimulation intensities of 200–300 μA.

are applied at 100 Hz. The induced "ripple-like" oscillations are recorded within 2 s after the application of HFS. The induced oscillatory activity in the *str. pyr.* is detectable starting at 100 μA and is maximal at 200–300 μA (Fig. 4). After the initial positive-going component, a slow negative-going component is detected in the *str. pyr.*, when spontaneous high-frequency oscillatory activity is generated.

3.6. Induction of LTP by a Protocol Eliciting High-Frequency Oscillations

The protocol used for the induction of oscillations results in an induction of LTP in the stimulated pathway (Fig. 5). During baseline recording, the stimulation intensity is set to elicit fEPSPs with 50% of the supramaximal intensity. During an HFS of 20 pulses at 100 Hz, the stimulation intensity is fivefold higher than

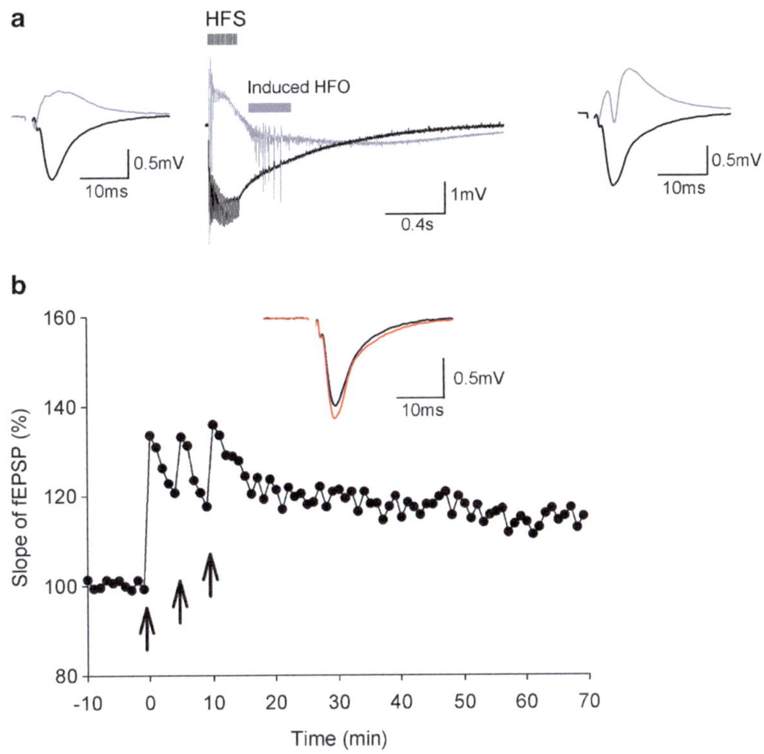

Fig. 5. HFS-induced oscillations and associated LTP. (**a**) Representative fEPSPs before (*left*) and after the induction of LTP (*right*), which were recorded in the *stratum pyramidale (gray)* and the *stratum radiatum (black),* are presented as in Fig. 2. To induce LTP, three trains of HFS were applied with 5-min intervals. Each HFS contained 20 pulses at 100 Hz. Baseline responses were monitored every 20 s. The stimulation intensity during HFS was set to be five times higher than during baseline recording. The HFS applied at this stimulation strength induced oscillatory activity. (**b**) A profile of HFS-induced LTP. (*Upper panel*) Superimposed representative fEPSPs before (*black trace*) and after the induction (*red trace*) of LTP that were shown individually in (**a**). (*Lower panel*) Measurement of fEPSP slopes before and after LTP induction. Three trains of HFS were delivered every 5 min (indicated by *arrows*). Significant STP (1–2 min) and LTP (50–60 min) of the fEPSPs is detected after TBS. The mean slope of the fEPSPs recorded 0–10 min before TBS is defined as 100%.

during baseline recording (because such a stimulation intensity is the most effective for inducing oscillations). After HFS, the stimulation intensity is reset to the intensity that has been used for the baseline recordings. To induce LTP, three trains of HFS are delivered at 5-min intervals. The recording of fEPSPs for 1 h reveals that robust LTP can be induced by this protocol (Fig. 5).

4. Notes

4.1. Mice for LTP Recordings

LTP can be easily induced in slices derived from mice of different ages. However, the age should be within a well-defined interval, because some forms of LTP exhibit age-dependence, and there can be a gene-dependent interaction between age and synaptic plasticity in genetically modified mice (44). Two ages are particularly attractive: 3 weeks (if the extracellular recordings will be complemented with a patch-clamp study) and 3 months (if the electrophysiological recordings will be complemented by a behavioral analysis of cognitive function). The level of LTP significantly depends on the genetic background of the mouse strain that is used in the experiments. Therefore, the use of wild-type littermate controls is highly recommended, particularly if the background of the genetically modified mice is mixed or the mice were not inbred for more than six generations. If the mice originate from different parents, it is important that they are raised under identical conditions. If the animal facility is located in a building that is different from where the experiments are performed, the mice should be brought into the laboratory 1 day before the start of experiment and kept in a quiet place to minimize the possible effect of stress on synaptic plasticity. Acute or chronic stresses can easily alter LTP, and therefore, investigators should minimize their effects as much as possible before slice preparation.

4.2. Brain Slicing

The blade should be cleaned in acetone for several minutes and then washed in water. The blade should be fixed securely and adjusted in the blade holder to ensure that the blade is at the optimal angle relative to the brain tissue. The age of the animals plays an important role in determining the thickness of the slices being prepared. We used 350-μm-thick hippocampal slices for 2- to 3-month-old mice. For animals younger than 1 month, 400 μm-thick slices are used. For animals older than 3 months (up to 24 months), 300-μm-thick slices are used.

4.3. Construction and Use of the Incubation Chamber

A chamber for the incubation of brain slices can be made from a plastic vessel (10 cm in diameter and height), inside which a plastic cylinder (5 cm in diameter, 3 cm in height) is glued in the middle of the wall. The slices are placed at the bottom of the cylinder, which can be made of nylon stockings glued to the edges of the cylinder. The vessel is filled with ACSF to the top of the cylinder. The diffuser (connected to oxygenation tubing) is placed at the bottom of the vessel (outside of cylinder), and the gas pressure is adjusted to saturate the solution but avoid any movement of the floating slices. At the end of each slice preparation, all of the instruments and the vibratome should be washed and dried to avoid agar soiling.

4.4. Slice Handling in the Recording Chamber

The slice is placed on the recording platform and covered with a nylon mesh. The mesh is placed on top of slice such that a hole cut in the mesh is located above the hippocampus. The mesh is fixed with two small metal loads to prevent the slice from floating. The recording chamber is continuously perfused (3 ml/min) with freshly prepared ACSF via a tubing system that connects a reservoir containing ACSF to the recording chamber (using the force of gravity). A configuration, when a hippocampus is supplied with ACSF from both sides, which is believed to be more efficient for keeping the hippocampal slice viable during long-lasting recordings than the traditional method, in which ACSF is mostly supplied from the top. Before entering the recording chamber, the ACSF is warmed by a mounted-line heater, which keeps the fluid at a constant 30°C temperature. (Even a slight fluctuation in temperature can significantly affect the fEPSP.) The ACSF is drawn from the recording chamber with the peristaltic pump in a timely manner and delivered back into plastic reservoir, thereby achieving circulated perfusion. The ACSF is oxygenated with 95% O_2 and 5% CO_2 for 30 min prior to all of the recordings and continuously during all of the recordings.

4.5. Mastering Baseline Recording

The stability of baseline fEPSPs is "the must" for LTP recording and its subsequent analysis. There should be no tendency for fEPSPs to run up or down. Otherwise, one should wait until the fEPSP responses have stabilized and then remeasure the stimulus–response curve. For the stability of the fEPSP recordings, it is important to have a constant ACSF flow rate, oxygenation and temperature during the recording of one slice and for different slices and animals. The time for the baseline recordings depends on the time-course of the LTP recordings. If early LTP is studied for not more than 1 h, it is sufficient to record the baseline for 10–20 min. In case of late LTP recordings (recordings for 4–8 h after the induction of LTP), a stable baseline should be monitored from 30 min to 1 h.

Acknowledgments

This work was supported by Italian Institute of Technology and by grants from the Deutsche Forschungsgemeinschaft (DI 702/6-1 to A.D.) and the San Paolo "Programma in Neuroscienze" (to A.D.).

References

1. Bliss TV, Lomo T (1973) Long-lasting potentiation of synaptic transmission in the dentate area of the anaesthetized rabbit following stimulation of the perforant path. J Physiol 232 (2):331–356
2. Lynch MA (2004) Long-term potentiation and memory. Physiol Rev 84(1):87–136
3. Andersen P, Morris R, Amaral D et al (2007) The hippocampus book. Oxford University Press, New York
4. Chapman PF, Kairiss EW, Keenan CL et al (1990) Long-term synaptic potentiation in the amygdala. Synapse 6(3):271–278
5. Rogan MT, Stäubli UV, LeDoux JE (1997) Fear conditioning induces associative long-term potentiation in the amygdala. Nature 390(6660):604–607
6. Dityatev AE, Bolshakov VY (2005) Amygdala, long-term potentiation, and fear conditioning. Neuroscientist 11(1):75–88
7. Crepel F, Jaillard D (1991) Pairing of pre- and postsynaptic activities in cerebellar Purkinje cells induces long-term changes in synaptic efficacy in vitro. J Physiol 432:123–141
8. Jörntell H, Hansel C (2006) Synaptic memories upside down: bidirectional plasticity at cerebellar parallel fiber-Purkinje cell synapses. Neuron 52(2):227–238
9. Laroche S, Jay TM, Thierry AM (1990) Long-term potentiation in the prefrontal cortex following stimulation of the hippocampal CA1/subicular region. Neurosci Lett 114 (2):184–190
10. Alonso A, de Curtis M, Llinás R (1990) Postsynaptic Hebbian and non-Hebbian long-term potentiation of synaptic efficacy in the entorhinal cortex in slices and in the isolated adult guinea pig brain. Proc Natl Acad Sci U S A 87 (23):9280–9284
11. Artola A, Singer W (1987) Long-term potentiation and NMDA receptors in rat visual cortex. Nature 330(6149):649–652
12. Iriki A, Pavlides C, Keller A et al (1989) Long-term potentiation in the motor cortex. Science 245(4924):1385–1387
13. Uhlhaas PJ, Singer W (2010) Abnormal neural oscillations and synchrony in schizophrenia. Nat Rev Neurosci 11(2):100–113
14. Engel AK, Moll CK, Fried I et al (2005) Invasive recordings from the human brain: clinical insights and beyond. Nat Rev Neurosci 6 (1):35–47
15. Buzsáki G (2006) Rhythms in the brain. Oxford University Press, New York
16. Buzsáki G (2002) Theta oscillations in the hippocampus. Neuron 33(3):325–340
17. Buzsáki G (2004) Large-scale recording of neuronal ensembles. Nat Neurosci 7(5):446–451
18. Lisman J, Buzsáki G (2008) A neural coding scheme formed by the combined function of gamma and theta oscillations. Schizophr Bull 34(5):974–980
19. Buzsàki G, Eidelberg E (1983) Phase relations of hippocampal projection cells and interneurons to theta activity in the anesthetized rat. Brain Res 266(2):334–339
20. O'Keefe J (1993) Hippocampus, theta, and spatial memory. Curr Opin Neurobiol 3 (6):917–924
21. Moser EI, Kropff E, Moser MB (2008) Place cells, grid cells, and the brain's spatial representation system. Annu Rev Neurosci 31:69–89
22. Winson J (1978) Loss of hippocampal theta rhythm results in spatial memory deficit in the rat. Science 201(4351):160–163
23. Bikbaev A, Manahan-Vaughan D (2007) Hippocampal network activity is transiently altered by induction of long-term potentiation in the dentate gyrus of freely behaving rats. Front Behav Neurosci 1:7
24. Bikbaev A, Manahan-Vaughan D (2008) Relationship of hippocampal theta and gamma oscillations to potentiation of synaptic transmission. Front Neurosci 2(1):56–63
25. Hyman JM, Wyble BP, Goyal V et al (2003) Stimulation in hippocampal region CA1 in behaving rats yields long-term potentiation when delivered to the peak of theta and long-term depression when delivered to the trough. J Neurosci 23(37):11725–11731
26. Hölscher C, Anwyl R, Rowan MJ (1997) Stimulation on the positive phase of hippocampal theta rhythm induces long-term potentiation that can be depotentiated by stimulation on the negative phase in area CA1 in vivo. J Neurosci 17(16):6470–6473
27. Pavlides C, Greenstein YJ, Grudman M (1988) Long-term potentiation in the dentate gyrus is induced preferentially on the positive phase of theta-rhythm. Brain Res 439(1–2):383–387
28. Greenstein YJ, Pavlides C, Winson J (1988) Long-term potentiation in the dentate gyrus is preferentially induced at theta rhythm periodicity. Brain Res 438(1–2):331–334
29. Orr G, Rao G, Houston FP et al (2001) Hippocampal synaptic plasticity is modulated by theta rhythm in the fascia dentata of adult

and aged freely behaving rats. Hippocampus 11 (6):647–654
30. Poe GR, Nitz DA, McNaughton BL et al (2000) Experience-dependent phase-reversal of hippocampal neuron firing during REM sleep. Brain Res 855(1):176–180
31. Huerta PT, Lisman JE (1995) Bidirectional synaptic plasticity induced by a single burst during cholinergic theta oscillation in CA1 in vitro. Neuron 15(5):1053–1063
32. Huerta PT, Lisman JE (1993) Heightened synaptic plasticity of hippocampal CA1 neurons during a cholinergically induced rhythmic state. Nature 364(6439):723–725
33. Hájos N, Pálhalmi J, Mann EO et al (2004) Spike timing of distinct types of GABAergic interneuron during hippocampal gammaγ oscillations in vitro. J Neurosci 24(41):9127–9137
34. Csicsvari J, Jamieson B, Wise KD et al (2003) Mechanisms of gamma oscillations in the hippocampus of the behaving rat. Neuron 37 (2):311–322
35. Traub RD, Whittington MA, Colling SB et al (1996) Analysis of gamma rhythms in the rat hippocampus in vitro and in vivo. J Physiol 493 (Pt 2):471–484
36. Pöschel B, Heinemann U, Draguhn A (2003) High frequency oscillations in the dentate gyrus of rat hippocampal slices induced by tetanic stimulation. Brain Res 959(2):320–327
37. Goutagny R, Jackson J, Williams S (2009) Self-generated theta oscillations in the hippocampus. Nat Neurosci 12(12):1491–1493
38. Singer W, Gray CM (1995) Visual feature integration and the temporal correlation hypothesis. Annu Rev Neurosci 18:555–586
39. Colgin LL, Denninger T, Fyhn M et al (2009) Frequency of gamma oscillations routes flow of information in the hippocampus. Nature 462 (7271):353–357
40. Wespatat V, Tennigkeit F, Singer W (2004) Phase sensitivity of synaptic modifications in oscillating cells of rat visual cortex. J Neurosci 24(41):9067–9075
41. Morellini F, Lepsveridze E, Kähler B et al (2007) Reduced reactivity to novelty, impaired social behavior, and enhanced basal synaptic excitatory activity in perforant path projections to the dentate gyrus in young adult mice deficient in the neural cell adhesion molecule CHL1. Mol Cell Neurosci 34(2):121–136
42. Kochlamazashvili G, Senkov O, Grebenyuk S et al (2010) Neural cell adhesion molecule-associated polysialic acid regulates synaptic plasticity and learning by restraining the signaling through GluN2B-containing NMDA receptors. J Neurosci 30(11):4171–4183
43. Claiborne BJ, Xiang Z, Brown TH (1993) Hippocampal circuitry complicates analysis of long-term potentiation in mossy fiber synapses. Hippocampus 3(2):115–121
44. Eckhardt M, Bukalo O, Chazal G et al (2000) Mice deficient in the polysialyltransferase ST8SiaIV/PST-1 allow discrimination of the roles of neural cell adhesion molecule protein and polysialic acid in neural development and synaptic plasticity. J Neurosci 20 (14):5234–5244
45. Muller D, Wang C, Skibo G et al (1996) PSA-NCAM is required for activity-induced synaptic plasticity. Neuron 17(3):413–422
46. Remondes M, Schuman EM (2003) Molecular mechanisms contributing to long-lasting synaptic plasticity at the temporoammonic-CA1 synapse. Learn Mem 10(4):247–252
47. Evers MR, Salmen B, Bukalo O et al (2002) Impairment of L-type Ca^{2+} channel-dependent forms of hippocampal synaptic plasticity in mice deficient in the extracellular matrix glycoprotein tenascin-C. J Neurosci 22 (16):7177–7194
48. Hanse E, Gustafsson B (1992) Long-term potentiation and field EPSPs in the lateral and medial perforant paths in the dentate gyrus in vitro: a comparison. Eur J Neurosci 4 (11):1191–1201
49. Berzhanskaya J, Urban NN, Barrionuevo G (1998) Electrophysiological and pharmacological characterization of the direct perforant path input to hippocampal area CA3. J Neurophysiol 79(4):2111–2118
50. Bukalo O, Fentrop N, Lee AY et al (2004) Conditional ablation of the neural cell adhesion molecule reduces precision of spatial learning, long-term potentiation, and depression in the CA1 subfield of mouse hippocampus. J Neurosci 24(7):1565–1577
51. Cremer H, Chazal G, Carleton A et al (1998) Long-term but not short-term plasticity at mossy fiber synapses is impaired in neural cell adhesion molecule-deficient mice. Proc Natl Acad Sci U S A 95(22):13242–13247

Neuromethods (2012) 67: 149–171
DOI 10.1007/7657_2011_5

Published online: 27 November 2011

Use of Dynamic-Clamp as a Tool to Reveal the Computational Properties of Single Neurons Embedded in Cortical Circuits

Alberto Bacci, Antonio Pazienti, and Michele Giugliano

Abstract

Dynamic clamp is a technique that combines computer modeling with experimental electrophysiology and is used to examine how specific ion channels modulate a variety of single-cell activities, by artificially emulating the response properties of specific ionic conductances during an electrophysiological recording. This is accomplished by continuously and instantaneously varying the current that is injected into a recorded neuron as a function of a computer-generated conductance and of the difference between its apparent reversal potential and the actual fluctuations of the membrane potential of the recorded cell. Dynamic clamp is often used to model voltage-independent, voltage-dependent, and synaptic ion currents and is very useful to study how cortical neurons compute and integrate diverse synaptic currents or sequences of synaptic inputs into specific spike-train outputs. Cortical networks are composed of highly heterogeneous cell types, and it is often difficult to dissect individual aspects of signal propagation between neurons and how they contribute to shape network activities underlying several cortical functions. Here, we describe some implementations of the dynamic clamp technique useful to studying the contribution of different elements of cortical circuits to the generation of single cell spike outputs.

Key words: Cortical networks, Dynamic clamp, Electrophysiology, Synaptic transmission

1. Introduction

In the mammalian brain, the cerebral cortex is the final destination of all sensory input and the site where incoming information is integrated into elaborate cognitive functions. This is accomplished by the concerted activity of large populations of cortical neurons that form complex, intertwined networks. One of the hallmarks of cortical networks is their highly heterogeneous cell population—especially of locally projecting GABAergic interneurons (1–7)—and this rich diversity allows for complex and fine control of cortical activities (3, 7–10). In addition to the multitude of cell types, each cortical neuron receives a huge number of synapses, whose biological and electrophysiological properties are also extremely diverse (2, 11, 12). Synaptic activity is fundamental for integrating and filtering incoming information or dictating postsynaptic neuronal spike timing, therefore providing a tight temporal code used by each neuron to perform sophisticated computational operations (3, 6–10, 13, 14). However, the heterogeneity of cortical cell types and the high diversity of their firing properties as well as the strength, dynamic range, spatial extent, anatomical localization, and molecular components of synaptic connections between them (2, 15, 16) increase the difficulty to dissect the role played by each specific element of the microcircuit during cortical operations.

Therefore, several fundamental questions are still open and represent the focus of intense investigation, such as the following: what is the functional role of cortical neuron heterogeneity? How does each neuronal element control and generate specific network activities? How do neurons compute information originating from diverse presynaptic partners, to produce complex firing behaviors? Are different network activities the result of specific properties of synaptic connections that cortical neurons form and receive? What is the role of the differential effects of specific neuromodulators on both intrinsic excitability and synaptic activity in different neuronal types?

Although conventional voltage- and current-clamp approaches to these questions are frequently used in cellular electrophysiology, and they can be indeed applied to address some specific aspects of cortical network function, they suffer from several limitations. For example, injection of constant current steps into a neuron held in current clamp elicit a firing pattern that is useful to characterize some general properties of that particular cell (6, 17–20), but should not be used to infer its temporal code during physiological cortical activities. Neurons in the intact brain in fact respond to a continuous noisy and irregular bombardment of a myriad of simultaneous synaptic inputs with highly intermittent and dynamic spike patterns (21–23). Likewise, stimulations of single neurons by injecting current waveforms designed to mimic single or sequences of synaptic responses are not accurate. Indeed, in reality, the point-by-point strength of a synaptic input changes as a function of the point-by-point fluctuations of the membrane potential, whose value at each given time determines the instantaneous driving force of that particular synaptic ion channel. This is particularly important, for example, if one wishes to study the impact of GABA-mediated responses on neuronal firing. Indeed, although the reversal potential of GABAergic responses in mature animals varies across different neuron types (24–26) and locations (27, 28), it is generally close to the resting membrane potential, thereby having a shunting effect at rest. As a consequence, the natural neuronal hyperpolarizations or depolarizations will determine the amplitude and direction of GABAergic currents that will ultimately depend on the instantaneous dynamic changes of the membrane potential.

The dynamic clamp offers the advantage of providing a neuron with an artificial (i.e., computer generated) and realistic stimulus that features all the appropriate waveform details. The amplitude of these waveforms is instantaneously computed and updated taking into account membrane potential fluctuations, and thus the driving force of one or more synaptic conductances. In particular, the dynamic clamp is used to simulate voltage-gated and voltage-independent conductances, excitatory and inhibitory synapses between neurons and to recreate in vitro high-conductance states typical of in vivo conditions (21). The dynamic clamp can be successfully used in any cell that can be recorded electrophysiologically, including excitable or nonexcitable cells in culture, as well as neurons in acute brain slices and/or in in vivo preparations (29, 30).

In this chapter we focus on the use on this technique as a tool to study the input–output properties of single neurons embedded in a network and receiving a rich diversity of synaptic connections with different properties, originating form a multitude of cell types, and differently responding to several neuromodulators. Whereas, in principle, the dynamic clamp can be successfully used in neurons recorded in vivo (31, 32), the large majority of the studies defining the integration properties of single neurons in a network used the dynamic clamp in acute brain slices. This preparation is metabolically deprived and is more quiescent, as it receives less ongoing synaptic activity than it does in vivo, but it offers many advantages in terms of simultaneous multielectrode recordings, visualization, and pharmacological manipulations, while keeping most of the microcircuit cytoarchitecture intact.

1.1. Working Principles of the Dynamic Clamp

The dynamic clamp can be defined as a closed-loop stimulation protocol, and it can take several forms (Fig. 1). Here, we focus on the conductance-clamp, as a computer-controlled way to effectively alter the local membrane conductance of a neuron. This is more different than injecting intracellularly an arbitrary current stimulus, and it is accomplished by using a hybrid circuit formed by a computer and a real neuron. Under the conditions of this dynamic feedback loop, the recorded neuron receives a current stimulus I that is instantaneously updated realistically as a function of the membrane potential V_m. By doing so, the computer mimics in real-time the response properties of an artificial population of ion channels characterized by a conductance g and an apparent reversal potential E_{rev}.

Since all ion channels obey Ohm's law, the current I flowing through a channel is:

$$I(t) = g(t) \cdot (V_m(t) - E_{rev}) \tag{1}$$

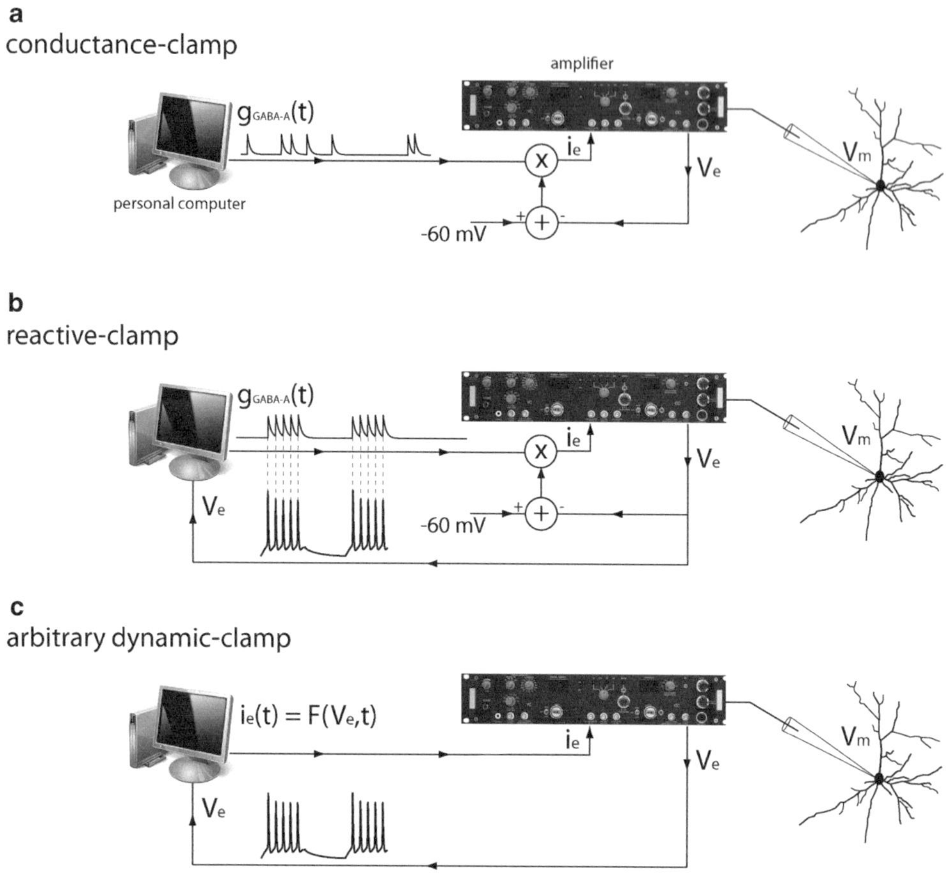

Fig. 1. The figure presents three protocols, involving nonconventional current injection. The simplest (**a**), known as conductance-clamp (53), replaces the current-clamp technique with a more realistic mean to mimic artificial synaptic inputs to a cell (e.g., here mediated by synthetic $GABA_A$ receptors, activated at random times). This technique is in principle the easiest to implement, as it requires a signal multiplication to combine the instantaneous membrane potential with an analog waveform, synthesized by a computer AD/DA board and having the meaning of a synaptic conductance. In (**b**), the timing of individual action potentials fired by the recorded neuron, is employed in real-time to trigger the repeated activation of a synthetic conductance (e.g., as in an autapse) (33). This requires the computer to monitor the membrane potential and perform a threshold-crossing test to identify the spike times. In (**c**), the most general concept of a current injection whose amplitude depends instantaneously on the membrane potential and on its previous history is sketched and identified as arbitrary dynamic-clamp (Sharp et al. (52)).

As opposed to conventional current-clamp stimulation, under conductance-clamp, the current flowing through the artificial ion channels changes dynamically, depending on the value of the membrane potential V_m—through the instantaneous difference E_{rev} and V_m or *driving force*—as well as on the channels conductance g (1). This conductance may vary in time according to the membrane potential (i.e., as in voltage-gated ion channels) or to the presence of a ligand (i.e., as in ionotropic synaptic receptors).

Assume that we wish to inject a realistic synaptic input into a neuron and record how this simulated synaptic stimulus changes the postsynaptic membrane potential in real time. The first requirement is the mathematical description of the synaptic conductance waveform, $g(t)$, resulting macroscopically from the complex conformational modifications induced by the neurotransmitter binding to postsynaptic receptors. In the case of a single synaptic event generated by release of neurotransmitter (e.g., glutamate or GABA) triggered by a single presynaptic action potential, each unitary synaptic waveform can be modeled as the algebraic sum (2) or the product (3) of two or more exponential functions (say n), whose variables are the peak amplitude a, the rise and decay time constants (τ_R and τ_D, respectively).

$$g(t) = \sum_{i=1}^{n} a_i \cdot (e^{-t/\tau_{Ri}} - e^{-t/\tau_{Di}}) \tag{2}$$

or

$$g(t) = \sum_{i=1}^{n} a_i \cdot (1 - e^{-t/\tau_{Ri}}) \cdot (e^{-t/\tau_{Di}}) \tag{3}$$

where in a realistic $g(t)$ waveform, the number of exponentials, n, represents several biological factors, such as receptor subunit composition, geometry of the synapse, diffusion properties of the neurotransmitter, etc. (2) and (3) describe unitary synaptic waveforms, but it is also possible to model the synaptic input resulting from a complex temporal pattern of presynaptic action potentials, generated by one or more neuron types (e.g., glutamatergic and GABAergic). In these cases, the injected conductance will be modeled as the superposition of several $g(t)$ with varying amplitudes and times of occurrence: a realistic representation of the composite synaptic input that a neuron receives during network activity.

A $g(t)$ waveform is synthesized accurately on the basis of real data previously acquired, and of a specific E_{rev} for that particular ion channel one aims at emulating (e.g., $E_{rev} = 0$ mV, for glutamatergic synapses). A conductance stimulus is then computer-synthesized and delivered to the recorded neuron under dynamic-clamp, either at explicit fixed times during the experiment, or it can be autonomously triggered by neuronal activity itself, i.e., as in a case of an artificial autapse, when the membrane potential crosses a certain threshold (as in (33)). Section 1 details how to iteratively synthesize realistic synaptic conductance waveforms. During the recording, the computer detect the neuronal V_m in real-time, and creates a current whose amplitude changes instantaneously based on the modeled $g(t)$, and as a function of its driving force. In a typical dynamic clamp setup, a computer (or analog device; see below) is used to detect V_m and update the current input that is imposed to a neuron continuously, while a second computer is used to display and log the neuronal response on disk.

For example, if one wishes to emulate a synaptic stimulus that is realistically fast, perisomatic GABAergic current, one first needs to model a waveform $g(t)$ with a fast (<1 ms) rise-time, a peak (~6–10 nS), and an exponential decay (ranging between 6 and 20 ms, depending on presynaptic cell type one aims at emulating). One then sets the reversal potential E_{GABA} at ~−60 mV. We record from a neuron, and inject a current that is a function described in (1). If the actual V_m is more negative than −60 mV, as in the case of more hyperpolarized resting potentials or under endogenous hyperpolarizing synaptic inputs, the injected current will be depolarizing. If, however, during the waveform injection, V_m depolarizes—as under endogenous excitatory synaptic inputs—and the cell fires an action potential, the injected current will dynamically become hyperpolarizing (after the neuron crosses −60 mV) and when V_m sharply depolarizes during the rising phase of the action potential its driving force ($V_m - E_{GABA}$) will promptly increase. At this point, the computer will update the current to be injected as a sharp hyperpolarizing current, whose amplitude and kinetics will depend on the much increased driving force and the actual level of $g(t)$, which can be at its peak or already decaying, depending on its waveform kinetics. The injected stimulus, therefore, will dynamically change and have profound effects on the spike waveform and particularly on its after-potential.

The dynamic clamp stimulation protocol as described above requires a real-time and uninterrupted measurement of the membrane potential and a fast capability of continuously computing and updating the current amplitude that is delivered to the recorded cell. If sampling rate and computation velocity are fast enough, the emulated ion channel population will have the same impact as if real ion channels were located where the voltage sampling and current injection take place (i.e., the pipette tip). In principle, given the neuronal membrane time constants, to achieve a quasi real-time performance the update rate should be between 20 and 50 kHz. The maximal sampling and update rates will largely depend on the hardware acquisition board, used to interface computer, electrophysiological amplifier, and neuron. Most importantly, however, the computer operating system will decisively determine the maximal performance achievable. Some operating systems, such as Windows (Microsoft, Redmond, USA), are multitasking, i.e., they distribute processor time between different computer tasks. This leads to a nondeterministic execution of user software code, due to the frequent interrupts (or hiccups) internally generated by the computer hardware and software. These interruptions will create gaps in dynamic clamp computations and ultimately precluding tight real-time performances. To circumvent this problem, it is possible to use a deterministic real-time operating system that does not hiccup and can reach appropriate update rates, at the expense of lack of multitasking for the duration of the dynamic clamp stimulation protocol. Alternatively, some artifices may be devised to compensate for Windows unexpected jitters or delays mentioned above. Finally, in other hardware implementations of the dynamic clamp the personal computer may be replaced by a dedicated embedded processor, running a custom real-time operating system as in DSP-based hardware. In Sects. 2.3–2.7, we detail how to assemble different dynamic clamp setups using various software and hardware implementations.

1.2. Advantages and Limitations of the Dynamic Clamp

In order to study the physiology of a complex biological system as the cortical network, it is often useful to perturb it in a highly controlled manner, while probing the response of different elements of the cortical circuit. The dynamic clamp offers this opportunity at the level of realistic, artificial ion conductances. Novel conductances (i.e., ion channels and/or synaptic inputs) that were not present or were not active can be introduced in a real neuron. Similarly, existing conductances can be ablated, while artificially canceling their specific responses in real-time. The same manipulations are often impossible to obtain by means of molecular biology or pharmacological blockade, leaving dynamic clamp as a precious tool to help understanding the mechanisms underlying specific aspects of intrinsic cellular excitability and information processing between neurons (2). Moreover, the dynamic clamp combines computational neuroscience with "wet-hands" experiments, allowing testing complex theoretical hypotheses experimentally (34). For instance, only when the model employed to generate $g(t)$ [e.g., (2) and (3)] is sufficiently realistic, can the neuronal responses induced by the dynamic clamp stimulation coincide with those observed under naturalistic cortical activation. Similarly, it is possible to improve and extend the complexity of the model, until the observed responses match their naturalistic counterparts, literally leading to reverse-engineering of cortical cellular and synaptic physiology.

To unequivocally assess the role played, for example, by a certain synaptic waveform or the reversal potential of a ligand-gated ion channel one can use complex genetic approaches (35) or specific pharmacological agents. However, whereas genetic alterations represent an extremely powerful tool to probe the molecular events underlying a multitude of phenomena, they often suffer from several limitations, such as developmentally regulated compensations, lack of cellular specificity, and interactions with complex cellular processes that are difficult to control. Importantly, genetic alterations, even if sophisticated enough to circumvent the above-mentioned limitations, do not allow monitoring the instantaneous reactions to "on the fly" alterations of specific voltage-gated ion channels or synaptic receptors.

Likewise, pharmacology is extensively used to quickly and reversibly alter neuronal excitability and/or synaptic function, but it is strongly dependent on specificity of drugs, as well as their penetration, diffusion and pharmacokinetic properties. A major issue of pharmacological applications is that they often block or activate specific ion channels or receptors globally. This prevents the fine dissection of how a neuron responds to alterations of an individual set of synapses. Moreover, global pharmacological alterations limit the evaluation of certain activities, such as the temporal control of neuronal firing or any other question relying on a network that should be as intact as possible. The dynamic clamp allows altering different properties of specific injected conductances (e.g., the reversal potential, amplitude, decay and/or rise times) by quickly toggling different components of the modeled waveform.

In summary, using dynamic clamp one can easily change the modeled synaptic conductance and examine how neuronal behavior changes in response to different manipulations including (1) changes in synaptic strength (by changing maximum conductance peak), (2) changes in the reversal potential, (3) mimic the action of drugs affecting the biophysics of that particular synaptic channel (by changing the synthetic conductance waveform), (4) modify short-term synaptic plasticity (such as depression or facilitation), (5) alter the interplay between inhibitory and excitatory connections. Along this line, the dynamic clamp makes it possible to recreate the impact of both inhibitory and excitatory synaptic waveforms, modeling physiologically relevant synaptic patterns and therefore mimicking, in vitro, the synaptic bombardment that a neuron receives in vivo. This application of the dynamic clamp was instrumental in determining the impact on neuronal integration properties of the strong background synaptic noise as measured in vivo (21).

In addition to modeling and testing intrinsic and synaptic conductances, the dynamic clamp can be used to create hybrid circuits formed by computers and real neurons, allowing testing hypotheses that would otherwise require either extremely complex experimental or oversimplified theoretical approaches. Indeed, by coupling a real neuron with one or more model neurons, it is possible to recreate physiologically relevant firing patterns and reveal the contribution to these firing activities of specific synaptic or intrinsic conductances in the "presynaptic" model neuron(s) (29, 36–39).

Despite the advantages mentioned above, the dynamic clamp suffers from some drawbacks, most of which are technical limitations. The first problem is inherent with all electrophysiological approaches, such as, for example, quality and stability of the recording, artifacts introduced by the pipette's electrical properties, whole-cell dialysis, rundown of ion channel functions in the recorded cell, and instrument artifacts (see Sect. 2.7 below). Moreover, as previously discussed, the dynamic clamp crucially relies on (real-time) measurement of membrane potential coupled to fast computation and injection of the modeled conductance into the recorded cell. This requirement might not be fulfilled in case of very fast conductances, and it largely depends on specific hardware and software implementation choices (discussed in Sects. 2, 2.4 and 2.5). In addition, some real-time applications require comprehensive expertise in hardware configuration and software programming, which might discourage many users.

One important issue of dynamic clamp experiments is space-clamp, i.e., the inability to electrically control a large portion of the recorded cell. In fact, the effects of the emulated ion channels are confined near the recording (and injection) site, which is in most cases the soma. Perisomatic synaptic transmission can be effectively modeled with the dynamic clamp, but modeling activity resulting from ion channels or receptors that are naturally expressed far from the recording electrode becomes problematic. As in voltage-clamp experiments, this problem becomes more serious if the recorded cell is a neuron with an elaborate dendritic tree.

Finally, simulation and injection of any synthetic conductance is powerful to assess electrical responses, neglecting however other potentially important biochemical functions of ion channels. Indeed, the opening of real ion channels and receptors is associated to the transmembrane flow of a specific ion, often leading to numerous metabolic effects. Changes in specific (e.g., calcium) intracellular ion concentrations and alterations of intracellular second messengers and enzymatic pathways are all physiological phenomena known to have short- and long-term effects on several neuronal functions, including cell excitability and synaptic responsiveness, and they obviously cannot be mimicked by the electrical channel properties alone.

2. Materials and Methods

In the following sections, we provide some examples of different experimental implementations. As for the materials and the experimental procedures necessary to obtain and perform electrophysiological recordings from acute brain slices we wish to refer to the chapter by Varela et al., in this volume (69), and other significant contributions (40–44).

2.1. How to (Iteratively) Synthesize Realistic Synaptic Conductance Waveforms

As an alternative to the explicit descriptions of (2) and (3), a minimal model can be considered to effectively implement, even under nonideal real-time conditions, the generation of a realistic synaptic conductance time-course. Let us consider for simplicity the conductance $g(t)$ associated to the repeated activation of ionotropic postsynaptic receptors, as the concentration of their ligand changes abruptly as in a train of pulses (see Fig. 1a). Assuming a two-state kinetic scheme describes the conformational states of the receptors (i.e., open and closed), the conductance change will abruptly increase up to a peak amplitude a and exponential decay with time constant τ (45). This is reminiscent of, e.g., glutamate or GABA release in the synaptic cleft and its effect on AMPA- or $GABA_A$-receptors conductance, although receptors desensitization and saturation have been neglected here for simplicity (45–47). Then, indicating as $t_1, t_2, t_3, \ldots$ the times of presynaptic activation, the time course of $g(t)$ can be computed as the solution of a linear differential equation:

$$\tau \frac{\mathrm{d}}{\mathrm{d}t} g = -g + \sum_i a\,\delta(t - t_i) \tag{4}$$

Equation (4) can be easily implemented and translated into an exact update rule, to be used iteratively at each time-step when the dynamic-clamp is performed:

$$g(t + \Delta t) = g(t)\mathrm{e}^{\Delta t/\tau} \tag{5}$$

In formula (5), Δt is the time passed since the last update of the variable g and is not required to be infinitesimal or constant. In other words, should the update-rate (i.e., the sampling-rate) of the dynamic-clamp implementation be nonuniform (e.g., as in Windows non-real-time implementations, due to glitches of the real-time schedule, see Sect. 5), (5) allows to compensate the unexpected latencies and correctly synthesize the next sample to be injected by the amplifier: $I = g(t)\ (E_{\mathrm{AMPA}} - V_{\mathrm{e}}(t))$.

The formula (5) allows one to account for the spontaneous decay rate of the conductance activation [e.g. as the τ_{D} of (2) and (3)]. In order to mimic the very quick increase in the conductance, immediately following a presynaptic action potential [e.g., as those occurring over τ_{R} in (2) and (3)], an additional rule must be added. Thus, whenever a presynaptic event occurs (e.g., generated as random events by the computer, as in Fig. 1a, or triggered by external events, as in Fig. 1b) $g(t)$ will be increased of a fixed quantity a, which indirectly mimics both the synaptic efficacy and the presynaptic probability of release:

$$g(t_i) \rightarrow g(t_i) + a \tag{6}$$

In the case of a number N of independent synapses of the same type (e.g., $AMPA_R$-mediated), impinging on the very same neuron and thus contributing to its overall synaptic input, the formula (6) can be still employed. However, in this scenario $g(t)$ now represents the overall conductance, under the assumption of linear superposition of the N contributions. In this case, (6) must be modified to account for distinct amplitude peaks $a_1, a_2, \ldots a_j$ associated to distinct synapses (e.g., approximating distinct electronic attenuation factors, synaptic efficacies, probabilities of release, etc.):

$$g(t_i) \to g(t_i) + a_j. \tag{7}$$

The simple algorithm represented by (5) and (6) [or (5)–(7)] requires a minimal computational load (i.e., one multiplication per time-step and one addition whenever a presynaptic activation event occurs). It can be easily generalized to the case of multiple receptor types (e.g., $I = g_{AMPA}(t))\ (E_{AMPA} - V_e(t)) + g_{GABA}(t)$ $(E_{GABA} - V_e(t))$. In that case, one must simulate a distinct variable $g(t)$ per synapse type, each with its own parameters (i.e., a, τ, and reversal potential E).

Finally, in the case of random presynaptic activation at frequency f (e.g., as in background synaptic activity of high conductance states), simulating a large number N of synapses of the same type can be achieved replacing (5) and (6) by a single equation and thus by a single update rule per time-step, regardless of the choice for f. This corresponds to employing a continuous noise waveform (48), instead of the individual *jumps* in $g(t)$. In this case, an exact iterative formula is also available (49):

$$g(t + \Delta t) = m + (g(t) - m)\, e^{\Delta t/\tau} + s\sqrt{1 - e^{-2\Delta t/\tau}}\, R \tag{8}$$

and it requires the generation of a pseudo-random number R, Gauss distributed (50), at each iteration. Finally, in (8), $m = Nfm_a$ and $s^2 = Nf(m_a^{\,2} + s_a^{\,2})$, with m_a the mean of the heterogeneous peaks amplitudes $\{a_j\}$, and s_a their standard deviation.

2.2. How to Synthesize Short-Term Depressing Synaptic Conductance Waveforms

Equation (6) [or (7)] can be replaced by a very simple time-dependent iterative formula, accounting for the exhaustion of synaptic efficacy and probability of release due to short-term homosynaptic depression (33), thereby resulting in history-dependent peak amplitude of the conductance (51). If one calls A the absolute maximal peak amplitude and U the fraction of synaptic resources used at activation event by a real synapse, one must add another variable r representing the readily releasable synaptic resources and replace (6) by:

$$g(t_i) \to g(t_i) + AU\, r(t_i) \tag{9}$$

Now, at each presynaptic event, the new variable r has also to be updated (i.e., decreased) as:

$$r(t_i) \to r(t_i)\,(1 - U) \tag{10}$$

Equation (5) must be then accompanied by an iterative exact update rule for the new variable r, which accounts for the recovery from depression time constant τ_D. When no presynaptic event occurs, r slowly reaches its resting value, here set to 1.

$$r(t+\Delta t) = (r(t)-1)e^{\Delta t/\tau D}+1. \tag{11}$$

2.3. An End-User Perspective on Dynamic-Clamp Implementation

Although the introduction of dynamic-clamp dates back to the early 1990s (52, 53), this technique is still neither sufficiently standardized nor commercially available as a complete turnkey system (Fig. 2). Nevertheless, especially in the last few years, open-source and public domain software is being continuously contributed and maintained by few labs around the world (e.g., (54–60)). In addition, the International Neuroinformatics Coordinating Facility[1] is increasingly providing infrastructures towards easy-to-use resources for the neuroscience community (61). As a consequence, in the near future the end-users will require only limited computer or system integration skills, depending inversely on the degree of customization required by his/her specific application.

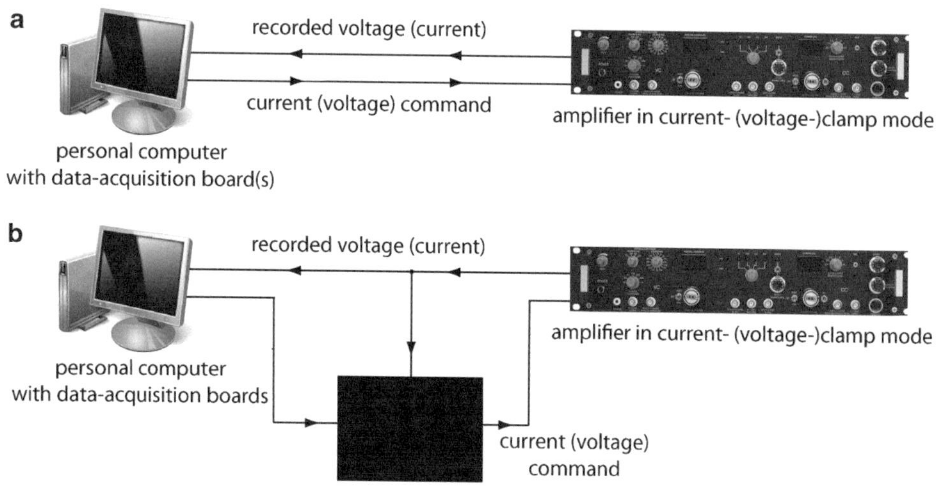

Fig. 2. Comparison between (**a**) a conventional PC-based data-acquisition system for cellular electrophysiology and one (**b**) including dynamic-clamp. The personal computer is equipped with a data-acquisition board(s) capable of analog-to-digital (A/D) conversion of the amplifier outputs (e.g., the pipette voltage, under the current-clamp configuration) and of digital-to-analog (D/A) generation for a stimulation protocol, fed into the amplifier command-input (e.g., the injected current, under the current-clamp configuration). The diagrams show that dynamic-clamp can be generally implemented as an external add-on to a preexisting setup, making possible to keep the original system and software for data-logging and analysis, as well as making the dynamic-clamp system portable across setups. The extra component for dynamic-clamp (i.e., the *black box*) could be a stand-alone hardware, a dedicated additional PC incorporating special boards, or it can partly or completely be incorporated in the preexisting personal computer and original data-acquisition system.

[1] http://www.incf.org.

2.4. Dedicated Hardware for Dynamic-Clamp

The available implementations of a dynamic-clamp system can be roughly subdivided into two broad and partly overlapping categories: (1) those taking advantage of dedicated hardware and (2) those requiring a software running on a (additional) personal computer, equipped with conventional data-acquisition boards (and with an ad hoc operating system). Only a subset of the software implementations is however advisable for beginners with no computer-skill. In fact, the other implementations require some programming skills or confidence in hardware configuration and nonmainstream operating systems installation. This specific consideration is likely to change as soon as more users approach dynamic-clamp.

The simplest and cheapest dedicated hardware component for dynamic clamp consists in an analog, four-quadrant electronic multiplier (see Fig. 1a, b). While such electronic circuit is today a solid standard, sold by manufacturing companies as a cheap monolithic integrated-circuit (e.g., the AD834 from Analog Devices Inc., Norwood, MA, USA and see its application notes[2]), it requires some printed circuit boards construction and assembly skills to be smoothly interfaced with a standard amplifier and data-acquisition boards (Fig. 2b). In addition, the use of an analog multiplier is limited to the most basic conductance-clamp protocols (see Fig. 1a), as it only allows combining an off-line computer-generated waveform (i.e., as in current-clamp) with the instantaneously recorded voltage (i.e., measured from the amplifier output). The output of the multiplier is then fed into current-command of the amplifier. Although analog circuits enable the fastest signal processing, an alternative solution is largely preferred for implementing more complex protocols and for its increased versatility. This alternative is available as a stand-alone hardware or as expansion boards to a host PC. It consists of dedicated digital electronic systems using programmable embedded processors, digital-signal-processors (DSPs), or field-programmable gate arrays (FPGAs). Examples are the ADWIN systems (Keithley Instruments Inc., Cleveland, OH, USA), the Mstar-DAP (Microstar Laboratories, Bellevue, WA, USA), the Toro-16 or the DSP-M67 + A4D4 boards (Innovative Integration, Simi Valley, CA, USA), and the DSP products by Sheldon Instruments (San Diego, CA, USA) and by National Instruments (Austin, TX, USA). These systems should be considered as reconfigurable electronic devices, or as programmable miniature computers, which also incorporate data-acquisition hardware capable of analog-to-digital (A/D) conversion and of digital-to-analog (D/A) signal generation. These systems usually require a connection to a host computer to be (initially) configured, and their programming takes place as in a kind of *upload* of a set of instructions from the host computer to

[2] http://www.analog.com/static/imported-files/data_sheets/AD834.pdf.

the embedded system. Programming or scripting languages are then available to the user, to define how the system will perform on-the-fly mathematical operations on the individual A/D data samples (e.g., the pipette voltage, from the amplifier output) and on internally generated responses (e.g., the conductance of a synthesized voltage-gated synthetic current). The complexity of these computations are of course limited by the chosen A/D-D/A sampling rate and by the total time required by mathematical operations (i.e., number of dedicated-CPU cycles): if the last take too long to be executed, then the sampling-rate has to be lowered accordingly. These operations must be specified in the details to numerically solve a set of differential equations (e.g., the kinetics of an artificial voltage-gated ion conductance—(52)) and to precisely specify how to output D/A data samples (e.g., the injected current). Therefore, these systems definitely require some programming effort to specify in the details which equations to use and how to numerically solve them [e.g. (4) and (5)], while incorporating in real-time the acquired inputs (e.g., the pipette voltage). With the recent increase in demand for these technologies, programming DSP, FPGA, and embedded processors became more user-friendly than in the past. In most of the cases in fact, a simplified programming language (e.g., a subset of C++, a custom Basic, Python, etc.) or a standard (visual) software environment (e.g., LabView, from National Instruments, Austin, TX, USA) are available.

As an example of the inner-loop programming required in a simple conductance-clamp application, we list below a minimal code snippet written in ADWIN Basic (courtesy of M. Wijnant). All the commands and keywords of the following example are custom keywords of the particular programming language. They are nevertheless reported here without extensive explanation, to provide the reader with a flavor of the limited complexity of programming a DSP. The DSP is instructed to execute this code at every sampling step, and only the lines reported in bold-face should be extended or replaced for alternative protocols and applications, while the rest of the code remains the same and might be considered as a template or a skeleton for developing one's own dynamic-clamp protocol. This code is schematized as a flowchart in Fig. 3.

EVENT:

Start_Conv(AI_chan0)	*Start A/D-conversion for input channel 0 (from the amplifier)*
Wait_EOC (AI_channel0)	*Wait until a single sample is acquired*
W=Read_ADC (AI_chan0)	*Get the value of the sample, as 16-bit integer*
V=((W/32768)-1)*10	*Convert the sample back into real number (in V)*

(continued)

(continued)

V=V/amplyGAIN	*Convert the sample into pipette voltage (in mV)*
Start_Conv(AI_chan1)	*Start A/D-conversion for the channel 1 (from the PC D/A)*
Wait_EOC(AI_chan1)	*Wait until a single sample is acquired*
W=Read_ADC (AI_chan1)	*Get the value of the sample, as 16-bit integer*
G=((W/32768)-1)*10	*Convert the sample back into real number (in V)*
G=G/gFACTOR	*Convert the sample into a conductance (in mS)*
I = G(−60−V)*	*The simplest conductance-clamp protocol possible: generating the current to be injected (see Figs. 1a and 3)*
I=I/ amplyCMD_GAIN	*Convert output into a command to the amplifier*
W=((I/10)+1)*32768	*Prepare the output sample into a 16-bit integer*
Write_DAC (A0_chan0,W)	*Set the value of the sample, as 16-bit integer*
Start_DAC(A0_chan0)	*Start output now*
Start_Conv(A0_chan0)	*Start A/D-conversion*
Wait_EOC(A0_chan0)	*Wait until the single sample is generated*

In summary, while the programming of these systems is generally straightforward even for nonexpert users, the most complex implementation of channel-kinetics or hybrid model neuronal networks, as well as the seamless integration and synchronization of the dynamic-clamp loop with the existing data-acquisition hardware (Fig. 2), may require slightly more efforts. Examples of these implementation details range from the choice of a numerical method for solving specific model equations [e.g., (6) and (7)], the design of a graphical user interface, the ad hoc generation of extra TTL-pulses to trigger data-acquisition systems, and so on and so forth.

2.5. Ad Hoc Software for Dynamic-Clamp, Under Windows

While it is certainly possible to use a single PC for experiment management, data visualization/logging as well as for dynamic-clamp, the concurrent software applications reduce the overall performance of the hardware thereby limiting the maximal sampling rates or potentially causing glitches in the execution of the dynamic-clamp operations loop. Consumer operating systems (OS), in addition, do not strictly allow real-time execution of user programs resulting in a nonpredictable execution of the dynamic-clamp loop code. Therefore, it is desirable to design the dynamic-clamp system as a separate, dedicated PC, as outlined in Fig. 2. The choice of the operating system and its compatibility with (preexisting) data-acquisition boards are the most important

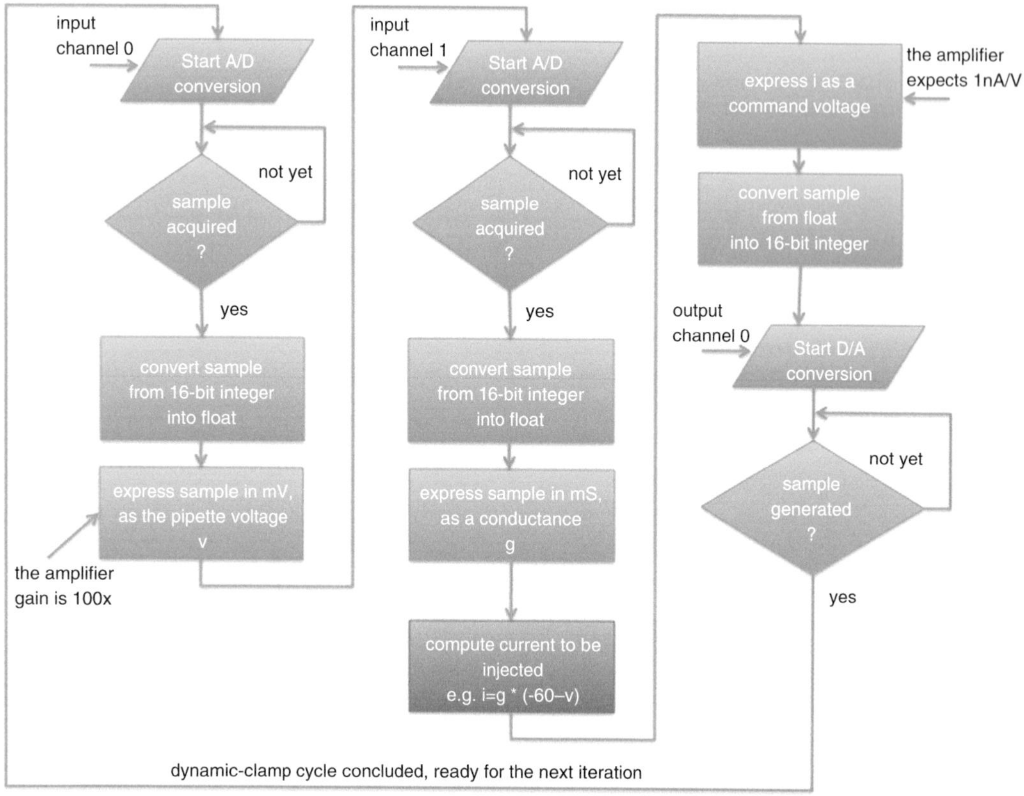

Fig. 3. Diagram illustrating a simple code as an example of for inner-loop programming required by a simple conductance-clamp application using a programmable DSP to generate real-time dynamic clamp.

issues for correctly implementing dynamic-clamp. We distinguish between dynamic-clamp software implementations running on native real-time OS (e.g., Linux-RTAI[3]; Lab-View Real-Time, National Instruments; MATLAB xPC Toolbox, The Mathworks, Natick, USA) and on conventional OS (e.g., Microsoft Windows, Mac OS X, Linux). Modified real-time Windows kernels have also been proposed recently (see for example INtime from tenAsys Corporation, Beaverton, OR, USA; and OnTime Software, Littleton, MA, USA) and might be excellent future alternatives. In all cases, choosing the operating system requires prior guarantee that proper driver support for the (preexisting) data-acquisition boards exists. For the (real-time) Linux OS, the open-source project COMEDI[4] is a notable initiative, providing free drivers, libraries and tools, supporting a large variety of commercial data-acquisition boards, otherwise unsupported by their manufacturers.

If the operating system of choice is not real-time, such as Microsoft Windows, a few tricks can be used to obtain a quasi-real-time

[3] https://www.rtai.org/.

[4] http://www.comedi.org/.

performance, while keeping track and compensating for the extra latencies and unexpected delays. These tricks consist in adapting the mathematical operations and the numerical methods to solve the differential equations at each step of the dynamic-clamp into a variable integration/sampling time-step [e.g., (5), (6), (8), and (11)]. While this is not possible for every application, it reveals to be a convenient and cheap solution that makes use of the familiar Windows system. For instance, the free software[5] available for Windows-based dynamic-clamp (56–58, 60) takes advantage of a real-time counter, available in specific data-acquisition boards (i.e., the Digidata 1200 series, by the former Axon Instruments) or accessible inside a modern Pentium-processor (i.e., the Time Stamp Counter register). These are excellent high-resolution, low-overhead ways of getting timing information.

While these approaches are successful for simulating hybrid circuits and (reactive) synaptic stimulations, the dynamic-clamp implementation of, for example, fast-inactivating voltage-dependent sodium conductances definitely requires higher dynamic-clamp sampling/integration rates and cannot be afforded under Windows.

2.6. Ad Hoc Software for Dynamic-Clamp, Under Real-Time OS

In the very recent years few notable proposals[6] have been made. Most of them are fully functional software written to work under the real-time extension Linux RTAI operating system (55, 59). They explicitly take advantage of the COMEDI project drivers, include a graphical user interface and even a simple way to add plug-in custom code and therefore extend their functionalities. The major obstacle from the user perspective is the correct installation and configuration of the operating system and of the COMEDI drivers. In fact, although it is highly documented in the many tutorials made available, at this very moment it is imperative to recompile the Linux kernel, after incorporating the real-time extensions, for the specific PC of choice by the user, requiring some confidence with Linux and the kernel configuration. Very recently, however, the RTXI[7] software is available as a *live* CD installation: in other words, the user can download and burn the content of a CD and use it to boot the fully functional real-time operating system. This is an excellent opportunity to both explore the software provided as well as to implement one's custom routines, bypassing the difficulties of the Linux real-time installation.

[5] http://stg.rutgers.edu/software, http://stdpc.sourceforge.net, http://www.qub.buffalo.edu/wiki/index.php/Dynamic_Clamp.

[6] http://www.hornlab.neurobio.pitt.edu/G%20clamp%20v2.html, http://www.rtxi.org, http://relacs.sourceforge.net.

[7] http://www.rtxi.org.

We conclude mentioning the free availability of the dynamic-clamp software develop to run on the LabView Real-Time operating system (54), which is a commercial extension provided by National Instruments for real-time applications. Along the same line, although not specifically addressing dynamic-clamp experiment, C. Zrenner et al. proposed for the first time the use of MATLAB (The Mathworks, Natick, USA) and of an ad hoc toolbox to design a real-time system (70). The additional toolbox works basically as LabView Real-Time OS and must be installed on the target PC, which receives from a host machine its program and the specification on how it must interface the acquisition hardware.

2.7. Active Electrode Compensation Versus Bridge-Balance

As opposed to (1), in the previous section as well as in the figures we have distinguished between the actual voltage V_e recorded by the pipette and the true membrane potential V_m. As in voltage-clamp, dynamic-clamp experiments require ideally the use of a two-electrode configuration. In fact, injecting an arbitrary current i_e and monitoring the membrane voltage simultaneously by a single-electrode is known to induce considerable artifacts in the measurements of the cell response properties (62). Indeed, the current i_e passing through the very same pipette makes V_e differing from V_m, by the voltage-drop U_e across the pipette.

$$V_e = V_m + U_e \tag{12}$$

U_e is in fact the unavoidable result of the electrical filtering properties of the electrode, common to sharp- and patch-pipette electrodes (63). Standard techniques are available, and nowadays incorporated in most of electronic amplifiers, to subtract U_e while compensating for the electrode capacitance (i.e., by negative capacitance correction) and for access resistance (i.e., by bridge-balance). These methods assume that the pipette acts as an R–C circuit in series to the membrane (Fig. 4c), and subtract from V_e both an offset and the exponential transients, resulting from the resistive and capacitive components of the pipette in response to i_e, respectively (64). In practice, however, the electrical equivalent circuit model of a sharp- or of a patch-electrode requires a distributed-parameters description (63), and it is more complex than a simple R–C. Moreover, in the case of strong, high-frequency stimuli, or when the characterization of fast neuronal response properties is of interest, the correction provided by the conventional methods is completely inadequate (65, 66).

Particularly, in closed-loop cellular electrophysiology techniques such as voltage-clamp and dynamic-clamp, where the injected current depends instantaneously on the measured voltage, conventional bridge-balance and capacitance compensation alone may dramatically fail. Then, inaccuracies in the estimate of the V_m and the

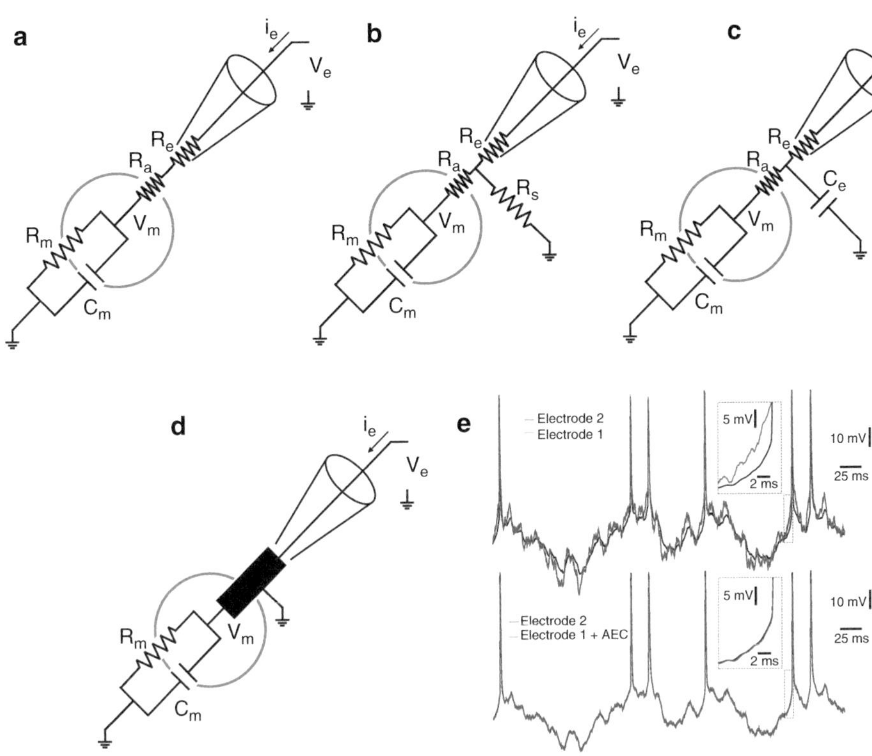

Fig. 4. Injecting an arbitrary current i_e and monitoring the membrane voltage V_m simultaneously is known to induce considerable artifacts in the pipette measurement V_e. Although compensation methods exist for accounting and subtracting the filtering effect of the electrode, no matter which lumped-parameters equivalent circuit model is employed or assumed (**a–d**), the correction is never complete. In closed-loop stimulation protocols such as voltage-clamp and dynamic-clamp, this represents a severe limitation. (**e**) A two-electrode (*open loop*) current injection and recordings (*upper panel*) are compared with a single-electrode injection and recording (*lower panel*) that is implementing the active electrode compensation (AEC) defiltering: when a fluctuating current is injected by electrode 1 and recorded simultaneously by electrode 1 and electrode 2 (*upper trace*), the severe distortion and artifacts induced in the electrode 1 recordings are apparent. By (off-line) employing the AEC on electrode 1 (*lower panel*), the single-electrode recordings can be hardly distinguished from the control recordings obtained by electrode 2 [(**e**) is modified and reproduced with permission from *Journal of Neurophysiology*, © 2008, The American Physiological Society].

artifacts induced by the partial compensation are reinjected, inducing further undesired voltage responses and sometimes leading to escalating instabilities (66, 67).

However, the single-electrode configuration for dynamic-clamp is particularly attractive for its simplicity and has no alternative in in vivo intracellular recordings (68). Very recently, Brette et al. proposed a successful computer-based technique for the off-line estimation of the pipette electrode filtering as well as its (online) subtraction (66). The method, known as the Active Electrode Compensation (AEC), assumes that a linear, yet unknown, electrical circuit can capture the filtering properties of the electrode (Fig. 4d). Then, U_e is expressed as the convolution between the current injected i_e and an arbitrary kernel $k(n)$, which is neither constant—as in the bridge-balance compensation (i.e., $U_e = R\ i_e$)—nor corresponding to a single-exponential—as in negative capacitance compensation. For digitally sampled signals, the convolution reduces to a sum, which expresses the effects of the past current injected samples on the voltage response U_e:

$$\begin{aligned} U_e(n) &= \sum_{i=0}^{M} k(i)\, i_e(n-i) \\ &= k_0\, i_e(n) + k_1\, i_e(n-1) + k_2\, i_e(n-2) + \cdots \\ &\quad + k_M\, i_e(n-M) \end{aligned} \tag{13}$$

The kernel $k(n)$ therefore captures the entire signal filtering between the digital output of the computer D/A board to the pipette tip. The identification of the kernel (i.e., the numbers k_0, k_1, k_2, …, k_M) must of course precede the application of the dynamic-clamp protocols, and requires injecting subthreshold white-noise currents and cross-correlating this inputs i_e with the recorded output V_e. Software for the full implementation of the AEC method is publicly available by Brette et al. for a variety of programming languages,[8] and makes relatively easy to implement the AEC for (online) compensation (Fig. 4e).

Acknowledgments

We thank Pablo Méndez, Simone Pacioni, and Silvia Marinelli for critically reading the manuscript. This work was supported by the Giovanni Armenise-Harvard Foundation: Career Development Award (A. Bacci); European Commission: Marie Curie International Reintegration Grant (A. Bacci); European Research Council (ERC) under the European Community's 7th Framework Program (FP7/2007-2013)/ERC grant agreement No 200808; the Belgian InterUniversity Attraction Pole (grant n. IUAP P6/29, M. Giugliano), the University of Antwerp (NOI-BOF2009, M. Giugliano), the Flanders Research Foundation (grant n. G.0836.09, M. Giugliano), and the Royal Society (2009/R4, M. Giugliano); A. Bacci is the 2007/2008 NARSAD Henry and William Test Investigator and M. Giugliano is the Emilie Francqui Foundation Professor.

[8] http://cns.iaf.cnrs-gif.fr/hr-cortex.html, http://audition.ens.fr/brette/HRCORTEX/index.html.

References

1. McBain CJ, Fisahn A (2001) Interneurons unbound. Nat Rev Neurosci 2:11–23
2. Markram H, Toledo-Rodriguez M, Wang Y et al (2004) Interneurons of the neocortical inhibitory system. Nat Rev Neurosci 5:793–807
3. Klausberger T, Somogyi P (2008) Neuronal diversity and temporal dynamics: the unity of hippocampal circuit operations. Science 321:53–57
4. Kawaguchi Y, Kubota Y (1997) GABAergic cell subtypes and their synaptic connections in rat frontal cortex. Cereb Cortex 7:476–486
5. Maccaferri G, Lacaille JC (2003) Interneuron diversity series: hippocampal interneuron classifications-making things as simple as possible, not simpler. Trends Neurosci 26:564–571
6. Ascoli GA, Alonso-Nanclares L, Anderson SA et al (2008) Petilla terminology: nomenclature of features of GABAergic interneurons of the cerebral cortex. Nat Rev Neurosci 9:557–568
7. Freund TF, Buzsaki G (1996) Interneurons of the hippocampus. Hippocampus 6:347–470
8. Klausberger T, Magill PJ, Marton LF et al (2003) Brain-state- and cell-type-specific firing of hippocampal interneurons in vivo. Nature 421:844–848
9. Hajos N, Palhalmi J, Mann EO et al (2004) Spike timing of distinct types of GABAergic interneuron during hippocampal gamma oscillations in vitro. J Neurosci 24:9127–9137
10. Somogyi P, Klausberger T (2005) Defined types of cortical interneurone structure space and spike timing in the hippocampus. J Physiol (Lond) 562:9–26
11. Jonas P, Bischofberger J, Fricker D, Miles R (2004) Interneuron diversity series: fast in, fast out-temporal and spatial signal processing in hippocampal interneurons. Trends Neurosci 27:30–40
12. Freund TF (2003) Interneuron diversity series: rhythm and mood in perisomatic inhibition. Trends Neurosci 26:489–495
13. Miles R, Toth K, Gulyas AI et al (1996) Differences between somatic and dendritic inhibition in the hippocampus. Neuron 16:815–823
14. Freund TF, Katona I (2007) Perisomatic inhibition. Neuron 56:33–42
15. Reyes A, Lujan R, Rozov A, Burnashev N et al (1998) Target-cell-specific facilitation and depression in neocortical circuits. Nat Neurosci 1:279–285
16. Scanziani M, Gahwiler BH, Charpak S (1998) Target cell-specific modulation of transmitter release at terminals from a single axon. Proc Natl Acad Sci USA 95:12004–12009
17. Bacci A, Rudolph U, Huguenard JR, Prince DA (2003) Major differences in inhibitory synaptic transmission onto two neocortical interneuron subclasses. J Neurosci 23: 9664–9674
18. Beierlein M, Gibson JR, Connors BW (2003) Two dynamically distinct inhibitory networks in layer 4 of the neocortex. J Neurophysiol 90:2987–3000
19. Wainer BH, Bolam JP, Freund TF, Henderson Z et al (1984) Cholinergic synapses in the rat brain: a correlated light and electron microscopic immunohistochemical study employing a monoclonal antibody against choline acetyltransferase. Brain Res 308:69–76
20. McCormick DA, Connors BW, Lighthall JW, Prince DA (1985) Comparative electrophysiology of pyramidal and sparsely spiny stellate neurons of the neocortex. J Neurophysiol 54:782–806
21. Destexhe A, Rudolph M, Pare D (2003) The high-conductance state of neocortical neurons in vivo. Nat Rev Neurosci 4:739–751
22. Kondgen H, Geisler C, Fusi S et al (2008) The dynamical response properties of neocortical neurons to temporally modulated noisy inputs in vitro. Cereb Cortex 18:2086–2097
23. Arsiero M, Luscher HR, Lundstrom BN, Giugliano M (2007) The impact of input fluctuations on the frequency-current relationships of layer 5 pyramidal neurons in the rat medial prefrontal cortex. J Neurosci 27:3274–3284
24. Martina M, Royer S, Pare D (2001) Cell-type-specific GABA responses and chloride homeostasis in the cortex and amygdala. J Neurophysiol 86:2887–2895
25. Vida I, Bartos M, Jonas P (2006) Shunting inhibition improves robustness of gamma oscillations in hippocampal interneuron networks by homogenizing firing rates. Neuron 49:107–117
26. Bartos M, Vida I, Jonas P (2007) Synaptic mechanisms of synchronized gamma oscillations in inhibitory interneuron networks. Nat Rev Neurosci 8:45–56
27. Khirug S, Yamada J, Afzalov R et al (2008) GABAergic depolarization of the axon initial segment in cortical principal neurons is caused by the Na-K-2Cl cotransporter NKCC1. J Neurosci 28:4635–4639
28. Szabadics J, Varga C, Molnar G et al (2006) Excitatory effect of GABAergic axo-axonic cells in cortical microcircuits. Science 311:233–235
29. Prinz AA, Abbott LF, Marder E (2004) The dynamic clamp comes of age. Trends Neurosci 27:218–224

30. Wilders R (2006) Dynamic clamp: a powerful tool in cardiac electrophysiology. J Physiol 576:349–359
31. Sohal VS, Zhang F, Yizhar O, Deisseroth K (2009) Parvalbumin neurons and gamma rhythms enhance cortical circuit performance. Nature 459:698–702
32. Nagtegaal AP, Borst JG (2010) In vivo dynamic clamp study of I(h) in the mouse inferior colliculus. J Neurophysiol 104:940–948
33. Bacci A, Huguenard JR (2006) Enhancement of spike-timing precision by autaptic transmission in neocortical inhibitory interneurons. Neuron 49:119–130
34. Wolfart J, Debay D, Le Masson G et al (2005) Synaptic background activity controls spike transfer from thalamus to cortex. Nat Neurosci 8:1760–1767
35. Wang DD, Kriegstein AR (2008) GABA regulates excitatory synapse formation in the neocortex via NMDA receptor activation. J Neurosci 28:5547–5558
36. Oprisan SA, Prinz AA, Canavier CC (2004) Phase resetting and phase locking in hybrid circuits of one model and one biological neuron. Biophys J 87:2283–2298
37. Le Masson G, Renaud-Le Masson S, Debay D, Bal T (2002) Feedback inhibition controls spike transfer in hybrid thalamic circuits. Nature 417:854–858
38. Bucher D, Prinz AA, Marder E (2005) Animal-to-animal variability in motor pattern production in adults and during growth. J Neurosci 25:1611–1619
39. Sieling FH, Canavier CC, Prinz AA (2009) Predictions of phase-locking in excitatory hybrid networks: excitation does not promote phase-locking in pattern-generating networks as reliably as inhibition. J Neurophysiol 102:69–84
40. Castaneda-Castellanos DR, Flint AC, Kriegstein AR (2006) Blind patch clamp recordings in embryonic and adult mammalian brain slices. Nat Protoc 1:532–542
41. Sakmann B, Neher E (1995) Single-channel recording. Plenum, New York
42. Bischofberger J, Engel D, Li L, Geiger JR, Jonas P (2006) Patch-clamp recording from mossy fiber terminals in hippocampal slices. Nat Protoc 1:2075–2081
43. Davie JT, Kole MH, Letzkus JJ et al (2006) Dendritic patch-clamp recording. Nat Protoc 1:1235–1247
44. Debanne D, Boudkkazi S, Campanac E et al (2008) Paired-recordings from synaptically coupled cortical and hippocampal neurons in acute and cultured brain slices. Nat Protoc 3:1559–1568
45. Destexhe A, Mainen Z, Sejnowski T (1998) Kinetic models of synaptic transmission. In: Segev I, Koch C (eds) Methods in neuronal modeling. MIT, Cambridge, MA, pp 1–25
46. Giugliano M, Bove M, Grattarola M (1999) Fast calculation of short-term depressing synaptic conductances. Neural Comput 11:1413–1426
47. Lytton WW (1996) Optimizing synaptic conductance calculation for network simulations. Neural Comput 8:501–509
48. Destexhe A, Rudolph M, Fellous JM, Sejnowski TJ (2001) Fluctuating synaptic conductances recreate in vivo-like activity in neocortical neurons. Neuroscience 107:13–24
49. Gillespie DT (1996) Exact numerical simulation of the Ornstein–Uhlenbeck process and its integral. Phys Rev E Stat Phys Plasmas Fluids Relat Interdiscip Topics 54:2084–2091
50. Press WH, Flannery BP, Teukolsky SA, Vetterling WT (1993) Numerical recipes in C: the art of scientific computing. Cambridge University Press, Cambridge
51. Markram H, Tsodyks M (1996) Redistribution of synaptic efficacy between neocortical pyramidal neurons. Nature 382:807–810
52. Sharp AA, O'Neil MB, Abbott LF, Marder E (1993) Dynamic clamp: computer-generated conductances in real neurons. J Neurophysiol 69:992–995
53. Robinson HP, Kawai N (1993) Injection of digitally synthesized synaptic conductance transients to measure the integrative properties of neurons. J Neurosci Methods 49: 157–165
54. Kullmann PH, Wheeler DW, Beacom J, Horn JP (2004) Implementation of a fast 16-bit dynamic clamp using LabVIEW-RT. J Neurophysiol 91:542–554
55. Raikov I, Preyer A, Butera RJ (2004) MRCI: a flexible real-time dynamic clamp system for electrophysiology experiments. J Neurosci Methods 132:109–123
56. Rabbah P, Nadim F (2005) Synaptic dynamics do not determine proper phase of activity in a central pattern generator. J Neurosci 25: 11269–11278
57. Nowotny T, Szucs A, Pinto RD, Selverston AI (2006) StdpC: a modern dynamic clamp. J Neurosci Methods 158:287–299
58. Milescu LS, Yamanishi T, Ptak K et al (2008) Real-time kinetic modeling of voltage-gated ion channels using dynamic clamp. Biophys J 95:66–87

59. Benda J, Gollisch T, Machens CK, Herz AV (2007) From response to stimulus: adaptive sampling in sensory physiology. Curr Opin Neurobiol 17:430–436
60. Hughes SW, Lorincz M, Cope DW, Crunelli V (2008) NeuReal: an interactive simulation system for implementing artificial dendrites and large hybrid networks. J Neurosci Methods 169:290–301
61. Bjaalie JG, Grillner S (2007) Global neuroinformatics: the International Neuroinformatics Coordinating Facility. J Neurosci 27:3613–3615
62. Molleman A (2003) Patch-clamping: an introductory guide to patch clamp electrophysiology. Wiley, New York
63. Geddes LA (1972) Electrodes and the measurement of bioelectric events. Wiley, New York
64. Wilson CJ, Park MR (1989) Capacitance compensation and bridge balance adjustment in intracellular recording from dendritic neurons. J Neurosci Methods 27:51–75
65. Badel L, Lefort S, Brette R et al (2008) Dynamic I–V curves are reliable predictors of naturalistic pyramidal-neuron voltage traces. J Neurophysiol 99:656–666
66. Brette R, Piwkowska Z, Monier C et al (2008) High-resolution intracellular recordings using a real-time computational model of the electrode. Neuron 59:379–391
67. Preyer AJ, Butera RJ (2009) Causes of transient instabilities in the dynamic clamp. IEEE Trans Neural Syst Rehabil Eng 17:190–198
68. Okun M, Naim A, Lampl I (2010) The subthreshold relation between cortical local field potential and neuronal firing unveiled by intracellular recordings in awake rats. J Neurosci 30:4440–4448
69. Varela C, Llano DA, Theyel BB "An Introduction to in vitro slice approaches for the study of neuronal circuitry" in Neuronal network analysis: concepts and experimental approaches, Neuromethods Book Series, Springer
70. Zrenner C, Eytan D, Wallach A, Thier P, Marom S (2010) A generic framework for real-time multi-channel neuronal signal analysis, telemetry control and sub-millisecond latency feedback generation. Front. Neurosci. 4:173

Neuromethods (2012) 67: 173–187
DOI 10.1007/7657_2011_20

Published online: 1 December 2011

Fast Perfusion Methods for the Study of Ligand-Gated Ion Channels

Andrea Barberis

Abstract

The elementary information transfer between two neurons is represented by the generation of a synaptic current in the post-synaptic element due to the activation of post-synaptic receptors by a neurotransmitter quantum. The duration and the amplitude of such signals are largely determined by the properties of post-synaptic receptors and the profile of neurotransmitter concentration sensed by post-synaptic receptors. To date, the knowledge about the properties of post-synaptic receptors activated in synaptic conditions has been limited by the difficulty to control and reproduce the synaptic neurotransmitter exposures. In this chapter, it is shown how to build and optimize devices capable to deliver neurotransmitter pulses approaching those occurring at the synapse. In addition, the role of the neurotransmitter concentration profile at the synaptic cleft in shaping post-synaptic currents is emphasized.

Key words: Synaptic transmission, Post-synaptic receptors, Synaptic cleft, Theta-glass, Neurotransmitter release, Neurotransmitter concentration

1. Introduction

Within neuronal networks, neurons receive a variety of excitatory and inhibitory inputs. It has been shown that projections of specific classes of principal cells and interneurons contact their target neurons in precise subneuronal compartments forming synapses with distinct kinetic properties (1, 2). Such precise temporal and spatial compartmentalization of synaptic inputs supports the idea that concerted network activity strongly relies on the dynamic interplay between excitation and inhibition (3–5) thus highlighting the influence of the synaptic current kinetics on the network functioning. A major limitation in studying the factors determining the kinetics and the amplitude and of synaptic currents is that the profile of the neurotransmitter concentration activating post-synaptic receptors in the cleft cannot be manipulated. This prevents the possibility to assess the precise relationship between neurotransmitter released and the shape of synaptic currents. Experimental data and computer model simulations have clarified that at both

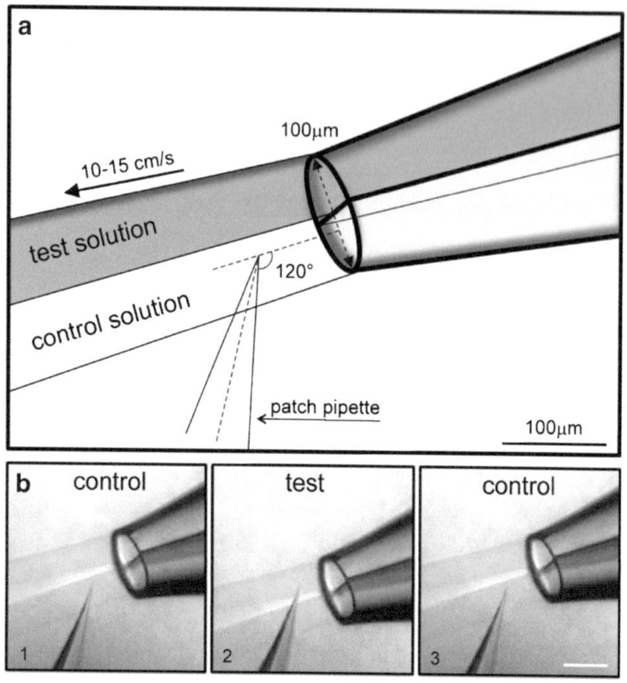

Fig. 1. (**a**) Schematic representation of the theta-barrel based ultra-fast perfusion system. The *control solution* and the neurotransmitter-containing solution (*test solution*) continuously flow in the theta-pipette at the speed of 10–15 cm/s generating a thin solution interface. The patch pipette axis forms a 120° angle with the solution stream axis in order to reduce the mechanic stress. (**b**) Sequence of the theta-pipette translation: (1) The patch pipette is in the *control solution* stream (2) the piezo electric translator quickly moves the theta-tube so that the patch pipette is perfused with the *test solution* (3) the theta-tube is back to the original position.

glutamatergic and GABAergic synapses, after the release of a single synaptic vesicle, the neurotransmitter peaks at high concentration and lasts in the cleft for ~100 µs (6, 7). The neurotransmitter synaptic exposure time is often faster than the time needed by the receptors to fully activate, implying that post-synaptic receptors activate in conditions of nonequilibrium. In such conditions, the amplitude and time course of the synaptic current are dictated by a complex interplay between receptor gating properties and neurotransmitter concentration time course in the cleft. In order to understand the behavior of fast ligand-gated ion channels activated in synaptic conditions, therefore, it is important to study the currents evoked by exogenous neurotransmitter pulses approaching the duration of synaptic neurotransmitter exposures. The basic requirement to deliver brief and controlled agonist pulses is to apply (and remove) the agonist-containing solution in ultra-short time. During the last three decades, neuroscientists have developed several devices attempting to minimize such solution exchange time. Important advancement in the optimization of fast perfusion systems has been obtained by

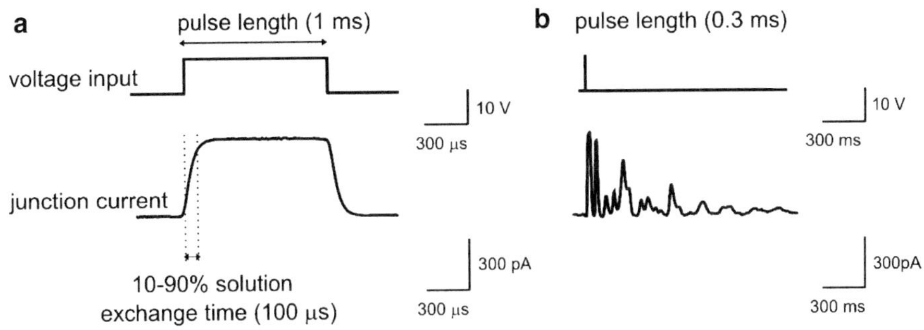

Fig. 2. Test of the solution exchange time and theta-pipette stability monitored with open-tip liquid junction current. (**a**) *Upper panel*: input voltage pulse applied to the piezo electric translator (1 ms, 10 V) *lower panel*: profile of the solution exchange measured at the open tip patch pipette following 1 ms piezo translator pulse. Note that the 10–90% rise time of the transition between control solution and test solution occurs in ~100 μs. Although, for reasons of clarity, the piezo pulse and the junction current are aligned, the piezo pulse precedes the junction current by ~1.5 ms. This delay is mainly due to the time needed to reach the solution interface (**b**) Example of critical vibrations induced by piezo pulses <1 ms. *Upper panel:* input voltage pulse applied to the piezo electric translator (0.3 ms, 10 V) *Lower panel:* pipette solution exchange profile in response to 0.3 ms piezo translator pulse showing oscillations of the theta-pipette solution interface across the patch pipette tip. Note the different scale bar with respect to **a**.

(1) creating a thin interface between the control and neurotransmitter-containing solutions to be moved over the sample and (2) limiting the surface of the sample to be perfused (8, 9). A stable interface between two solutions can be easily created by exploiting theta-glass tubes (so-called because the profile of the capillary mouth recalls the Greek letter "theta" see Fig. 1a). Briefly, a theta-barrel forms two independent channels continuously conveying the control solution and the neurotransmitter-containing solution, respectively (Fig. 1a). The patch pipette, with an excised-out patch on its tip, is positioned in front of the control stream. By means of a fast piezo electric translator, the theta-barrel is switched in a way that the patch pipette tip crosses the solution interface and jumps in the neurotransmitter stream. Subsequently, the theta is switched back to the original position bringing the patch pipette tip in the control solution stream (Fig. 1b). Using this device it is possible to obtain solution exchange time of ~100 μs, while the shortest pulse duration achievable is typically ~1 ms (Fig. 2a). Such fast agonist exposure is still longer than that occurring at the synapse during synaptic transmission but represents an excellent tool to study post-synaptic receptors activated in a time range relevant for synaptic transmission. In addition, since currents mediated by fast-ligand gated channels display submillisecond rise time (~300 μs), the exchange time of ~100 μs is adequate to satisfactory resolve the activation kinetics of post-synaptic receptors. Exploiting this system, therefore, it is possible to apply neurotransmitter pulses at precise concentration and duration. Researchers usually refer to this technique as *ultra-fast perfusion*, *concentration jumps*, *or liquid filaments* (8, 9). The study of the

behavior of ligand-gated receptors challenged by brief agonist applications emphasizes the role of the neurotransmitter pulse timing in shaping synaptic current. During synaptic transmission, the neurotransmitter reaches high concentration (1–3 mM) at both glutamatergic and GABAergic synapses (6, 7). Such concentrations are saturating when applied in equilibrium conditions. Nonetheless, it has to be borne in mind that, at the synapse, post-synaptic receptors contact the neurotransmitter for only 50–100 μs (7). In these conditions, 1–3 mM GABA and glutamate are no longer saturating (7, 10). This indicates, for instance, that the degree of activation of post-synaptic receptors (hence the amplitude of synaptic currents) depends upon the duration of the "synaptic pulse" in the cleft (11). In addition to the influence on current amplitude, the neurotransmitter timing also modulates the kinetics of synaptic currents by unmasking specific features of post-synaptic receptor gating. The influence of the neurotransmitter time course on the current kinetics strongly relies on the fine tuning of fast receptor desensitization (12). In this context, the use of ultra-fast perfusion has been crucial to demonstrate that "synaptic-like" agonist pulses (1 ms) induce a significant accumulation of desensitization raising the possibility that fast desensitization could participate in synaptic transmission. Although desensitization is a nonconductive state, it plays a central role in dictating the kinetics of ligand-gated receptors (13). The deactivation kinetics of NMDA and $GABA_A$ receptors, for instance, is slowed down by multiple reopenings of receptors previously accumulated in the desensitized state (14, 15). In contrast, in AMPA/kainate receptors (KARs), desensitization contributes to fast deactivation kinetics (16). The role of fast desensitization in determining the properties of synaptic currents in response to different agonist exposure time is exemplified by the behavior of post-synaptic KARs at mossy fibers-CA3 synapses in the hippocampus. Synaptic currents mediated by KAR (KAR-EPSCs) show slow decay time and play an important role in determining the hippocampal network properties (17). The slow decay of synaptic current significantly differs from the extremely fast kinetics of current mediated by recombinant KAR expressed in heterologous systems. By exploiting ultra-fast perfusion system, it has been possible to elucidate that brief ("synaptic-like") glutamate exposures elicit slow decaying currents, while longer glutamate applications elicit fast decay kinetics (12) (Fig. 3d). This behavior has been shown to be attributable to the presence of two different binding sites on GluK2/GluK5 KARs, one with high affinity and poor desensitization and one with low affinity and strong desensitization (Fig. 3d) (18). During synaptic exposure, due to the short interaction of neurotransmitter with the post-synaptic receptors, the high affinity/low desensitizing site is preferentially activated yielding slow decaying currents. One of the most important advantages of ultra-fast perfusion system is the possibility to record the peak amplitude

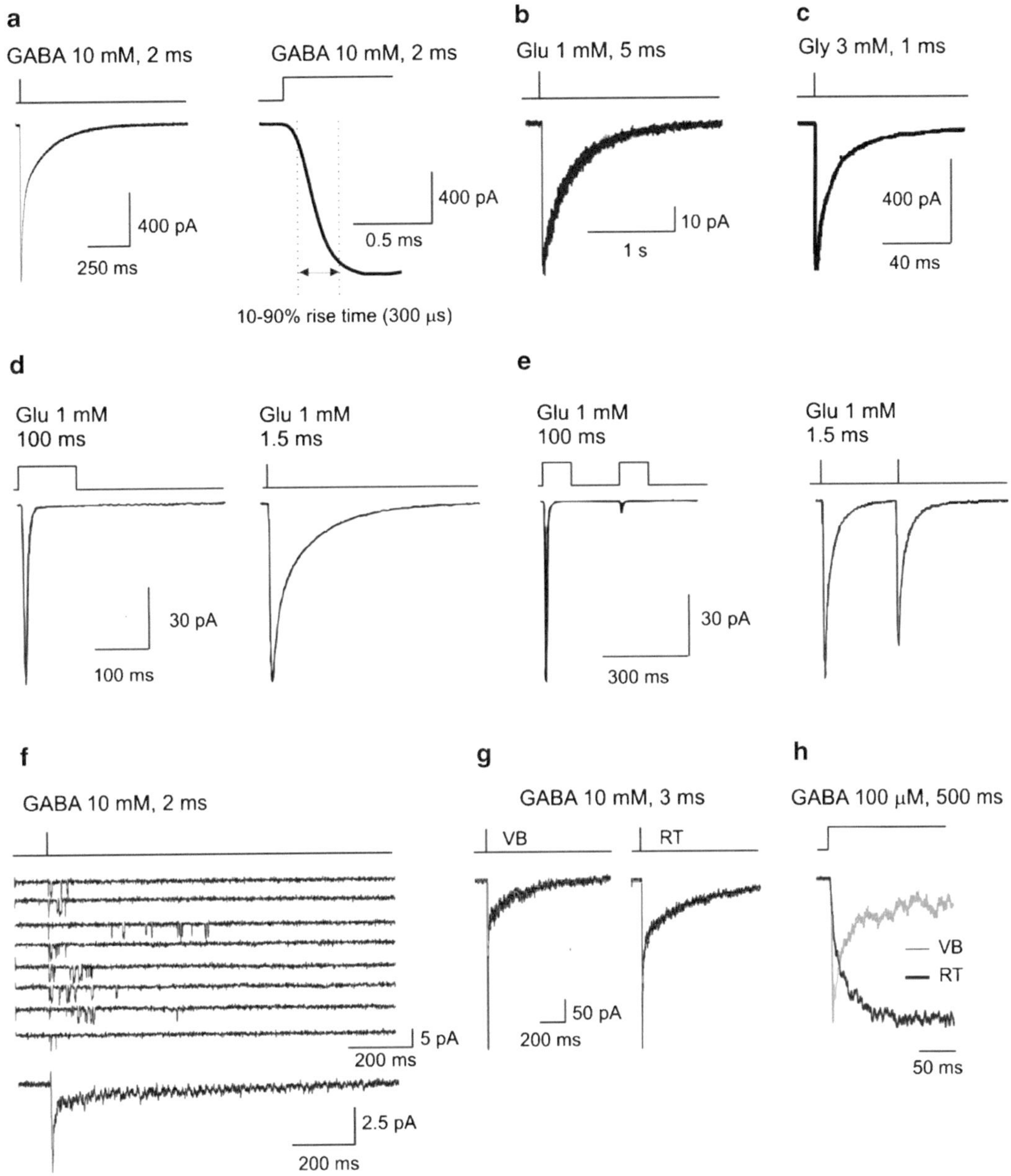

Fig. 3. Examples of traces recorded using the ultra-fast perfusion. (**a**) *Left panel*: $GABA_AR$-mediated currents evoked by brief (2 ms) pulses of saturating GABA (10 mM). Currents were recorded in excised-out patches pulled from HEK293-cells transfected with $\alpha1\beta2\gamma2$ $GABA_AR$ subtype. *Right panel*: same trace as in **a** shown in expanded time scale in order to appreciate the fast current onset (10–90% rise time: 300 μs). (**b**) NMDAR-mediated current (GluN1-GluN2C subtype) in response to 1 mM glutamate and 0.5 mM glycine (5 ms pulse). Adapted from (32). (**c**) Currents mediated by GlyRs (1 ms glycine 3 mM). Adapted from (33). (**d**) Short (1.5 ms) and long (100 ms) pulses applied to recombinant kainate receptors (GluK2/GluK5 subtype). Note that long pulses elicit fast-decaying currents while short ("synaptic-like") pulses unmask slow deactivation kinetics. Adapted from (12) (**e**) Desensitization properties of GluK2/GluK5 receptors. *Left panel*: two consecutive 100 ms pulses (300 ms gap) show strong induction of desensitization. *Right panel*: delivering of 1.5 ms pulse pairs (to the same patch as in *left panel*) induces poor receptor desensitization. Adapted from (12). (**f**) Example of $GABA_AR$-mediated currents evoked by short pulses of saturating GABA resolved at single-channel level. *Upper panel*: seven sweeps showing the stochastic single channel activation. *Bottom panel*: macroscopic current obtained by averaging the seven single channel sweeps shown in the *upper panel*. Adapted from (34). (**g**) Example of current obtained from patches pulled from ventrobasal (VB, *left panel*) and reticular (RT, *right panel*) thalamic nuclei. Adapted from (19). (**h**) Currents recorded in patches pulled from ventrobasal and reticular nuclei shown normalized and expanded time scale. Note the slower onset kinetics of patches pulled from reticular nucleus with respect to that of ventrobasal patches. Adapted from (19).

of the current with minimal distortions. Using traditional slow perfusion system, indeed, the current onset is often mixed with fast desensitization implying that the peak amplitude is systematically underestimated. This leads, for instance, to misinterpretation of receptor affinity/efficacy obtained from dose–response studies (13). By pulling patches from neurons in culture or in acute/organotypic slices ultra-fast perfusion technique allows mapping the kinetics and the amplitude of synaptic currents in different neuron subtypes (19). As shown in Fig. 3g, h, for instance, $GABA_AR$-mediated currents obtained from patches pulled from either ventrobasal or reticular thalamic nuclei show markedly different kinetics, paralleling the difference in kinetics of IPSCs recorded in the same brain regions (19). Altogether, this system is suitable for detecting changes in number and/or gating properties of postsynaptic receptors after, for instance, the application of drugs or plasticity protocols. Technically, it is possible to expand the use of fast perfusion in in-vivo experiments by pulling patches from, e.g., deep layers of the cortex otherwise difficult to be studied with optogenetic approaches for the poor light penetration. The ultrafast perfusion system is the only perfusion system providing the "concentration clamp," meaning that is possible to precisely control the concentration profile "seen" by the patch. On the other hand, the major drawback of this system is that the best performances are obtained in outside-out excised patch. In this experimental configuration, the channels are probably devoided of intracellular factors that might be important for their proper functioning. The rapid application of controlled neurotransmitters pulse at individual intact synapses is, in fact, a major technological challenge. Laser photolysis of caged neurotransmitters, for instance, is believed to deliver "synaptic-like" neurotransmitter pulses restricted at the synapse. Using two-photon laser excitation, indeed, it is possible to focally uncage a small volume of neurotransmitter in spots comparable to the synapse size (~250 nm). The enormous advantage of this method is to elicit currents at individual putative synapses. In addition, the versatility of light illumination allows stimulating with complex spatial and temporal patterns. The main disadvantage of this technique, however, is that the neurotransmitter pulse concentration and duration cannot be fully controlled. In particular, it is impossible to quickly remove the agonist. The laser excitation, indeed, generates a high neurotransmitter concentration spot that, subsequently, spreads in uncontrolled way in the extra-synaptic space. It could be objected that the transmitter uncaging system would mimic the intrinsic mechanisms of synaptic release, as neurotransmitter diffuses in the synaptic cleft from a concentrated spot (the synaptic vesicle). At the synapse, however, the synaptic vesicle has diameter ~20 nm and contains neurotransmitter at concentration ~100–150 mM while, with the uncaging technique, low millimolar concentration of neurotransmitter diffuses from a 250 nm spot. The situation is certainly very different and might account for some kinetic differences between uncaging-evoked currents (uPSCs) and mPSCs. In keeping with this, current evoked by laser photolysis of caged-GABA showed slower rise rime (20) and markedly slower decay kinetics (20, 21). Glutamatergic current elicited by glutamate photorelease, in contrast, closely matches mEPSCs (22) although some groups reported slower decay kinetics of uEPSCs with respect to mEPSCs (23). To this concern, it has to be stressed that the similarity between uEPSCs and mEPSCs could be only apparent, due to the fact that desensitization and deactivation kinetics of AMPA receptors are relatively similar. In other words, the slower glutamate exposure obtained with laser uncaging (with respect to the synaptic glutamate release) would be masked by fast AMPA receptor desensitization. In addition, caged compounds (especially the GABA-caged compounds) are not completely inert in their caged form, acting as competitive antagonists for $GABA_A$ receptors (20). A possible competition between the photoreleased GABA and the GABA-caged could further account for the slower kinetics of uIPSCs with respect to mIPSCs (20). Another system capable of fast neurotransmitter delivery restricted to submicron areas is the micro-iontophoretic device. High neurotransmitter concentration (1–5 M) is ejected from a small pipette (~50–100 mΩ) at submillisecond pulses creating a cloud of neurotransmitter with size ~500 nm (24). Similarly to photoreleased agonists, the neurotransmitter diffuses in uncontrolled way and cannot be quickly removed. With respect to the uncaging technique, however, this approach has the advantage to apply the actual neurotransmitter and not modified molecules that are potential receptor blockers. In addition to the uncaging technique, the use of light to study post-synaptic receptors could be theoretically exploited by generating light-sensitive ligand-gated receptors. This has been made possible by conjugating ligand-gated receptors with photoisomerizable compounds (e.g., azobenzene) coupled to the agonist molecule of interest (25, 26). In brief, light triggers azobenzene isomerization with consequent interaction of the agonist with its binding site. Interestingly, specific illumination wavelength can selectively induce the *cis-* or *trans-* isomerization determining the agonist binding and unbinding, respectively. The azobenzene isomerization therefore represents a "reversible cage" allowing precise control of the pulse length timing. The considerable advantage with respect to uncaging is that the stimulus duration is precisely controlled and spatially restricted to the illuminated area. Although this technique is extremely powerful to control cell excitability, it remains to be established to what extent receptor gating induced by light is equivalent to that operated by "regular" agonist binding (27). The generation of modified ligand-gated receptors activated by light with the same properties of native receptors activated by regular neurotransmitter of will represent a fundamental breakthrough in the study of the properties, function, and physiology of ligand-gated receptors. In the next sections, we show how to build and optimize the theta-barrel-based ultra-fast perfusion.

2. Materials

1. Theta capillaries (Hilgenberg, Germany). Outer diameter: 2.0 mm, Inner diameter: 1.4 mm, Septum thickness: 0.1 mm. Capillary length: 10 cm
2. P-1000 micro-pipette puller (Sutter Instruments, Rome, Italy)
3. Diamond-tip engraving pen (Sigma-Aldrich, Milano, Italy)
4. Tubings: (1) Inner diameter: 1.6 mm Outer diameter: 3.2 mm; (2) Inner diameter: 0.4 mm Outer diameter: 0.8 mm (Tygon, Harvard Apparatus, Holliston, MA, USA)
5. Syringe infusion pump SP220IZ (WPI, Berlin, Germany)
6. Piezo translator P-245.50 (Physik Instrumente, Karlsruhe, Germany)
7. High-Voltage Piezo Amplifier (HVPZT) E-470 (Physik Instrumente, Karlsruhe, Germany)

3. Methods

3.1. Theta-Pipette Preparation

Theta-glass capillaries are commercially available from different dealers (Hilgenberg, Germany; Sutter, CA, USA; Clark Electromedical Instruments, England; WPI, FL, USA). Standard theta-tubing sizes are: outer and inner diameter 2.0 and 1.4 mm, respectively, and septum thickness ~0.2 mm. It is advisable, however, to request custom-sized septum thickness of ~0.1 mm (9). Before pulling a pipette, it is important to check the symmetry of the theta-channels. In 10–15% of theta-pipettes, indeed, the septum might be slightly off-center generating two barrels with different size. These pipettes generate significant distortion of the solution stream and should be discarded. In order to prepare theta-pipettes, the capillary is first thinned to the desired diameter (~100–150 μm) by using a standard pipette puller. Subsequently, the pipette is cut at the lowest pipette diameter by scratching the glass with a diamond tip and gently bending the capillary. Horizontal pullers, allowing precise control of the pulling force and velocity are preferred with respect to the vertical pullers. Generally, the aforementioned procedure allows the preparation of two pipettes with a blunt tip end, an important requirement to avoid turbulences in the solution streams. However, it is not infrequent to break the glass in an irregular way. Pipettes showing pronounced irregularities in their tip cannot be used. Once obtained a proper theta-pipette it is necessary to connect both channels to the solutions. To this end, small polyethylene tubing is inserted in both theta-ending channels until reaching the theta-tip. Some users prefer to use small quartz capillaries that, generally, are smaller and stiffer than polyethylene tubing, easing the insertion procedure. The disadvantage is that quartz capillaries are more fragile and less flexible than polyethylene ones. Subsequently, once the tubings are inserted into the theta-barrels, the theta-pipette back is sealed by filling the theta-channels with two component epoxy resins or cyanoacrylic glue. It is advisable to leave as little dead volume as possible. Epoxy resins need to be injected into the theta back using syringes, while cyanoacrylic glues will easily fill the theta-barrels by capillarity. Both glues are assumed not to release any bioactive compound.

3.2. Creating the Solution Streams

In order to flow the control and the neurotransmitter solutions into the theta-barrels, different approaches have been adopted. Some users exploit gravity. The flow rate, in this case, is determined by the elevation of the solution pools with respect to the application pipette. This experimental setup has the fundamental problem that the solution speed would change according to the solution level. Moreover, even small differences in the solution pathway might significantly change the relative speed between the two channels with consequent alterations of the solution interface. In contrast, a stable and balanced solution stream can be easily obtained by using infusion syringe pumps. We use an SP220IZ from WPI (Germany) but other similar devices guarantee the same performances. Syringe pumps allow creating a constant flow along the whole experiment and, importantly, the flow rate can be readily changed according to the experimental conditions. Typically, the flow rate is set to 6–10 ml/h (solution speed 10–15 cm/s, through a theta-mouth diameter 100–150 μm). This speed allows satisfactory quality of the solution interface and should not be critical for the patch stability. Different configurations from outside out patch including nucleated patch and whole cell recording (discussed below) will require a markedly lower solution rate (~4–6 and ~2–4 ml/h, respectively). Importantly, solution must be carefully filtered, not only because debris in the solution would ruin the patch, but also because the theta-pipette would severely block (in particular the small ones). In addition, it is necessary to avoid formation of air bubbles in the solution stream as they would be fatal for the patch. Heating up the solutions and letting the temperature to relax back to room temperature greatly minimizes the risks of air bubbles in the solution (9). Additional bath perfusion is very important in order to avoid neurotransmitter accumulation in the recording chamber. This is particularly important using high agonist concentrations. Bath perfusion must be oriented in the same direction of the theta-glass solution stream in order to prevent bending and turbulence of the solution interface. To mount the theta-pipette onto the piezo electric translator, we build small Plexiglas blocks with a hole that fits the theta-pipette. Then, through a second hole in the Plexiglas the complex Plexiglas + theta-pipette is screwed directly to the piezo translator. It is important that in the working position the theta-septum is oriented perpendicularly to the microscope stage. To achieve that, the theta in the Plexiglas hole is kept loose until the pipette is placed in the working position so it can be rotated and adjusted. Once the correct septum position is reached, the theta-pipette is glued to the Plexiglas block. The piezo translator is then connected to an XYZ micromanipulator on the recording rig. In order to minimize the mechanical pressure exerted by the solution stream on the patch, the angle formed by the theta and patch pipette should be around 120° (Fig. 1a). This configuration also reduces the risk of collision with debris expelled from the theta-pipette (9, 28).

3.3. Checking the Theta-Pipette Performance

The quality of a theta-pipette is evaluated on both the speed and reliability of the exchange between the control and the test solutions (Fig. 2a). This test can be done by performing open-tip recordings of liquid junction currents. To this end, standard Ringer's solution (control) and 10% of standard Ringer's solution (test) flow in each of the two theta-channels, respectively. Subsequently, a test patch pipette (filled with Ringer's solution) is positioned in front of the control solution stream. During the displacement of the theta-pipette by the piezo electric translator, as soon as the test solution contacts the patch pipette, a liquid junction current is detected by patch clamp amplifier. The junction current then precisely reports the solution concentration temporal profile "seen" be the patch pipette tip. Typically, both the rise time and the relaxation of the junction current (indicating the time needed to exchange the solution around the open tip) are ~100 μs (Fig. 2a). It is important to apply positive pressure to the test patch pipette in order to avoid artifacts due to partial dilution of the solution on the tip of the recording pipette. In addition to the solution exchange time, it is also crucial to assess whether the duration and shape of the junction current matches the input pulse to the piezo. Distortions of the junction current pulse due to unwanted pipette oscillations are discussed in the next sections. In order to obtain reliable pulses, the positioning in front of the theta it is crucial. The distance from the theta-mouth (*y*-axis), is typically 100 μm, while the distance from the septum (*x*-axis) and the *z*-axis positioning has to be determined for every theta-pipette. Usually, the best exchange speed and reliability is obtained by positioning the patch pipette very close to the solution interface, although this does not represent a rule. It has to be pointed out that solution exchange time measured in open tip recording might differ from that obtained in excised-out patch configuration. The solution exchange speed around an excised patch can be assessed by recording the depolarization-induced current following application of high-potassium solution (9). In our hands, the open tip recordings and the excised patch recordings pulled from hippocampal cultured neurons slightly differed (10–20% slower in excised patch with respect to open tip).

3.4. Finding the Best Recording Configuration

As discussed in the previous sections, the best performances in terms of solution exchange speed are obtained in excised patch configuration. However, if the patch contains a small number of receptors making the current undetectable, for instance, nucleated-patch could represent a preferable configuration. In nucleated-patch configuration, indeed, the membrane surface is ~25 times larger than in excised-patch thus increasing the probability to include functional channels. In order to obtain nucleated patches, during the procedure of patch excision, it is necessary to apply a slight negative pressure to the patch pipette. In most cases, the nucleus will leave the neuron and will be included in the patch. In this configuration, the exchange time is limited by the time needed for the solution to cover the membrane surface (300–500 μs). This exchange time is still reasonably fast, although it is not indicated for studies aiming to resolve the fast gating processes of the receptor. It is also possible to perform fast perfusion in the whole-cell configuration. A lifted cell (typically transfected HEK cells) is positioned in front of the theta-pipette with solution exchange times ~1–3 ms. In this configuration, the exchange time heavily affects the current kinetics. However, even in the whole-cell configuration, theta-glass-based perfusion will perform better than most of older perfusion systems (including multibarrel systems, puffs, etc.). As a general remark, the degree of exchange speed has to be chosen according to the kinetics of the process to be studied. For instance, ligand-gated receptors activated by low agonist doses display slow activation rate. In this condition, the slowdown of the solution exchange due to the nucleated or whole-cell configuration is not rate-limiting.

3.5. Application of Multiple Solutions

Many experiments require the application of more than two solutions. In particular, studying the effect of drugs or dose–response curves requires applications of many different solutions to the same patch. To this end, it is possible to build a system of valves controlling multiple inflows into both theta-barrels. This setup certainly assures versatility but has some drawbacks. First of all, the complete solution replacement in the theta-barrel takes from 20 to 60 s according to the distance of the valves from the theta-pipette. This can be critical with excised patches exhibiting short lifetime. Secondarily, in case of use of very lipophilic and/or high affinity drugs, it is difficult to establish when the solution has been completely washed out. Some researchers therefore prefer to use duplet or triplets of theta-barrel pipettes. In this configuration, during the experiment, the patch pipette is moved from one to another theta, minimizing the switching time and avoiding solution mix.

3.6. Piezoelectric Devices

The theta-pipette has to be moved in and out the test solution in a very brief time. In order to achieve this, it is necessary to use a piezo translators showing high resonance frequency. We use the Physik Instrumente P-245.50 with excursion of 80 μm and resonance frequency 8 KHz (9). In order to perform an entire cycle in the millisecond range, it is imperative to drive the piezoelectric translator with a high-voltage piezo amplifier (HVPZT). We use an E-470 (from Physik Instrumente) yielding an output voltage of 1,100 V, with peak output power of 550 W. Using these devices, it is possible to translate the theta-glass pipette in less than 50 μs. Shorter translation times can be detrimental since they can generate vibrations.

3.7. Practical and Theoretical Limits of Solution Exchange Speed Around the Patch

At the interface between two solutions with different concentrations, there is a concentration gradient due to the fact that one solution partially diffuses into the other, respectively. Theoretically, in the case of solutions flowing in the theta-barrels, assuming zero septum thickness, the sharpest concentration gradient is obtained at $t = 0$ when the solutions first contact after leaving the theta-mouth. At $t > 0$, the gradient will dissipate according to: (29)

$$C(x, t) = 0.5 \cdot \mathrm{erfc}[x/(4Dt)^{0.5}] \tag{1}$$

where x is the distance, D is the diffusion coefficient, and t is the time. At distance from the theta-mouth $d = 1$ μm, stream flow speed $v = 100$ cm/s, the two solutions contact for a time $t = v/d = 1$ μs. For diffusion coefficient $D = 10^{-5}$ cm^2/s, according to (1), such "contact time", generates a 10–90% gradient length of ~0.1 μm. Moving such interface across a patch with radius $r = 1$ μm, at piezo translation speed of 100 cm/s, the 10–90% of the exchange time will be achieved in ~1 μs (28). Considering that both the flow and translation speed reported above cannot be reasonably further increased, this is the best exchange time value achievable (28). In practice, good experimental values are: 10–90% solution interface length ~15–25 μm and 10–90% solutions exchange time ~80–100 μs. The difference from the ideal values is given by the fact that (1) the septum thickness is ~10–20 μm (2) the maximum flow rate tolerated by the patch is 10–15 cm/s and (3) the theta-translation speed is ~50 cm/s. Model simulations revealed that at septa thickness >1 μm, the solution gradient does not increase monotonically with the distance from the theta-mouth. Due to the presence of stagnation solution layers in front of the septum, the point at which the interface is the thinnest (also called "sweet point") is located few tens of microns away from the theta-mouth (and not right at the end of the theta-pipette as expected in ideal conditions) (28). The position of the sweet point depends on both the septum thickness and the solution flow rate. In typical experimental conditions (septum thickness 10–20 μm and flow rate 10–15 cm/s), it lays ~50–100 μm away from the theta-mouth (9, 28). The sharpness of the solution interface at the sweet point critically depends on the flow rate. As stated above, excised-out patches tolerate 10–15 cm/s as maximal solution flow rate. Very stable patches, however, can be perfused at even higher flow rates (20–30 cm/s). This improves the solution exchange-time up to ~60 μs. In addition, the septum thickness can be significantly thinned by pulling smaller theta-tip endings (diameter ~50 μm). Using both high speed flow and small theta-pipettes, it is possible to reach solution exchange-time to ~40 μs. This represents the very best realistic performance achievable using such devices. It has to be pointed out however that small theta-pipettes might show severe vibration artifacts (see below). The operator, therefore, has to find out the best compromise between speed and stability.

4. Notes

4.1. Cancelation of Vibration Artifacts

Solution exchange time in the range of 40–80 μs would theoretically allow the application of square-like neurotransmitter pulses with durations in the submillisecond range. In practice, however, the pulse length is limited by the theta-capillary and piezo electric translator oscillations. Imposing submillisecond pulses, indeed, the system reacts with uncontrolled vibrations determining unwanted transitions of the solution interface across the patch pipette with consequent distortions of the pulse from the expected concentration profile (Fig. 2b). For this reasons, using the equipment and the procedures illustrated so far, it is very difficult to obtain reliable pulses with controlled duration <1 ms. However, sometimes oscillations appear even at pulses length >1 ms. In this case, some practices can decrease the probability of the system to vibrate. First of all, bigger pipettes vibrate much less than smaller ones. Theta-pipettes with mouth diameter in the range of 100–150 μm are relatively stable at pulses >1 ms. Furthermore, filtering the input TTL to the piezo amplifier may help in eliminating vibrations by inducing a smoother theta-barrel shift. On the contrary, stiffening the theta-pipette shaft with metal or plastic shields does not eliminate vibrations. If it is envisaged to deliver double pulses or pulse trains, it is not sufficient to check for system stability at single

1 ms pulses. More complex protocols might unmask critical harmonics of the system otherwise undetected using single pulse protocols. Noteworthy, some research groups optimized a challenging procedure allowing the cancelation of the system vibrations (30, 31). In brief, the oscillation waveform induced by imposing submillisecond pulse (monitored at the patch pipette by using liquid junction potential) is used to appropriately model a transfer function that, once applied as voltage input waveform to the piezo electric actuator, generates a vibration-free pipette displacement. The detailed procedure of the transfer function modeling is behind the purposes of this book and is comprehensively described in (30). Using this method, it is possible to deliver reliable "square-like" pulses of ~200 μs duration (31).

Acknowledgments

I wish to thank Enrica M. Petrini and Jerzy W. Mozrzymas for critical reading of the manuscript. This work was supported by the Fet Proactive7 grant to Fabio Benfenati (Italian Institute of Technology, Neuroscience and Brain Technology, Italy) and AB.

References

1. Freund TF, Katona I (2007) Perisomatic inhibition. Neuron 56:33–42
2. Karayannis T, Elfant D, Huerta-Ocampo I et al (2010) Slow GABA transient and receptor desensitization shape synaptic responses evoked by hippocampal neurogliaform cells. J Neurosci 30:9898–9909
3. Klausberger T, Somogyi P (2008) Neuronal diversity and temporal dynamics: the unity of hippocampal circuit operations. Science 321:53–57
4. Spruston N (2008) Pyramidal neurons: dendritic structure and synaptic integration. Nat Rev Neurosci 9:206–221
5. Klausberger T (2009) GABAergic interneurons targeting dendrites of pyramidal cells in the CA1 area of the hippocampus. Eur J Neurosci 30:947–957
6. Clements JD, Lester RA, Tong G et al (1992) The time course of glutamate in the synaptic cleft. Science 258:1498–1501
7. Mozrzymas JW, Zarnowska ED, Pytel M et al (2003) Modulation of GABA(A) receptors by hydrogen ions reveals synaptic GABA transient and a crucial role of the desensitization process. J Neurosci 23:7981–7992
8. Franke C, Hatt H, Dudel J (1987) Liquid filament switch for ultra-fast exchanges of solutions at excised patches of synaptic membrane of crayfish muscle. Neurosci Lett 77 (2):199–204
9. Jonas P (1995) Fast application of agonists to isolated membrane patches. Single-channel recording. In: Sakmann B, Neher E (eds) Single-channel recording. Plenum press, New York, pp 231–243
10. Popescu G, Robert A, Howe JR et al (2004) Reaction mechanism determines NMDA receptor response to repetitive stimulation. Nature 430:790–793
11. He L, Wu XS, Mohan R et al (2006) Two modes of fusion pore opening revealed by cell-attached recordings at a synapse. Nature 444:102–105
12. Barberis A, Sachidhanandam S, Mulle C (2008) GluR6/KA2 kainate receptors mediate slow-deactivating currents. J Neurosci 28:6402–6406

13. Mozrzymas JW, Barberis A, Mercik K et al (2003) Binding sites, singly bound states, and conformation coupling shape GABA-evoked currents. J Neurophysiol 89:871–883
14. Lester RA, Jahr CE (1992) NMDA channel behavior depends on agonist affinity. J Neurosci 12:635–643
15. Jones MV, Westbrook GL (1995) Desensitized states prolong GABAA channel responses to brief agonist pulses. Neuron 15:181–191
16. Partin KM, Patneau DK, Winters CA et al (1993) Selective modulation of desensitization at AMPA versus kainate receptors by cyclothiazide and concanavalin A. Neuron 11:1069–1082
17. Sachidhanandam S, Blanchet C, Jeantet Y et al (2009) Kainate receptors act as conditional amplifiers of spike transmission at hippocampal mossy fiber synapses. J Neurosci 29:5000–5008
18. Mott DD, Rojas A, Fisher JL et al (2010) Subunit-specific desensitization of heteromeric kainate receptors. J Physiol 588:683–700
19. Mozrzymas JW, Barberis A, Vicini S (2007) GABAergic currents in RT and VB thalamic nuclei follow kinetic pattern of alpha3- and alpha1-subunit-containing GABAA receptors. Eur J Neurosci 26:657–665
20. Trigo FF, Papageorgiou G, Corrie JE et al (2009) Laser photolysis of DPNI-GABA, a tool for investigating the properties and distribution of GABA receptors and for silencing neurons in situ. J Neurosci Methods 181:159–169
21. Matsuzaki M, Hayama T, Kasai H et al (2010) Two-photon uncaging of gamma-aminobutyric acid in intact brain tissue. Nat Chem Biol 6:255–257
22. Matsuzaki M, Ellis-Davies GC, Nemoto T et al (2001) Dendritic spine geometry is critical for AMPA receptor expression in hippocampal CA1 pyramidal neurons. Nat Neurosci 4: 1086–1092
23. DiGregorio DA, Rothman JS, Nielsen TA et al (2007) Desensitization properties of AMPA receptors at the cerebellar mossy fiber granule cell synapse. J Neurosci 27:8344–8357
24. Heine M, Groc L, Frischknecht R et al (2008) Surface mobility of postsynaptic AMPARs tunes synaptic transmission. Science 320:201–205
25. Gorostiza P, Isacoff EY (2008) Optical switches for remote and noninvasive control of cell signaling. Science 322:395–399
26. Gorostiza P, Isacoff EY (2008) Nanoengineering ion channels for optical control. Physiology (Bethesda) 23:238–247
27. Numano R, Szobota S, Lau AY et al (2009) Nanosculpting reversed wavelength sensitivity into a photoswitchable iGluR. Proc Natl Acad Sci U S A 106:6814–6819
28. Sachs F (1999) Practical limits on the maximal speed of solution exchange for patch clamp experiments. Biophys J 77:682–690
29. Carslaw H (1959) Conduction of heat in solids. Clarendon, Oxford
30. Stilson S, McClellan A, Devasia S (2001) High-speed solution switching using piezo-based micropositioning stages. IEEE Trans Biomed Eng 48:806–814
31. Moffatt L, Hume RI (2007) Responses of rat P2X2 receptors to ultrashort pulses of ATP provide insights into ATP binding and channel gating. J Gen Physiol 130:183–201
32. Dravid SM, Prakash A, Traynelis SF (2008) Activation of recombinant NR1/NR2C NMDA receptors. J Physiol 586:4425–4439
33. Pitt SJ, Sivilotti LG, Beato M (2008) High intracellular chloride slows the decay of glycinergic currents. J Neurosci 28:11454–11467
34. Barberis A, Mozrzymas JW, Ortinski PI et al (2007) Desensitization and binding properties determine distinct alpha1beta2gamma2 and alpha3beta2gamma2 GABA(A) receptor-channel kinetic behavior. Eur J Neurosci 25:2726–2740

Part III

Electrophysiological Approaches In Vivo

Neuromethods (2012) 67: 191–203
DOI 10.1007/7657_2011_21

Published online: 4 December 2011

Electrophysiological Approaches for Studying Neuronal Circuits In Vivo

George Dragoi

Abstract

Most of our understanding of the brain activity and its relation to behavior comes from the electrophysiological studies of neuronal activity at different levels. Extracellular recording of spiking and local field potential activity gives the most comprehensive picture of the brain activity at the neuronal ensemble level within a region as well as across brain areas. Juxtacellular recording approach is best suited for investigating the activity of individual classes of neurons that are further characterized in relation to their morphology, molecular content, and axonal projection profile. Intracellular recordings uncover intrinsic and synaptic subthreshold dynamics of individual neurons that are not detectible using the other two techniques. Electrical and optical stimulation of neurons complements the recording techniques and are used to artificially control the neuronal activity with incredibly high temporal resolution. The current work reviews these four approaches for studying neuronal activity in vivo and provides an overview of the most recent advances and trends in this field.

Key words: Local field potential, Juxtacellular recordings, Intracellular recordings, Electrical stimulation, Optogenetics

1. Introduction

The living brain is essentially an electrical organ at the interface with the external world. Consequently, most of our knowledge regarding the neuronal correlates of behavior comes from studying the electrical activity of the brain using a variety of electrophysiological techniques. Among them, the oldest and most frequently used to study the living brain is the in vivo extracellular recording. The recent years in the neurophysiology of extracellular recordings have seen a trend toward miniaturization, increased neuronal sampling, and a diversity of combinations between electrophysiology and other neurotechniques. In addition, intracellular recordings, classically used only in anesthetized preparations have been performed successfully in behaving animals (1, 2) further expanding the repertoire of electrophysiological approaches for studying neuronal circuits in awake behaving animals.

The large variety of in vivo electrophysiological approaches can be classified using several criteria. Depending on the location of the electrode in relation to the recorded cellular activity, the techniques can be extracellular, when the electrode is outside the cell, juxtacellular, when the electrode is outside but very near to the cell, or intracellular, when the tip of the electrode is contiguous with the internal milieu of the cell. Classically, in vivo intracellular and juxtacellular recordings have been performed in anesthetized animals to minimize the electrode movement while extracellular recordings have been targeting the neuronal activity in behaving animals. Extracellular techniques can be used to record spikes from multiple or single cell activity in which case they measure the overall output of the cell or to record local field potentials that are generated by the sum of the presynaptic inputs and cellular firing in the sampled area. Based on their effect on the spontaneous activity of the brain, electrophysiological approaches can be passive, as is generally the case for recording, or active, when extrinsic electrical current or light are applied to interfere with (i.e., to excite or to block) the ongoing brain activity. For a fast read-out and online feedback, stimulation is used in combination with recording techniques.

2. Extracellular Recordings of Neuronal Spiking Activity

Extracellular recordings of neuronal spiking activity measure large fast changes in the flow of ions in the extracellular milieu of the cell. In vivo simultaneous extracellular and intracellular recordings from the same neuron in the CA1 area of the rat hippocampus have shown that the extracellular changes in voltage around the neuron during action potentials are best approximated by the first derivative of the intracellular changes in voltage (3). Since both processes generally happen at the submillisecond timescale, in vivo electrophysiology has the best temporal resolution of all the techniques for assessing neuronal activity. Unlike intracellular technique, extracellular recording of spiking activity is blind to the subthreshold depolarization that contributed to the generation of the recoded action potential (see below the section on recording of local field potentials).

Historically, extracellular recordings were performed using single electrodes implanted in one region of the brain in behaving animals. Fundamental advances in our understanding of the neural code were made possible using this approach, among them being the discovery of "place cells" in the rat hippocampus (4), orientation-sensitive cells in the cat visual cortex (5), "face cells" in the monkey inferotemporal cortex (6), or reward prediction signal in the monkey dopamine cells (7). In all of these cases, neurons were recorded one at a time and temporal relationship among neurons from the same region or from connected areas of the brain was ignored. The improvements in the extracellular recording technique over the single site, single area electrodes developed later on in several directions with the primary goal of enabling the study of neuronal interaction and organization within and across brain regions.

First, the need to better separate multiple single cells recorded from the same electrode led to the development of multiple site electrodes originally created by twisting two or four thin wires together. Subsequently, the wires were glued together into one macroscopic electrode by partially melting their insulation using heat thus producing the stereotrode and the tetrode (8), respectively. The tetrode in particular became the standard electrode for extracellular recordings in the hippocampus and the connected areas, where it significantly improved the separation and the overall yield of multiple single neurons by offline clustering procedures (Fig. 1). Currently, tetrodes can be machine manufactured using thin, flat silicon probes where the four recording sites are distributed along the shank of the electrode in a diamond shape with about 20 μm along its sides (Neuronexus probes). An eight site per shank "octotrode" version is also available (9).

A second improvement concerned an increase in the total number of electrodes and recording sites that allowed for simultaneous recording of multiple single units from the same, two, or multiple brain areas in behaving animals. Light "hyperdrives" made of 3D-printed plastic materials (10) can now carry as many as 32 or more independently moveable tetrodes that can be distributed over several regions of the brain. The use of similar hyperdrives opened for the first time the possibility of recording and studying the activity of large ensembles of neurons (11) from selected areas as well as the interaction between anatomically or functionally connected brain regions (12). Phenomena like the offline replay of spatial trajectories in the CA1 area of the hippocampus (13) and its coordination with similar replay activity observed in the upstream primary visual cortex (14) could be discovered in part due to the existence of this technology. An alternative to the hyperdrive is the multiple shanks multiple sites silicon probes (15) that can simultaneously record neuronal activity from multiple nearby and connected areas with the advantage of being light-weight, slim-design causing less brain tissue damage, and requiring no manufacturing time and effort for the end-user. On the other hand, the advantages of the tetrode wires over the silicon-based ones are the independent control of each electrode's position and depth in the brain, the low-cost manufacturing, and the control and flexibility over their construction time and availability.

3. Considerations on the Analysis of Large Recording Datasets

The advancements in the extracellular recording techniques resulted in the simultaneous recording of a large number of cells which allowed researchers to study the temporal relationship between

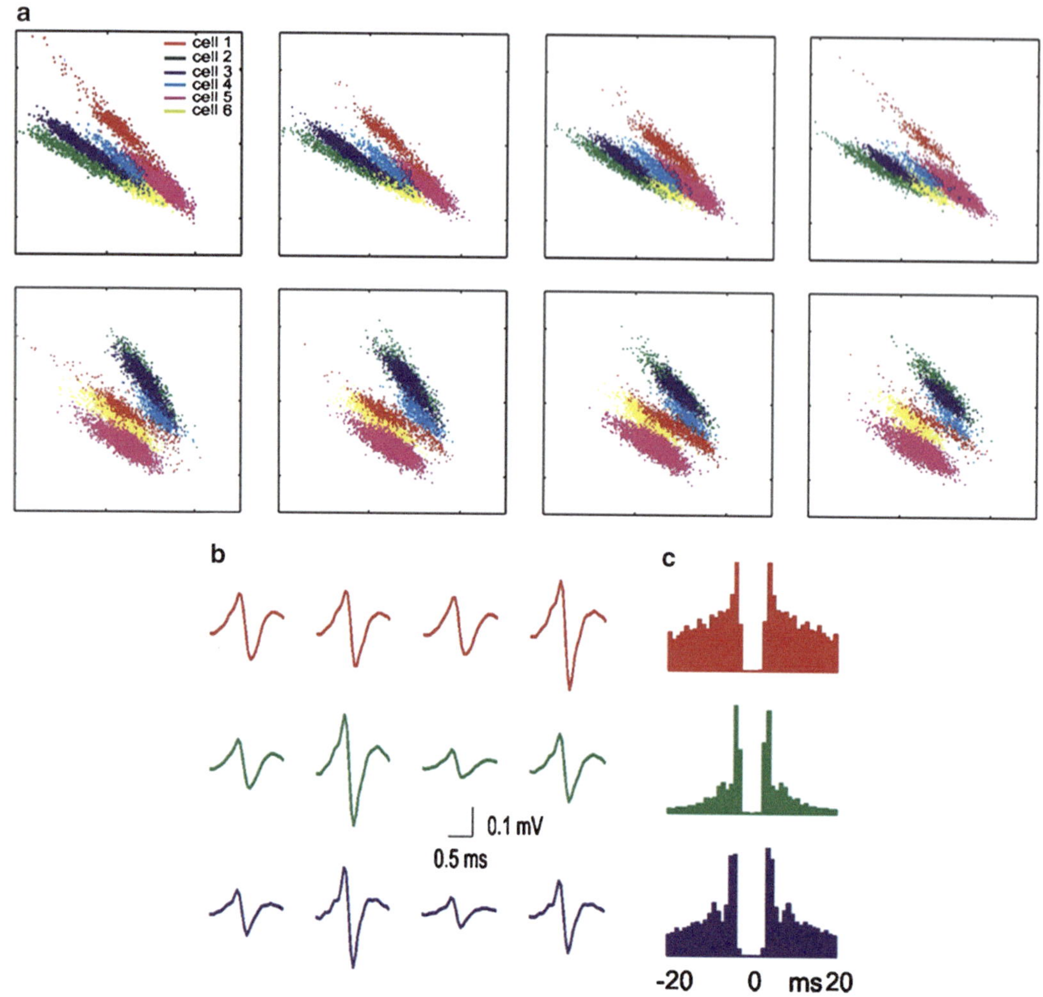

Fig. 1. Tetrode recording of multiple single neurons from the rat hippocampus. (**a**) Multiple single cell clusters (color-coded) obtained by triangulation of the spike amplitudes recorded on tetrode sites 1 versus 2 (*top* four panels) and 3 versus 4 (*bottom* four panels) during four successive run sessions (columns). (**b**) Average waveforms of three simultaneously recorded cells (rows, color-coded) on each of the four wires of the tetrode (columns). (**c**) Autocorrelograms corresponding to the cells in (**b**) Reprinted and adapted from [Neuron], (26), Copyright (2003) with permission from Elsevier.

ensemble of cells within and across brain areas. In order to isolate single cells out of the recorded pool of multiunit spikes, several offline clustering programs were developed. All clustering programs were based on the multidimensional triangulation of the features (mainly amplitude) of the voltage change as recorded by individual wires of stereotrodes or tetrodes (11). Semiautomatic routines that diminish the subjectivity, time, and effort of the human operator have also been developed to assist researchers with clustering (16), but to date the clustering process has not been fully automated. Instead, a number of postclustering analyses have been developed to indicate the relative quality of each cell cluster (17).

4. Extracellular Recording of Local Field Potentials

Every neuron receives a large number of excitatory and inhibitory synaptic inputs that interact with the intrinsic dynamics of its membrane potential to generate action potentials in an all-or-none fashion. In addition to the detection of action potentials as described in the above sections, extracellular electrodes are used to record the population of excitatory and inhibitory potentials generated locally in the form of local field potential. A popular approach uses the same physical electrode to record both multiple single units and local field potentials that can later be correlated using temporal and phase analyses (see also chapters "Investigating Sleep Homeostasis with Extracellular Recording of Multiunit Activity from the Neocortex in Freely-Behaving Rats" and "Analysis of Hippocampal Memory Replay Using Neural Population Decoding"). Using this approach, O'Keefe and Recce discovered the phenomenon of phase precession of hippocampal place cell spikes relative to the theta oscillation (18). A further development in the local field potential recording technique was made possible by using the silicon linear probes in which 16–32 recording sites are equally spaced along the shank of the electrode thus simultaneously recording the neuronal activity across several layers within a brain region (19, 20) (Fig. 2). Based on the difference in voltage transients across adjacent recording sites, multiple phase-shifted electrical dipoles can be detected during various oscillatory activities in the region. The dipoles are formed by pairs of electrical sinks and sources localized at different depths in the brain region sampled by the electrode. Electrical sinks generally reflect a local depolarization in response to excitatory presynaptic inputs. Electrical sources indicate either local inhibitory inputs or return

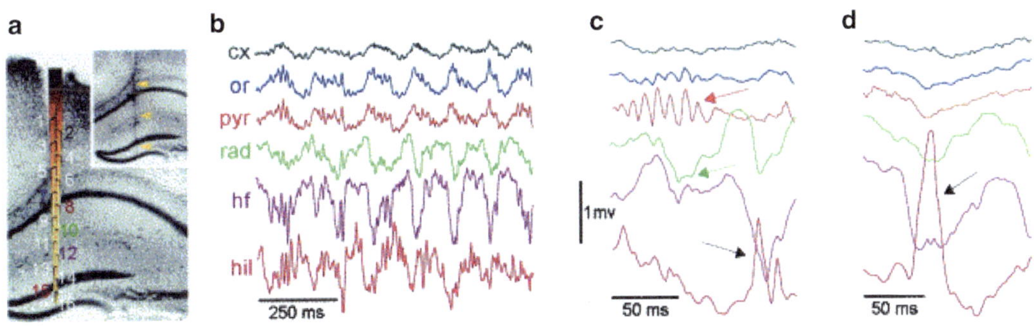

Fig. 2. Depth profile of hippocampal local field potential activity in the mouse. (**a**) Location of a 16-site silicon probe (100-μm spacing) in the CA1-dentate gyrus axis. Inset, *yellow arrowheads* indicate the actual location of the probe in the brain. (**b**) Theta and gamma oscillations simultaneously recorded from different brain regions and layers as animals ran in a wheel (cx, neocortex; or, stratum oriens; pyr, pyramidal layer; rad, stratum radiatum; hf, hippocampal fissure; hil, hilus). (**c**) Fast field oscillation ("ripple," *red arrow*) in the pyramidal cell layer associated with a negative sharp wave in stratum radiatum (*green arrow*). *Black arrow* points to a dentate spike with polarity reversal across the dentate granule cell layer. (**d**) An isolated dentate spike. Reprinted from [Neuroscience], (20), Copyright (2003) with permission from Elsevier.

currents and together with an associated electrical sink form an electric dipole. The detection of active sinks superimposed on the anatomical location of their maximum amplitude is a powerful way of detecting layer-specific inputs in a specific brain region in vivo arriving at particular times or phases of a know oscillatory pattern. Using silicon-based linear probes implanted in the hippocampus, it became possible to describe the dynamic spatial–temporal profile of the main oscillatory patterns in the CA1 area, namely the 6–12 Hz theta oscillation (21), the 40–100 Hz gamma oscillation (22), and the 150–200 Hz ripple oscillation (23) during associated sharp waves. More recently, probes with linear arrangements of recording sites are used to detect layer-specific neuronal spiking activity in the neocortex together with the corresponding local field potentials (9) (see also chapter "Determination of Cortical Circuit Function Using Current Source-Density (CSD) Analysis In Vivo").

5. Electrical and Optical Stimulation of Neuronal Activity

Extracellular electrodes passively record neuronal spiking activity and local field potentials in the absence of any electrical interference with the brain activity. However, simple observation of electrophysiological phenomena is only correlational and insufficient to extract causal relationships between phenomena. In order to gain control over neuronal activity at the spiking or local field potential level, electrical neuronal stimulation or optical stimulation of genetically engineered neurons have been used successfully in the past and the more recent years, respectively. In particular, the long-term effects of patterned electrical stimulation on the brain activity have been a significant driving force in the field of learning and memory since the discovery of long-term potentiation by Bliss and Lomo (24) at the perforant path to dentate gyrus synapse in the rabbit hippocampus. In vivo experiments have subsequently addressed the role of electrically induced long-term potentiation on learning (25) and tuning properties of single neurons (26). The field of epilepsy research has also benefited from the method of electrical stimulation as a model for induction of seizures (27). Electrical stimulation of axon terminals, also known as antidromic stimulation, in conjunction with extracellular recordings at the cell body has been proven useful in determining the axonal projection and connectivity pattern of the recorded neuron (28). In the more recent years, the discovery of genetically engineered opsins has resulted in a new research paradigm, the field of optogenetics (29) (see Part IV in the present volume) which has promised to revolutionize the neuroscience research. The expression of channelrhodopsin (30) into specific classes of neurons and its activation with blue light followed by neuronal excitation and spiking has taken over the field of electrical stimulation. On the other hand, the expression of halorhodopsin (31) molecules in selected neurons followed by their temporary inhibition upon yellow light stimulation has in most cases replaced the need for crude, irreversible brain lesions.

6. Combination of Extracellular Recordings with Other Techniques

The high temporal precision of the in vivo extracellular recording technique has made it the most accurate method for measuring the neuronal activity. However, its invasive nature has made it incompatible with the measurement of neuronal activity in normal human subjects where, instead, functional magnetic resonance imaging (fMRI) technique has been successfully employed. In an attempt to translate the changes in blood oxygenation levels measured by fMRI into their electrophysiological equivalent, simultaneous fMRI and extracellular recordings have been performed in monkeys. The results indicate that increases in the fMRI BOLD signal are most correlated with the local field potential activity, primarily with the local oscillatory activity in the gamma range (32), rather than with the spiking activity of the imaged neurons (33). This somehow surprising finding indicates that the fMRI technique primarily measures the overall strength of the input into the imaged region rather than its output in the form of spiking activity. More recently, the optogenetic techniques have been used in vivo in combination with behavioral measurements in order to assess the role of specific circuits of particular regions of the brain in determining or controlling animal's behavior (34). However, with very few exceptions, the in vivo application of optogenetic techniques has not simultaneously monitored the effects that light activation had on the electrical activity of the specific neuronal circuits at the time of the particular behavior. In order to demonstrate that the resulting behavior at the time of optical stimulation is due to the hypothesized changes in neuronal activity, simultaneous extracellular recordings and optical stimulation will need to be performed. In cases when stimulation and recording take place in the same brain region, one can use a combined "optotrode" made by gluing together an optic fiber and an extracellular recording electrode.

7. Wireless and Multiplexed Recordings

The increase in the number of extracellular recording electrodes and sites led to an increase in the volume of the wire tether used to transmit the neuronal electrical signals to an external amplifier. This increase could interfere with animal's sleep and exploratory

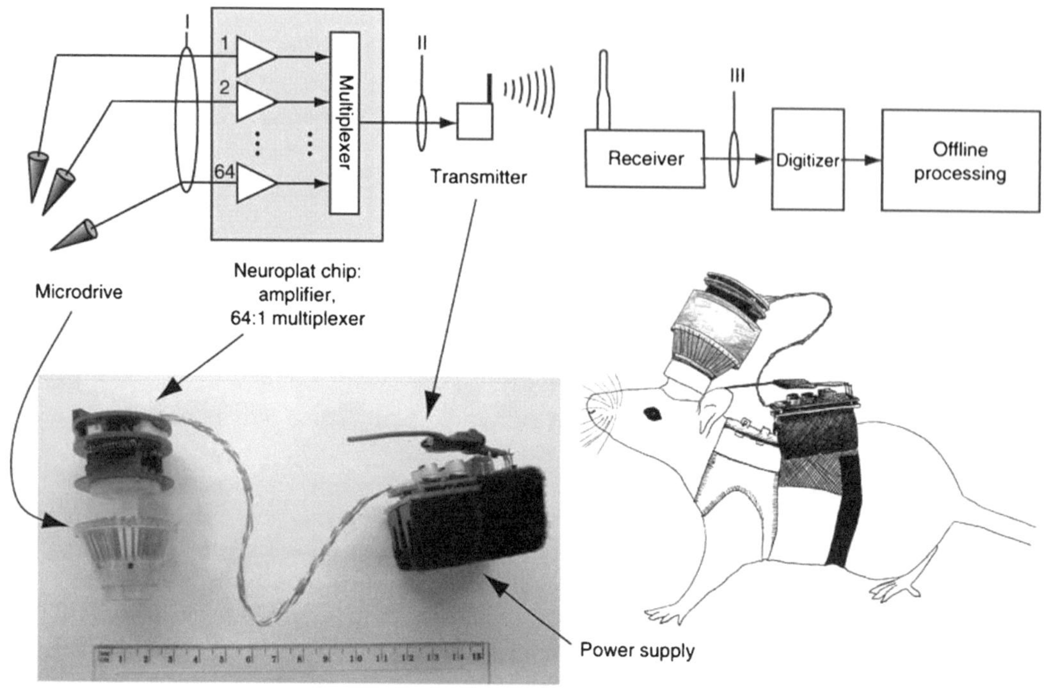

Fig. 3. Wireless recording system. *Top*, Schematic of the system. *Bottom*, System overview, showing all components (*left*) and the complete system worn by the rat (*right*). Reprinted by permission from Macmillan Publishers Ltd.: [Nature Neuroscience], (35), Copyright (2011).

behavior and potentially bias the overall electrophysiological results. A first step in decreasing the bulk of wires connected on the animal's head was achieved by multiplexing the signal from 32 independent recording channels into one wire channel (32:1 ratio) used to transmit the electrical signal to the external amplifier (Triangle Biosystems). A more thorough solution consists in the total elimination of the tether by the use of wireless transmission of up to 64 channels of data (35) (Fig. 3) to a receiver placed 3–4 m away from the implanted animal (i.e., telemetry). The preamplifier is left on the animal's head for the entire duration of the recording and is powered by a long-lasting battery placed either on the animal's head or on a pack around its body. Telemetry allows the animals to more freely move throughout the environment with the possibility of them engaging into social and more natural exploratory behaviors that were previously not possible in a tethered recording system.

8. Juxtacellular Recording and Stimulation of Single Neurons

One of the aims of electrophysiology is to be able to identify and morphologically characterize the neuron whose activity is recorded. To this end, this can be achieved using the juxtacellular recording (36) followed by the stimulation and labeling of the recorded neuron. The juxtacellular recording is essentially a form of extracellular recording in which a glass electrode filled with an intracellular tracer, such as neurobiotyn or biocytin, is placed in close proximity of one neuron. Following completion of the in vivo juxtacellular recording, electrical current is passed through the recording electrode to intensely stimulate the neuron and through electroporation to pass the tracer across the cell membrane inside the neuron. The tracer fills the neuron whose morphology, dendritic arborization, and axonal trajectory can be further reconstructed histologically. The strength of this method consists in the correlation that can be established between the electrophysiological and histological features of the neurons, to which additional labeling for particular neuronal markers is generally being added. Using this technique, it was possible to define the cell types and the projection pattern from the basal forebrain cholinergic system into the neocortex (37). An important caveat is that juxtacellular technique could only be applied to characterize one neuron per brain region per animal, making this technique laborious compared with the most modern techniques of extracelluar recording when the activity of several tens of neurons could be simultaneously monitored. In addition, in cases where two or more neurons were accidentally labeled with the tracer, the identity of the recorded neuron could not be established with certainty thus diminishing the usability of the associated electrophysiological recording. Most juxtacellular recordings were performed in anesthetized or head-fixed animals in order to reduce the electrode movement relative to the neuron. However, more recently the technique has been applied to freely moving animals (38).

9. Intracellular In Vivo Recordings

Extracellular and juxtacellular recording techniques measure the all-or-none output of the neurons in the form of action potentials. However, these techniques cannot measure and distinguish between the intrinsic subthreshold activity of the neuron and the inputs that the neuron receives in the form of excitatory or inhibitory postsynaptic potentials both of which contribute to the generation of action potentials. The aim of the in vivo intracellular recording techniques is to fill in this gap by recording the subthreshold intrinsic activity and the postsynaptic potentials preceding and leading to the action potential generation while still enabling stimulation and labeling of the recorded neuron (see chapter "Intracellular Whole-Cell Patch-Clamp Recordings of Cortical Neurons in Awake Head-Restrained Mice"). Classically, the use of in vivo intracellular recording techniques was limited to anesthetized animals partially due to the reduced electrode movement with respect to the recorded neuron and partially due to the relatively large dimensions of the intracellular recording device. Using this approach it could be determined for instance that both inhibitory and excitatory postsynaptic potentials participated in the generation of the 150–200 Hz ripple oscillation in the CA1 area of the hippocampus (39), an epoch associated with sequence replay during sleep (13) and resting (40). While gaining access to the inside of the neuron in vivo is a remarkable step in itself, the presence of the anesthetic can make it difficult to generalize the results to a freely behaving animal situation. Moreover, certain neuronal activities are brought about only by specific behaviors in which the freely moving animals engage, an example being the case of hippocampal place cells active during spatial exploration in rodents. Recent technical advances have partially overcome the previous limitations in two ways. A group of experiments first established a strongly sealed, whole cell patch recording that was later stabilized by gluing the glass pipette to the skull. Subsequently, the animals were allowed to freely move in space for a few minutes while the intracellular activity was recorded with a miniaturized device (1). Unprecedented records of changes in the subthreshold intracellular membrane potential and of postsynaptic potentials were obtained from the neocortex and hippocampal putative place cells while rats engaged in spatial exploration. A second group of experiments used an awake, head-fixed preparation in mice trained to run on an air-filled flexible ball which under computer-control maintained the head position fixed while animals ran and took turns in a virtual environment context. The intrinsic and synaptic subtreshold and spiking activity of virtual place cells as well as phase precession could be recorded as animals ran on the ball through the virtual environment (2). In both sets of cases, however, the success rate of intracellular recording remained low while the overall duration of recording sessions was from a few minutes to less than 1 h in the best case. These caveats of using the in vivo intracellular recordings in freely moving animals together with the limited number of cells recorded per animal constitute limitations to a more general use of this method in the neuroscience research.

10. Conclusions

The electrophysiological approach to study the brain activity in vivo encompasses a range of very successful techniques that despite being among the oldest in the field of neuroscience are continuously improved to satisfy new levels of understanding of the brain function. Among the most remarkable trends observed in

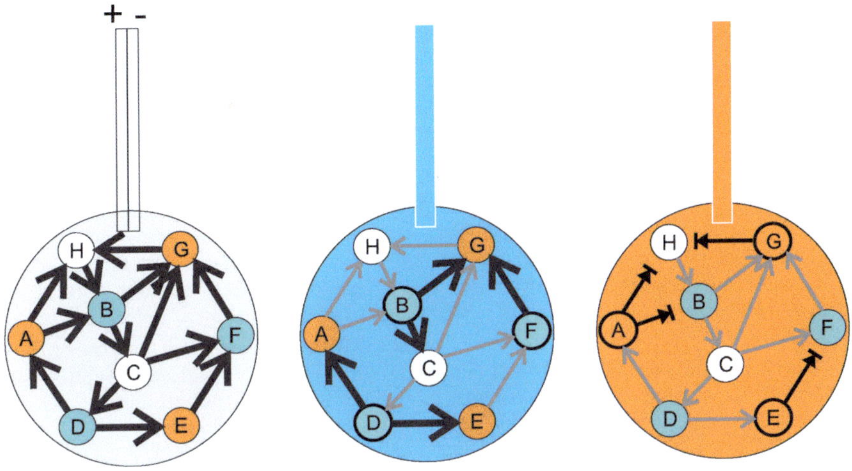

Fig. 4. Comparison between electrical and optogenetic stimulation. Electrical stimulation (*left*) activates all the cells in the vicinity of the electrode. Optogenetic excitation (as with a *blue* light-activated channelrhodopsin, *middle*) or inhibition (as with a *yellow* light-activated halorhodopsin, *right*) is selective to the cells expressing the particular type of opsin.

electrophysiology are the extracellular recordings of larger ensembles of neurons from single or connected brain areas, the use of juxtacellular and intracellular recordings in awake behaving animals, and the newly born field of optogenetics that combines the light-induced stimulation of genetically specified classes of neurons (Fig. 4) with recordings and measurements of behavior performance in freely moving animals. This rather impressive technological improvement stimulated the development of new data analysis methods that required larger number of neurons, such as the Bayesian decoding of behavior (41, 42), the design of new behavioral paradigms adapted to intracellular recordings in awake animals, such as the virtual reality maze (2), and the engineering of more selective and potent (re)agents in the field of optogenetics (29). The next few years will certainly witness important discoveries and advancement in our understanding of the brain function thanks to the broad availability and use of all these technological tools.

References

1. Lee AK, Manns ID, Sakmann B et al (2006) Whole-cell recordings in freely moving rats. Neuron 51(4):399–407
2. Harvey CD, Collman F, Dombeck DA et al (2009) Intracellular dynamics of hippocampal place cells during virtual navigation. Nature 461(7266):941–946
3. Henze DA, Borhegyi Z, Csicsvari J et al (2000) Intracellular features predicted by extracellular recordings in the hippocampus in vivo. J Neurophysiol 84(1):390–400
4. O'Keefe J, Dostrovsky J (1971) The hippocampus as a spatial map. Preliminary evidence from unit activity in the freely-moving rat. Brain Res 34(1):171–175
5. Hubel DH, Wiesel TN (1959) Receptive fields of single neurones in the cat's striate cortex. J Physiol 148:574–591
6. Gross CG, Rocha-Miranda CE, Bender DB (1972) Visual properties of neurons in inferotemporal cortex of the Macaque. J Neurophysiol 35(1):96–111

7. Schultz W, Dayan P, Montague PR (1997) A neural substrate of prediction and reward. Science 275(5306):1593–1599
8. Gray CM, Maldonado PE, Wilson M et al (1995) Tetrodes markedly improve the reliability and yield of multiple single-unit isolation from multi-unit recordings in cat striate cortex. J Neurosci Methods 63(1–2):43–54
9. Bartho P, Hirase H, Monconduit L et al (2004) Characterization of neocortical principal cells and interneurons by network interactions and extracellular features. J Neurophysiol 92 (1):600–608
10. Kloosterman F, Davidson TJ, Gomperts SN et al (2009) Micro-drive array for chronic in vivo recording: drive fabrication. J Vis Exp (26). pii: 1094. doi: 10.3791/1094
11. Wilson MA, McNaughton BL (1993) Dynamics of the hippocampal ensemble code for space. Science 261(5124):1055–1058
12. Siapas AG, Wilson MA (1998) Coordinated interactions between hippocampal ripples and cortical spindles during slow-wave sleep. Neuron 21(5):1123–1128
13. Lee AK, Wilson MA (2002) Memory of sequential experience in the hippocampus during slow wave sleep. Neuron 36 (6):1183–1194
14. Ji D, Wilson MA (2007) Coordinated memory replay in the visual cortex and hippocampus during sleep. Nat Neurosci 10(1):100–107
15. Harris KD, Csicsvari J, Hirase H et al (2003) Organization of cell assemblies in the hippocampus. Nature 424(6948):552–556
16. Harris KD, Henze DA, Csicsvari J (2000) Accuracy of tetrode spike separation as determined by simultaneous intracellular and extracellular measurements. J Neurophysiol 84(1):401–414
17. Schmitzer-Torbert N, Jackson J, Henze D et al (2005) Quantitative measures of cluster quality for use in extracellular recordings. Neuroscience 131(1):1–11
18. O'Keefe J, Recce ML (1993) Phase relationship between hippocampal place units and the EEG theta rhythm. Hippocampus 3 (3):317–330
19. Csicsvari J, Henze DA, Jamieson B et al (2003) Massively parallel recording of unit and local field potentials with silicon-based electrodes. J Neurophysiol 90(2):1314–1323
20. Buzsaki G, Buhl DL, Harris KD et al (2003) Hippocampal network patterns of activity in the mouse. Neuroscience 116(1):201–211
21. Kocsis B, Bragin A, Buzsaki G (1999) Interdependence of multiple theta generators in the hippocampus: a partial coherence analysis. J Neurosci 19(14):6200–6212
22. Csicsvari J, Jamieson B, Wise KD et al (2003) Mechanisms of gamma oscillations in the hippocampus of the behaving rat. Neuron 37 (2):311–322
23. Buzsaki G, Horvath Z, Urioste R et al (1992) High-frequency network oscillation in the hippocampus. Science 256(5059):1025–1027
24. Bliss TV, Lomo T (1973) Long-lasting potentiation of synaptic transmission in the dentate area of the anaesthetized rabbit following stimulation of the perforant path. J Physiol 232 (2):331–356
25. Moser EI, Krobert KA, Moser MB et al (1998) Impaired spatial learning after saturation of long-term potentiation. Science 281 (5385):2038–2042
26. Dragoi G, Harris KD, Buzsaki G (2003) Place representation within hippocampal networks is modified by long-term potentiation. Neuron 39(5):843–853
27. Buzsaki G, Hsu M, Horvath Z et al (1992) Kindling-induced changes of protein kinase C levels in hippocampus and neocortex. Epilepsy Res Suppl 9:279–283
28. Tepper JM, Martin LP, Anderson DR (1995) GABAA receptor-mediated inhibition of rat substantia nigra dopaminergic neurons by pars reticulata projection neurons. J Neurosci 15 (4):3092–3103
29. Deisseroth K (2011) Optogenetics. Nat Methods 8(1):26–29
30. Boyden ES, Zhang F, Bamberg E et al (2005) Millisecond-timescale, genetically targeted optical control of neural activity. Nat Neurosci 8(9):1263–1268
31. Zhao S, Cunha C, Zhang F et al (2008) Improved expression of halorhodopsin for light-induced silencing of neuronal activity. Brain Cell Biol 36(1–4):141–154
32. Niessing J, Ebisch B, Schmidt KE et al (2005) Hemodynamic signals correlate tightly with synchronized gamma oscillations. Science 309 (5736):948–951
33. Logothetis NK, Pauls J, Augath M et al (2001) Neurophysiological investigation of the basis of the fMRI signal. Nature 412(6843):150–157
34. Tsai HC, Zhang F, Adamantidis A et al (2009) Phasic firing in dopaminergic neurons is sufficient for behavioral conditioning. Science 324 (5930):1080–1084
35. Szuts TA, Fadeyev V, Kachiguine S et al (2011) A wireless multi-channel neural amplifier for freely moving animals. Nat Neurosci 14 (2):263–269
36. Pinault D (1996) A novel single-cell staining procedure performed in vivo under electrophysiological control: morpho-functional

features of juxtacellularly labeled thalamic cells and other central neurons with biocytin or Neurobiotin. J Neurosci Methods 65(2):113–136

37. Duque A, Balatoni B, Detari L et al (2000) EEG correlation of the discharge properties of identified neurons in the basal forebrain. J Neurophysiol 84(3):1627–1635
38. Isomura Y, Harukuni R, Takekawa T et al (2009) Microcircuitry coordination of cortical motor information in self-initiation of voluntary movements. Nat Neurosci 12(12):1586–1593
39. Ylinen A, Bragin A, Nadasdy Z et al (1995) Sharp wave-associated high-frequency oscillation (200 Hz) in the intact hippocampus: network and intracellular mechanisms. J Neurosci 15(1 Pt 1):30–46
40. Foster DJ, Wilson MA (2006) Reverse replay of behavioural sequences in hippocampal place cells during the awake state. Nature 440 (7084):680–683
41. Zhang K, Ginzburg I, McNaughton BL et al (1998) Interpreting neuronal population activity by reconstruction: unified framework with application to hippocampal place cells. J Neurophysiol 79(2):1017–1044
42. Davidson TJ, Kloosterman F, Wilson MA (2009) Hippocampal replay of extended experience. Neuron 63(4):497–507

Neuromethods (2012) 67: 205–218
DOI 10.1007/7657_2011_6

Published online: 25 November 2011

Determination of Cortical Circuit Function Using Current Source-Density Analysis In Vivo

Michael J. Higley

Abstract

The recording of local field potentials (LFPs) has become a major tool in the analysis of electrical signaling in the brain. By yielding information about correlated activity in small neuronal populations, LFPs have been used to study mechanisms of sensory encoding, motor planning, cognitive function, and the occurrence of pathological activity patterns seen in epilepsy. However, the relatively poor spatial resolution of LFP signals often impedes characterization of the cellular processes underlying their generation. Thus, current source-density (CSD) analysis has been developed as a means to enhance the discrimination of extracellular current sinks and sources that are produced by neural synaptic activity. CSD analysis has been used by a number of groups to construct functional maps of the flow of information through cortical networks in vivo. Moreover, the relative technical ease of acquiring data for CSD computation has made this an attractive approach for neuroscientists in a diverse array of fields. Here, we describe the basic theory underlying CSD analysis and give a methodological overview of its application in the rodent neocortex.

Key words: Current source-density, Local field potential, Electrophysiology, In vivo, Laminar, Cortex, Silicon probe

1. Introduction

Brain function is subserved by the electrical signaling properties of individual neurons that receive and integrate analog inputs in the form of synaptic potentials and generate output in the form of action potentials. A basic experimental goal in neuroscience research is to record this electrical activity, both to understand its relationship to behavior and cognition and also to elucidate the cellular mechanisms underlying its generation.

One of the oldest approaches used to characterize neuronal activity is the recording of extracellular field potentials, first described in 1875 by Richard Caton (1). Field potentials are signals generated by the synchronous changes in current flowing across the membranes of individual or groups of neurons, and therefore, through the extracellular space. Events contributing to these potentials include fast sodium action potentials, synaptic activity, dendritic calcium "spikes" in neuronal dendrites, and large voltage-dependent intrinsic oscillations in single neurons (2, 3). The signals recorded can take a variety of forms, including fast impulses recorded in peripheral nerves, event-related potentials in localized brain regions, and the complex patterns of the electroencephalogram (EEG) spanning the cortical mantle (2, 3).

In particular, the lower frequency (<100 Hz) signals recorded by microelectrodes either inserted into the brain or positioned at the surface, termed local field potentials (LFPs), have been of great interest. LFPs are thought to reflect the linearly summed postsynaptic potentials from neighboring groups of neurons (2–4), and therefore provide an estimation of activity across local populations. Traditionally, LFPs have been used to characterize the spontaneous oscillatory activity of the brain in both healthy conditions and during pathological states such as seizures (2). However, an increasing number of studies have used LFPs to investigate the cellular mechanisms of sensory encoding, motor planning, memory, and decision-making (5–11). In addition, the population activity measured by LFPs appears to be related to the BOLD response measured by fMRI (12), providing a cellular correlate for imaging studies. LFPs have also emerged as a candidate signal for the design and control of neural prostheses (13).

One challenge to the interpretation of LFP data is the relatively poor spatial resolution inherent in these signals. Volume conduction of extracellular potentials through the brain parenchyma significantly impairs determination of the site of origin for LFPs. One solution has been to perform current source-density (CSD) analysis. This approach employs the recording of voltage gradients through geometrically arranged arrays of microelectrodes to estimate extracellular current flow in and out of a proscribed volume of tissue. In this chapter, we briefly review the interpretation of LFP data and the computation of the CSD. We also illustrate some uses of CSD analysis and provide an experimental outline for pursuing this approach in the rodent cortex.

1.1. Origin and Interpretation of Local Field Potentials

Neuronal activity involves current flow across a lipid bilayer membrane, resulting in a potential difference between the intracellular and extracellular spaces. In most cases, this membrane potential is nonuniformly polarized (e.g., the local depolarization of a region of axon produced by a propagating action potential or the depolarization of the dendritic arbor relative to the cell body following synaptic excitation). These potential gradients produce current flow from one part of the cell to another, necessarily generating equal extracellular currents flowing in the opposite direction. At points where net current is directed inward across the membrane, the location is termed a *sink*. Where current is net outward, the location is termed a *source*. The flow of current through the finite conductivity of the extracellular fluid establishes potential gradients that can be recorded by local electrodes. Thus, LFPs are potentials between two points in extracellular space rather than across a membrane.

Determination of the exact generators of extracellular current flow underlying the LFP is still an area of active research (2). However, in most situations, excitatory synaptic activity mediated by glutamatergic receptors is thought to be the most significant contributor (2, 3). Although fast sodium action potentials generate large transmembrane currents that can be recorded as extracellular potentials, their short duration (~1 ms) limits temporal summation. Furthermore, the low-pass filtering characteristics of the extracellular space attenuate spatial summation of higher frequency events (2). Inhibitory currents through $GABA_A$ receptors may also play a role in the generation of LFPs, particular during highly synchronous activation of GABAergic interneurons (14). However, their contribution is believed to be small because the chloride reversal potential is close to the resting membrane potential. Nevertheless, strong perisomatic shunting inhibition may enhance dendritic inward currents by preventing membrane depolarization and the resulting reduction in excitatory driving force.

Thus, the *magnitude* of the LFP is determined by the total amount of current flowing in or out of a small volume of brain tissue. Consequently, the *time course* of the LFP will closely follow the kinetics of current flow rather than changes in the membrane potential per se. In fact, for brief events shorter than the time constant of the cell (~10–20 ms), transmembrane current is largely due to capacitative charging and is proportional to the first derivative of membrane potential.

To consider an example, we can imagine a glutamatergic synaptic conductance being opened in the apical dendrite of a cortical pyramidal neuron (Fig. 1). We refer to the site of the synapse as an "active" sink because it is generated by a conductance change. Current flowing into the cell at the synapse is balanced by passive current flow out of the cell at some distance away (a "passive" source). This current loop establishes a potential dipole, measurable by an extracellular recording electrode.

From Fig. 1, we can see that the magnitude of the recorded LFP depends on the placement of the electrode relative to the active sink. Specifically, the magnitude of the signal decreases roughly as $1/r$, where r is the distance between the electrode and the sink (15). Furthermore, as dictated by the superposition principle, the LFP recorded at any one site is proportional to the linear summation of all currents generated in vicinity of the electrode (16). A square millimeter of brain tissue can contain ~10^4 neurons and several orders of magnitude greater numbers of synapses, each of which may be active approximately once per second, producing a churning sea of extracellular current flow. LFP data do not allow the discrimination of weak currents

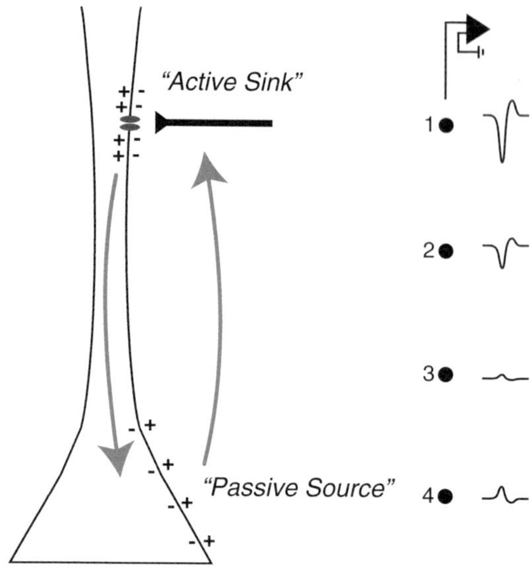

Fig. 1. Schematic illustration of the electrical dipole established by synaptic inputs to the apical dendrite of a cortical pyramidal neuron. Glutamatergic excitation generates an active current sink (inward transmembrane current), while a corresponding passive source (outward transmembrane current) forms at some distance away in the cell body. The direction of current flow is shown by the *curving arrows*. The local field potential (LFP) generated by this synaptic current and recorded by an extracellular electrode varies as a function of distance from the active current sink, as illustrated by the *traces at the right*. Thus, experimental recordings of LFP data have relatively poor spatial resolution and cannot distinguish low amplitude nearby signals from larger more distant signals.

generated by nearby synapses from strong (or highly synchronous) currents generated some distance away. Thus, LFPs give us information that some temporal pattern of synaptic activity has occurred, but the spatial resolution of the underlying current sinks and sources is relatively poor.

1.2. Current Source-Density Analysis

CSD analysis provides a means for enhancing the spatial resolution of current sinks and sources underlying recorded LFPs. Stated directly, the CSD is a scalar quantity that measures the *net* amplitude of extracellular current flowing into and out of neuronal tissue at a given point in space. The additional spatial information is derived from the recording of LFPs at multiple locations and analyzing their spatial distribution. This can be performed by sequentially moving a single microelectrode, but is most often done using a single multisite recording array, allowing simultaneous recording of LFPs over a large volume of tissue.

The utility of CSD analysis is illustrated in Fig. 2 (see Sect. 3 for the computation of the CSD). Here, LFPs were recorded from

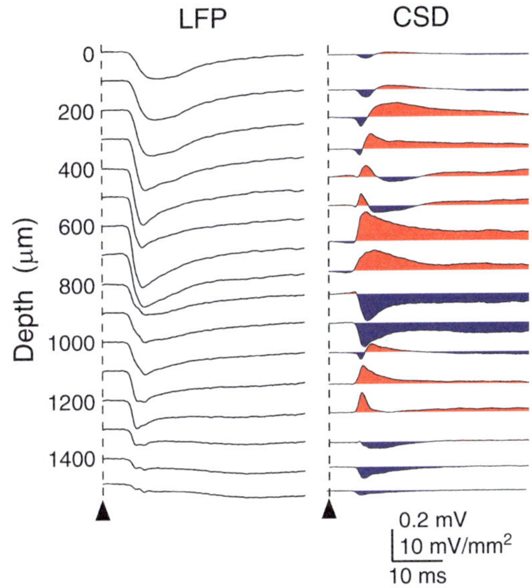

Fig. 2. Example of sensory-evoked LFP data and the corresponding current sinks and sources. The LFPs (*left traces*) were recorded by a 16-channel silicon probe inserted into the barrel cortex of an anesthetized rat. Deflection of a single somatotopically aligned whisker (*black triangle*) produced a complex set of voltage traces at multiple locations throughout the cortical depths. The CSD traces at the *right*, computed using methods described in the text, elucidate the synaptic currents underlying generation of the LFPs. Consistent with known anatomy, whisker deflection evokes an early thalamocortical sink in mid-layers, followed by delayed sinks in superficial and deeper layers. Thus, CSD analysis provides a functional map for the flow of excitation in cortical circuits.

a 16-site silicon-based laminar recording array inserted into the barrel cortex of an anesthetized rat. Recording sites are separated by 100 μm, and the entire probe spans most of the cortical thickness. The barrel cortex is organized in a columnar pattern, with neurons in vertical register being most strongly driven by deflection of the same somatotopically aligned "principal" whisker (6, 17, 18). Here, average LFPs at each site were recorded following deflection of the principal whisker (left traces). Note that the spatial locations of synaptic inputs evoked by whisker deflection are difficult to identify from the LFP data. However, the computed CSD for each site is shown at the right, with positive values (shown in red) representing current sinks and negative values (shown in blue) representing current sources. These traces reveal that whisker deflection produces early excitation (current sink) at a depth of 500–700 μm, corresponding to cortical layer 4. Subsequently, delayed excitation appears in superficial and deeper cortical layers. These data provide a picture of the intracolumnar flow of sensory information and agree with the known anatomy of thalamocortical and intracortical projections (19–21). Thus, the power of CSD analysis is the ability to generate a functional spatiotemporal map of population synaptic activity in vivo. In this respect, the CSD provides a unique source of information that cannot be gained from traditional intracellular recordings, which measure the integrated synaptic responses at the cell body, or extracellular single unit recordings, which reveal only suprathreshold activity.

CSD analysis has been successfully applied to the analysis of spontaneous and evoked neuronal activity in structures throughout the brain, including the cortex, cerebellum, hippocampus, retina, and brainstem. A thorough discussion of these works is beyond the scope of this chapter. However, to illustrate some of the possible uses of the CSD, three examples are presented here.

First, as shown above, CSD analysis has been used by multiple groups to map the flow of intra- and intercolumnar excitation in somatosensory (6, 7, 18), auditory (5), and visual (22) cortices. In combination with anatomical knowledge, this approach has been particularly valuable in determining the contribution of ascending (thalamocortical) versus horizontal (intracortical) inputs to sensory encoding. In the case of the rat whisker system, Higley and Contreras (6) used computation of the CSD along with pharmacological inactivation of intracortical circuits to show that horizontal inputs from adjacent cortical regions do not contribute significantly to suppressive interactions between sensory stimuli. By contrast, Happel et al. (5) showed that for the primary auditory cortex, long-range horizontal connections terminating in superficial layers are critical to the generation of spectral tuning receptive fields. Similarly, Chen et al. (22) used laminar activation profiles in ventral areas V4 and IT of the macaque visual system to conclude that inputs from dorsal stream pathways and nonspecific thalamic nuclei modulate neuronal activity in these areas.

Second, CSD analysis has been used to determine the functional terminal arborization of a single presynaptic neuron in vivo. In an elegant study, Swadlow et al. (7) recorded the average LFPs triggered in the rat barrel cortex by a simultaneously recorded single unit in the somatotopically aligned somatosensory thalamus. Computation of the underlying active sinks revealed excitatory inputs to layers 4 and 6, precisely matching the anatomical projections of thalamocortical neurons. In a later study, they further showed that this functional map was unaltered by changes in behavioral state, despite changes in overall patterns of cortical activity (23).

Third, CSD analysis has also been used to analyze the anatomical substrates of spontaneously occurring oscillations in the rodent cortex and hippocampus. Kandel and Buzsaki (24) recorded LFPs in the neocortex of both awake and anesthetized rats during natural sleep spindles and epileptiform high-voltage spike-and-wave activity. Comparing the corresponding sinks and sources to those evoked by sensory stimulation, the authors concluded that these spontaneous oscillations are largely mediated by intracortical circuits.

1.3. Derivation of the CSD

We now give a brief overview of the computation of the CSD. The reader is referred to additional resources for a more in depth mathematical treatment of these concepts (3, 16, 25).

The relationship between the transmembrane current, i_m, (e.g., that is produced by synaptic activation) and the concurrent three-dimensional extracellular field potential, Φ, is given by:

$$i_m = K \times \left[\frac{\partial^2 \Phi}{\partial x^2} + \frac{\partial^2 \Phi}{\partial y^2} + \frac{\partial^2 \Phi}{\partial z^2}\right]$$

where K is a proportionality constant that includes the conductivity of the extracellular fluid. A practical difficulty in applying this relationship is the necessity of measuring the LFP in three orthogonal axes.

However, if the volume of tissue being analyzed is laminar, consisting of stacked two-dimensional planes of uniform anatomy, we can assume that the change in potential within each of these planes is zero, and therefore:

$$\frac{\partial^2 \Phi}{\partial x^2} = \frac{\partial^2 \Phi}{\partial y^2} = 0$$

and the extracellular current is directly proportional to the second spatial derivative of the LFPs recorded along the axis perpendicular to the laminar structure.

In the case of the neocortex (and possibly the hippocampus and cerebellum), there is reason to believe this assumption is justified. Cortical laminae comprise multiple cell-body and dendritic layers, each with approximately uniform synaptic inputs. Furthermore, the neocortex is typically arranged as a columnar structure, with intracolumnar diameters of several hundred microns (26). Recent studies have suggested that >95% of the LFP signal originates within 250 μm of the recording site, suggesting that most currents are generated by synapses in the same column and exhibit similar patterns of spontaneous and evoked activity (27, 28). Thus, assuming a uniform laminar signal under these conditions is reasonable. Nevertheless, in their studies of the cerebellum, Nicholson and Llinas (29) recommended a full three-dimensional analysis using multiple arrays of electrodes arranged in a three-dimensional configuration. Furthermore, Pettersen et al. (30) developed an alternative analysis, termed inverse CSD (iCSD), where laminar heterogeneity and differential spatial distribution of conductivity are explicitly considered and which may yield more accurate calculations under some circumstances.

In the simplified case, LFP measurements are taken from a finite number of sampling points (e.g., the 16 channels of a silicon-based laminar probe). From these discrete points, the second spatial derivative can be calculated using any one of standard difference formulas. We have adopted the following procedure from Freeman and Nicholson (25).

Take LFP measurements at three neighboring points, V_0, V_a, and V_b, such that:

$$V_0 = V(z, t)$$
$$V_a = V(z - 2\Delta z, t)$$
$$V_b = V(z + 2\Delta z, t),$$

where z is the location along the recording axis and Δz is the spacing between consecutive recording sites. A simple approximation of the second spatial derivative, discussed by Freeman and Nicholson (25), is given by:

$$\frac{\partial^2 V(t)}{\partial z^2} = i_m(t) \approx \frac{V_b + V_a - 2V_0}{(2\Delta z)^2}$$

Here, $i_m(t)$ is called the current source-density function.

A number of other more complicated estimations for the second spatial derivative are possible. However, a direct comparison of the computed CSD values for several cases did not find a considerable improvement in the accuracy of the calculation (25).

It may also be apparent that, if n separate LFP channels are recorded, only n-4 are available for the CSD computation, reducing the area over which the CSD can be derived. However, we and other authors have taken advantage of the observation that, in many cases, the recorded potentials do not vary significantly above and below the first and last recording site, respectively. In this situation, Vaknin et al. (31) showed that extrapolating the LFP data at the two extremes produced an accurate calculation of the underlying CSD, and this procedure was used to generate the CSD traces from the corresponding LFP data in Fig. 2.

2. Materials

2.1. Surgical Equipment

- Experimental animals (rats)
- Stereotaxic frame with anesthesia head holder and arms for electrode holders (Model#930, #1460, #1929-B, David Kopf, Inc.).
- Heating blanket
- Surgical tools (forceps, scissors, scalpel, etc.)
- High-speed dental drill
- Sterile saline
- Anesthetic (Isoflurane)

- Local anesthetic (Lidocaine 1%)
- Surgical scrub (Betadine)
- Hair trimmer
- Dental acrylic

2.2. Electrophysiological Recordings

- 16-channel silicon-based laminar probe with A-type connector (Part# A1×16-3mm100-413, Neuronexus Technologies, Inc.)
- Connecting cable (28 ga insulated wire)
- Silver wire (~32 ga)
- Two 8-channel differential amplifiers (Model HiZ×8, FHC, Inc.), custom built with low cutoff filter settings of 0.1 or 1.0 Hz for LFP recordings

2.3. Data Acquisition and Analysis

- Power1401 Acquisition System (Cambridge Electronic Design, Inc.)
- Spike2 acquisition software (Cambridge Electronic Design, Inc.)
- Standard computer (highest processing power available is recommended)
- IgorPro (Wavemetrics, Inc.)
- DataAccess (Bruxton, Inc.)

3. Methods

3.1. Hardware

In order to record LFP data at multiple sites through the cortical depth, the 16-channel silicon-based laminar recording array from Neuronexus Technologies provides a high quality signal in a compact and economical form factor (Fig. 3). Each recording site is equally spaced at 100 μm intervals, which is sufficient to span the rat somatosensory cortex. The size of the recording site can be specified at either 177 or 413 or 703 μm^2. For recording LFPs, we recommend either of the larger-sized probes. Thus, the appropriate part number is A1×16-3mm100-413.

A variety of printed circuit board connectors ("Probe Packages") are also available. For acute recordings, we recommend the A16 version. This connector consists of a standard 2 × 8 dual-inline-pin (DIP) block, with a pin length of 5.5 mm and pin spacing of 2.5 mm. Cables can be directly soldered to these pins, or matching connectors can be obtained from any small electronics dealer.

An amplifier with sufficient channels is also required. We have successfully used two 8-channel HiZ×8 differential amplifiers from FHC, Inc. However, note that the standard low cutoff filter

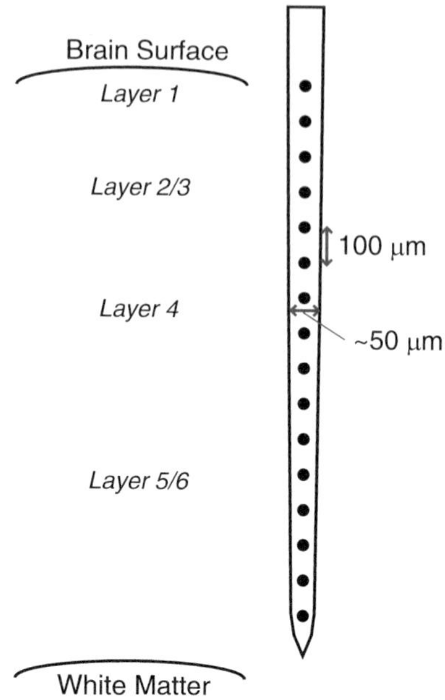

Fig. 3. Schematic illustration of the 16-channel silicon probe, available from Neuronexus, Inc., typically used to record LFP data in laminar structures, to be used for subsequent computation of the CSD. The probe is suitable for spanning the depth of the rat neocortex.

settings (100–500 Hz) are insufficient for LFP data. It is necessary to request a custom circuit with a low cutoff of 0.1 or 1 Hz. The headstage consists of eight pairs of pin inputs for multiple bipolar recordings. In this case, each recording site from the probe is directed to an individual positive terminal and a ganged reference signal (see below) is directed to each negative terminal (Fig. 4).

3.2. Data Acquisition

For recording LFP data, a sampling rate of 10 kHz is sufficient (if you wish to simultaneously record multiunit activity from the probe recording sites, this rate should be >20 kHz). Recording 16 channels at this rate requires considerable bandwidth for data acquisition. The Power1401 from Cambridge Electronic Design (Cambridge, UK) is a data acquisition board that is suitable for these recordings, and allows, in different configurations, up to 64 channels of waveform data acquisition. The Power 1401 requires the use of Spike2 software, also from Cambridge Electronic Design.

3.3. Analysis

Spike2 saves acquired data in a proprietary format. Although some analysis is possible with Spike2, the increased flexibility of dedicated analysis software warrants exporting the raw data. We highly recommend IgorPro from Wavemetrics as an easy-to-learn and flexible environment for waveform data analysis.

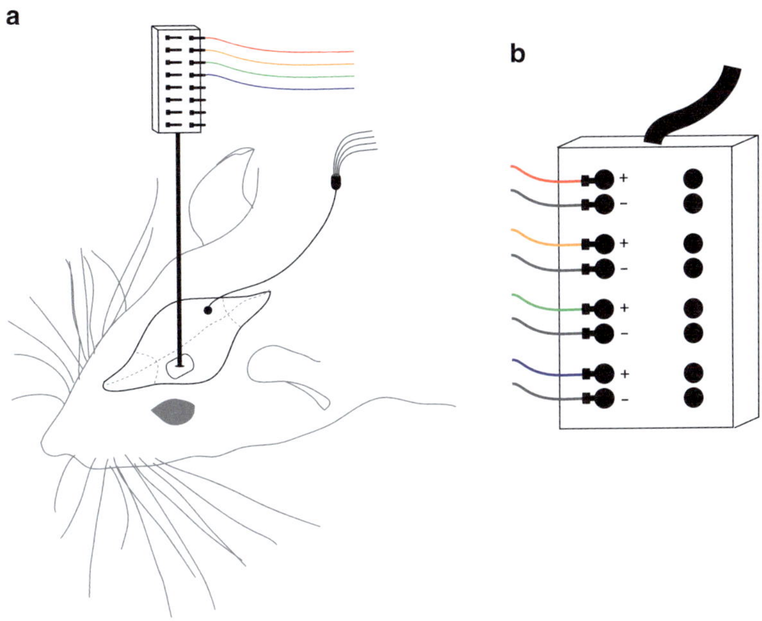

Fig. 4. Schematic illustration of the recording setup used to obtain LFP and CSD data from the rat cortex. The probe is inserted into the cortex of an anesthetized rat, positioned normal to the brain surface. Each channel of the probe is connected to an individual positive terminal on the headstage of a differential amplifier (Only four wires are shown in the diagram). Reference signals are provided by a silver wire, inserted between the dura and the inner skull surface and connected to a cable bus directed to the negative terminals of the headstage.

To export data from Spike2 to IgorPro, we recommend DataAccess by Bruxton. This software adds several user-accessible routines to IgorPro, allowing the importing of Spike2 files. Alternatively, DataAccess has a simple graphical interface for managing data conversion.

Once the data has been moved to IgorPro, standard analysis methods can be used to generate averaged LFP traces around selected time-points. From these averages, the CSD can be computed algebraically as detailed in the previous section.

3.4. Surgical Setup

The experimental approach described briefly here is performed under isoflurane anesthesia. The reader is referred to other excellent chapters in the present volume by Cardin, Crochet, and Vyazovskiy et al. for additional detailed information on surgical protocols.

3.5. Electrode Placement

1. Induce anesthesia and position the rat into the stereotaxic apparatus. During initial surgical manipulations, isoflurane should be set to ~2%. After incisions and craniotomy have been completed and the animal's depth of anesthesia stabilizes, isoflurane can be reduced to ~0.5%.

2. Expose the skull, and make a small rectangular craniotomy over the appropriate recording site (~200–300 μm per side). The craniotomy should be large enough to accommodate positioning the electrode but no larger. The detached piece of bone can be removed with forceps, exposing the dura. Flushing with sterile saline will remove any accumulated blood. It may take 2–3 min for small severed vessels to coagulate.
3. At this point, the grounding wire should be placed. This is typically done be making a small burr hole through the skull at some distance (>2 mm) from the recording site. Insert a thin, chlorided silver wire through the burr hole to rest between the dura and interior skull surface. Silver wire can be readily chlorided by soaking in bleach for 1–2 h, followed by rinsing with water. Use of poorly chlorided reference electrodes may generate a large voltage drop that may saturate the amplifier headstage. The exterior end of the wire should be connected to a ganged grounding bus, consisting of 16 individual wires that will serve as the references for each recording channel (Fig. 3). This silver wire can be held in place on the skull surface by application of a small amount of dental acrylic.
4. Using a fine syringe tip (30 ga) or fine dissecting scissors, make a small incision in the dura to allow electrode penetration.
5. When handling the probe, only touch the printed circuit board. Position the probe in the stereotaxic manipulator and attach the connector to the headstage. This will minimize post-implant movement. Using the manipulator, orient the probe so that its principal axis is normal to the brain surface. Gradually lower the probe into the brain. If buckling occurs, wait for the probe to work its way into the tissue (may take several minutes). Continue lowering the probe until the dorsal-most recording site is even with the brain surface.
6. Electrode insertion often results in temporary spreading depression, where neural activity around the electrode is silenced for a period of time. Thus, it is recommended to wait for ~30 min following electrode placement to begin recordings.
7. Probes can be reused for multiple experiments if rinsed gently with hydrogen peroxide or enzymatic contact lens cleaner followed by distilled water at the conclusion of the recording session.

References

1. Caton R (1875) The electrical currents of the brain. Br Med J 2:278
2. Ebersole JS, Pedley TA (eds) (2003) Current practice of clinical electroencephalography. Lippincott Williams & Wilkins, Philadelphia
3. Mitzdorf U (1985) Current source-density method and application in cat cerebral cortex: investigation of evoked potentials and EEG phenomena. Physiol Rev 65:37–100
4. Buzsaki G (2004) Large-scale recording of neuronal ensembles. Nat Neurosci 7:446–451
5. Happel MF, Jeschke M, Ohl FW (2010) Spectral integration in primary auditory cortex attributable to temporally precise convergence of thalamocortical and intracortical input. J Neurosci 30:11114–11127
6. Higley MJ, Contreras D (2007) Cellular mechanisms of suppressive interactions between somatosensory responses in vivo. J Neurophysiol 97:647–658
7. Swadlow HA, Gusev AG, Bezdudnaya T (2002) Activation of a cortical column by a thalamocortical impulse. J Neurosci 22:7766–7773
8. Henrie JA, Shapley R (2005) LFP power spectra in V1 cortex: the graded effect of stimulus contrast. J Neurophysiol 94:479–490
9. Kreiman G, Hung CP, Kraskov A et al (2006) Object selectivity of local field potentials and spikes in the macaque inferior temporal cortex. Neuron 49:433–445
10. Pesaran B, Pezaris JS, Sahani M et al (2002) Temporal structure in neuronal activity during working memory in macaque parietal cortex. Nat Neurosci 5:805–811
11. Scherberger H, Jarvis MR, Andersen RA (2005) Cortical local field potential encodes movement intentions in the posterior parietal cortex. Neuron 46:347–354
12. Logothetis NK, Pauls J, Augath M et al (2001) Neurophysiological investigation of the basis of the fMRI signal. Nature 412:150–157
13. Andersen RA, Musallam S, Pesaran B (2004) Selecting the signals for a brain-machine interface. Curr Opin Neurobiol 14:720–726
14. Ylinen A, Bragin A, Nadasdy Z et al (1995) Sharp wave-associated high-frequency oscillation (200 Hz) in the intact hippocampus: network and intracellular mechanisms. J Neurosci 15:30–46
15. Bedard C, Kroger H, Destexhe A (2006) Model of low-pass filtering of local field potentials in brain tissue. Phys Rev E Stat Nonlin Soft Matter Phys 73:051911
16. Johnston D, Wu SM (1997) Foundations of cellular neurophysiology. MIT Press, Cambridge
17. Simons DJ, Woolsey TA (1979) Functional organization in mouse barrel cortex. Brain Res 165:327–332
18. Di S, Baumgartner C, Barth DS (1990) Laminar analysis of extracellular field potentials in rat vibrissa/barrel cortex. J Neurophysiol 63:832–840
19. Bernardo KL, Woolsey TA (1987) Axonal trajectories between mouse somatosensory thalamus and cortex. J Comp Neurol 258: 542–564
20. Lubke J, Roth A, Feldmeyer D et al (2003) Morphometric analysis of the columnar innervation domain of neurons connecting layer 4 and layer 2/3 of juvenile rat barrel cortex. Cereb Cortex 13:1051–1063
21. Shepherd GM, Svoboda K (2005) Laminar and columnar organization of ascending excitatory projections to layer 2/3 pyramidal neurons in rat barrel cortex. J Neurosci 25:5670–5679
22. Chen CM, Lakatos P, Shah AS et al (2007) Functional anatomy and interaction of fast and slow visual pathways in macaque monkeys. Cereb Cortex 17:1561–1569
23. Stoelzel CR, Bereshpolova Y, Swadlow HA (2009) Stability of thalamocortical synaptic transmission across awake brain states. J Neurosci 29:6851–6859
24. Kandel A, Buzsaki G (1997) Cellular-synaptic generation of sleep spindles, spike-and-wave discharges, and evoked thalamocortical responses in the neocortex of the rat. J Neurosci 17:6783–6797
25. Freeman JA, Nicholson C (1975) Experimental optimization of current source-density technique for anuran cerebellum. J Neurophysiol 38:369–382
26. Mountcastle VB (1998) The cerebral cortex. Harvard University Press, Cambridge
27. Katzner S, Nauhaus I, Benucci A et al (2009) Local origin of field potentials in visual cortex. Neuron 61:35–41
28. Xing D, Yeh CI, Shapley RM (2009) Spatial spread of the local field potential and its laminar variation in visual cortex. J Neurosci 29:11540–11549

29. Nicholson C, Llinas R (1975) Real time current source-density analysis using multi-electrode array in cat cerebellum. Brain Res 100:418–424
30. Pettersen KH, Devor A, Ulbert I et al (2006) Current-source density estimation based on inversion of electrostatic forward solution: effects of finite extent of neuronal activity and conductivity discontinuities. J Neurosci Methods 154:116–133
31. Vaknin G, DiScenna PG, Teyler TJ (1988) A method for calculating current source density (CSD) analysis without resorting to recording sites outside the sampling volume. J Neurosci Methods 24:131–135

Neuromethods (2012) 67: 219–235
DOI 10.1007/7657_2011_7

Published online: 25 November 2011

Intracellular Whole-Cell Patch-Clamp Recordings of Cortical Neurons in Awake Head-Restrained Mice

Sylvain Crochet

Abstract

Membrane potential dynamics resulting from the integration of thousands of synaptic inputs and intrinsic membrane properties underlie the generation of action potential in neurons of the central nervous system. The investigation of membrane potential dynamics is, therefore, of major importance to the understanding of brain function. This level of neuronal activity can only be assessed by measuring differences of potential between the inside and the outside of a neuron, i.e., intracellular recording. In mammals, this approach has been so far mainly restricted to reduced preparations in vitro and more recently to the intact brain in anesthetized animals. Such preparations do not reproduce the complexity and the diversity of the brain activities observed in behaving animals and are, therefore, of limited interest to the understanding of complex brain processing and cognitive functions. Recently, we have developed an approach that enables intracellular recordings of cortical neurons in awake behaving mice. The mechanical stability of the brain being the main technical issue, it has been successfully overcome by (1) using "blind" whole-cell patch-clamp technique conferring higher stability in the initial phase of the recording, (2) implanting mice with light metal posts that enable painless and stable fixation of the head, (3) habituating the animal to avoid large and brisk body movements during the recording session, and (4) reducing the size of the craniotomy to minimal to prevent large brain pulsations and edema. This technique has been successfully applied to the investigation of cortical sensory processing during active sensing in the mouse whisker system and has been expanded to simultaneous double intracellular recordings or combined with other recording techniques, such as local field potentials or two-photon microscopy.

Key words: Patch clamp, Intracellular recording, In vivo, Cortex, Behaving animal

1. Introduction

Information is encoded in the central nervous system by the active generation of action potentials (APs) in neurons, which are interconnected through chemical synapses. When an AP is generated in one neuron, it propagates along its axon to the terminals, where it releases neurotransmitter into the synaptic cleft. The neurotransmitter binds with its receptors on postsynaptic neurons, opening ion channels that generate unitary postsynaptic potentials (uPSPs). Depending upon which neurotransmitter is released, these uPSPs can be excitatory or inhibitory in nature. Due to the largely interconnected nature of most neuronal networks in the central nervous system, neurons have to integrate uPSPs from thousands of presynaptic neurons (1, 2). The integration of excitatory and inhibitory inputs is done across space (location of the synapse on the dendritic arbor) and time (summation, short-term plasticity) resulting in membrane potential (Vm) subthreshold fluctuations (3–6). When these fluctuations bring the Vm close to spike threshold, voltage-gated sodium channels open and generate an AP that impact several other postsynaptic neurons.

Recording APs with extracellular electrodes in intact animals has been the primary way of assessing neuronal code and brain functions since the very beginning of electrophysiology by correlating neuronal discharge with sensory stimuli or animal behavior (7–9). Indeed, APs are high-amplitude (40–60 mV) voltage deflections that can be easily recorded a few micrometers away from the cell using even relatively big metal electrodes (tens of μm in diameter). Extracellular recordings are, thus, less sensitive to movements between the electrode and the recorded neuron and can be performed in awake head-restrained (10–13) or even freely moving behaving animals (14–16). More recently, the parallel development of electrodes allowing for better spike detection/sorting and multisite recordings (silicon probes), as well as high-capacity multichannel acquisition systems, enabled simultaneous recording of large neuronal population in freely moving behaving animals (17–20).

If recording neuronal discharge is of critical importance to the understanding of brain functions, the synaptic mechanisms leading a neuron to fire an AP are also of equal importance to understand how the neuronal code is elaborated. However, the recording of subthreshold Vm fluctuations is of much higher difficulty because it can be achieved only by recording the difference of potential between the inside and the outside of a neuron, which requires a direct access of the tip of the electrode to the inside of the neuron (intracellular recording). To do so, several technical difficulties must be overcome. First, the tip of the electrode must be smaller than the soma of the neuron (for instance, the soma of a cortical neuron in the mouse is about 10–20 μm in diameter). Second, accessing the inside of a neuron implies breaking the integrity of the cellular membrane, which may cause leaks between the intracellular and extracellular spaces resulting in the depolarization of the cell and ultimately its death. It is essential, then, to establish a seal between the membrane of the recorded cell and the electrode to prevent any leak. Finally, because of the small size of the cell, any small movement between the electrode and the tissue may result in artifacts in the recorded Vm or in the electrode going out or through the cell.

Two approaches have been developed in parallel over the last decades to perform intracellular recordings: sharp electrode

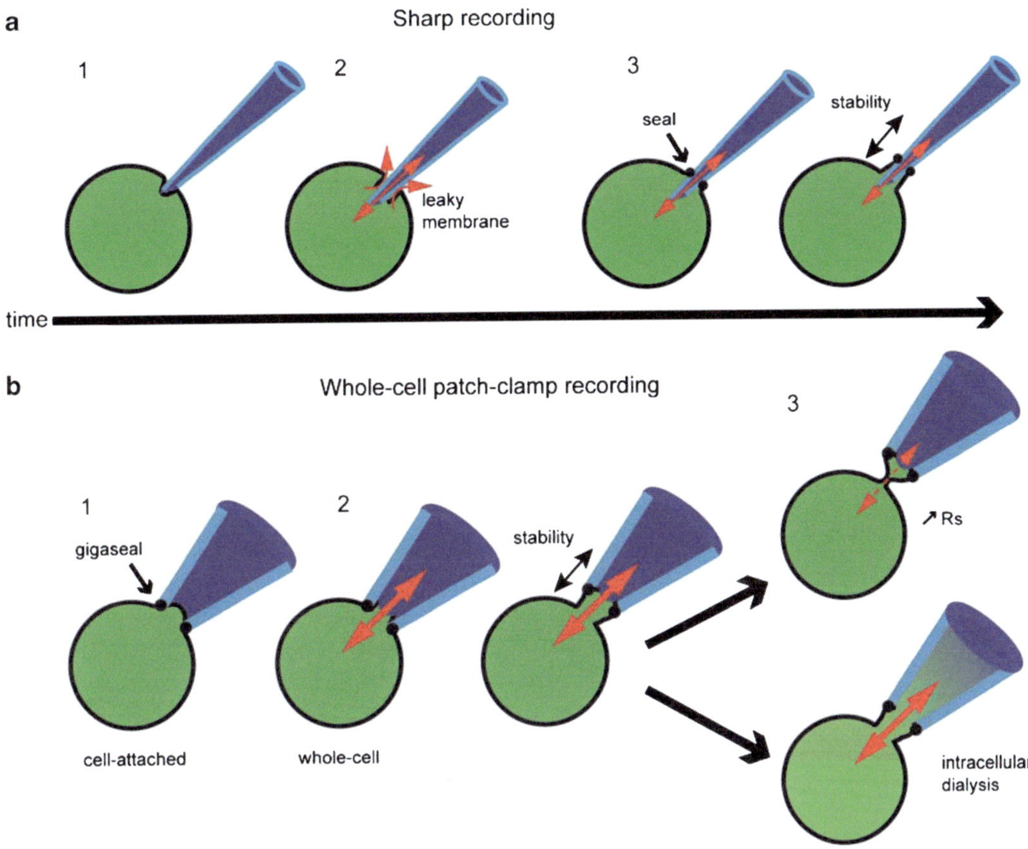

Fig. 1. Evolution of intracellular recordings over time. (**a**) Schematic representation of the evolution of sharp electrode intracellular recording. In 1, the tip of the glass pipette contacts the cell membrane. In 2, the pipette has been advanced further to penetrate the cell membrane, which results in leaky membrane and cell depolarization. This is compensated by injection of hyperpolarizing current through the pipette. In 3, over time and in the absence of movement between the cell and the electrode, a seal forms around the tip of the pipette and hyperpolarizing current can be removed. In this configuration, the recording is stable and sustains moderated tissue movements. (**b**) Schematic representation of the evolution of whole-cell patch-clamp recording. In 1, the tip of the glass pipette contacts the cell membrane and a gigaseal is established by gentle suction. In 2, a short negative pressure pulse applied to the pipette breaks the membrane to establish the whole-cell configuration. In this condition, thanks to the gigaseal and the elasticity of the cell membrane, the recording is stable and sustains moderated movements between the cell and the electrode. In 3, over time, whole-cell recordings tend to deteriorate due to increased access resistance (Rs) or the dialysis of the intracellular space.

recordings and whole-cell patch-clamp recordings. Sharp electrodes are thin glass electrodes with very small tip and high resistance (20–80 MΩ). The tip of the electrode has to physically penetrate the cell to access intracellular space (Fig. 1a). The penetration of the electrode produces a leak and a depolarization of the neuron, which is compensated by the injection of hyperpolarizing current through the pipette. Over time and in the absence of movement between the electrode and the cell, something similar to a seal forms and a stable recording can be obtained without current injection (Fig. 1a). For patch-clamp technique, glass electrodes are bigger, with lower resistance (4–6 MΩ) and smooth tip. The tip is approached against the cell membrane and a small suction is applied to the pipette to establish a seal of very high resistance (gigaseal) before opening the membrane to access intracellular space (Fig. 1b). The elasticity of the cell membrane attached to the pipette by the gigaseal renders the patch-clamp approach much less sensitive to tissue movements than the sharp electrode approach in the early stage of the recording (Fig. 1). The main issue with the patch-clamp approach is that in whole-cell configuration the important exchanges between the intracellular milieu and the pipette solution may wash out secondary messengers and therefore affect the physiology of the cell (intrinsic properties, establishment of plasticity, ...) (Fig. 1b). Another issue with patch-clamp recording is that the access resistance tends to deteriorate more rapidly due to the closure of the membrane. As a result, high-frequency components of the signal, such as APs, can be attenuated (low-pass filter) (Fig. 1b).

Due to different technical issues (mechanical stability, possibility to approach the neuron with the electrode under visual control), intracellular recordings have long been restricted to in vitro preparations (isolated cells, culture of neurons, or slices of brain tissue) before being adapted to intact anesthetized animals. The two main issues have been the stabilization of the preparation to prevent movements of the brain and the development of "blind" recording techniques. The stabilization can be achieved easily in anesthetized animal by different procedures, like muscular paralysis using curare, opening of the cisterna magna to prevent brain pulsation, pneumothorax and hip suspension, and covering the craniotomy with agarose. The blind recording technique consists in advancing the recording electrode by small steps into the brain tissue until a contact with a neuron can be detected. Both sharp electrodes and whole-cell patch-clamp techniques have been used in vivo with their own advantages and disadvantages. Intracellular sharp recordings are more sensitive to tissue movement during the phase of stabilization, but once stabilized last longer than whole-cell patch-clamp recordings on average (up to several hours) with more stable access resistance and less dialysis of the intracellular space (Fig. 1). Only very recently, intracellular recordings technique has been expanded to nonanesthetized animals during natural sleep–wake cycle or quiet wakefulness with either sharp electrodes (21–23) or patch-clamp technique (23, 24). But the application of intracellular recordings techniques to higher brain functions, such as active sensory perception, has not been possible so far due to the difficulty of performing intracellular recordings in awake behaving animals. However, studying sensory perception in awake behaving animals is of major interest because neuronal dynamics and sensory processing are dramatically affected by the animal's behavior (21, 25–28).

We describe here a methodological approach that we have developed allowing for intracellular recordings in awake behaving mice. We detail how the major difficulty, i.e., the movements between the cell and the electrode, can be overcome. We have achieved optimal stability of the recording thanks to the use of "blind" patch-clamp technique, which is less sensitive to movements than sharp electrodes; implantation of a light metal post on the head of the animal, enabling painless and stable head fixation (Fig. 2); a period of habituation of the mouse to the procedure of head fixation to avoid large and brisk body movement during the recording session; and reduction of the size of the craniotomy.

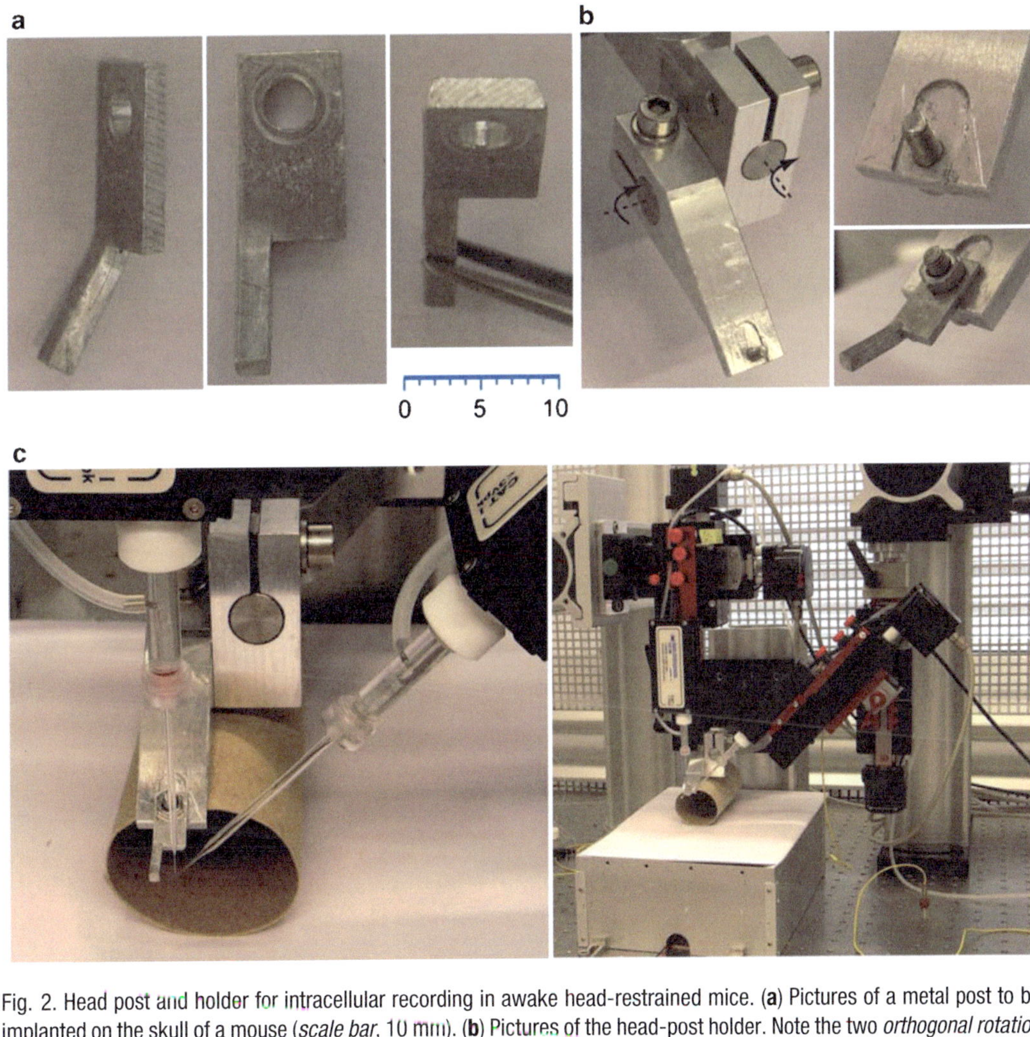

Fig. 2. Head post and holder for intracellular recording in awake head-restrained mice. (**a**) Pictures of a metal post to be implanted on the skull of a mouse (*scale bar*, 10 mm). (**b**) Pictures of the head-post holder. Note the two *orthogonal rotation axes* allowing the adjustment of the position of the mouse head. The head post is fixed thanks to a nut. (**c**) View of the intracellular setup configuration with two microdrives allowing for simultaneous positioning of two electrodes (local field potential + intracellular; juxtacellular+intracellular; dual whole-cell recordings).

Assessing Vm dynamics in awake behaving animals provides new insights into brain function, opening the possibility of dissecting the synaptic mechanisms that drive neuronal networks. We have applied this technique to the somatosensory system of the mouse, which constitutes a very attractive model of active sensing. Mice and rats have a very-well-organized system of vibrissae (or whiskers) with a topological organization conserved from the whisker pad to the primary sensory cortex, with each single-whisker activating neuron preferentially within a single barrel column (29, 30). Mice and rats use their whiskers to actively explore their environment by moving their whiskers back and forth in a rhythmic manner (whisking) and are able to perform very fine discrimination of texture or object position (14, 15, 31–34). We have combined intracellular recordings with high-speed video filming of whisker movements to correlate Vm dynamics with whisker behavior and active sensing (35). This technique has been further expanded in our laboratory to simultaneous dual whole-cell recordings and targeted whole-cell recordings of GABAergic interneurons using two-photon microscopy in awake mice, showing the impact of behavior on neuronal discharge and correlation of Vm fluctuations between pyramidal and GABAergic interneurons in the primary somatosensory cortex (25, 36). Other laboratories have also recently successfully implemented a very similar approach to perform whole-cell patch-clamp recordings in awake head-restrained rodents in the neocortex or hippocampus (23, 37).

The major weakness of the present approach are as follows: (1) We have noticed that 2–3 weeks after the implantation, the dura mater underneath the exposed bone in the recording chamber tends to grow thicker and becomes more vascularized—opening a clean craniotomy in this condition becomes much more difficult and the success rate of the patch-clamp recording is significantly reduced. (2) The behavioral tasks that can be used are obviously limited by the reduced motor outputs and the increased difficulty of training an animal, especially a mouse, while head fixed. But it should be noted that despite recent successful attempts, the success rate of intracellular recordings in freely moving animals so far is very low (about 15%) (38, 39). In addition, several laboratories have recently developed more sophisticated behavioral tasks in head-restrained rodents (34, 37). The use of intracellular whole-cell recording in awake head-restrained rodents is, therefore, a promising technique, which will certainly develop in the future, in combination with other electrophysiological or imaging approaches.

2. Materials

2.1. Experimental Animals of Interest

Our experiments have been conducted on 4–8-weeks-old C57Bl6J mice from Janvier (see Note 1). All experiments were carried out in accordance with the Swiss Federal Veterinary Office. Make sure that the described protocol conforms to institutional and national regulation.

2.2. Implantation

- Vapomatic anesthetic vaporizer (VIP3000 Isoflurane, MDS Matrx) and Attane Isoflurane (Provet).
- Stereotaxic device: Nose clamp (custom made) or small animal stereotaxic frame (Kopf).
- Heating blanket: DC Temperature Control Module, heating pad, and immersion heating rod (FHC).
- Surgical instruments: Fine scissors, disposable scalpel blades with handle, spatula.
- Head post (custom made): We have developed an L-shaped head post that enables maximum contact surface with the skull and good access to most of brain regions (Fig. 2a). The long part of the head post is positioned contralaterally to the region of interest and the short part over the cerebellum; the shape can be modified for particular applications (bilateral recordings, access to the cerebellum). An angle of ~140° between the part in contact with the skull and the part fixed to the holder (Fig. 2b) provides a better access to the craniotomy. The part fixed to the holder has a trapezoidal shape that ensures a good stability between the head post and the holder.
- Ophthalmic gel (Viscotears, Novartis).
- Topical antiseptic (Betadine).
- Hydrogen peroxide solution (2%).
- Cotton swabs.
- Cyanoacrylate glue (Loctite Super Attak 401, Henkel).
- Acrylic dental cement (Paladur pink powder 67407963 and Paladur liquid, Kaladent).
- Silicone sealant (Kwik-Cast, WPI).
- Carprofene (Pfizer).

2.3. Habituation

- Head-post holder (custom made): The head-post holder has been developed to provide maximum stability and flexibility with two orthogonal axes of rotation allowing for the adjustment of the position of the mouse head (Fig. 2b).

2.4. Craniotomy

- Ringer's solution containing (in mM): 135 NaCl, 5 KCl, 5 HEPES, 1 $MgCl_2$, and 1.8 $CaCl_2$ (adjusted to pH 7.3 with NaOH).
- Stereomicroscope (MZ95, Leica) with long working distance objective (plan ×1.0, 112 mm), magnification from 6.3 to 60 with 10×/21B eyepieces.
- Lightweight drill (Model EXL-M40, Osada) and small diameter drill bit (Komet Dental Drill Bit H1.204.005, CondorDental).
- Surgical instruments: Fine forceps (Dumont #5) and needles (U-100 Insulin, 30 G × 8 mm, BD Micro-Fine).

2.5. Recording

- Glass capillaries: Bo-glass capillaries, ends fire-polished 75 mm length OD = 2.0, ID = 1.4 mm (Hilgenberg).
- Pipette puller: P-97 Flaming/Brown Micropipette Puller (Sutter Instrument).
- Intracellular solution containing (in mM): 135 potassium gluconate, 4 KCl, 10 HEPES, 10 sodium phosphocreatine, 4 MgATP, 0.3 Na_3GTP (adjusted to pH 7.3 with KOH and to osmolarity 285 mOsm/L with distilled water). The intracellular solution contains ATP and GTP that degrade rapidly at room temperature. Stock solution must be prepared on ice and a 1-mL aliquot can be stored at −20°C for a few months. Just before the recording session, unfreeze the aliquot and add 3 mg/mL biocytin. Then, keep the solution on ice.
- Microdrive: Mini 4 axes, 4MRE/4MLE (Luigs & Neumann).
- Pipette pressure control: Manual seal sucker (Sigmann Elektronik).
- Intracellular amplifier: Multiclamp 700 amplifier (Axon Instruments).
- Recording system: Signals are filtered online at 10 kHz (bessel filter, Multiclamp) and digitized at 20 kHz by ITC-18 (Instrutech Corporation) under the control of IgorPro (Wavemetrics).
- High speed camera (MotionPro, Redlake), synchronized with electrophysiology through TTL pulse.
- Reference electrode: Ag/AgCl Electrode 2.0 mm diameter (WPI).
- Multichannel digital oscilloscope: TDS 2014 (Tektronic).

3. Methods

3.1. Implantation

Timing: 30–45 min per animal

1. Anesthetize the animal with isoflurane (3%) in oxygen. Once deeply anesthetized (deep and slow breathing), the mouse

remains unconscious for tens of seconds out of the induction box (see Note 2) and its head can be fixed into the nose clamp or the stereotaxic frame.

2. Place the animal onto the heating blanket and insert the rectal probe to maintain the body temperature at ~37°C. Decrease isoflurane concentration to 2%. Apply ophthalmic gel on the eyes to prevent ocular dryness.
3. Expose the skull by removing the skin from the cerebellum to the olfactory bulb.
4. To ensure a good anchoring of the head post, it is essential to carefully clean the skull. First, gently clean the exposed skull and surrounded cut skin with betadine applied with cotton swabs. Retract the membranes covering the skull. Then, clean the skull with peroxide solution. Dry the skull with cotton swabs. Use the edge of a scalpel blade to clean the whole surface of the skull free of periosteum. Finally, make shallow grooves every 0.5–1 mm on the surface of the skull with the tip of the scalpel blade, except where the recording chamber is going to be (see Note 3).
5. Cover the whole exposed skull with a thin layer of cyanoacrylate glue. After a few minutes, the L-shaped head post is glued onto the skull with the long bar in contact with the skull positioned contralaterally to the region of interest (the primary sensory cortex in our case) and the short part lying over the cerebellum (see Note 4).
6. After a few minutes, the head post is securely attached to the skull with acrylic dental cement, which is also used to build the recording chamber. Liquid dental cement is first applied to the whole surface of the skull (except where the recording chamber is going to be positioned) as well as over the parts of the head post in contact with the skull. Prepare dental cement into a consistency comparable with toothpaste and use it to build the walls of the recording chamber. The recording chamber is typically 5–8 mm in diameter and the chamber walls 2–3 mm high (see Note 5). Once the dental cement is solid, fill the recording chamber with silicone sealant.
7. Administrate a pain killer to the mouse (carprofene, i.p.).
8. Put the mouse in a clean cage and let it recover from surgery for 2–3 days. Control body weight right after and in the next days following implantation. The animal should not lose more than 10% of its body weight and then recover.

3.2. Habituation

Timing: 3–5 days per animal

1. Put the animal onto the platform and fix the head post to the holder. The body of the animal can be slightly restrained by a tub or cardboard walls (see Note 6).

2. At the end of each session, reward the animal with a few drops of sweet water, then release it, and put it back into its cage.
3. Control body weight during the habituation period. The animal should not lose weight.
4. The duration of each session should increase gradually, starting from 10 to 20 min up to 1–1.5 h. The number and the increment in time between each session should be adapted for each animal depending on its behavior and the experimental protocol (see Note 7).

3.3. Craniotomy

Timing: 30–40 min per animal

The craniotomy is a very critical step for the patch recording, especially when recording from superficial regions of the neocortex. The dura must be removed without bleeding in order to ensure a recording of good quality. Any trace of membrane or blood might prevent the establishment of the gigaseal. In mice, there is very little space between the pia and the dura mater. Opening the dura without damaging the superficial layer of the cortex is a delicate operation that requires some practice. The size of the craniotomy is also very important: the smaller it is, the more stable the recording is (see Note 8). This procedure must be conducted using a high-magnification stereomicroscope and a good illumination of the preparation.

1. Anesthetize the animal (see step 1, section “Implantation”).
2. Fixe the head post to the holder firmly and decrease isoflurane concentration to 2%.
3. Remove the silicone sealant from the recording chamber and clean it with warm (~37°C) Ringer’s solution.
4. Determine the location of the craniotomy (see Note 9).
5. Drill circularly around the region of interest to thin the bone. Start with circles of ~1.5 mm centered on targeted point and progressively reduce diameter as the bone gets thinner (see Note 10). Stop drilling before reaching the dura. The bone should be thinned on a region of 0.2–0.4 mm in diameter (Fig. 3).
6. Insert a fine needle between the thinned bone and the dura to open the craniotomy (Fig. 3).
7. Carefully clean if any bleeding occurs.
8. Incise the dura mater with the tip of a clean and new needle (Fig. 3). The opening of the dura can be seen by a leak of cerebrospinal fluid (see Note 11).
9. Fill the recording chamber with Ringer’s solution and seal the chamber with silicone sealant.

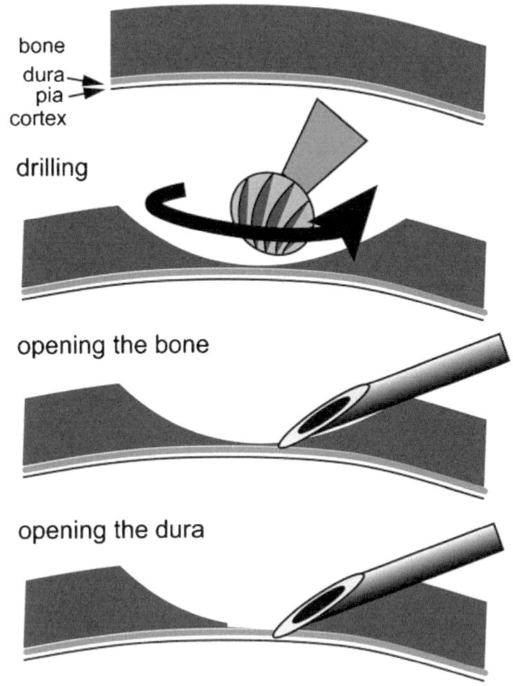

Fig. 3. Procedure for the craniotomy. Schematic representation of the procedure to open a craniotomy for cortical intracellular recording. The bone is thinned by drilling circularly and progressively reducing circle diameter. The bone is then open by inserting a thin sharp needle between the bone and the dura. After gentle cleaning to remove possible blood trace, a sharp needle is used to open the dura.

10. Put the animal in a separate cage for recovery (2–24 h) until the recording session.

3.4. Recording

Timing: 1–3 h per animal

1. Pull the pipettes (long tip, 4–7 MΩ).
2. Unfreeze one aliquot of intracellular solution and add the biocytin.
3. Place the animal on the recording setup and fix the head post to the holder.
4. Remove the silicone sealant and gently clean the craniotomy (see Note 12).
5. Position the reference electrode in or at the border of the recording chamber. A small piece of plasticine can be used to maintain the reference electrode in position.
6. Fill the patch pipette with fresh intracellular solution.
7. Insert the pipette in the pipette holder and apply positive pressure to the pipette (100–150 mb).

8. Under stereomicroscope guidance, position the tip of the pipette into the craniotomy, on the surface of the brain.
9. The blind patch technique consists in advancing the patch pipette by steps of 2 μm into the brain until the tip of the pipette contacts a cell, which can be seen as a sudden increase in the resistance of the patch pipette. This procedure is conducted in voltage-clamp mode by constantly applying 10-mV square test pulses at 20–50 Hz and monitoring the current deflection recorded by the patch pipette. A decrease in the deflection indicates an increase in the pipette resistance.
10. Using the wheel of the control pad, slowly advance the pipette into the brain by ~50 μm, then release the pressure to 50–70 mb, and check the resistance of the pipette (see Note 13). Advance the pipette by a few 2-μm steps. If the resistance of the pipette does not increase, the pipette can be slowly advanced to the region of interest using the wheel of the control pad. Constantly check the resistance of the pipette; the resistance may increase due to contact with a cell or a blood vessel but generally returns to baseline after tens of μm. If the resistance remains high, withdraw the pipette a bit and apply higher positive pressure until the resistance returns to baseline (see Note 14).
11. Once the recording site is reached, decrease the pressure to 20–30 mb and advance the pipette by 2-μm steps, until a contact occurs. A good contact appears as an abrupt approximately twofold increase in input resistance (Fig. 4a). Release the positive pressure. Immediately and gently apply mild negative pressure (suction) while simultaneously and progressively decreasing the holding potential to −70 mV, until a gigaseal is formed (Fig. 4a) (see Note 15). At this stage, the electrode capacitance can be compensated. Then, open the membrane by applying short and transient negative-pressure pulses to the pipette until the membrane breaks (Fig. 4a). Then, switch the amplifier to current-clamp mode (Fig. 4b).
12. The good quality of the recording is attested by mean Vm < −40 mV, overshooting APs with amplitude >40 mV. Low-amplitude, broad APs generally result from high access resistance, which may indicate that the membrane is not opened enough. Applying slight negative or positive pressure as well as moving the pipette one step back may correct for the bad access resistance, but this may also cause the loss of the recording.
13. Once a whole-cell recording of good quality is obtained, proceed with your protocol. An example of correlation between Vm dynamics from a layer 2/3 pyramidal neuron in the barrel cortex and whisker behavior is depicted in Fig. 4b–e.

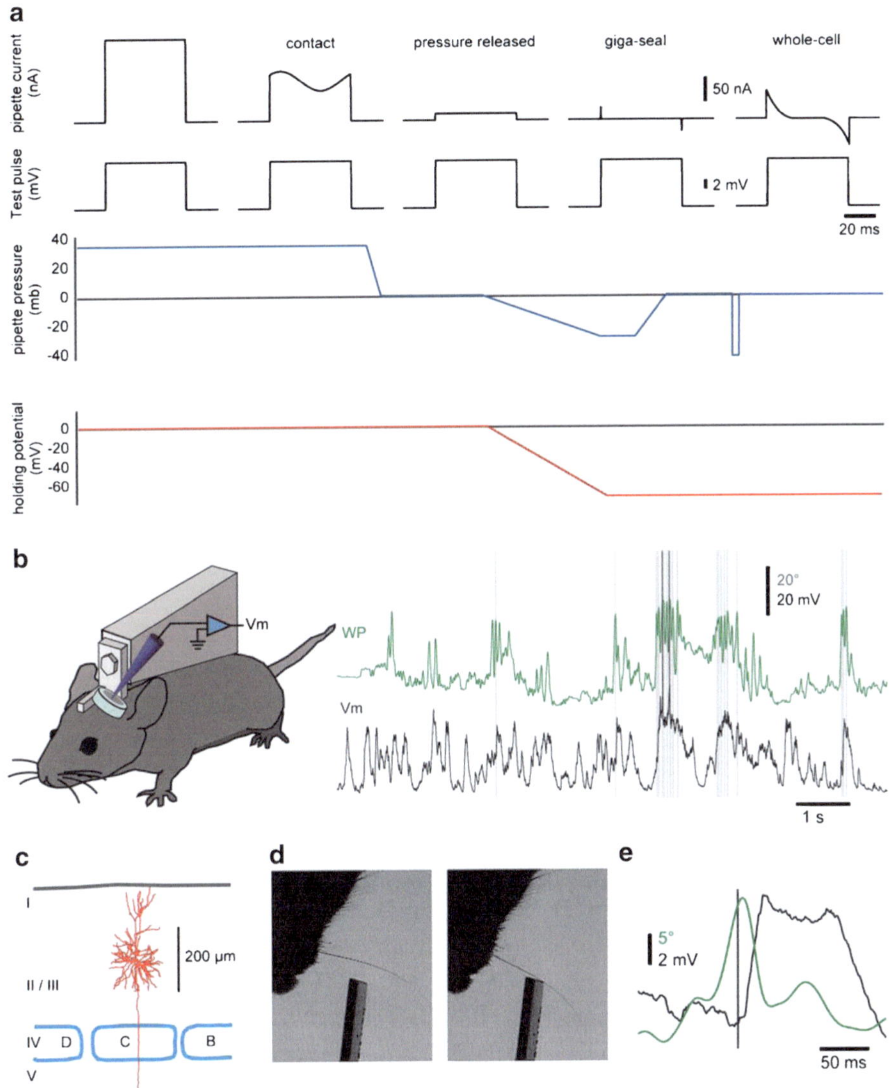

Fig. 4. Blind whole-cell patch-clamp recording in an awake behaving mouse. (**a**) Schematic representation of the successive phases leading to whole-cell configuration. In voltage clamp mode, 10-mV test pulses are applied to the pipette at 20–50 Hz while monitoring the current response on the oscilloscope. A constant 15–30-mb positive pressure is applied to the pipette, which is advanced in 2-μm steps into the brain. A contact with a cell is detected as a sudden increase in the pipette resistance (decrease in current in response to test pulse). The pressure is then released producing a further increase in the pipette resistance. A gentle suction (negative pressure) is applied to the pipette while the holding potential is progressively decreased to −70 mV, until a gigaseal is formed. After capacitance and transient compensation, short, negative-pressure pulses are applied to the pipette to break the cell membrane and establish the whole-cell recording. The amplifier can then be switched to current-clamp mode. (**b**) Schematic representation of the recording configuration in awake head-restrained mouse and an example of a whole-cell patch-clamp recording of a layer 2/3 pyramidal cell within C2 barrel column during a sequence of active touches. *Top trace*, whisker angular position (WP); *bottom trace*, membrane potential (Vm); *gray bars* indicate contacts between the whisker and the object. (**c**) Reconstructed dendrites and descending axon within barrel map (normal view). The recorded neuron was filled with biocytin and stained with an ABC kit and diaminobenzidine for post hoc anatomical identification and localization relative to barrel map. (**d**) Simultaneous high-speed filming of the whisker behavior allowed for identification of whisker–object contact time. (**e**) Averaged WP and Vm triggered by the onset of active contacts. (**e**) Averaged WP and Vm triggered by the onset of the active contacts (black line).

14. At the end of the recording, the pipette is slowly withdrawn and the animal can be Transcardially perfused with PFA solution for post hoc anatomical identification of the recorded neuron (Fig. 4c). To increase the chance of recovery of the cell anatomy, the pipette should preferentially be withdrawn while the recording is still of good quality.

4. Notes

1. A similar approach can be used with rats. However, rats are much stronger than mice and can easily break their implants during the habituation procedure. It is, thus, recommended to use anchoring screws during the implantation and to process very gently and progressively during the habituation (23, 40).
2. For short implantations (up to 45 nm), gas anesthesia can be substituted by a mixture of Ketamine and Xylazine administered by i.p.
3. The skull of the mouse is very soft, so particular attention should be paid not to apply excessive pressure with the cotton swabs or the scalpel blade.
4. The head post can be stereotaxically positioned using a modified arm mount micromanipulator (Kopf) to enable reproducible implantation.
5. The size of the recording chamber may be adapted to experimental needs. The critical point is to position and build the recording chamber in a way that enables a good access to the craniotomy, i.e., the needle should reach the craniotomy with an angle not higher than 40° to the surface of the brain (Fig. 3).
6. The use of any plastic materials should be avoided in close proximity of the animal. For example, if the animal is sitting on a plastic board, we observed very large electrical artifacts due to static charge when the animal is moving or whisking.
7. The habituation process can be largely adapted depending on the experimental protocol. Three to five sessions of habituation are usually necessary to perform recordings in an awake mouse. Recordings across different sleep–wake stages (wakefulness, slow-wave sleep, and REM sleep) are also possible but require prolonged habituation (2–3 weeks) with longer sessions (3–5 h) (41). It is also possible to train an implanted animal to perform simple tasks under head-restrained condition during the period that precedes the recording (34). However, there are two limiting factors to a long habituation process or training procedure: (1) when the animals are implanted for too long (more than 3 weeks), the dura underneath the exposed skull grows thicker

making the craniotomy more difficult; (2) it is also more difficult, though not impossible, to perform patch-clamp recordings in mice older than 6 weeks and our protocol for implantation might not be appropriate for mice younger than 4 weeks. Therefore, it is not recommended to wait more than 2–3 weeks between the implantation and the recording session.

8. For combination with two-photon microscopy, the size of the craniotomy must be larger (1–2 mm in diameter). To prevent tissue movements and edema, the brain is stabilized by covering the craniotomy with agarose and a thin glass coverslip.
9. The location of the craniotomy can be determined in different ways. Stereotaxic coordinates can be used; in this case, a stereotaxic frame must be used during the implantation and the targeted location can be marked during the implantation. A major advantage of the mouse is that the skull, when hydrated, is transparent enough to perform intrinsic imaging without opening or thinning. For experimenters interested in sensory perception, the exact location of the craniotomy can thus be functionally determined using intrinsic optical imaging (28, 35). In any case, it is recommended to choose a location avoiding big blood vessels.
10. The skull of the mouse is very thin and transparent when hydrated but becomes opaque when dry. It is easier to drill when the bone is dry, but it is recommended to regularly control the progression of the drilling by filling the chamber with Ringer's solution.
11. Alternatively, the opening of the dura can be done at the beginning of the recording session, just prior to the insertion of the pipette. This is actually recommended if the recording session occurs the day after the craniotomy.
12. Depending on the duration of the recovery period after the craniotomy, it might be necessary to reopen the dura with a fine needle.
13. If the resistance of the pipette is high after its insertion into the tissue or progressively increases when advancing it, it means that the tip of the pipette is obstructed probably because the craniotomy was not perfectly clean or the dura not completely removed. One can try to clean the tip of the pipette by increasing the positive pressure applied to the pipette and advancing the pipette by 50–100 μm. If the pipette cannot be cleaned, it must be changed and it is recommended to clean the craniotomy again.
14. The pressure used to advance the pipette into the tissue might be adjusted depending on which structure is targeted; reaching deep structures may require passing through areas rich in fibers or even through ventricles with high risk of obstructing the tip of the pipette.

15. Sometime, the establishment of a gigaseal is difficult despite the application of negative pressure to the pipette. Depolarizing the holding Vm to −80/−90 mV may help forming the seal. The holding Vm should then be set back to −70 mV before breaking the membrane to establish the whole-cell configuration.

References

1. Cragg B (1967) The density of synapses and neurones in the motor and visual areas of the cerebral cortex. J Anat 101:639–654
2. DeFelipe J, Farinas I (1992) The pyramidal neuron of the cerebral cortex: morphological and chemical characteristics of the synaptic inputs. Prog Neurobiol 39:563–607
3. Yuste R, Tank DW (1996) Dendritic integration in mammalian neurons, a century after Cajal. Neuron 16:701–716
4. Williams SR, Stuart GJ (2003) Role of dendritic synapse location in the control of action potential output. Trends Neurosci 26:147–154
5. Magee JC (2000) Dendritic integration of excitatory synaptic input. Nat Rev Neurosci 1:181–190
6. Reyes A, Lujan R, Rozov A et al (1998) Target-cell-specific facilitation and depression in neocortical circuits. Nat Neurosci 1:279–285
7. Hubel D (1959) Single unit activity in striate cortex of unrestrained cats. J Physiol 147:226–238
8. Hobson JA, McCarley RW, Wyzinski P (1975) Sleep cycle oscillation: reciprocal discharge by two brainstem neuronal groups. Science 189:55–58
9. Buzsaki G, Czeh G (1981) Commissural and perforant path interactions in the rat hippocampus. Field potentials and unitary activity. Exp Brain Res 43:429–438
10. Romo R, Salinas E (2001) Touch and go: decision-making mechanisms in somatosensation. Annu Rev Neurosci 24:107–137
11. Wiest MC, Bentley N, Nicolelis MA (2005) Heterogeneous integration of bilateral whisker signals by neurons in primary somatosensory cortex of awake rats. J Neurophysiol 93:2966–2973
12. Huetz C, Philibert B, Edeline JM (2009) A spike-timing code for discriminating conspecific vocalizations in the thalamocortical system of anesthetized and awake guinea pigs. J Neurosci 29:334–350
13. Ecker AS, Berens P, Keliris GA et al (2010) Decorrelated neuronal firing in cortical microcircuits. Science 327:584–587
14. von Heimendahl M, Itskov PM, Arabzadeh E et al (2007) Neuronal activity in rat barrel cortex underlying texture discrimination. PLoS Biol 5:e305
15. Krupa DJ, Wiest MC, Shuler MG et al (2004) Layer-specific somatosensory cortical activation during active tactile discrimination. Science 304:1989–1992
16. Sakai K, Crochet S (2001) Differentiation of presumed serotonergic dorsal raphe neurons in relation to behavior and wake-sleep states. Neuroscience 104:1141–1155
17. Csicsvari J, Henze DA, Jamieson B et al (2003) Massively parallel recording of unit and local field potentials with silicon-based electrodes. J Neurophysiol 90:1314–1323
18. Pastalkova E, Itskov V, Amarasingham A et al (2008) Internally generated cell assembly sequences in the rat hippocampus. Science 321:1322–1327
19. Peyrache A, Khamassi M, Benchenane K et al (2009) Replay of rule-learning related neural patterns in the prefrontal cortex during sleep. Nat Neurosci 12:919–926
20. Lin SC, Nicolelis MA (2008) Neuronal ensemble bursting in the basal forebrain encodes salience irrespective of valence. Neuron 59:138–149
21. Timofeev I, Grenier F, Steriade M (2001) Disfacilitation and active inhibition in the neocortex during the natural sleep-wake cycle: an intracellular study. Proc Natl Acad Sci USA 98:1924–1929
22. Mahon S, Vautrelle N, Pezard L et al (2006) Distinct patterns of striatal medium spiny neuron activity during the natural sleep-wake cycle. J Neurosci 26:12587–12595
23. Okun M, Naim A, Lampl I (2010) The subthreshold relation between cortical local field potential and neuronal firing unveiled by intracellular recordings in awake rats. J Neurosci 30:4440–4448

24. Petersen CC, Hahn TT, Mehta M et al (2003) Interaction of sensory responses with spontaneous depolarization in layer 2/3 barrel cortex. Proc Natl Acad Sci USA 100:13638–13643
25. Gentet LJ, Avermann M, Matyas F et al (2010) Membrane potential dynamics of GABAergic neurons in the barrel cortex of behaving mice. Neuron 65:422–435
26. Edeline JM, Dutrieux G, Manunta Y et al (2001) Diversity of receptive field changes in auditory cortex during natural sleep. Eur J Neurosci 14:1865–1880
27. Castro-Alamancos MA (2004) Absence of rapid sensory adaptation in neocortex during information processing states. Neuron 41:455–464
28. Ferezou I, Bolea S, Petersen CC (2006) Visualizing the cortical representation of whisker touch: voltage-sensitive dye imaging in freely moving mice. Neuron 50:617–629
29. Petersen CC (2007) The functional organization of the barrel cortex. Neuron 56:339–355
30. Brecht M (2007) Barrel cortex and whisker-mediated behaviors. Curr Opin Neurobiol 17:408–416
31. Carvell GE, Simons DJ (1990) Biometric analyses of vibrissal tactile discrimination in the rat. J Neurosci 10:2638–2648
32. Prigg T, Goldreich D, Carvell GE et al (2002) Texture discrimination and unit recordings in the rat whisker/barrel system. Physiol Behav 77:671–675
33. Knutsen PM, Pietr M, Ahissar E (2006) Haptic object localization in the vibrissal system: behavior and performance. J Neurosci 26:8451–8464
34. O'Connor DH, Clack NG, Huber D et al (2010) Vibrissa-based object localization in head-fixed mice. J Neurosci 30:1947–1967
35. Crochet S, Petersen CC (2006) Correlating whisker behavior with membrane potential in barrel cortex of awake mice. Nat Neurosci 9:608–610
36. Poulet JF, Petersen CC (2008) Internal brain state regulates membrane potential synchrony in barrel cortex of behaving mice. Nature 454:881–885
37. Harvey CD, Collman F, Dombeck DA et al (2009) Intracellular dynamics of hippocampal place cells during virtual navigation. Nature 461:941–946
38. Lee AK, Manns ID, Sakmann B et al (2006) Whole-cell recordings in freely moving rats. Neuron 51:399–407
39. Epsztein J, Lee AK, Chorev E et al (2010) Impact of spikelets on hippocampal CA1 pyramidal cell activity during spatial exploration. Science 327:474–477
40. Souliere F, Urbain N, Gervasoni D et al (2000) Single-unit and polygraphic recordings associated with systemic or local pharmacology: a multi-purpose stereotaxic approach for the awake, anaesthetic-free, and head-restrained rat. J Neurosci Res 61: 88–100
41. Takahashi K, Lin JS, Sakai K (2006) Neuronal activity of histaminergic tuberomammillary neurons during wake-sleep states in the mouse. J Neurosci 26:10292–10298

Neuromethods (2012) 67: 237–258
DOI 10.1007/7657_2011_22

Published online: 1 December 2011

Investigating Sleep Homeostasis with Extracellular Recording of Multiunit Activity from the Neocortex in Freely Behaving Rats

Vladyslav V. Vyazovskiy, Umberto Olcese, and Giulio Tononi

Abstract

Cortical activity during sleep and waking is traditionally investigated with electroencephalography (EEG). The most distinctive feature of neocortical activity during sleep is the occurrence of EEG slow waves, arising from quasi-synchronous periods of activity and silence among cortical neurons. The EEG slow waves are regulated homeostatically: they are larger and have a higher incidence following long waking periods and decrease as a function of time spent asleep. Since intense early sleep seems to be important for restoration, understanding the cellular mechanisms underlying homeostatic regulation of sleep slow waves may appear crucial for understanding sleep function. While macrooscillations recorded with the EEG arise from synchronous activity and silence of large populations of cortical neurons, at present intracellular recording techniques do not allow monitoring the state of more than just a few cells at a time across spontaneous sleep–wake cycle in unrestrained animals. Here, we review a method for chronic recording of extracellular LFP and multiunit activity from the neocortex in freely moving rats. This technique is most useful for addressing cellular mechanisms of sleep homeostasis because it allows monitoring the activity of many cells simultaneously for many hours. The description of the surgical procedure is complemented with a detailed account of spike sorting, which is a crucial step in processing and interpreting extracellular waveforms.

Key words: Extracellular recordings, Local-filed potentials, Multiunit activity, Rats, Neocortex, Sleep homeostasis, Spike sorting

1. Introduction

Cortical activity shows ongoing spontaneous fluctuations at multiple temporal and spatial scales that are tightly linked to vigilance state, movement, behavior, and the presence or absence of sensory stimuli (1–5). Above all, on a slow time scale of minutes and hours brain state fluctuates regularly between two distinct states—wakefulness and sleep (6).

Traditionally, sleep and waking in animals and humans have been investigated with electroencephalography (EEG). The advantages of the EEG technique are that the signal, recorded with macroelectrodes placed on the skull or on the surface of the cortex, represents large-scale cortical activity and that the recordings can be performed simultaneously from many cortical areas, and for many hours. With the advent of chronic EEG technique, it became possible to record sleep during an entire night, and perform extensive analysis of various aspects of brain electrical activity. Most importantly, chronic sleep EEG recording and analysis led not only to the discovery that sleep consists of two different stages—NREM sleep and REM sleep, but also that sleep is regulated homeostatically, as manifested in an increase of sleep EEG slow-wave activity (SWA) (0.5–4 Hz) after waking and its gradual decline across the night (7, 8). Crucially, both these phenomena would have remained unnoticed if long-term, stable recordings for many hours were not performed. However, there are also some limitations to the EEG-approach, such as volume conduction, poor spatial resolution and the impossibility to trace the origin of EEG waves or their spatiotemporal dynamics to specific cellular and network mechanisms.

These shortcomings were partially resolved by intracellular single-cell electrophysiology. This technique provided a relationship between cortical neuronal activity, EEG waves, and global brain states. Specifically, intracellular recordings revealed that, in contrast to waking and REM sleep—when cortical neurons are tonically depolarized most of the time (4)—virtually all cortical neurons during sleep engage in the slow oscillation, consisting of a depolarized UP state, where neurons show sustained firing, and a silent hyperpolarized DOWN state (4, 9, 10). Moreover, a close temporal relationship between intracellular slow oscillation and simultaneously recorded EEG slow waves was found (11, 12). While this approach has proven essential for our understanding of how individual neurons change their activity in relation to behavior and vigilance states, it has also some significant limitations. Specifically, intracellular recording techniques do not allow long-term stable recordings in freely moving rodents, and even in head-restrained animals it is possible to record only a few cells at a time (9).

Thus, in order to investigate cellular mechanisms underlying homeostatic sleep regulation, a technique is necessary that on one hand allows long-term stable recording in freely moving, unanesthetized animals, and on the other hand has sufficient temporal and spatial resolution to resolve spiking activity of individual neurons. Extracellular recordings from the cortex with microelectrodes appeared to be useful in bridging this gap between the EEG and single-cell recordings. The advantage of extracellular recording techniques is that they allow sampling activity from a large neuronal population, assure long-term stable recording, and often allow simultaneous recording of local field potentials (LFPs) from the same electrodes (13). This technique is uniquely suited to record from several cells for many hours and days, and it is thus extremely valuable for investigating cellular and network mechanisms underlying sleep homeostasis (5). In this chapter, we describe different methodological aspects of chronic extracellular recording of neuronal activity and LFPs during waking and sleep in freely moving rats. We also consider the most essential steps of its analysis with an emphasis on spike sorting algorithms, and discuss potential problems and limitations of the technique.

1.1. The Origin of Extracellular Signals in the Neocortex

Extracellular recording is one of the most widely used techniques in in vivo electrophysiology. The signal measured by an electrode placed in the cortex represents the mean extracellular field potential from the weighted average of all current sinks and sources in the local cortical compartment (14). Typically, the largest component of the current is that generated by action potentials (APs), which produce large transmembrane potentials in the vicinity of the neuron (15). Cortical pyramidal cells have several zones that are independently capable of generating APs, but the axon initial segment is the preferred site of spike initiation (16). The signals can be recorded as a voltage difference between the electrode placed in close proximity to neurons and a reference electrode. The high-frequency signal (~300–5,000 Hz) mainly reflects sum of the extracellular APs (multiunit activity, MUA) of all neurons from their soma, axons, and dendrites within a sphere whose radius is about 140–300 μm, with the electrode at its center (14, 17), and rapidly decays with distance (18–20). Although larger electrodes pick up activity from a larger neuronal population, they usually produce also a larger site of damage, potentially affecting the physiological activity of local neuronal assemblies. Moreover, if too many neurons are recorded simultaneously from the same site, they are difficult to discriminate (19).

There is a consistent relationship between spiking activity and LFPs (14). LFP is usually recorded as a voltage difference either between the microelectrode and distant referential electrode or, in a bipolar configuration, between two recording sites located close (<1–2 mm) to each other (21). The low frequency signal (<200–300 Hz) mostly represents slow events reflecting cooperative activity in neural populations and is mostly due to synaptic activity, e.g., the dendritic processing of synaptic inputs (14, 18, 22). LFPs are thought to reflect the activity of large populations of neurons in a sphere of one to several millimeters around the electrode and are believed to have local origin (23). Cortical pyramidal neurons generate extracellular currents that flow mostly parallel to their somatodendritic axis. A fundamental question of neuroscience is the relationship between LFPs and spiking activity. LFPs convey information relatively independent of spiking activity, although faster LFP frequencies, e.g., gamma-band (~30–100 Hz), can be significantly affected by neuronal discharges (13, 24). Moreover, cortical neurons usually have a preferred phase of firing in respect to delta (25) and theta (26) oscillations. Finally, in humans and monkey firing rates are correlated

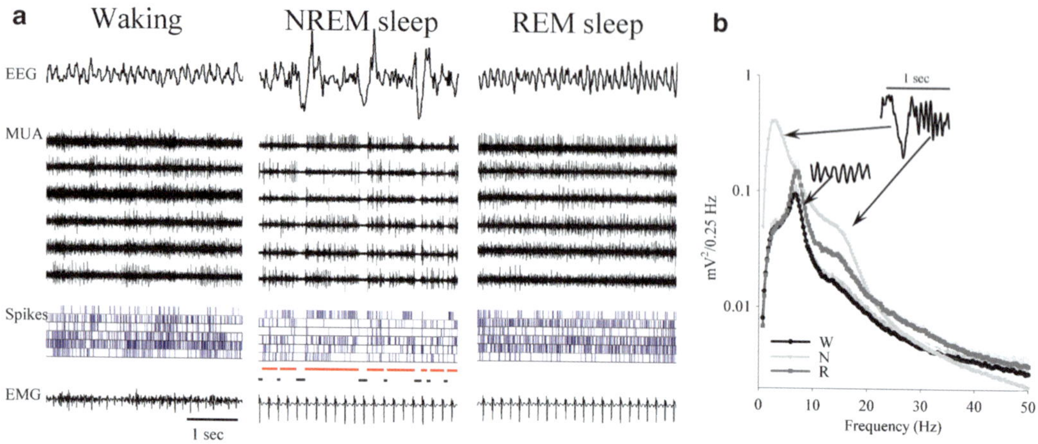

Fig. 1. EEG and multiunit activity in three basic vigilance states. (**a**) From top to bottom: Surface electroencephalogram (EEG) traces from the right barrel cortex during waking, NREM sleep and REM sleep in a representative rat during undisturbed baseline (positivity is upward). Raw multiunit activity (MUA) recorded simultaneously in the same rat from a microwire array placed in the left barrel cortex (6 individual channels are shown). Note high frequency tonic firing in waking and REM sleep, and the periods of population neuronal activity and silence in NREM sleep. Raster plots of spike activity for the same 6 channels (each vertical line is a spike). Note the close temporal relationship between the silent periods and the negative phase of EEG slow waves. Corresponding electromyogram (EMG). (**b**) Average EEG power spectra in NREM sleep, REM sleep and waking (mean + SEM, $n = 6$ rats). Insets: EEG slow wave followed by a spindle typical for NREM sleep and theta-activity typical for REM sleep and waking.

with shifts in the LFP power spectra (14, 27). It has been shown that the LFP in the vicinity of a cortical neuron correlates with its membrane potential (28–30). Thus, the correlation between spikes and LFP might reflects synchrony between the synaptic inputs (31).

1.2. EEG and Extracellular Activity in Waking and NREM Sleep

A fundamental difference between wakefulness and sleep is the extent to which the brain is engaged in the acquisition and processing of information. This "functional" difference between vigilance states is reflected in different patterns of neuronal activity and in the EEG (Fig. 1). Specifically, unlike in NREM sleep, when neocortical activity is organized largely "from inside" (4), cortical activity in waking is generated not only by ascending influences from specific waking-promoting areas (32, 33) and intracortical and cortico-subcortical interactions (34, 35), but also by behavior (2, 36, 37) and processing of incoming external stimuli (38). The wake EEG in rodents is distinguished by the absence of slow waves, and by pronounced theta (~7–9 Hz) activity (Fig. 1) presumably arising as a result of physical spread from the hippocampus due to volume conduction (26, 39–43).

Cortical neuronal firing activity is generally slower in NREM sleep compared to both wakefulness and REM sleep (4, 44–46). Moreover, independent firing of neurons in wakefulness is replaced

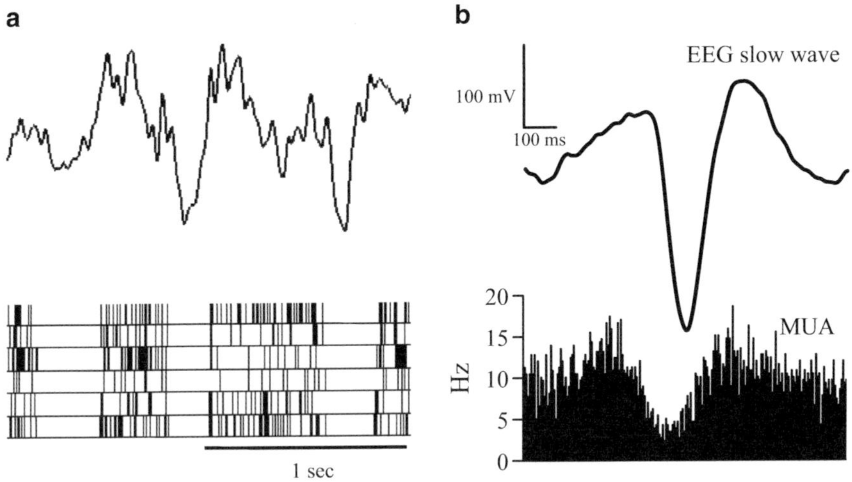

Fig. 2. Sleep slow waves are associated with generalized neuronal silence. (**a**) Individual trace of the EEG and raster plots of neuronal activity in NREM sleep in one representative rat (each *vertical line* is a spike). Note that during EEG slow waves neuronal activity is absent or reduced. (**b**) The profile of average neuronal firing rates (MUA) in NREM sleep aligned to the negative peak of the EEG slow wave.

with regular synchronous bursts of APs during sleep (45, 47–49). Intracellular recordings have shown that, during NREM sleep, virtually all cortical neurons engage in the slow (<1 Hz) oscillation, consisting of a depolarized UP state, when neurons show sustained firing, and a hyperpolarized DOWN state, characterized by neuronal silence (4, 10, 11, 50). Simultaneous EEG and intracellular or extracellular recordings have shown that slow waves are a reflection of near-synchronous transitions between UP and DOWN states in large populations of cortical neurons (4, 12, 44, 45, 49, 51–56). Furthermore, high amplitude slow waves are associated with a longer period of silence in MUA (5, 51) (Fig. 2).

The mechanisms underlying spontaneous cortical generation of the UP state have been investigated extensively but are not yet clearly understood. The cortical UP states can be triggered from the thalamus (57) or start spontaneously in the cortex (9, 52). It has been shown that in between UP states, the membrane potential is relatively hyperpolarized, and there is a marked decrease in synaptic activity (4). The beginning of the UP state is marked by the arrival of large barrages of postsynaptic potentials, which depolarize the neuron resulting in a burst of APs in the local network (28, 58). It has been postulated that the UP state is generated through local recurrent excitation that is balanced and controlled by the activity of GABAergic interneurons (28, 59, 60). An intriguing finding was that UP states are associated with discharge rates equal to or even exceeding those found in waking (4, 61). The rapid transition from the DOWN state into the UP state occurs

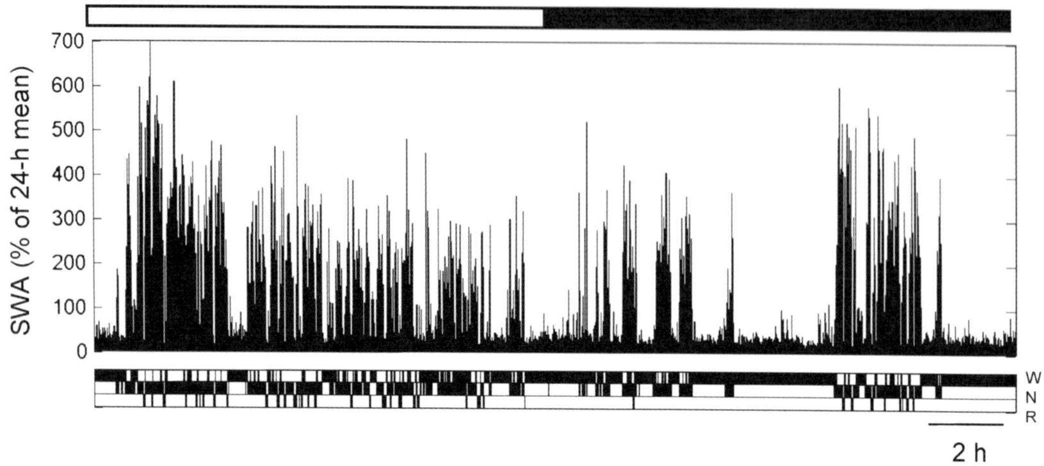

Fig. 3. Sleep SWA decreases homeostatically during sleep. Slow-wave activity (SWA, EEG power between 0.5 and 4 Hz) and corresponding hypnogram during an undisturbed baseline 24 h in one individual rat. Vertical bars depict consecutive 4-s values of SWA. Note that SWA decreases during 12 h light period and increases after long consolidated waking episodes.

synchronously throughout the local network (28) or can start in deep layers and propagate in other layers and horizontally with a delay of ~10–30 ms (59). The sequential propagation of the UP state in naturally sleeping animals is manifested in a precise timing of spike occurrence within the first ~100 ms after the onset of the UP state in which the average onset latency was ~50–60 ms (54).

1.3. Investigating Sleep Homeostasis with Extracellular Recordings of Cortical LFP and MUA

Sleep and wake have been traditionally investigated with EEG. EEG studies led to the notion that sleep is a global state, encompassing an entire brain (9). Indeed, the hallmark of sleep EEG—high-amplitude slow waves—are usually recorded near-simultaneously across distant cortical areas (62). Also sleep homeostasis, which was mostly characterized in terms of the temporal evolution of EEG power spectra in slow-frequency band, is usually regarded as a whole-brain phenomenon (7). Strong evidence accumulates, however, that in both humans and animals sleep regulation has a regional facet (36, 63–66), the homeostatic changes in sleep slow waves are related to cortical plasticity (21, 64, 67–74) and the shape of EEG slow waves is determined by specific cellular substrates (5).

It is now established that spontaneous brain activity changes not only as a function of vigilance state, behavior, or sensory processing, but is also strongly determined by preceding history (75). For example, sleep following extended wakefulness is characterized by intense cortical SWA, the NREM EEG power between 0.5 and 4 Hz, which decreases after periods of sleep (36, 76, 77). The time course of NREM SWA during undisturbed 24 h is illustrated on Fig. 3. There is also evidence for a regional regulation of slow waves (78, 79) and recent studies show that cortical areas that have been more active during waking show higher SWA relative to less engaged areas (64, 72, 74, 80), whereas areas that have been less active show reduced SWA (68). It has been shown recently that after wakefulness cortical firing in NREM sleep is characterized by short and frequent periods of population activity, where neurons fire intensely and synchronously, while the periods of neuronal silence are long and frequent (5).

Recording local extracellular LFPs with multiple microwire electrodes from several cortical areas thus appears a useful approach for addressing sleep regulation. Among the most important advantages of this technique is that it allows recording LFPs with high spatial resolution, and that it is not confounded by volume conduction as in the case of human scalp EEG. In addition, recording simultaneously extracellular activity from large assemblies of cortical neurons can clarify how the dynamics of single- and multiunit activity bring about the homeostatic changes in the LFPs.

For example, it is currently unclear what determines the regularity in the occurrence of prolonged intracellular UP and DOWN states and extracellular synchronous ON and OFF periods that ultimately give rise to the large number of high-amplitude slow waves typical of sleep after prolonged period of wakefulness. Our own hypothesis suggests that sleep SWA may reflect the average strength of cortical synapses, which would increase during wakefulness as a result of plastic processes, and decrease during sleep due to a sleep-dependent mechanism of synaptic downscaling (71, 81). According to this hypothesis, high average synaptic strength facilitates network synchronization during sleep, which would appear in the EEG as increased SWA. The progressive decrease in SWA during sleep, on the other hand, would reflect a decrease in network synchronization due to the weakening of cortical connections (81–83). Since in this case fewer neurons would fire in-phase with others, it should result in a less efficient recruitment of neurons into the UP state, more irregular firing rates, and a rare occurrence of prolonged synchronous DOWN states (5, 21, 70). Thus, understanding the network dynamics underlying synchronous slow oscillation in the neocortex of spontaneously sleeping animals should help to understand the mechanisms underlying the homeostatic changes in the sleep EEG.

2. Materials

Simultaneous recording of LFPs and MUA from a large number (a bundle or an array) of small electrodes (20–50 μm in diameter) usually assures minor cortical damage and a sufficient number of neurons to make the sample representative (19). It is believed that the low frequency band of the extracellular signal is much less affected by geometry of tissue-electrode interface, which is critical for single-unit recordings (84). While the shape and properties of the electrode play some role (15), it was argued that with proper recording equipment the impedance (which is inversely related to the size of an uninsulated surface area of the electrode) and geometry of microelectrode recording sites within typically used ranges for extracellular recordings do not appreciably affect LFP recordings (85). In our experiments, we used polyimide-insulated tungsten microwire arrays (16-ch: 2 rows with 8 channels, Tucker-Davis Technologies Inc, Alachua, FL) with the following dimensions: wire diameter 33 μm, electrode spacing 175–250 μm, row separation: horizontal, 375–500 μm; vertical, 500 μm.

The technique of implanting microwire electrode arrays generally follows the standard neurosurgical procedure in small rodents. We use the approach described in "Surgical implantation guidelines" (Neuronexus Technologies, Inc.) and (86) with some modifications. One day before surgery animals receive an i.p. dose of dexamethasone to suppress local immunological response (0.2 mg/kg) (87, 88). For anesthesia isoflurane (3–5% for induction, 1–2% for maintenance), or a ketamine/xylazine cocktail (ketamine 87 mg/kg–xylazine 13 mg/kg, i.p) can be used.

Apart from the conventional instruments and surgical tools used for neurosurgery in small rodents the following materials are required:

- Friedman-Pearson Rongeur (part number 16221-14, Fine Science Tools, CA, USA)
- Gelfoam (available from dental supplies companies)
- Vitrectomy scissors (2.2 mm straight blades, part number 15036-14, Fine Science Tools, CA, USA)
- Warm (40°C) agarose gel mixed with dexamethasone (0.2%)
- Two-component silicon gel (KwikSil; World Precision Instruments, FL, USA)

Before surgery all materials must be sterilized in ethylene oxide.

The adult (~3 month) male laboratory rat is a suitable species for the recording of neuronal activity during sleep. Sleep–waking architecture in several rat strains has been well characterized. Moreover, the techniques of sleep deprivation as well as a number of behavioral paradigms have been established for this species (65, 77, 89, 90). It is crucial in all sleep studies that the animals are kept in highly standardized conditions, are well adapted to the recording settings and well entrained to the light–dark cycle. In our laboratory, animals are kept on a 12:12 light–dark cycle while ambient temperature is kept constant at ~23°C. Prior to surgery animals must be well habituated to the recording cage, to the experimenter and to the experimental setting, e.g., novel objects used for sleep deprivation. At least 1 week is allowed for recovery after surgery, and experiments should be started only after the sleep–waking cycle had fully normalized. An example of the hypnograms of a fully entrained rat during a standard baseline 24 h is illustrated on Fig. 3.

3. Methods

3.1. Surgical Procedure

Under deep anesthesia, rats are set in the stereotaxic apparatus, skull is cleaned from skin and other tissues and the implantation areas are outlined with a pen. Depending on the aim of the experiment, suggested cortical areas are the barrel cortex (B: −2 to 3 mm, L: 5–6 mm) and/or the frontal cortex (B: +1–2 mm, L: 2–3 mm). An ~2 × 2 mm craniotomy is made using first a 1.4 mm drill bit and then a 0.7 mm drill bit, with the aid of a high-speed surgical drill. The hole is adjusted to the size of the array by removing the remaining fragments of the bone with a Friedman-Pearson Rongeur. In most cases, the removal of the dura does not cause bleeding, but if bleeding occurred it can be stopped with gelfoam soaked in sterile saline. The dura is pierced with a syringe needle of high gauge and dissected with vitrectomy scissors.

The electrode array is advanced into the brain tissue by penetrating the pia mater, making an effort to avoid vascular damage under constant visual control with magnification with a stereo microscope. The electrode arrays must be handled with care throughout the procedure and antistatic wrist strap must be worn at all times. Array insertion is achieved by advancing the electrode array until both the superficial and the deep rows of the arrays are under the cortical surface. The wires of the arrays are very easily bent and no effort should be applied if the remnants of the dura or bone fragments obstruct lowering the array in the cortical tissue. The final position of the array can be adjusted by withdrawing or lowering it slowly (~50 μm steps) until most recording channels show robust single- or multiunit activity. The final insertion depth for layer V is ~1.2 mm below the pial surface but may vary depending on cortical thickness. Rat brain atlas should be consulted for determining the coordinates of the region of interest and the final depth of the electrodes. Warm agarose gel is gently placed around the array to fill the craniotomy. At this stage, special care is taken to avoid displacing the array in the lateral direction. The two-component silicon gel is used to seal the craniotomy and protect the surface of the brain from dental acrylic. After ~10 min, required for the gel to polymerize, dental acrylic is gently placed around the electrode, fixing the array to the skull. The ground electrode is placed above the cerebellum and additional anchor screws are placed bilaterally in the frontal bone or occipital. EMG wires are placed in the neck muscles. The animal should be kept on a warm blanket until it recovers from anesthesia before transferring it to the recording cage.

3.2. Signal Recording, Processing, and Analysis

The experiment should begin at least after 1 week following surgery when the animal's behavior and sleep–wake cycle are fully normalized and signals are stable. It is recommended to collect spike data at minimal sampling rate of 25 kHz, after filtering between 300 and 5,000 Hz. LFPs and the EEGs can be recorded with sampling rate of 256 Hz, after filtering 0.1–128 Hz.

In the extracellularly recorded signal action potentials usually occur on the background of signals from distant low-amplitude MUA and electrical artifacts due to movement. Recording systems usually allow noise with a standard deviation of 10–20 μV, while spikes usually range from 30 to 100 μV, depending on the neuronal type, the distance between the cell and the recording site and on the impedance between brain tissue and the electrode. In order to increase signal-to-noise ratio of the recording, it is important to choose an appropriate referencing scheme. Common surgery procedures (91) require placing a ground screw, for example, above the cerebellum. However, large distance (several mm) between cortical arrays usually results in increased levels of noise due to motion artifacts (91, 92). A common solution of this problem is placing an additional reference electrode nearby the recording tips. Such reference electrode can either be a screw or an ad hoc recording site on the array itself.

It is not always possible to record raw data continuously at the full sampling rate (25 kHz or more) from multiple channels, making online spike detection necessary. In addition, the threshold-based technique is employed to extract spikes from background activity for further processing (15). Amplitude thresholds for online spike detection are set manually based on visual and auditory control and allow only crossings of spikes with high signal-to-noise ratio (at least 2, Fig. 4). Since extracellular MUA signals are usually asymmetric, the detection threshold is applied to the side where spike waveforms exhibit greater deflection (14). Other strategies for spike detection involve manually selecting certain characteristics of the signal or performing cross-correlation with template waveforms. Threshold methods have an advantage of being effective and computationally low-demanding, which is an important feature for online spike detection. Moreover, thresholding usually allows excluding the low-amplitude noise and most of high-amplitude artifacts related to chewing and grooming of the animal.

During our recording, whenever the recorded voltage exceeded a predefined threshold (usually ~20–30 μV), a segment of 46 samples (0.48 ms before, 1.36 ms after the threshold crossing) is extracted

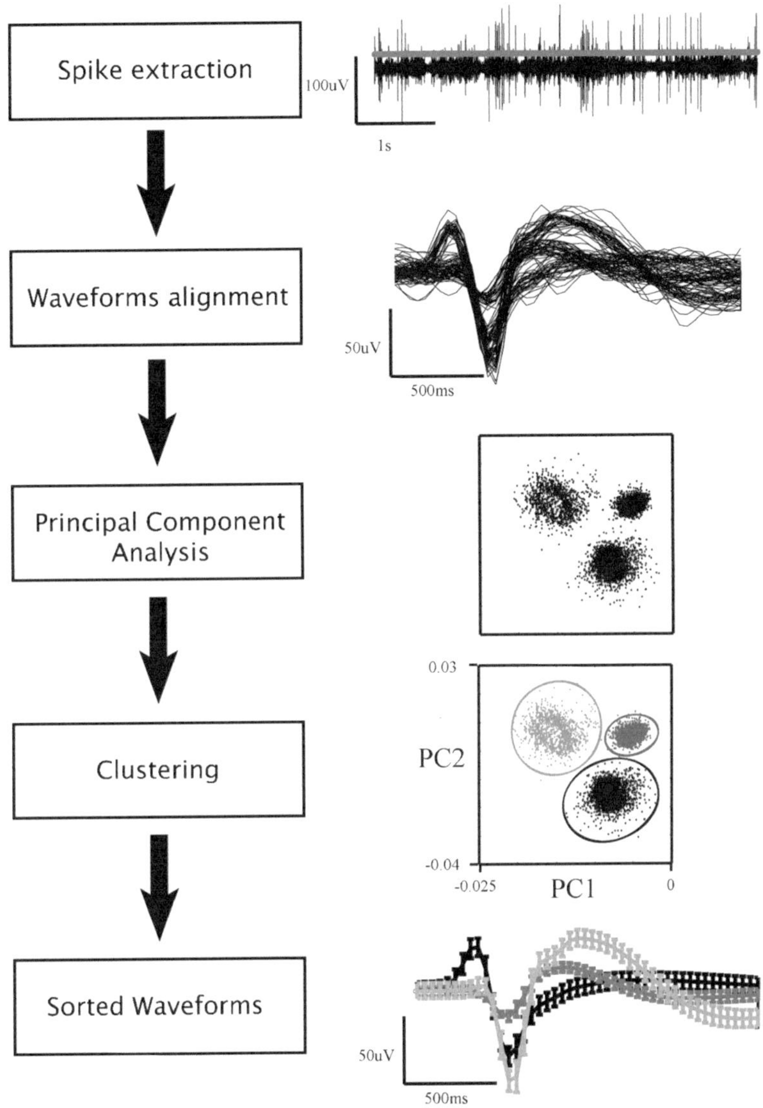

Fig. 4. Main steps in a spike sorting algorithm. Left, chart plot of the main steps of spike sorting. *Right*, from *top* to *bottom*: (1) 5-s trace of raw MUA recorded in from the barrel cortex. *Gray line*: amplitude threshold used for spike detection. (2) Individual waveforms ($n = 100$) extracted from the raw signal, aligned relative to their negative peak. For each spike, 46 consecutive data points are extracted (1.84 ms). (3) Scatter plot of the first two principal components (PCs) computed from 9,000 spikes taken from the same data set as in (2). Note that in this case two PCs are sufficient to describe most of the variance among different waveforms. (4) Clustering in the PCs space. The SMEM clustering algorithm discriminates three clusters depicted as *black, dark gray* and *light gray ellipses*. (5) Average waveforms ($\pm$ standard deviation) of action potentials belonging to each of the three clusters. Note that the clustering algorithm was able to successfully discriminate the different waveforms evident in (2).

and stored together with the corresponding timestamps. Spike data are then subjected to offline sorting procedure (see below). Subsequently, to avoid contamination of the sorted units, especially of small amplitude, by MUA, all individual neurons are carefully screened and units showing relatively high proportion (>1–2%) of short refractory periods (<2.5 ms) and no major peaks on the distribution of ISIs are discarded from the analysis.

The EEG and EMG signals, acquired together with spike data, are essential for correct staging of three basic vigilance states: waking, NREM sleep, and REM sleep. Off-line scoring of vigilance states based on the EEGs and EMG is performed by visual inspection of 4-s epochs. Waking is characterized by low voltage, high frequency EEG pattern and phasic EMG activity. Epochs of eating, drinking, and intense grooming are usually excluded, since during those periods LFP and MUA can be contaminated by muscle artifacts, affecting the spectral content of the signal and precluding reliable isolation of individual spikes (usually <5–10% of total time awake). NREM sleep is characterized by the occurrence of high amplitude slow waves and low tonic EMG activity (Fig. 1a). During REM sleep, the EEG is similar to that during waking, but only heart beats and occasional twitches are evident in the EMG signal. Epochs containing artifacts, predominantly during active waking, are excluded from spectral analysis.

3.3. Spike Sorting

A crucial step in the processing and interpreting extracellular signals is spike sorting. Usually, each electrode picks up activity from many cells within a sphere several hundred microns in diameter, and it is therefore necessary to employ algorithms to discriminate the contribution of single neurons and eliminate artifactual waveforms. Here, we review the main steps of processing spike data (Fig. 4). The filtering and spike extraction procedures are usually performed online and spike waveforms together with the corresponding time stamps are saved. Thresholding is a common way to extract spike waveforms from the continuous signal (see above). Prior to the feature extraction procedure the extracted waveforms are aligned typically to the largest value, which is usually the maximal amplitude of AP (15). It may be necessary to perform visual prescreening of waveforms in order to remove obviously artifactual signals and those cases where two or more spikes are superimposed.

Feature extraction. Spikes are commonly classified by their waveform features (15). If two cells are located at a different distance from the recording site, their corresponding extracellular APs will usually have different waveforms (19). Most techniques require to first extract features from the recorded waveforms because it is computationally very expensive to run clustering algorithm on a large number of data points per waveform and not all data points carry relevant information about spike shape (15).

Both manual and automated feature extraction techniques exist (15). Manual techniques require the measurement of biologically and visually relevant features, such as peak-to-peak amplitude, or half-spike width. Such parameters, although useful in characterizing different neuronal subtypes, do not optimally describe all the information contained in waveforms. Automated methods are most commonly employed for feature extraction, such as wavelet decomposition (93, 94) and principal component analysis (PCA) (95, 96). In PCA, an orthogonal linear transformation of the original data is performed, through singular value decomposition of the data matrix (97). After performing PCA, each spike (composed of n sampling values) will be described by n variables, called principal components, or PCs, each being a linear combination of the original sampling values. Each PC corresponds to an eigenvalue and an eigenvector from the singular value decomposition of the data matrix. The first PC is associated with the eigenvector on the direction of maximal variance of the dataset, and so on in decreasing order for the remaining PCs. The ratio between each eigenvalue and the sum of all eigenvalues represents the ratio of the total variance described by a single PC. For neuronal spikes, the first 2–4 PCs are usually able to describe over 70% of the total variance (15). Thus, with PCA, it is possible to reduce the number of data points used to describe a single spike and, at the same time, it is possible to quantify the amount of information which is lost during the process of dimensionality reduction (Fig. 4).

In Matlab, several functions are available to perform principal component analysis. We employed *runpca*, which is a part of the EEGLab toolbox (98). This function is applied to all the waveforms extracted from individual recording sites. The output of this function is the PCs and the corresponding eigenvalues. Then, a subset corresponding to the first n PCs can be extracted and used for subsequent steps of spike sorting.

Clustering procedure. Clustering is the most important step in spike sorting. Several kinds of clustering algorithms exist (99). Historically, the first strategy was visual discrimination of spikes by looking at the waveform shape or other features, although high variability is associated with such manual spike sorting (100). Subsequently, techniques have become more and more automated, although it is recommended that the whole process is supervised.

Clustering algorithms often employed in spike sorting are k-means (101), fuzzy c-means (102), and mixture of Gaussians fitting (103). Split and merge expectation maximization (SMEM, (104)) is an algorithm based on fitting a mixture of Gaussians to the data space that has several advantages over less sophisticated techniques. With respect to both k-means and fuzzy c-means, SMEM does not need the user to a priori select the number of clusters, or use heuristic measures of clustering quality, such as the Davies-Bouldin index (105). This algorithm is designed to find the correct number of clusters and characterize them by maximizing the log-likelihood function (103), a quantitative measure of goodness of fit. Moreover, in contrast to k-means clustering, clusters can overlap and each point has a probability of belonging to a cluster, instead of being a simple all-or-none function. The only drawback of the method, apart from it being computationally demanding, is that since clusters must have a Gaussian shape it requires model-based approach.

This algorithm is composed of two steps: expectation maximization step (EM) and a split and merge (SM) step. In the EM step, the parameters defining the considered k Gaussian distributions (i.e., n-dimensional mean values and variances) are fitted to the observed features by maximizing the expectation of the log-likelihood function for the complete data. Here, by complete data, we mean both observed and nonobserved data, where the former goes under the name of incomplete data. Thus, by using only features from the recorded data, it is possible to extrapolate the characteristics of the whole population data. In the SM step, Gaussians can be either split or merged according to some criteria and all the new cluster configurations, after being ranked, undergo an EM step. If the log-likelihood function for one of these configurations is larger than that at the previous step (by a factor larger than a certain value), the algorithm restarts with the new configuration, and otherwise it stops. Finally, each data point is assigned to the cluster that maximizes the posterior probability of having generated that sample (Fig. 4), i.e., each waveform is associated to the putative neuron which has the highest probability of having created it. With all these steps, it was demonstrated that the algorithm never stops in local maxima, but always reaches global maxima (104). In a simplified version, where only the split merge is implemented, at each step all clusters are—one at a time—split into two. If the log-likelihood function has increased in the new configuration, the new configuration is kept and the process restarts, otherwise the algorithm stops.

The algorithm is controlled by two parameters. These are two thresholds for determining when each EM step has reached convergence, and when to stop the SM procedure. The former threshold modulates the goodness of fit of each cluster, while the latter determines whether the algorithm tends to stop with fewer or more clusters than there actually are. Finally, a third parameter can be added for excluding artifacts—the samples for which the probability of belonging to one cluster falls below a certain value—is equivalent to removing the tails of Gaussian distributions. With long-term recordings, the algorithm may be too slow on standard workstations, and it should rather be run first only on a random sample of spikes (e.g., 20,000). In this case, the algorithm needs to be run multiple times, to make sure that the sample of spikes is representative of the global population, and the clustering is therefore performed correctly.

Several Matlab packages are available for performing the EM step and are found in the Matlab Central file repository (http://www.mathworks.com/matlabcentral/fileexchange). Although we implemented custom function for the SM step, some scripts can be found on the same Web site. These functions are to be applied on the PCs extracted from waveform data and return an index, which assigns each waveform to a cluster corresponding to an individual neuron.

Electrophysiological classification of neurons. Major neuronal subtypes can be discriminated based on extracellular waveform characteristics, specifically, the spike amplitude and the width at half-amplitude of the spike. Previous in vitro and in vivo studies classified neurons into three main categories: regular-spiking (RS), intrinsically bursting (IB), and fast-spiking (FS) neurons (4, 52, 106–111). It is generally accepted that broad APs occurring at low frequency are produced by pyramidal excitatory cells while narrow spikes occurring at faster frequency belong to GABAergic interneurons (106, 110, 112–121). It should be kept in mind, however, that the shape and amplitude of extracellularly recorded spikes are profoundly affected by axon geometry, inhomogeneities in the extracellular space, volume conduction or electrode-to-neuron distance (122–124), and the neuronal phenotype cannot be determined with certainty unless morphological analysis and/or immunohistochemistry is performed. In our recordings, neurons showed significant variability in their spike shape and firing characteristics but could roughly be subdivided into three groups (Fig. 5). Specifically, neurons with narrow APs and mostly short interspike intervals (ISIs) are likely putative fast-spiking neurons, neurons with broad spikes and relatively low firing rates are likely putative regular spiking, whereas putative bursting neurons also had broad spikes and had a major mode of ISIs distribution between 3 and 4 ms.

4. Notes

Having reviewed the main steps in data collection and processing, we should mention now several significant methodological and conceptual issues that should be taken into account while recording and interpreting extracellular signals. First, this technique does not permit to assess directly the intracellular states of neurons, because some neurons might undergo slow oscillation or exhibit subthreshold fluctuations of the membrane potential without emitting APs, rendering their activity undetectable.

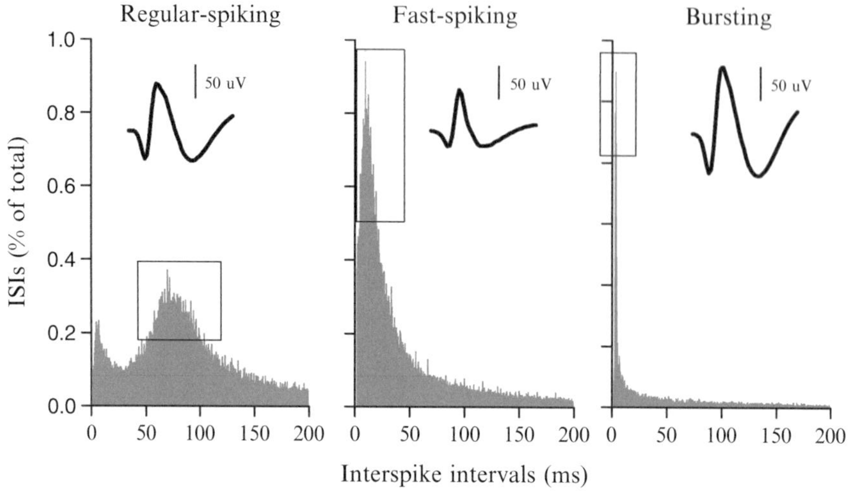

Fig. 5. Electrophysiological classification of neurons. Interspike intervals (ISIs) distribution and corresponding average waveforms for three representative neurons recorded from the barrel cortex of an individual rat belonging to three major firing phenotypes, putative regular spiking, putative fast spiking, and putative intrinsically bursting cell. The major modes of the ISIs distributions are boxed.

A common problem of chronically implanted microelectrodes is recording stability (95). It is primarily affected by the reactions occurring on the tissue–electrode interface, for example disruption of blood vessels and damage of neurons (84, 125). As a result, microglia and astrocytes form an encapsulation around the electrode for a considerable distance (100–200μm) that lead to the deterioration of the signal (126–128). As a result of the movements between arrays and tissue, inflammatory processes and formation of scar tissue the features of recorded waveforms may change over time (usually days or weeks). If such changes occur in the course of a single recording session, they may lead to the separation of waveforms belonging to a single neuron into two or more clusters. If two clusters have similar average waveforms and ISI distributions, but timestamps of one cluster mostly precede those of the second cluster, signal drifting is likely to have occurred. Therefore, it is necessary to verify signal stability across multiple recording sessions. A typical technique is to compare average waveforms, via template matching algorithms or comparing extracted features (5, 92, 95, 129, 130).

It is believed that none of the spike sorting techniques allows to discriminate spikes with 100% accuracy because of numerous factors, including similar wave shapes of simultaneously recorded neurons, overlapping spikes, background noise, or synchronous discharge of nearby neurons (15). If two spikes occur one after another with a time interval less than 1 ms, a spurious waveform can be recorded. While this problem is still unresolved, several techniques have been developed to identify the contribution of single spikes to the average waveform (96). The quality of isolation can be checked by measuring ISIs: if the spike waveform indeed belongs to a single unit, then there should be no ISI less than its refractory period, which is for most cells >1 ms (15).

Another potential confound is the distance between neurons and recording sites that can affect the shape of extracellularly recorded APs (19). To avoid cases when distinct neurons produce spikes with similar waveforms, tetrodes can be used where recording sites are organized in groups of four with average distance between recording sites of about 25 μm. All four recording sites detect signals coming from the same cell (19, 95). Since two neurons usually have different distances from the distinct sites, their APs are likely to have different features depending on the recording site, making it possible to discriminate them correctly. The SMEM algorithm has been successfully employed for tetrode recordings (95). Another technique requires the measurement of neuronal responses to stimulation or their functional tuning, since neurons producing similar spikes might have different functional properties (131).

References

1. MacLean JN, Watson BO, Aaron GB et al (2005) Internal dynamics determine the cortical response to thalamic stimulation. Neuron 48:811–823
2. Petersen CC, Hahn TT, Mehta M et al (2003) Interaction of sensory responses with spontaneous depolarization in layer 2/3 barrel cortex. Proc Natl Acad Sci USA 100:13638–13643
3. Sakata S, Harris KD (2009) Laminar structure of spontaneous and sensory-evoked population activity in auditory cortex. Neuron 64:404–418
4. Steriade M, Timofeev I, Grenier F (2001) Natural waking and sleep states: a view from inside neocortical neurons. J Neurophysiol 85:1969–1985
5. Vyazovskiy VV, Olcese U, Lazimy YM et al (2009) Cortical firing and sleep homeostasis. Neuron 63:865–878
6. Cirelli C, Tononi G (2008) Is sleep essential? PLoS Biol 6:e216
7. Borbély AA, Achermann P (2005) Sleep homeostasis and models of sleep regulation. In: Kryger MH, Roth T, Dement WC (eds) Principles and practice of sleep medicine. W. B. Saunders, Philadelphia, PA, pp 405–417
8. Tobler I (2005) Phylogeny of sleep regulation. In: Kryger MH, Roth T, Dement WC (eds) Principles and practice of sleep medicine. W. B. Saunders, Philadelphia, PA
9. Chauvette S, Volgushev M, Timofeev I (2010) Origin of active states in local neocortical networks during slow sleep oscillation. Cereb Cortex 20(11):2660–2674
10. Destexhe A, Contreras D, Steriade M (1999) Spatiotemporal analysis of local field potentials and unit discharges in cat cerebral cortex during natural wake and sleep states. J Neurosci 19:4595–4608
11. Amzica F, Steriade M (1998) Electrophysiological correlates of sleep delta waves. Electroencephalogr Clin Neurophysiol 107:69–83
12. Contreras D, Steriade M (1995) Cellular basis of EEG slow rhythms: a study of dynamic corticothalamic relationships. J Neurosci 15:604–622
13. Buzsáki G (2006) Rhythms of the brain. Oxford University Press, Oxford

14. Rasch MJ, Gretton A, Murayama Y et al (2008) Inferring spike trains from local field potentials. J Neurophysiol 99:1461–1476
15. Lewicki MS (1998) A review of methods for spike sorting: the detection and classification of neural action potentials. Network 9: R53–R78
16. Shu Y, Duque A, Yu Y et al (2007) Properties of action-potential initiation in neocortical pyramidal cells: evidence from whole cell axon recordings. J Neurophysiol 97:746–760
17. Holt GR, Koch C (1999) Electrical interactions via the extracellular potential near cell bodies. J Comput Neurosci 6:169–184
18. Mitzdorf U (1985) Current source-density method and application in cat cerebral cortex: investigation of evoked potentials and EEG phenomena. Physiol Rev 65:37–100
19. Buzsaki G (2004) Large-scale recording of neuronal ensembles. Nat Neurosci 7:446–451
20. Gold C, Henze DA, Koch C et al (2006) On the origin of the extracellular action potential waveform: a modeling study. J Neurophysiol 95:3113–3128
21. Vyazovskiy VV, Riedner BA, Cirelli C et al (2007) Sleep homeostasis and cortical synchronization: II. A local field potential study of sleep slow waves in the rat. Sleep 30:1631–1642
22. Mitzdorf U (1987) Properties of the evoked potential generators: current source-density analysis of visually evoked potentials in the cat cortex. Int J Neurosci 33:33–59
23. Katzner S, Nauhaus I, Benucci A et al (2009) Local origin of field potentials in visual cortex. Neuron 61:35–41
24. Csicsvari J, Jamieson B, Wise KD et al (2003) Mechanisms of gamma oscillations in the hippocampus of the behaving rat. Neuron 37:311–322
25. Whittingstall K, Logothetis NK (2009) Frequency-band coupling in surface EEG reflects spiking activity in monkey visual cortex. Neuron 64:281–289
26. Sirota A, Montgomery S, Fujisawa S et al (2008) Entrainment of neocortical neurons and gamma oscillations by the hippocampal theta rhythm. Neuron 60:683–697
27. Manning JR, Jacobs J, Fried I et al (2009) Broadband shifts in local field potential power spectra are correlated with single-neuron spiking in humans. J Neurosci 29:13613–13620
28. Haider B, Duque A, Hasenstaub AR et al (2006) Neocortical network activity in vivo is generated through a dynamic balance of excitation and inhibition. J Neurosci 26:4535–4545
29. Poulet JF, Petersen CC (2008) Internal brain state regulates membrane potential synchrony in barrel cortex of behaving mice. Nature 454:881–885
30. Rudolph M, Pospischil M, Timofeev I et al (2007) Inhibition determines membrane potential dynamics and controls action potential generation in awake and sleeping cat cortex. J Neurosci 27:5280–5290
31. Okun M, Naim A, Lampl I (2010) The subthreshold relation between cortical local field potential and neuronal firing unveiled by intracellular recordings in awake rats. J Neurosci 30:4440–4448
32. Villablanca JR (2004) Counterpointing the functional role of the forebrain and of the brainstem in the control of the sleep-waking system. J Sleep Res 13:179–208
33. Jones BE (2005) From waking to sleeping: neuronal and chemical substrates. Trends Pharmacol Sci 26:578–586
34. Miller DB, O'Callaghan JP (2006) The pharmacology of wakefulness. Metabolism 55: S13–S19
35. Boutrel B, Koob GF (2004) What keeps us awake: the neuropharmacology of stimulants and wakefulness-promoting medications. Sleep 27:1181–1194
36. Vyazovskiy VV, Ruijgrok G, Deboer T et al (2006) Running wheel accessibility affects the regional electroencephalogram during sleep in mice. Cereb Cortex 16:328–336
37. Gentet LJ, Avermann M, Matyas F et al (2010) Membrane potential dynamics of GABAergic neurons in the barrel cortex of behaving mice. Neuron 65:422–435
38. Sporns O, Tononi G, Edelman GM (2000) Connectivity and complexity: the relationship between neuroanatomy and brain dynamics. Neural Netw 13:909–922
39. Green JD, Arduini AA (1954) Hippocampal electrical activity in arousal. J Neurophysiol 17:533–557
40. Petsche H, Stumpf C (1960) Topographic and toposcopic study of origin and spread of the regular synchronized arousal pattern in the rabbit. Electroencephalogr Clin Neurophysiol 12:589–600
41. Whishaw IQ, Vanderwolf CH (1973) Hippocampal EEG and behavior: changes in amplitude and frequency of RSA (theta rhythm) associated with spontaneous and learned movement patterns in rats and cats. Behav Biol 8:461–484

42. Robinson TE (1980) Hippocampal rhythmic slow activity (RSA; theta): a critical analysis of selected studies and discussion of possible species-differences. Brain Res 203:69–101
43. Leung LW, Borst JG (1987) Electrical activity of the cingulate cortex. I. Generating mechanisms and relations to behavior. Brain Res 407:68–80
44. Murata K, Kameda K (1963) The activity of single cortical neurones of unrestrained cats during sleep and wakefulness. Arch Ital Biol 101:306–331
45. Noda H, Adey WR (1973) Neuronal activity in the association cortex of the cat during sleep, wakefulness and anesthesia. Brain Res 54:243–259
46. Hobson JA, McCarley RW (1971) Cortical unit activity in sleep and waking. Electroencephalogr Clin Neurophysiol 30:97–112
47. Verzeano M, Negishi K (1960) Neuronal activity in cortical and thalamic networks. J Gen Physiol 43(6):177–195
48. Noda H, Adey WR (1970) Firing of neuron pairs in cat association cortex during sleep and wakefulness. J Neurophysiol 33:672–684
49. Burns BD, Stean JP, Webb AC (1979) The effects of sleep on neurons in isolated cerebral cortex. Proc R Soc Lond B Biol Sci 206:281–291
50. Steriade M, Nunez A, Amzica F (1993) A novel slow (<1 Hz) oscillation of neocortical neurons in vivo: depolarizing and hyperpolarizing components. J Neurosci 13:3252–3265
51. Calvet J, Fourment A, Thiefry M (1973) Electrical activity in neocortical projection and association areas during slow wave sleep. Brain Res 52:173–187
52. Steriade M, Nunez A, Amzica F (1993) Intracellular analysis of relations between the slow (<1 Hz) neocortical oscillation and other sleep rhythms of the electroencephalogram. J Neurosci 13:3266–3283
53. Ji D, Wilson MA (2007) Coordinated memory replay in the visual cortex and hippocampus during sleep. Nat Neurosci 10:100–107
54. Luczak A, Bartho P, Marguet SL et al (2007) Sequential structure of neocortical spontaneous activity in vivo. Proc Natl Acad Sci USA 104:347–352
55. Mukovski M, Chauvette S, Timofeev I et al (2007) Detection of active and silent states in neocortical neurons from the field potential signal during slow-wave sleep. Cereb Cortex 17(2):400–414
56. Molle M, Yeshenko O, Marshall L et al (2006) Hippocampal sharp wave-ripples linked to slow oscillations in rat slow-wave sleep. J Neurophysiol 96:62–70
57. Crunelli V, Hughes SW (2009) The slow (<1 Hz) rhythm of non-REM sleep: a dialogue between three cardinal oscillators. Nat Neurosci 13:9–17
58. Hasenstaub A, Shu Y, Haider B et al (2005) Inhibitory postsynaptic potentials carry synchronized frequency information in active cortical networks. Neuron 47:423–435
59. Sanchez-Vives MV, McCormick DA (2000) Cellular and network mechanisms of rhythmic recurrent activity in neocortex. Nat Neurosci 3:1027–1034
60. Fanselow EE, Connors BW (2010) The roles of somatostatin-expressing (GIN) and fast-spiking inhibitory interneurons in UP-DOWN states of mouse neocortex. J Neurophysiol 104:596–606
61. Webb AC (1976) The effects of changing levels of arousal on the spontaneous activity of cortical neurones: I. Sleep and wakefulness. Proc R Soc Lond B Biol Sci 194:225–237
62. Massimini M, Huber R, Ferrarelli F et al (2004) The sleep slow oscillation as a traveling wave. J Neurosci 24:6862–6870
63. Huber R, Deboer T, Tobler I (2000) Topography of EEG dynamics after sleep deprivation in mice. J Neurophysiol 84:1888–1893
64. Huber R, Ghilardi MF, Massimini M et al (2004) Local sleep and learning. Nature 430:78–81
65. Vyazovskiy VV, Borbely AA, Tobler I (2002) Interhemispheric sleep EEG asymmetry in the rat is enhanced by sleep deprivation. J Neurophysiol 88:2280–2286
66. Vyazovskiy VV, Tobler I, Winsky-Sommerer R (2007) Alteration of behavior in mice by muscimol is associated with regional electroencephalogram synchronization. Neuroscience 147:833–841
67. De Gennaro L, Fratello F, Marzano C et al (2008) Cortical plasticity induced by transcranial magnetic stimulation during wakefulness affects electroencephalogram activity during sleep. PLoS One 3:e2483
68. Huber R, Ghilardi MF, Massimini M et al (2006) Arm immobilization causes cortical plastic changes and locally decreases sleep slow wave activity. Nat Neurosci 9:1169–1176
69. Krueger JM, Rector DM, Roy S et al (2008) Sleep as a fundamental property of neuronal assemblies. Nat Rev Neurosci 9:910–919
70. Riedner BA, Vyazovskiy VV, Huber R et al (2007) Sleep homeostasis and cortical synchronization: III. A high-density EEG study

of sleep slow waves in humans. Sleep 30:1643–1657

71. Tononi G, Cirelli C (2006) Sleep function and synaptic homeostasis. Sleep Med Rev 10:49–62
72. Vyazovskiy V, Borbely AA, Tobler I (2000) Unilateral vibrissae stimulation during waking induces interhemispheric EEG asymmetry during subsequent sleep in the rat. J Sleep Res 9:367–371
73. Vyazovskiy VV, Cirelli C, Pfister-Genskow M et al (2008) Molecular and electrophysiological evidence for net synaptic potentiation in wake and depression in sleep. Nat Neurosci 11:200–208
74. Vyazovskiy VV, Tobler I (2008) Handedness leads to interhemispheric EEG asymmetry during sleep in the rat. J Neurophysiol 99:969–975
75. Borbely AA (1982) A two process model of sleep regulation. Hum Neurobiol 1:195–204
76. Achermann P, Borbely AA (2003) Mathematical models of sleep regulation. Front Biosci 8: s683–s693
77. Tobler I, Borbely AA (1986) Sleep EEG in the rat as a function of prior waking. Electroencephalogr Clin Neurophysiol 64:74–76
78. Cajochen C, Foy R, Dijk DJ (1999) Frontal predominance of a relative increase in sleep delta and theta EEG activity after sleep loss in humans. Sleep Res Online 2:65–69
79. Oleksenko AI, Mukhametov LM, Polyakova IG et al (1992) Unihemispheric sleep deprivation in bottlenose dolphins. J Sleep Res 1:40–44
80. Kattler H, Dijk DJ, Borbely AA (1994) Effect of unilateral somatosensory stimulation prior to sleep on the sleep EEG in humans. J Sleep Res 3:159–164
81. Tononi G, Cirelli C (2003) Sleep and synaptic homeostasis: a hypothesis. Brain Res Bull 62:143–150
82. Hill S, Tononi G (2005) Modeling sleep and wakefulness in the thalamocortical system. J Neurophysiol 93:1671–1698
83. Esser SK, Hill SL, Tononi G (2007) Sleep homeostasis and cortical synchronization: I. Modeling the effects of synaptic strength on sleep slow waves. Sleep 30:1617–1630
84. Schwartz AB, Cui XT, Weber DJ et al (2006) Brain-controlled interfaces: movement restoration with neural prosthetics. Neuron 52:205–220
85. Nelson MJ, Pouget P (2010) Do electrode properties create a problem in interpreting local field potential recordings? J Neurophysiol 103:2315–2317
86. Kralik JD, Dimitrov DF, Krupa DJ et al (2001) Techniques for long-term multisite neuronal ensemble recordings in behaving animals. Methods 25:121–150
87. Spataro L, Dilgen J, Retterer S et al (2005) Dexamethasone treatment reduces astroglia responses to inserted neuroprosthetic devices in rat neocortex. Exp Neurol 194:289–300
88. Zhong Y, Bellamkonda RV (2007) Dexamethasone-coated neural probes elicit attenuated inflammatory response and neuronal loss compared to uncoated neural probes. Brain Res 1148:15–27
89. Hanlon EC, Faraguna U, Vyazovskiy VV et al (2009) Effects of skilled training on sleep slow wave activity and cortical gene expression in the rat. Sleep 32:719–729
90. Huber R, Tononi G, Cirelli C (2007) Exploratory behavior, cortical BDNF expression, and sleep homeostasis. Sleep 30:129–139
91. Lemon R, Prochazka A (1984) Methods for neuronal recording in conscious animals. Wiley, Chichester
92. Nicolelis MAL (2008) Methods for neural ensemble recordings, 2nd edn. CRC Press, Boca Raton, FL
93. Hulata E, Segev R, Ben-Jacob E (2002) A method for spike sorting and detection based on wavelet packets and Shannon's mutual information. J Neurosci Methods 117:1–12
94. Quiroga RQ, Nadasdy Z, Ben-Shaul Y (2004) Unsupervised spike detection and sorting with wavelets and superparamagnetic clustering. Neural Comput 16:1661–1687
95. Tolias AS, Ecker AS, Siapas AG et al (2007) Recording chronically from the same neurons in awake, behaving primates. J Neurophysiol 98:3780–3790
96. Zhang PM, Wu JY, Zhou Y et al (2004) Spike sorting based on automatic template reconstruction with a partial solution to the overlapping problem. J Neurosci Methods 135:55–65
97. Jolliffe IT (2002) Principal component analysis, 2nd edn. Springer, New York, NY
98. Delorme A, Makeig S (2004) EEGLAB: an open source toolbox for analysis of single-trial EEG dynamics including independent component analysis. J Neurosci Methods 134:9–21
99. Tan P-N, Steinbach M, Kumar V (2006) Introduction to data mining, 1st edn. Pearson Addison Wesley, Boston, MA

100. Wood F, Black MJ, Vargas-Irwin C et al (2004) On the variability of manual spike sorting. IEEE Trans Biomed Eng 51:912–918
101. Hartigan JA (1975) Clustering algorithms. Wiley, New York, NY
102. Bezdek JC, Ehrlich R, Full W (1984) Fcm—the fuzzy C-means clustering-algorithm. Comput Geosci 10:191–203
103. Bishop CM (2006) Pattern recognition and machine learning. Springer, New York, NY
104. Ueda N, Nakano R, Ghahramani Z et al (2000) SMEM algorithm for mixture models. Neural Comput 12:2109–2128
105. Davies DL, Bouldin DW (1979) Cluster Separation Measure. IEEE Trans Pattern Anal Mach Intell 1:224–227
106. Connors BW, Gutnick MJ (1990) Intrinsic firing patterns of diverse neocortical neurons. Trends Neurosci 13:99–104
107. Kawaguchi Y, Kubota Y (1997) GABAergic cell subtypes and their synaptic connections in rat frontal cortex. Cereb Cortex 7:476–486
108. Bartho P, Hirase H, Monconduit L et al (2004) Characterization of neocortical principal cells and interneurons by network interactions and extracellular features. J Neurophysiol 92:600–608
109. Bruno RM, Simons DJ (2002) Feedforward mechanisms of excitatory and inhibitory cortical receptive fields. J Neurosci 22:10966–10975
110. McCormick DA, Connors BW, Lighthall JW et al (1985) Comparative electrophysiology of pyramidal and sparsely spiny stellate neurons of the neocortex. J Neurophysiol 54:782–806
111. Gray CM, McCormick DA (1996) Chattering cells: superficial pyramidal neurons contributing to the generation of synchronous oscillations in the visual cortex. Science 274:109–113
112. Diaz-Mataix L, Artigas F, Celada P (2006) Activation of pyramidal cells in rat medial prefrontal cortex projecting to ventral tegmental area by a 5-HT1A receptor agonist. Eur Neuropsychopharmacol 16:288–296
113. Diester I, Nieder A (2008) Complementary contributions of prefrontal neuron classes in abstract numerical categorization. J Neurosci 28:7737–7747
114. Gonzalez-Burgos G, Krimer LS, Povysheva NV et al (2005) Functional properties of fast spiking interneurons and their synaptic connections with pyramidal cells in primate dorsolateral prefrontal cortex. J Neurophysiol 93:942–953
115. Jung MW, Qin Y, McNaughton BL et al (1998) Firing characteristics of deep layer neurons in prefrontal cortex in rats performing spatial working memory tasks. Cereb Cortex 8:437–450
116. Mallet N, Le Moine C, Charpier S et al (2005) Feedforward inhibition of projection neurons by fast-spiking GABA interneurons in the rat striatum in vivo. J Neurosci 25:3857–3869
117. Povysheva NV, Gonzalez-Burgos G, Zaitsev AV et al (2006) Properties of excitatory synaptic responses in fast-spiking interneurons and pyramidal cells from monkey and rat prefrontal cortex. Cereb Cortex 16:541–552
118. Puig MV, Ushimaru M, Kawaguchi Y (2008) Two distinct activity patterns of fast-spiking interneurons during neocortical UP states. Proc Natl Acad Sci USA 105 (24):8428–8433
119. Tierney PL, Degenetais E, Thierry AM et al (2004) Influence of the hippocampus on interneurons of the rat prefrontal cortex. Eur J Neurosci 20:514–524
120. Tseng KY, Mallet N, Toreson KL et al (2006) Excitatory response of prefrontal cortical fast-spiking interneurons to ventral tegmental area stimulation in vivo. Synapse 59:412–417
121. Valenti O, Grace AA (2009) Entorhinal cortex inhibits medial prefrontal cortex and modulates the activity states of electrophysiologically characterized pyramidal neurons in vivo. Cereb Cortex 19 (3):658–674
122. Kole MH, Letzkus JJ, Stuart GJ (2007) Axon initial segment Kv1 channels control axonal action potential waveform and synaptic efficacy. Neuron 55:633–647
123. Pattle RE (1971) The external action potential of a nerve or muscle fibre in an extended medium. Phys Med Biol 16:673–685
124. Richerson S, Ingram M, Perry D et al (2005) Classification of the extracellular fields produced by activated neural structures. Biomed Eng Online 4:53
125. Schwartz AB (2004) Cortical neural prosthetics. Annu Rev Neurosci 27:487–507
126. Polikov VS, Tresco PA, Reichert WM (2005) Response of brain tissue to chronically implanted neural electrodes. J Neurosci Methods 148:1–18
127. Biran R, Martin DC, Tresco PA (2005) Neuronal cell loss accompanies the brain tissue

response to chronically implanted silicon microelectrode arrays. Exp Neurol 195:115–126

128. Turner JN, Shain W, Szarowski DH et al (1999) Cerebral astrocyte response to micromachined silicon implants. Exp Neurol 156:33–49

129. Greenberg PA, Wilson FA (2004) Functional stability of dorsolateral prefrontal neurons. J Neurophysiol 92:1042–1055

130. Nicolelis MA, Dimitrov D, Carmena JM et al (2003) Chronic, multisite, multielectrode recordings in macaque monkeys. Proc Natl Acad Sci USA 100:11041–11046

131. Greenberg PA, Wilson FAW (2004) Functional stability of dorsolateral prefrontal neurons. J Neurophysiol 92:1042–1055

Neuromethods (2012) 67: 259–282
DOI 10.1007/7657_2011_8

Published online: 9 December 2011

Analysis of Hippocampal Memory Replay Using Neural Population Decoding

Fabian Kloosterman

Abstract

Large-scale recording of neural activity in waking animals provides us with a window on computations performed in the brain during behavior. In order to better understand these computations, population decoding techniques are used to study what information about the external environment is represented by neural ensemble activity. Decoding approaches are especially appealing when neural activity does not reflect the actual sensory experience or motor output, but is generated by internal cognitive processes, such as planning, memory, and decision making. This chapter describes the background of Bayesian decoding and outlines the general procedures for decoding ensemble activity during cognitive processes. The power of this method is demonstrated by applying it to recordings of hippocampal place cell activity in order to identify and characterize hippocampal memory replay.

Key words: Neural ensemble recording, Spike train decoding, Hippocampus, Memory replay

1. Introduction

Simultaneous recording of the activity of large numbers of neurons in animals that perform a behavioral task has proven to be an important tool to understand the neurophysiological basis of perception, memory, and cognition. Multiple single cell activity in waking animals is commonly recorded using multi-electrode recording techniques (1–3) or, more recently, imaging of activity-dependent fluorescent markers (4–6). Using these approaches, it is possible to study how information about sensory stimuli, intended motor actions and behavior is encoded in the temporal pattern of action potentials ("spikes") emitted by individual neurons in an ensemble. Unlike methods that analyze the activity of single cells or averaged activity (i.e., field potentials) in a brain area, these methods can address questions about how information is distributed in a population of neurons and what the role is of interactions between neurons in the population.

The analysis of multiple simultaneously recorded spike trains is a challenging problem and the development of statistical tools for this problem is an active area of research (7). Decoding algorithms offer one approach to study how neuronal ensembles represent information about a stimulus (7–9). These algorithms have been used successfully in a variety of brain systems (10–15). For example, Georgopoulos et al. (11) showed that in rhesus monkeys the movement direction of an arm could be estimated from a population of motor cortical neurons. Later studies applied the decoding approach to the hippocampus (10, 14) and sensory areas (13, 15).

Decoding analysis is an especially attractive approach when neuronal activity does *not* reflect the actual experience or physical stimuli, but instead is generated by internal cognitive processes (16). This mismatch occurs for example in memory recall, where neurons that were previously active during an experience are reactivated later in time (17, 18), even though the stimuli of the original experience are not present at the time of recall. By decoding the spike activity measured during a memory recall event, it should be possible to identify and characterize the covert stimulus being recalled. In principle, other cognitive processes can be studied using neural population decoding as well, such as planning, visual imagery, and decision making.

Early demonstrations of "reading out" cognitive operations are provided by studies of the motor cortex in monkeys (19, 20). In one study, monkeys had to move their arm in a direction that deviated by 90° from the direction of a target light. In this case, the representation of movement direction in a population of motor cortical neurons changed gradually from the direction of the target light to the direction of the final movement, providing evidence at the neural level that mental rotation was used to solve this task (19).

More recently, decoding approaches were used to study spatial memory processes in the hippocampus of rodents (21–23). Hippocampal pyramidal neurons fire when a rat moves through a specific region of the environment (24, 25). Each neuron has its own spatial receptive field, or place field, and at every location in the environment a large number of neurons are active, suggesting that the hippocampus contains a map of space (25, 26). Previously, it was found that hippocampal firing patterns that occur when an animal follows a path through space, are reactivated (or replayed) during the so-called offline states, such as sleep (27, 28) and periods of quiescence (29–31). These studies directly compared the sequential activation of hippocampal neurons during the experience ("online" state) with firing sequences during the offline state. One important limitation of this method is that it relies on individual neurons having unimodal spatial receptive fields, which is not the case in large environments (32). A decoding approach, however, does not suffer from this limitation as it makes optimal use of all spatial information contained in the hippocampal spike trains (21).

Another advantage of decoding is that ensemble activity can be described in the stimulus domain and therefore a direct comparison to the actual experience of the animal is possible. As such, decoding provides a more general framework for understanding the computations and cognitive processes in offline states.

This chapter describes a general method for decoding ensemble activity during cognitive processes. The power of this method is demonstrated by applying it to hippocampal ensemble recordings to identify and characterize memory replay.

2. Mathematical Preliminaries

There are a number of different approaches to the problem of neural decoding (11, 14, 25, 33–35); however, in this chapter only decoding approaches based on Bayesian statistical methods are discussed (10, 12, 14). Neural decoding can be divided into two stages: an encoding stage and a decoding stage (Fig. 1a). The encoding stage, i.e., the transformation from stimulus to spike trains, can be described by a mathematical model that is informed by the experimental measurements of spike activity and knowledge of the stimuli that gave rise to the spike trains. Additional

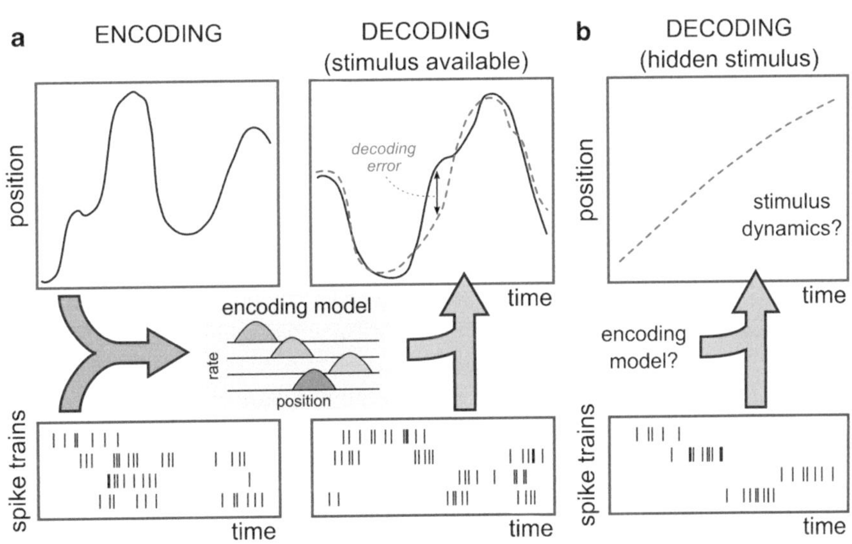

Fig. 1. (**a**) Schematic overview of encoding (*left*) and decoding (*right*) during online states, when an experimenter has access to both the stimulus (in this case, a rat's position on a track) and the neuronal activity. The relationship between the stimulus and observed spike trains can be described by an encoding model, as illustrated by the spatial tuning curves in the *gray box*. The same model can then be used to create a prediction of the stimulus (*dashed line*) at other time points. Stimulus statistics and dynamics, if known, can aid the decoding process. Comparison of the decoded stimulus with the true stimulus provides information about the performance of the decoding process. (**b**) In offline states decoding may reveal a hidden stimulus. One difficulty is that both the exact encoding model and the hidden stimulus dynamics are unknown.

information about the history of the activity in the ensemble can be incorporated in the model as well. The model can then be used to predict spike activity for a given a set of stimuli, or, conversely, to derive information about the stimuli from observed neuronal activity in the decoding process. In the next two sections, a general mathematical background is given of encoding and decoding in a Bayesian framework. The last section discusses important issues that arise when applying decoding methods to study hidden cognitive processes (Fig. 1b).

2.1. Encoding

To generate an encoding model, the activity of a neuron is described as a random point process, or a set of discrete events (spikes), that occur in continuous time (36). A point process is fully characterized by its generalized rate function $\lambda(t|H(t))$, which specifies the cell's firing rate in spikes per second at time t, given the history H of a set of covariates up to time t. It is possible to construct a model of a neuron's generalized rate function that incorporates both spiking history of the neuron and the activity of the population in addition to external stimuli (37). However, often a simplifying assumption is made, where the activity in a neuronal population constitutes an ensemble of conditionally independent Poisson processes. According to this assumption, the probability that a neuron emits a spike at a particular time is conditionally independent of any previous spike event in the same neuron or any other neuron in the population. Thus, while the activity of any two cells can still be correlated, the assumption postulates that the correlation is solely due to a common effect of the external stimulus on the neurons. Given the assumption of Poisson firing statistics, the rate function of a neuron can now be written purely as a function of the time-varying stimulus $x(t)$: $\lambda(t|H(t)) = \lambda[x(t)]$. Here, $\lambda[x]$ represents the classical tuning curve of a neuron, which relates average firing rate to the stimulus x. (For the present discussion, it is assumed that the stimulus is a discrete variable and this is indicated by the use of square brackets). In practice, the tuning curve can be defined parametrically by fitting a model to the data, or non-parametrically by constructing a histogram.

For any Poisson process, the response of a neuron in a given time window $[t, t+\Delta)$ can be summarized by the number of spikes, ignoring the precise temporal pattern. The expected number of spikes in this time window is $\lambda[x(t)]\Delta$ if the firing rate function over the course of the time window does not change. This applies to a homogeneous Poisson process (i.e., $\lambda[x]$ is constant for all x) or, by approximation, if the stimulus changes at a slower time scale than the duration Δ of the time window (see Note 1). For a given stimulus x, the probability that a neuron responds with n spikes is given by the Poisson equation:

$$P[n|x] = \frac{1}{n!}(\lambda[x]\Delta)^n e^{-\Delta\lambda[x]} \tag{1}$$

$P[n|x]$ is called the conditional probability and when computed for all possible responses n and stimulus values x, it fully characterizes the encoding process. This means that from the set of probabilities $P[n|x]$ and an observed stimulus record, a spike train can be generated that, statistically speaking, is indistinguishable from the true spike train, provided that the encoding model and its assumptions are correct. $P[n|x]$ is also called the likelihood function if it is considered as a function of the stimulus x for a fixed response n.

For a population of C neurons that are conditionally independent, the joint likelihood can be expressed as the product of their individual likelihoods:

$$P[\mathbf{n}|x] = \prod_{c=1}^{C} P[n_c|x] \tag{2}$$

Here, $\mathbf{n}$ represents the response of the population constructed from the vector of single cell spike counts, i.e., $\mathbf{n} = \{n_1, n_2, \ldots, n_C\}$. When combined with (1), the joint likelihood can be written as:

$$P[\mathbf{n}|x] = \left\{ \prod_{c=1}^{C} \frac{(\Delta\lambda_c[x])^{n_c}}{n_c!} \right\} e^{-\Delta \sum_{c=1}^{C} \lambda_c[x]} \tag{3}$$

2.2. Decoding

While the encoding model indicates how stimuli are translated into a set of spike trains, an inversion of the model can be used to infer the value of the stimulus that likely produced observed spike trains—a process called decoding (Fig. 1a). The relation between encoding and decoding is given by Bayes' theorem, which is optimal if the encoding model is correct:

$$P[x_t|\mathbf{n}_t] = \frac{P[\mathbf{n}_t|x_t]P[x_t]}{P[\mathbf{n}_t]} \tag{4}$$

$P[x_t|\mathbf{n}_t]$ is the posterior probability distribution which contains all information about the stimulus x at time t that can be learned from the observed neuronal response $\mathbf{n}_t$. The posterior is calculated from the joint likelihood $P[\mathbf{n}_t|x_t]$ (see Sect. 1), the prior probability distribution $P[x_t]$ and the response probability $P[\mathbf{n}_t]$.

The prior gives information about the stimulus statistics that is independent of the observed response. For example, suppose that stimulus is the animal's position on a linear track and the animal is known to visit one location more often than another. This preference for certain locations can be captured in the prior. It is possible to choose a prior that is uniform, in which case it does not provide additional information about the stimulus x and the posterior probability distribution is proportional to the likelihood. The response probability $P[\mathbf{n}_t]$ can be treated as a normalization factor, i.e., $P[\mathbf{n}_t] = \sum_{x_t} P[\mathbf{n}_t|x_t]P[x_t]$, such that the posterior distribution is a true probability distribution and sums to 1.

The posterior distribution is often summarized by a single (point) estimate of the stimulus $\hat{x}$. Three commonly used estimates are the posterior mode ($\arg\max_x P[x_t|\mathbf{n}_t]$), the posterior mean ($\sum_x xP[x_t|\mathbf{n}_t]$) and the posterior median. The confidence in the estimate is high if most probability in the posterior is concentrated around $\hat{x}$, but low if the probability is spread out over many stimulus values. Different measures can be used to express the level of confidence, such as the probability at the mode, the posterior variance ($\sum_x (x - \hat{x}_t)^2 P[x_t|\mathbf{n}_t]$), inter-quartile range or credible interval.

Decoding as defined in (4) is memoryless, which means that the estimate is computed using the neuronal activity at time t, but it does not depend on past activity. However, many stimuli evolve in time, and knowledge about the stimulus dynamics can improve the decoding (Fig. 1a). A Bayesian filter approach can be used to recursively compute the posterior distribution at time t by combining the estimate at the previous time step with the newly observed neuronal activity. Letting $\mathbf{N}_t$ be the past and present population activity up to time t and $\mathbf{n}_t$ the present activity only, then using Bayes' theorem the posterior can be written as:

$$P[x_t|\mathbf{N}_t] = P[x_t|\mathbf{n}_t, \mathbf{N}_{t-1}] = \frac{P[\mathbf{n}_t|x_t, \mathbf{N}_{t-1}]P[x_t|\mathbf{N}_{t-1}]}{P[\mathbf{n}_t|\mathbf{N}_{t-1}]} \tag{5}$$

Decoding now becomes a two step process—first the prior $P[x_t|\mathbf{N}_{t-1}]$ defines a prediction of the most likely next value of the stimulus based solely on past activity. The prediction is computed from the posterior at the previous time step according to:

$$P[x_t|\mathbf{N}_{t-1}] = \sum_{x_{t-1}} P[x_t|x_{t-1}]P[x_{t-1}|\mathbf{N}_{t-1}] \tag{6}$$

Here, $P[x_t|x_{t-1}]$ represents a transition probability distribution that maps the last estimate to the set of most likely next stimulus values. The transition probability distribution can be defined by a stochastic model of the stimulus dynamics. For example, the stimulus can be modeled as a random walk process in order to include a continuity constraint: $x_t = x_{t-1} + N(0, \sigma)$, where $N(0, \sigma)$ represents a zero-mean normal distribution with standard deviation σ. In this case, the transition probability distribution can be written as $P[x_t|x_{t-1}] = N(x_t - x_{t-1}, \sigma)$.

In the second step of decoding, the prediction is updated with the new data observed at time t through the likelihood function using (5). The resulting posterior distribution can then be used in the prediction step for the next estimate.

2.3. Decoding Covert Stimuli

For the present discussion, it is useful to differentiate between an online state in which neuronal populations are mainly driven by external stimuli, and an offline state in which activity patterns are (primarily) generated internally by cognitive processes. The aim then is to study the cognitive processes in the offline state by decoding a covert stimulus from observed neuronal activity (Fig. 1b). By definition there is no ground truth value to which the estimated stimulus in the offline state can be compared and thus it is not possible to measure the performance of the decoding. Instead, the only possible "validation" is provided by any unexpected structure or correlations in the decoded stimulus estimates. In addition, neuronal activity is the *only* information available during the offline state—both the encoding model (i.e., how the hidden stimulus is represented by the neuronal population) and the dynamics of the hidden stimulus are unknown.

The unknown encoding model for the offline state may be estimated from observations in the online state, based on the fundamental assumption that the population activity in both states represents the same physical variable. However, it is possible that some spiking dynamics, such as propensity to burst, firing rates and the relationship to ongoing oscillations, differ between online and offline states. Thus, online encoding models that incorporate these factors may not be applicable to offline states. Rather a reduced model that just captures the relationship between stimulus and response would be more appropriate.

It is important to consider which prior information about the stimulus statistics and dynamics should be included in the decoding process. One option is to use the statistics of the known stimulus in the online state. However, there is no reason to assume that this prior information about the real stimulus applies to the hidden stimulus in the offline state. In fact, this could be considered a hypothesis that needs to be tested. Davidson et al. (21) used a noninformative prior initially, making the decoding process a form of likelihood estimation. The resulting stimulus estimates were then subjected to analysis in order to describe the temporal structure and to construct a model of the hidden stimulus dynamics.

3. Decoding Hippocampal Memory Replay

The decoding techniques described here can be applied to hippocampal spike trains to estimate the position of an awake, freely moving animal in its environment. The demonstration here is based on the study performed by Davidson et al. (21), in which the activity of dozens of hippocampal units were recorded while a rat explored a long narrow track. In the following sections, a brief overview is given of the experimental design and recording procedures, followed by a description of the encoding stages and decoding stages. In particular, the decoding stage not only shows how to estimate an animal's real position in the online state or virtual position in the offline state, but also how to validate the decoding performance. Finally, the estimated virtual position is analyzed to discover any (temporal) structure.

3.1. Experimental Design and Data Acquisition

3.1.1. Overview

Hippocampal cell activity is recorded using arrays of 4-wire bundles (tetrodes (3)) carried by a small micro-drive (38–40) that is surgically implanted onto a rat's head. At the same time, the rat's position and head direction are monitored using an overhead video tracking system that detects the location of two light emitting diodes that are mounted on the rat's implant. For more detailed descriptions of neural ensemble recording techniques, the reader is referred to the relevant literature (3, 41).

In the experiment, rats run on a 10-m long, narrow sinuous track and receive food reward at either end of the track after completing a full lap. The linear position of the rat is measured along the track, relative to one of the ends. Spatially selective firing of hippocampal pyramidal neurons is particularly strong when a rat is actively moving through the environment (24, 25). Hence, the online state is defined as those times in which the rat is actively exploring the track with a minimum running speed of 15 cm/s. In contrast, the offline state is defined as periods of immobility (running speed <5 cm/s), when rats are eating, grooming or are otherwise disengaged, and activity in the hippocampus switches to a state characterized by intermittent population bursts (42–44). In these population bursts, hippocampal spiking patterns do not reflect the rat's actual position in the environment, but replay behavioral relevant spike sequences (29–31).

3.1.2. Spike Sorting

On a single tetrode, the spikes from dozens of neurons are recorded simultaneously. In order to analyze this activity, spikes must first be sorted into separate clusters (or "units") that correspond to putative single neurons (3). Spike sorting relies on the distinctive spike amplitude profiles for each neuron, which is the result of the unique location of a neuron relative to the tip of the tetrode. Other waveform features (e.g., spike width) can also be used to assist spike sorting, in particular to distinguish pyramidal neurons with wide waveforms from interneurons that have narrow waveforms.

Variations in single cell spike waveform features, for example because of noisy background activity, can result in overlapping clusters for neighboring neurons. This overlap becomes more significant for neurons which are located further away from the tetrode tip and have smaller amplitude spikes. For most analyses, well-isolated clusters that pass several isolation criteria (e.g., clear refractory period, high cluster quality measures (45)) are required. Even small contaminations of clusters with spikes from a different neuron could lead to spurious temporal ordering effects (46).

Population decoding, however, puts less stringent requirements on the isolation quality of units. Even units that are not well-isolated can be included, as long as the spike amplitude signature remains stable (21, 45). For example, if the spikes from two hippocampal place cells cannot be separated and are combined into a single unit, the resulting spatial tuning curve is an average of the tuning curves of the two place cells. Nevertheless, the combined unit still provides information about the location of the animal, albeit less than when the place cells were considered separately. In this way, the maximum amount of information can be extracted from the ensemble of neurons surrounding each tetrode.

3.1.3. Sampling of Online and Offline States

An accurate estimate of the encoding model requires a sufficient sampling of neuronal activity over the full range of positions on the track during the online state. Practically, this means that a rat should complete at least 5–10 laps. Due to the use of a speed threshold to define the online state (see Sect. 3.1), positions on the track where the rat stops consistently (such as reward locations and the end of the track) may be under-sampled and would have to be excluded from analysis.

In a simple task on a linear track, the training state of the rat affects the amount of online and offline data that can be collected (see Note 2). Working with untrained rats is advantageous because naive animals tend to pause often for long intervals, providing an abundance of candidate replay events. A well-trained rat, in contrast, will spend very little time in the offline state because it runs back and forth quickly between the two ends of the track to collect food rewards without much pausing. On the other hand, a well-trained animal provides a large sampling of data from the online state while a naive rat tends to run only a few laps at best (especially on a long track) and produces only limited sampling of the online state. For these reasons, Davidson et al. (21) analyzed early recording sessions (day 3 or 4) in which the rats completed several laps, but behavior was still variable enough to include many immobility epochs with candidate replay events.

3.2. Encoding

3.2.1. Selection of Encoding Model

Previous studies of encoding of spatial information in hippocampal neural ensembles used models that included history of spiking and the phase of the local theta oscillation in addition to the positional covariate (10, 47). These encoding models cannot be applied to the decoding of hippocampal memory replay because (1) theta oscillations are only present during active exploration, but not in the offline state and (2) there is no reason to assume that spiking dynamics are the same in the online and offline states. A reasonable alternative is an encoding model that relates the firing of a neuron only to the rat's position, but which omits influences of firing history or activity of other neurons. For this case, as detailed in

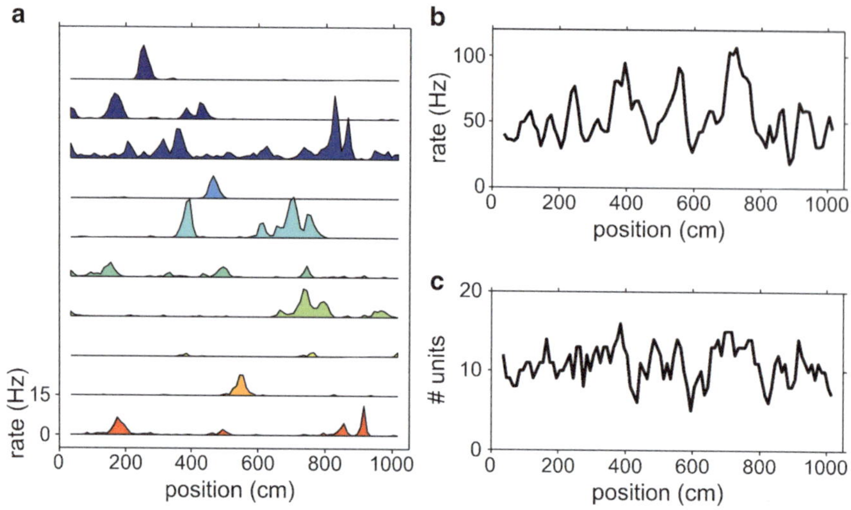

Fig. 2. (**a**) Spatial tuning curves of 10 of 46 recorded hippocampal units in one recording session. On a 10-m long track, many hippocampal units have multiple place fields. (**b**) Mean firing rate of the population of isolated units as a function of position on the track. (**c**) Number of active units as a function of position on the track. For each position, only units that have a mean firing rate higher than 1 Hz at that position are included.

Sect. 1, building an encoding model corresponds to computing the spatial tuning curve $\lambda_c[x]$ for each cell from the neural activity and position record during online epochs (see Note 5).

3.2.2. Construction of Spatial Tuning Curves

To construct a spatial tuning curve for each recorded unit, the physical length of the track is discretized into equal sized bins (e.g., 10 cm). An average firing rate histogram over all position bins is computed for each unit over the complete recording session (see Note 3 about stability of place fields). For this, the number of spikes emitted while the rat was located in any given position bin is divided by the total time spent at that position, i.e., $\lambda_c[x] = \frac{\text{nspikes}_c[x]}{\text{occupancy}[x]}$. Only the data obtained during the online state (i.e., when the rat moves faster than 15 cm/s) is used to compute the tuning curve. Optionally, spatial tuning curves can be smoothed to reduce the effect of binning (see Note 4). Figure 2a shows examples of spatial tuning curves of a small number of hippocampal units. Note that each unit is active in specific locations on the track and that some units have multiple place fields.

In some cases, two neurons that were never active together in the online state (i.e., two neurons with completely nonoverlapping spatial firing fields), nonetheless fire simultaneously in the offline state. When this occurs, the likelihood in (3) will be 0 everywhere and the posterior density is undefined. To remedy this problem, a small background value (e.g., 0.01 Hz) can be added to the tuning curves.

3.2.3. Selection of Units

In principle, any neuron whose firing shows a consistent relation to the animal's position can be used for decoding. Most decoding studies focus on hippocampal pyramidal neurons because of the high spatial information carried in their spike trains. Notably, hippocampal interneurons also show a consistent spatial modulation of their firing rate (48), and decoding could potentially benefit from their contribution. However, little is known about the participation of interneurons in hippocampal memory replay and the contribution of interneurons to the decoding of offline states remains an open question. In the present discussion, interneurons, identified by their high mean firing rate (>5 Hz) and narrow spike width (<0.25 ms, peak-to-trough), are not included.

3.3. Decoding

3.3.1. Decoding Position in Online and Offline States

As described in Sect. 2, the decoding process produces a posterior probability distribution containing any information about the stimulus that can be gained from the neuronal activity. The likelihood function and the prior probability distribution are essential for computing the posterior see (4). A noninformative prior is used for decoding position in offline states (see Sect. 3) and thus the posterior at time t is proportional to the likelihood function:

$$P[x_t|\mathbf{n}_t] = K_t P[\mathbf{n}_t|x_t] = K_t \left\{ \prod_{c=1}^{C} \lambda_c[x]^{n_{c,t}} \right\} \mathrm{e}^{-\Delta \sum_{c=1}^{C} \lambda_c[x]} \quad (7)$$

Here, K_t is a constant that includes all terms that are independent of x and which makes sure that the posterior distribution sums to 1.

In practice, time is divided into a set of nonoverlapping windows of duration Δ and a vector of spike counts for all units $\mathbf{n}_t = \{n_{1,t}, n_{2,t}, \ldots, n_{C,t}\}$ is constructed for each time window. Next, the spike counts are used to compute the posterior distribution $P[x_t|\mathbf{n}_t]$ according to (7). (Note that when none of the units emits a spike, the posterior is proportional to the exponential term). The optimal choice of Δ is determined by the temporal dynamics of the stimulus that is being decoded. During online states, Δ should be large enough such that sufficient data is available for accurate decoding and small enough to capture the behavioral dynamics. A value of 200–500 ms is a reasonable choice (14, 21). In Fig. 3a, the result of decoding is shown for a 70-s long time window during which the rat ran from one end of the track to the other. During online states, when the rat travels down the track, the decoding error is small. On the other hand, decoding performance degrades when the rat stops.

During offline states, the underlying dynamics of the hidden stimulus are unknown and an appropriate value for Δ can only be determined empirically. Previous studies indicated that hippocampal sequence replay in offline population bursts is sped up by a factor of 10–20× (27–30) and Davidson et al. (21) determined

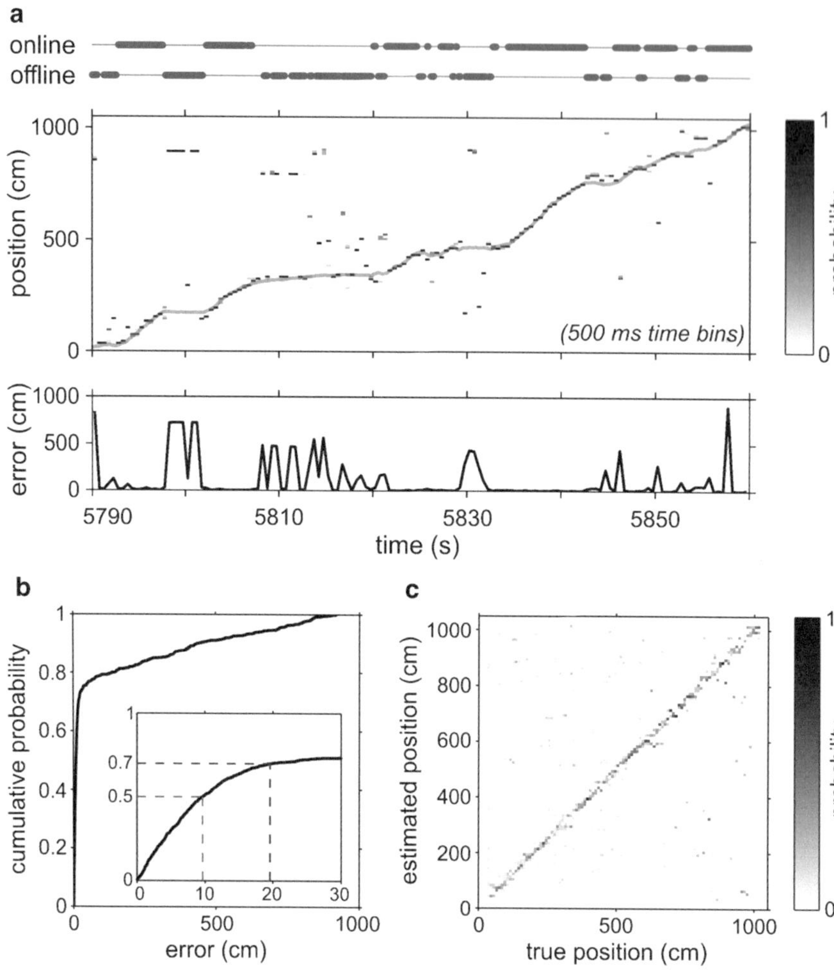

Fig. 3. (**a**) Position decoding during a 70-s long epoch in which the rat ran from one end of the track to the other end while stopping frequently. *Top*: Online and offline epochs. *Middle*: Posterior probability distributions of position computed in 500-ms nonoverlapping time bins. The animal's true running path (*semitransparent line*) is overlaid on top of the posteriors. Notice that good correspondence between the positions with high probability (*black*) and the animal's path. *Bottom*: Error between the maximum a posteriori estimate and the animal's true location. Note that during online epochs when the rat is running, the error is low and that during offline epochs, when the rat pauses, the error is high. (**b**) Cumulative distribution of decoding errors during online epochs, evaluated using a cross-validation procedure. (**c**) Confusion matrix showing the distribution of estimated positions for each true position of the rat on the track. To allow for easy comparison between the distributions for each true position (regardless of how often the animal has visited that location), the conditional distributions $p(\hat{x}|x)$ are shown, rather than the joint probability distribution $p(x,\hat{x})$.

that a value of 20 ms for the time bin duration gave a good trade-off between temporal resolution and low noise estimates.

3.3.2. Evaluating Decoding Performance

For each data set to be decoded, it is important to determine how well the stimulus can be estimated from the neuronal ensemble. During offline states, there is no ground truth to which the decoded position can be compared. Thus, the performance of the encoding model can only be assessed in the online state. If insufficient spatial information is present in the population of neurons, the encoding model cannot be used to accurately predict the rat's position on the track. In this case, it is also unlikely that the virtual position in the offline state can be successfully decoded.

To make a quick assessment of the available spatial information, both the population firing rate (i.e., the sum of all tuning curves $\sum_{c=1}^{C} \lambda_c[x]$) and the number of active neurons in the online state can be plotted as a function of position (Fig. 2b, c). For locations on the track that have low coverage, i.e., where the population firing rate is low and/or few units are active, decoding is unlikely to succeed.

In another test of decoding performance, the estimated position is compared to the rat's true position on the track in a cross-validation procedure. In this procedure, one part of the data in the online state ("training data") is used to construct the encoding model while the remainder of the online data ("test data") is used to decode position. Decoding error is then defined as the absolute difference between the estimated position and the rat's physical location. In Fig. 3b, the cumulative distribution of errors is shown for a typical experiment. Notably, the median error is smaller than the spatial bin size of 10 cm (see inset in Fig. 3b). Seventy percent of the estimates are still within 20 cm of the true location. For the 30% of the remaining estimates, the error is more or less uniformly distributed between 20 cm and a maximum error of ~900 cm, as indicated by the approximately linear increase of the cumulative probability.

The error distribution gives an indication of the overall decoding performance, but it does not provide information about the distribution of errors for specific positions on the track. By treating decoding as a classification problem, this additional information can be obtained from the (normalized) confusion matrix (49), which is the joint distribution $p(x, \hat{x})$ of the true position on the track x and the estimated position $\hat{x}$ (Fig. 3c). The confusion matrix can be constructed using either the full posterior distributions or the point estimates derived from the posterior distributions (see Sect. 2). In the case of perfect decoding, all power in the confusion matrix is concentrated along the diagonal. High values located off the diagonal are evidence of a systematic misclassification of the true position in the decoding. Conversely, a uniform distribution of estimated positions reflects a general inability to estimate the true position. In Fig. 3c, the highest probabilities are around the diagonal indicating that for most locations on the track the estimated positions are generally close to the rat's true position.

From the confusion matrix another decoding performance measure can be calculated: the mutual information between the true position and the estimated position (8). This quantity measures (in bits) how much information the estimate $\hat{x}$ shares with the

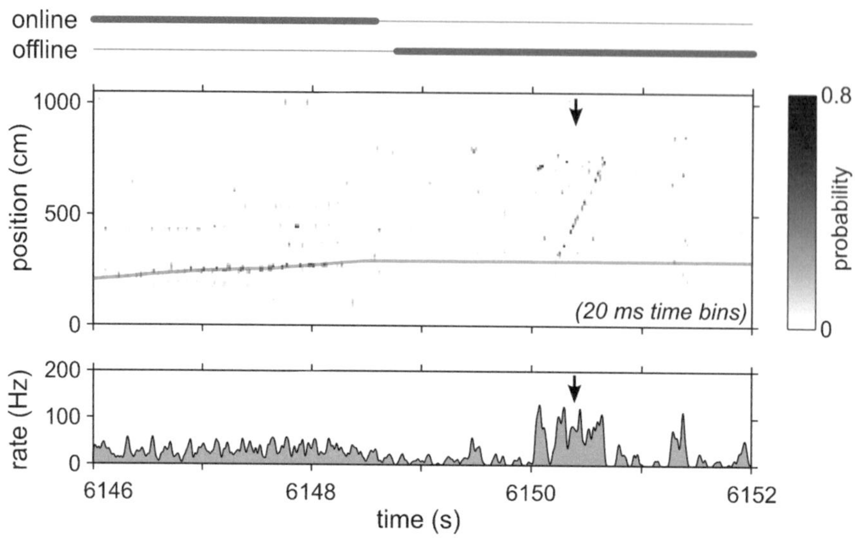

Fig. 4. Position decoding in a 6-s long window around a transition from online to offline state. *Top*: Online and offline epochs. *Middle*: Posterior probability distributions of position computed in 20-ms nonoverlapping time bins. The animal's true path (*semitransparent line*) is overlaid on top of the posteriors. In the online epoch, positions with high probability follow the animal's true path. In the offline epoch, decoding is poor except between 6,150 and 6,151 s, when position estimates appear to describe a trajectory away from the rat's physical location (*arrow*). *Bottom*: Multi-unit activity including all isolated and nonisolated spikes, showing the occurrence of population bursts in the offline epoch.

true position x. In perfect decoding, when x and $\hat{x}$ are identical, the mutual information is equal to the entropy of the animal's true position; in the worst case, when x and $\hat{x}$ are completely independent, then the mutual information is equal to 0. It should be noted that in practice the mutual information is positively biased due to finite sampling (50).

$$I = \sum_{\hat{x}} \sum_{x} p(x, \hat{x}) \log_2 \left(\frac{p(x, \hat{x})}{p(x)p(\hat{x})} \right) \tag{8}$$

3.4. Analysis of Position Estimates

3.4.1. Overview

The final step is to perform statistical analyses on the time-series of position estimates that are obtained by applying decoding to offline epochs. Figure 4 shows the result of decoding during an epoch in which the rat paused part-way the track. During most of the time when the rat is immobile, firing rate is low and decoding results in noisy posterior distributions. Occasionally, however, many neurons will fire in brief population bursts (Fig. 4, bottom panel). Memory replay can be observed during some of the population bursts, as is evident from the systematic deviation of the decoded position from the rat's physical location (Fig. 4, arrow).

In the next sections, three examples of analysis that can be applied to decoded positions in offline states are discussed. In all three cases, the analysis focuses exclusively on the set of population bursts while the rat is immobile on the track. The first two analyses deal with the discovery and description of temporal structure in the hidden position stimulus. The third analysis looks at the relation between single cell activity and the population representation of the hidden position stimulus. Other types of analyses are possible, for example correlations of the decoded position with the animal's actual behavior, with ongoing oscillations in the local field potential or with neuronal activity in other brain regions—but these are not discussed here.

3.4.2. Spatial–Temporal Structure I: Average Distance Traveled

An analysis that can be used to test if, on average, estimated positions in population bursts are temporally correlated is to quantify the relation between the average distance traveled and the time that has elapsed. This relationship between time and distance can provide information about the dynamics of the hidden position stimulus. For example, a linear relationship between time and distance traveled could indicate the presence of constant velocity trajectories. Alternatively, a square root relationship between time and distance could indicate that the dynamics are better described by a random walk model.

The probability distribution of the difference between two position estimates is obtained by cross-correlation of their posterior probability distributions. If $P_t[x]$ and $P_{t+\text{lag}}[x]$ represent two posterior probability distributions at time t and t + *lag*, then their cross-correlation is given by: $P[\mathrm{d}x] = \sum_x P_t[x]P_{t+\text{lag}}[x + \mathrm{d}x]$. The average probability distribution $P_{\text{lag}}[\mathrm{d}x]$ of the distance traveled for each time lag is computed by averaging the cross-correlations for all pairs of position estimates that are separated in time by *lag* seconds. The expected value of $P_{\text{lag}}[\mathrm{d}x]$ and its variance can be computed from the sample shuffle distribution that is obtained by repeatedly randomizing the posterior distributions in time and recalculating the cross-correlation.

Figure 5a shows the average distribution of the distance traveled for a single time lag of 100 ms, computed during offline population bursts only. Notice the distinct peaks at 80 and −80 cm, indicating that on average the estimated position has moved 80 cm in either direction along the track over 100 ms. In Fig. 5b, the distance distributions for a range of time lags is shown. Notice the linear relationship between distance traveled and time lag.

3.4.3. Spatial–Temporal Structure II: Model Fit

The previous analysis indicated that *on average* the decoded position during hippocampal offline population bursts is linearly related to time. Also, the example shown in Fig. 4, suggests that hippocampal activity within *single bursts* represent a trajectory through space (even though the animal is paused). In order to describe the dynamics of the hidden position stimulus in individual population bursts, a model of a constant speed trajectory will be fitted to the data (see Note 6).

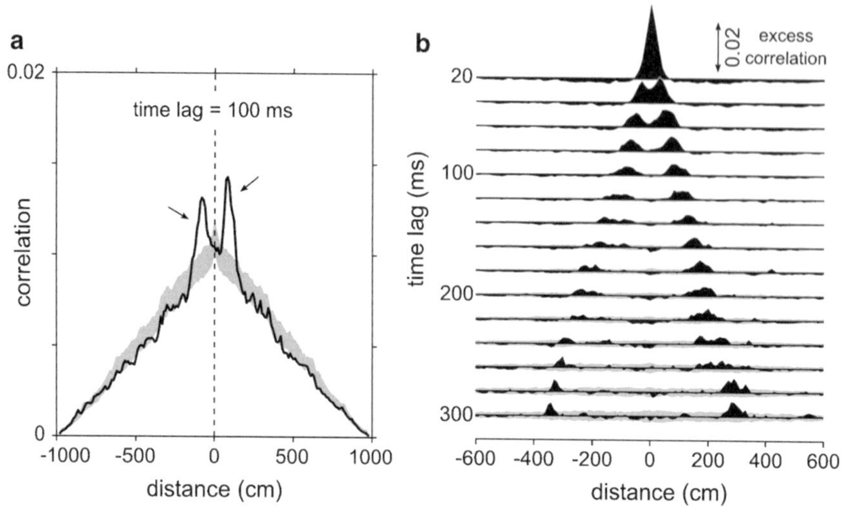

Fig. 5. (**a**) Cross-correlation between all pairs of posterior probability distributions that have a lag of 100 ms. Only posteriors during population bursts in offline epochs are included. Notice the increased correlation at ± 80 cm, indicating that position estimates at 100 ms lags are on average separated by 80 cm in either direction on the track. *Gray area* represents 95% confidence interval of the expected correlation under the null hypothesis that there is no temporal structure in the position estimates. The expected correlation is computed by repeatedly randomizing the posteriors in time. (**b**) Cross-correlations for lags ranging from 20 to 300 ms. For visual clarity, the expected correlation has been subtracted to only show the excess correlation. *Gray area* represents 95% confidence interval of the expected correlation.

Karlsson and Frank (22) drew samples from the posterior distributions and performed simple linear regression on these samples. However, this approach is sensitive to outliers and assumes that the errors are normally distributed. Instead, Davidson et al. (21) developed a method that is not sensitive to outliers and also uses the complete posterior distribution. This method is inspired by the radon transform which can be used to detect lines in images (51). The goal is to find a trajectory that best fits the posterior probabilities. A trajectory is characterized by its two parameters *velocity* and *intercept* and can be written as:

$$y(t) = \text{velocity} \times t + \text{intercept} \tag{9}$$

For each trajectory, a score can be computed that reflects the probability that the decoded positions are within a chosen distance z from the trajectory (see Fig. 6a):

$$\text{Score} = \frac{1}{NT} \sum_{t} \sum_{-z \leq x_t - y(t) < z} P[x_t | \mathbf{n}_t] \quad z \in \left\{ \tfrac{1}{2}\delta x, \tfrac{3}{2}\delta x, \tfrac{5}{2}\delta x, \ldots \right\} \tag{10}$$

Here, NT is the number of time windows in the candidate replay event for which a posterior distribution has been computed. The value of z dictates the size of the region around the trajectory

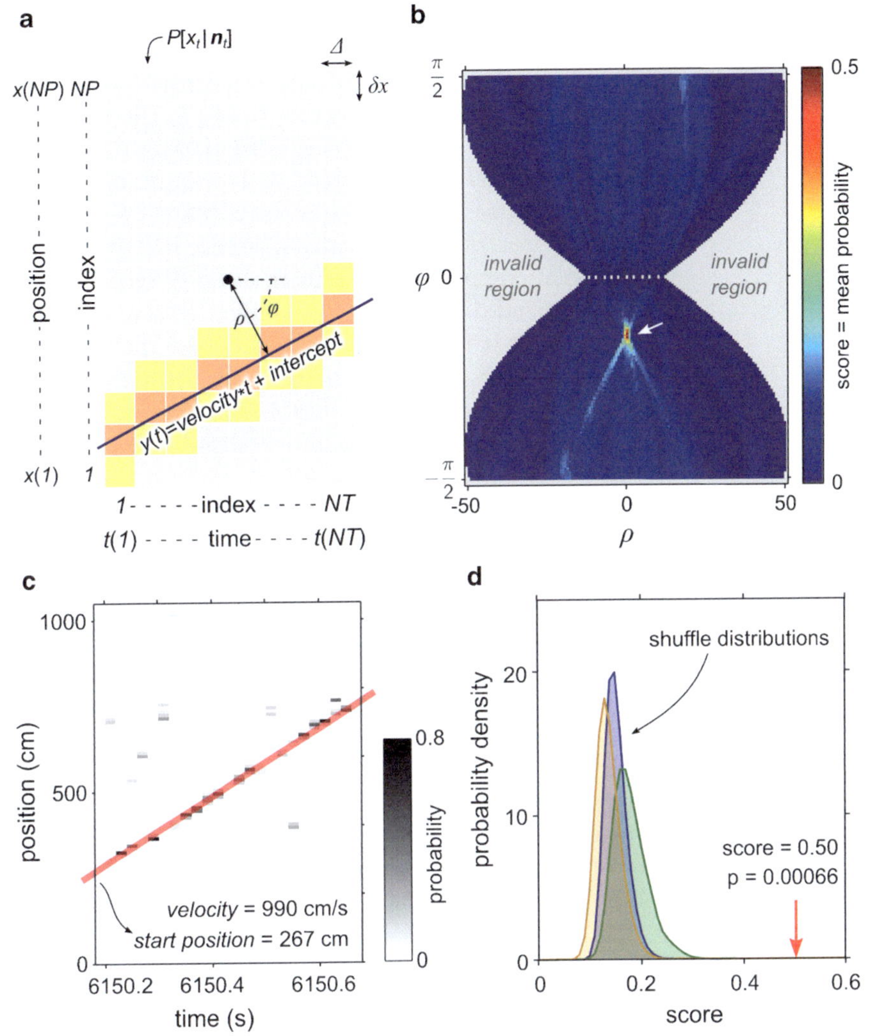

Fig. 6. (**a**) Diagram explaining the algorithm to detect trajectories in individual population bursts. Each *column* represents a posterior population distribution. The score of a trajectory (*blue line*) is computed by averaging the probabilities associated with the positions on the trajectory (*orange bins*). Optionally, a larger region around the trajectory can be included to compute the score (i.e., *orange + yellow bins*). The trajectory can be parameterized either by velocity and intercept or by the length (ρ) and angle (ϕ) of the normal vector. (**b**) Result of densely sampling trajectories for the population burst shown in Fig. 4. The score of each trajectory is plotted as a function of its ϕ and ρ parameters. Parameter combinations in the *gray areas* do not represent valid trajectories. The trajectory with the highest score (*white arrow*) is selected as the best fit for the spatial–temporal structure in the posterior probability distributions. (**c**) Best fitting trajectory from (**b**) plotted on *top* of the posterior probability distributions computed for the population burst in Fig. 4. (**d**) Comparison of the score of the best fitting trajectory in (**b**, **c**) and the score distributions obtained after three randomization schemes. From *left* to *right* the three different randomizations are: pseudo burst, position cycle, and unit permutation.

that is used to compute the score. A larger value of z allows for the detection of trajectories with small local variations in velocity. The higher the score, the better the trajectory fits the spatial–temporal pattern in the posterior probability distributions.

The general procedure is to densely sample trajectories from the set of all possible trajectories to find the parameters *velocity* and *intercept* that maximize the score in (10) (Fig. 6b, c; see Note 7 for implementation details).

The significance of the trajectory fitted to the decoded position in a single population bursts can be tested using Monte Carlo methods. The original data is randomized repeatedly and every time the trajectory fitting procedure is applied and a score is computed. The actual score is then compared to the distribution of scores obtained using the randomized data and a Monte Carlo p-value can be computed according to: $p = (r + 1)/(n + 1)$, where r is the number of randomizations resulting in a score higher than the real score and n is the total number randomizations performed. Davidson et al. (21) used three different randomizations and repeated each 1,500 times:

1. *Position cycle.* The position bins in each posterior distribution are circularly shifted by a random amount. The general shape of the posteriors is maintained, but any temporal correlation is lost. This randomization procedure corrects for chance alignment of position estimates.
2. *Unit permutation.* Each unit is randomly assigned to another spatial tuning curve. This randomization controls for patterns generated solely by spiking dynamics (for example, prolonged bursting of a single unit). Problems arise if the mean firing rates differ significantly between units and a high-rate unit is assigned to the tuning curve of a low-rate unit or vice versa. In that case, the randomization procedure should be modified so that only the shapes of the tuning curves are randomized, but not the mean firing rate.
3. *Pseudo burst.* For each population bursts, a new "pseudo" burst with the same number of time bins is constructed by randomly drawing posterior distributions from all available population bursts. This randomization procedure corrects for any systematic preference in the decoding for certain positions on the track.

In Fig. 6d, the score distributions of the three randomization procedures are shown for the population burst in Fig. 4. Since a fitted trajectory should pass all three tests in order to be considered significant, the final p-value is the maximum of all three Monte Carlo p-values.

3.4.4. Relation to Single Cell Activity: Virtual Tuning Curves

Analogous to the spatial tuning curves constructed using the measured position in online epochs (see Sect. 2) a "virtual" tuning curve can be constructed using the result of decoding in the offline epochs. Virtual tuning curves are one way to assess whether the activity of a single cell in offline epochs is coherent with the position represented by the population.

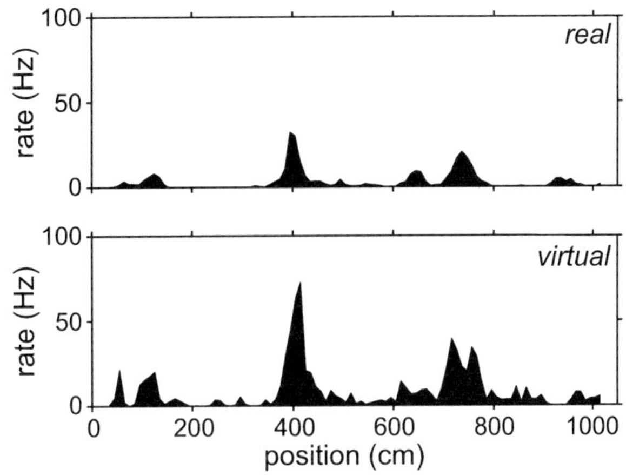

Fig. 7. Comparison between the real spatial tuning curve of unit as computed during online states (*top*) and the virtual spatial tuning curve for the same unit as computed from all significant ($p < 0.05$) replay trajectories in the offline state (*bottom*). Notice that the two curves are similar, except that the peak rates in the virtual tuning curve are higher.

To compute the virtual spatial tuning curve, one can use either the complete posterior probability distribution, point estimates (see Sect. 2), or a model fitted to the decoded position (see Sect. 3.3). Optionally, times at which there is high uncertainty about the estimated position may be excluded from the analysis. Importantly, to avoid any circularity in the analysis, the cell for which a virtual tuning curve is computed should be left out of the population of cells used to perform the decoding of position.

An example of the true and virtual tuning curves of a hippocampal pyramidal cell is shown in Fig. 7. The virtual curve was computed using the posterior modes in population bursts that contained significant replay trajectory according to the method described in the previous section. For this unit, a high correlation exists between the true and virtual spatial tuning curves. Notice, however, that the maximum rate in the virtual tuning curve is 2–3× higher than in the real tuning curve. This suggests that the encoding model may need to be modified to include a rate scaling factor (see Note 5).

4. Notes

1. If the stimulus does change quickly over the course of the window duration Δ, , the number of expected events in that time window is given by the integral $\int_t^{t+\Delta} \lambda[x(t)]\mathrm{d}t$.

2. The dependence of the amount of offline state on the rat's training status may vary with the experimental conditions and behavioral task. For example, in a t-maze alternation task with a delay phase before each trial, the total time spent in the offline state may be higher even after many training sessions. However, the general requirement for sufficient sampling of both online and offline states still holds.

3. By calculating the spatial tuning curves as averages over the total recording session, it is assumed that spatial receptive fields are stable in time. One way to check the stability is to compute tuning curves for the first half and second half of the recording session separately and compute the correlation between the two tuning curves. A low correlation indicates that the spatial tuning curves changes over the course of the recording session. In this case, an adaptive encoding model may be more appropriate (52).

4. A smoothed spatial tuning curve for a cell can be computed using the following equation:

$$\lambda_c[x] = \frac{\text{nspikes}_c[x]}{\text{occupancy}[x]} = \frac{\sum_{n=1}^{N_c} G\left(\frac{x_n^c - x}{h}\right)}{\frac{1}{Fs}\sum_{t=1}^{T} G\left(\frac{x_t^{\text{rat}} - x}{h}\right)} \tag{11}$$

Here, N_c is the number of spikes for unit c and x_n^c is the corresponding position of the rat on the track when the nth spike was emitted. The path of the animal x_t^{rat} is sampled at Fs samples per second for a total of T position samples in the online epochs. G is a smoothing kernel and h is a smoothing parameter. A range of smoothing kernels can be used, but commonly a normal kernel is chosen. In this case, h represents the standard deviation of the normal density function. Note that when a box kernel is chosen, with $2h$ equal to the spatial bin size, then (11) reduces to a simple nonsmoothed histogram.

5. By constructing the encoding model from the spatial tuning curves computed in the online state, it is tacitly assumed that the firing rates in the online and offline states are similar. In reality, the mean population firing rate in offline population bursts is 2–3× higher than the mean population firing rate in the online state. This information can be incorporated in the encoding model by scaling the spatial curves with a constant factor α. Equation (7) then becomes:

$$P[x_t|\mathbf{n}_t] = K_t\left\{\prod_{c=1}^{C} \lambda_c[x]^{n_{c,t}}\right\} e^{-\Delta\alpha\sum_{c=1}^{C}\lambda_c[x]} \tag{12}$$

The effect on the posterior distribution is modest for a range of scaling factors, especially if $\sum_{c=1}^{C} \lambda_c[x]$ is close to uniform.

6. The spatial–temporal structure in the posterior probability distributions can also be fitted with a stochastic model using Bayesian filtering, as described in Sect. 2. Analogous to the method described for fitting a deterministic model in Sect. 3.3, a score can be computed by comparing how well the fitted trajectory matches the likelihood estimates calculated using noninformative priors. This score is then compared to distributions of scores obtained after shuffling the original data and refitting the stochastic model.

7. The velocity parameter in (8) can take any value between $-\infty$ and $+\infty$, which is inconvenient for sampling purposes. Therefore in the following algorithm, a different form of the trajectory equation is used in which the parameters are bounded. First, let us present the series of posteriors as a matrix with NT columns or time bins (bin size is Δ s) and NP position bins (bin size is δx cm) (Fig. 6a). Each time bin is associated with an index (ci) and the absolute time at the center of the time bin (t). Likewise, each position bin is associated with an index (ri) and position on the track (x). Further, the center of the matrix can be defined in both index and time-position coordinates: $(ci_{\text{mid}}, ri_{\text{mid}}) = \left(\frac{NT+1}{2}, \frac{NP+1}{2}\right)$ and $(t_{\text{mid}}, x_{\text{mid}}) = \left(\frac{t(NT)+t(1)}{2}, \frac{x(NP)+x(1)}{2}\right)$. The relation between the two coordinate systems is given by: $t = t_{\text{mid}} + \Delta(ci - ci_{\text{mid}})$ and $x = x_{\text{mid}} + \delta x(ri - ri_{\text{mid}})$. Using these definitions, the equation for a trajectory can be written in the more convenient "normal form," using index coordinates and with the origin at the center of the matrix:

$$(ci - ci_{\text{mid}})\cos(\phi) + (ri - ri_{\text{mid}})\sin(\phi) = \rho \tag{13}$$

The parameters ϕ and ρ represent the angle of inclination of the normal vector and the length of the normal vector, respectively (see Fig. 6a; note that ρ can take on negative values as well). All possible trajectories that intersect with the matrix have parameters that lie in a compact space defined by the following conditions: $\phi \in \left[-\frac{1}{2}\pi, \frac{1}{2}\pi\right]$ and $\rho \in \left[-\frac{1}{2}D, \frac{1}{2}D\right]$, where D is the length of the matrix diagonal: $D = \sqrt{(NT-1)^2 + (NP-1)^2}$. Next, this parameter space is densely sampled and for each sampled trajectory and its corresponding parameters, (13) is used to compute the (nearest integer) position bin index for every time bin. These indices are then used to compute the score for the sampled trajectory by averaging the probabilities in the corresponding bins (orange bins in Fig. 6a). For those time bins where the sampled trajectory falls outside the length of the track, the median of the posterior distribution in that time bin is added to the score. This procedure ensures that there exists no intrinsic bias against partial trajectories.

To make the algorithm robust to small deviations or noise in the posterior distributions, the probability of a wider region around the sampled trajectory can be included into the score (i.e., yellow region in Fig. 6a; see also Sect. 3.3). This can be implemented by preconditioning the matrix, such that the probability in each position bin is replaced by the sum of itself and its n neighbors. This can be easily accomplished by convolving each posterior distribution with a length $2n + 1$ vector of all ones. Finally, out of all sampled (ϕ, ρ) parameters, the trajectory with the highest score is selected as the one that best fits the underlying posterior distributions (Fig. 6b). The velocity and intercept of this trajectory can then be determined using the following relationships: $\text{velocity} = -\cot(\phi)\frac{\delta x}{\Delta}$ and $\text{intercept} = \cot(\phi)\frac{\delta x}{\Delta}t_{\text{mid}} + \frac{\rho}{\sin(\phi)}\delta x + x_{\text{mid}}$.

Acknowledgments

I would like to thank Audrey Chang, Zhe Chen and Matt Wilson for their comments on an earlier version of this manuscript. Many thanks to Tom Davidson for the discussions about the methods presented in this chapter and for generously providing me with hippocampal unit data to prepare some of the figures.

References

1. Fujisawa S, Amarasingham A, Harrison MT et al (2008) Behavior-dependent short-term assembly dynamics in the medial prefrontal cortex. Nat Neurosci 11:823–833
2. Nicolelis MAL, Dimitrov D, Carmena JM et al (2003) Chronic, multisite, multielectrode recordings in macaque monkeys. Proc Natl Acad Sci USA 100:11041–11046
3. Gray CM, Maldonado PE, Wilson M et al (1995) Tetrodes markedly improve the reliability and yield of multiple single-unit isolation from multi-unit recordings in cat striate cortex. J Neurosci Methods 63:43–54
4. Lütcke H, Murayama M, Hahn T et al (2010) Optical recording of neuronal activity with a genetically-encoded calcium indicator in anesthetized and freely moving mice. Front Neural Circuits 4:9
5. Murayama M, Pérez-Garci E, Lüscher HR et al (2007) Fiberoptic system for recording dendritic calcium signals in layer 5 neocortical pyramidal cells in freely moving rats. J Neurophysiol 98:1791–1805
6. Sawinski J, Wallace DJ, Greenberg DS et al (2009) Visually evoked activity in cortical cells imaged in freely moving animals. Proc Natl Acad Sci USA 106:19557–19562
7. Brown EN, Kass RE, Mitra PP (2004) Multiple neural spike train data analysis: state-of-the-art and future challenges. Nat Neurosci 7:456–461
8. Paninski L, Pillow J, Lewi J (2007) Statistical models for neural encoding, decoding, and optimal stimulus design. Prog Brain Res 165:493–507
9. Quiroga RQ, Panzeri S (2009) Extracting information from neuronal populations:

information theory and decoding approaches. Nat Rev Neurosci 10:173–185

10. Brown EN, Frank LM, Tang D et al (1998) A statistical paradigm for neural spike train decoding applied to position prediction from ensemble firing patterns of rat hippocampal place cells. J Neurosci 18:7411–7425
11. Georgopoulos AP, Schwartz AB, Kettner RE (1986) Neuronal population coding of movement direction. Science 233:1416–1419
12. Sanger TD (1996) Probability density estimation for the interpretation of neural population codes. J Neurophysiol 76:2790–2793
13. Stanley GB, Li FF, Dan Y (1999) Reconstruction of natural scenes from ensemble responses in the lateral geniculate nucleus. J Neurosci 19:8036–8042
14. Zhang K, Ginzburg I, McNaughton BL et al (1998) Interpreting neuronal population activity by reconstruction: unified framework with application to hippocampal place cells. J Neurophysiol 79:1017–1044
15. Boloori AR, Jenks RA, Desbordes G et al (2010) Encoding and decoding cortical representations of tactile features in the vibrissa system. J Neurosci 30:9990–10005
16. Johnson A, Fenton AA, Kentros C et al (2009) Looking for cognition in the structure within the noise. Trends Cogn Sci 13:55–64
17. Reijmers LG, Perkins BL, Matsuo N et al (2007) Localization of a stable neural correlate of associative memory. Science 317:1230–1233
18. Gelbard-Sagiv H, Mukamel R, Harel M et al (2008) Internally generated reactivation of single neurons in human hippocampus during free recall. Science 322:96–101
19. Georgopoulos AP, Lurito JT, Petrides M et al (1989) Mental rotation of the neuronal population vector. Science 243:234–236
20. Georgopoulos AP, Crutcher MD, Schwartz AB (1989) Cognitive spatial-motor processes. 3. Motor cortical prediction of movement direction during an instructed delay period. Exp Brain Res 75:183–194
21. Davidson TJ, Kloosterman F, Wilson MA (2009) Hippocampal replay of extended experience. Neuron 63:497–507
22. Karlsson MP, Frank LM (2009) Awake replay of remote experiences in the hippocampus. Nat Neurosci 12:913–918
23. Johnson A, Redish AD (2007) Neural ensembles in CA3 transiently encode paths forward of the animal at a decision point. J Neurosci 27:12176–12189
24. O'Keefe J, Dostrovsky J (1971) The hippocampus as a spatial map. Preliminary evidence from unit activity in the freely-moving rat. Brain Res 34:171–175
25. Wilson MA, McNaughton BL (1993) Dynamics of the hippocampal ensemble code for space. Science 261:1055–1058
26. O'Keefe J, Nadel L (1978) The hippocampus as a cognitive map. Oxford University Press, Oxford
27. Lee AK, Wilson MA (2002) Memory of sequential experience in the hippocampus during slow wave sleep. Neuron 36:1183–1194
28. Ji D, Wilson MA (2007) Coordinated memory replay in the visual cortex and hippocampus during sleep. Nat Neurosci 10:100–107
29. Foster DJ, Wilson MA (2006) Reverse replay of behavioural sequences in hippocampal place cells during the awake state. Nature 440:680–683
30. Diba K, Buzsáki G (2007) Forward and reverse hippocampal place-cell sequences during ripples. Nat Neurosci 10:1241–1242
31. Csicsvari J, O'Neill J, Allen K et al (2007) Place-selective firing contributes to the reverse-order reactivation of CA1 pyramidal cells during sharp waves in open-field exploration. Eur J Neurosci 26:704–716
32. Fenton AA, Kao HY, Neymotin SA et al (2008) Unmasking the CA1 ensemble place code by exposures to small and large environments: more place cells and multiple, irregularly arranged, and expanded place fields in the larger space. J Neurosci 28:11250–11262
33. Seung HS, Sompolinsky H (1993) Simple models for reading neuronal population codes. Proc Natl Acad Sci USA 90:10749–10753
34. Salinas E, Abbott LF (1994) Vector reconstruction from firing rates. J Comput Neurosci 1:89–107
35. Zemel RS, Dayan P, Pouget A (1998) Probabilistic interpretation of population codes. Neural Comput 10:403–430
36. Daley DJ, Vere-Jones D (2003) An introduction to the theory of point processes. Springer, New York
37. Truccolo W, Eden UT, Fellows MR et al (2005) A point process framework for relating neural spiking activity to spiking history, neural ensemble, and extrinsic covariate effects. J Neurophysiol 93:1074–1089
38. Kloosterman F, Davidson TJ, Gomperts SN et al (2009) Micro-drive array for chronic in vivo recording: drive fabrication. J Vis Exp 26(pii):1094
39. Nguyen DP, Layton SP, Hale G et al (2009) Micro-drive array for chronic in vivo recording: tetrode assembly. J Vis Exp 26 (pii):1098

40. Yamamoto J, Wilson MA (2008) Large-scale chronically implantable precision motorized microdrive array for freely behaving animals. J Neurophysiol 100:2430–2440
41. Buzsáki G (2004) Large-scale recording of neuronal ensembles. Nat Neurosci 7:446–451
42. Buzsáki G, Horváth Z, Urioste R et al (1992) High-frequency network oscillation in the hippocampus. Science 256:1025–1027
43. Csicsvari J, Hirase H, Mamiya A et al (2000) Ensemble patterns of hippocampal CA3-CA1 neurons during sharp wave-associated population events. Neuron 28:585–594
44. Ylinen A, Bragin A, Nádasdy Z et al (1995) Sharp wave-associated high-frequency oscillation (200 Hz) in the intact hippocampus: network and intracellular mechanisms. J Neurosci 15:30–46
45. Schmitzer-Torbert N, Jackson J, Henze D et al (2005) Quantitative measures of cluster quality for use in extracellular recordings. Neuroscience 131:1–11
46. Quirk MC, Wilson MA (1999) Interaction between spike waveform classification and temporal sequence detection. J Neurosci Methods 94:41–52
47. Jensen O, Lisman JE (2000) Position reconstruction from an ensemble of hippocampal place cells: contribution of theta phase coding. J Neurophysiol 83:2602–2609
48. Ego-Stengel V, Wilson MA (2007) Spatial selectivity and theta phase precession in CA1 interneurons. Hippocampus 17: 161–174
49. Quiroga RQ, Reddy L, Koch C et al (2007) Decoding visual inputs from multiple neurons in the human temporal lobe. J Neurophysiol 98:1997–2007
50. Treves A, Panzeri S (1995) The upward bias in measures of information derived from limited data samples. Neural Comput 7:399–407
51. Toft PA (1996) The radon transform – theory and application. PhD thesis, Technical University of Denmark. http://petertoft.dk/PhD. Accessed on 2007
52. Eden UT, Frank LM, Barbieri R et al (2004) Dynamic analysis of neural encoding by point process adaptive filtering. Neural Comput 16:971–998

Neuromethods (2012) 67: 283–302
DOI 10.1007/7657_2011_9

Published online: 13 January 2012

High-Speed Videography of Embodied Active Sensing in the Rodent Whisker System

Jason T. Ritt

Abstract

Self-motion is often a major component of sensation. Comprehensive understanding of natural sensory processing in neural systems thus requires neural recording be coupled to high-resolution observation of behavior. The rodent vibrissa (whisker) system has several advantages as a model for network function during sensory behaviors, but only recently have inputs at the whisker level received serious study. This chapter reviews methods for capturing biological motion via high-speed videography, with emphasis on tracking rodent whiskers or other similar processes such as insect antennae.

Key words: Active touch, Vibrissa, Digital video, Computer vision, Biomechanics

1. Introduction

1.1. Sensation as a Behavior

A typical description of sensory processing consists of single neuron tuning curves and network level transformations of activity. To place this description in a functional context, one must know what sensory inputs an individual receives as it navigates and investigates its surroundings. In natural settings, characterization of these inputs must account for how embodiment and active sensing dominate the information initially translated into neural activity. *Embodiment* refers to how peripheral sensors connect the world to the brain, i.e., how energy of different kinds is translated into neural activity at the periphery. *Active sensing* refers to how these sensors are used to sample the world, e.g., how an animal positions them, sweeps them, tenses them, or focuses them during sensory acquisition. Such behavioral choices select the information to be perceived, and delimit the scope of later computation performed by the nervous system. Active sensing depends on embodiment, as the physical structure of transduction determines how a sensor should be moved to acquire the specific information sought.

As one example, not all eyes are created equal. We acquire visual information by first choosing where to look, making eye motions a central part of the filter stream. Parameters that vary with task and viewing conditions, such as fixation times, may induce changes to visual input statistics that induce adaptation or plasticity (1–5). By contrast, the eyes of some raptors have two foveas, a sideways pointing region for monocular prey tracking in spiral flight, and a forward pointing region for close-range binocular prey capture (6). Head motions and flight patterns, rather than saccades, dominate the gaze direction in an ecologically adapted fashion. At a lower level of complexity, fiddler crabs have fixed eye stalks with panoramic vision and minimal control of gaze direction, appropriate for their ecological requirements of interacting with ground level conspecifics while avoiding overhead predation by birds (7).

1.2. Preneural Computation

In some cases a sensory computation might be considered "preneural" (8), that is, occurring in the transduction process itself instead of via central neural transformations. Sound source localization provides a clear example. The standard Jeffress model of parallel delay lines, derived from the avian brainstem, requires substantial neural computation with finely tuned neural elements (9). By contrast, some lizards and other animals have mechanically coupled "pressure difference" ears, such that simple subtraction of the vibration amplitude of the two ears could provide a reliable estimate of sound source localization, with minimal requirements on neural processing (10).

Physical contact provides a more extensive set of possible interactions than light or sound transmission, including compression and expansion, lateral sheer, twisting, and temperature changes. Complementing the diversity of processes is diversity in mechanical components, including skin, nails, and hairs. The detailed biomechanics of touch transducers determine the information they relay, making them ideal for investigation of neural consequences of embodiment. For example, human subjects iteratively refine tactile exploration of objects (see e.g., (11, 12)), from a whole hand "haptic glance" that quickly assesses global features, to squeezing that assesses hardness or lateral fingertip scanning that assesses surface texture, to engaging in highly targeted motions to find specific features (e.g., a pencil tip). These behavioral choices manipulate the object–sensor physical interaction according to informational demands. However, not all contexts reveal significant effects of active touch (13, 14), and its computational impact remains a subject of debate.

1.3. Embodied Active Sensing in the Vibrissa System

The rodent vibrissa (whisker) tactile system offers an ideal platform to connect embodied active sensing to processing in networks of neurons. Cortical "barrels" provide unique anatomical landmarks for structure–function relationships (15). The externalized and discrete vibrissae can be directly observed and manipulated during natural sensing. Rats demonstrate remarkable perceptual capabilities and rich active sensing behaviors with their whiskers (e.g., see (16–26)). Many of these behaviors, especially large amplitude rhythmic "whisking" of vibrissae along a nearly anterior–posterior direction, can be tracked in surrogate by EMG of facial muscles (27). There is increasing interest in and study of mice (28) to exploit available genetic techniques, such as neural control through cell type specific stimulation (e.g., by optogenetics (29); see also Chap. "Combining Optical Stimulation with Extracellular Electrophysiology in Behaving Mice" by Siegle) and selective physiological measurement of distinct neuron classes (e.g., by 2-photon imaging).

1.4. Vibrissa Motions and Coding Questions

In the first detailed quantification of whisker motions, Simons and Carvell (17, 30) devoted significant effort to whisker tracking during texture discrimination, though by design their resolution was limited to determining the "macroscopic" self-motion of whisking, and they did not resolve putatively information bearing "micromotions" resulting from interaction with the textures. From the macromotions and the known spacing of the gratings, they calculated the probable number of grating ridges contacted per second, and found it well in excess of what single cortical neurons could follow with plausible spike rates. One of the primary research questions thus became what features of the textures were mechanically conducted by the whiskers to the primary afferents in the face, serving as a code for texture. Since whiskers are not rigid, it cannot be assumed that the motions at the base are a simple measure of the profile of the texture, and in this sense, the inputs to the whisker system remained unknown. In the last decade, multiple labs have attacked this problem and extensions to other whisker based tactile tasks.

Vibrissae are keratinous hairs, similar to but stiffer and thicker than other mammalian body hair. Each of the large "macrovibrissae" originates in a specialized follicle containing afferent nerve endings (31, 32), and extends out up to ~50 mm. Macrovibrissae taper from diameters ~0.15 mm near the face down to ~0.01 mm at the tip. Most vibrissae exhibit some degree of curvature at rest, which can vary during whisking (33).

Vibrissae are elastic, that is, if deflected and let go they return to their original resting shape. Most phenomena discussed in relation to vibrissa sensing depend implicitly or explicitly on elasticity. A prominent example, particularly in the context of texture sensing, is stick–slip motion (25, 34–41), which results from interaction of friction at a point of contact with an elastic restoring force (42). During a stick, the contact point (whisker tip) stays fixed while the shaft builds up potential energy in bending; when the elastic force overcomes the frictional force, a ballistic slip of the tip occurs to some new stopping point, where the cycle can repeat.

Distinct from but related to stick–slip phenomena is resonant motion. In this case, elasticity combined with vibrissa mass induces a preferred frequency of motion, strongly influenced by length, and moderately sensitive to contact boundary conditions, diameter, and other factors. There are at least two ways resonance could alter vibrissa motions. Ringing, in which the shaft of the vibrissa oscillates around a resting position with decaying amplitude, is commonly observed during texture contact, often in response to an impulsive stick following a large slip (25, 34–36, 39). Rings are also seen when vibrissae "spring off" from contact into air (25, 35, 43). A more subtle effect, whose significance is immersed in current debate, is biasing of complicated motion patterns during contact (25, 36, 38, 39), for example, altering the timing of stick–slip motions even in the absence of purely periodic motion. A key uncertainty is the degree of whisker damping.

A second elastic phenomenon to receive substantial attention is bending and associated reaction forces, particularly in the context of localization tasks (18, 19, 21, 26, 43–47). For example, a rigid lever contacting a fixed object would stop, resulting in high force but low angular deflection at the lever base. However, a flexible vibrissa under the same condition allows motion to continue as the vibrissa bends, resulting in a large angular deflection but low force at the time of contact. The magnitudes and rates of change of forces and deflections are interrelated and could be used to determine radial distance of contact (46). Bending also produces force pushing into the follicle tangent to the vibrissa axis (47), an effect of current interest but unsettled importance.

What aspects of micromotions and forces are coded by neurons and might underlie perception? There is broad agreement that coarse texture discriminations (rough vs. smooth) can be performed with only average quantities (average micromotion size or velocity, or number of slips per second) and that neurons are highly velocity sensitive (25, 37–41). Rats can detect single deflections (22), but pattern recognition capabilities are less clear. Some evidence suggests rats cannot distinguish distinct deflection trains of equal energy (48), but to date no study directly links textural micromotions to psychophysical performance.

It is important to account here for sensing behaviors. For example, Carvell and Simons found whisking behaviors varied with task, and measurements made in one condition may not be universal to all conditions. Rodents may regulate whisker touches to suppress most bending (21), so theories built on differential bending across whiskers during fixed whisking patterns may not capture the true function in behaving animals. An interesting question is how animals perform the regulation. Part of the "sensation" in such a closed loop system could come from knowing which motor program (e.g., whisk amplitude) to perform to achieve a target sensory feedback (e.g., minimally superthreshold contact). Unlike in efference copy models, self-motion is not "subtracted out" from the sensory signal, but is itself the representation of object information. This hypothesis remains an open possibility.

Any mechanical property that alters patterns of micromotions or reaction forces is potentially significant for coding questions, including how properties vary across the whisker pad. The pad is highly stereotyped between animals, with a gradient of length from long posterior to shorter anterior whiskers (44). From the animal's perspective, it is likely the vibrissa array is a unified organ as opposed to a collection of independent hairs, and large scale structure could play a role analogous to parts of the hand in primate touch. Mechanical differences across whiskers should be expected to affect stick–slip patterning. For example, longer whiskers have lower resonant frequencies, and may bend more during sticks, leading to larger slips. Some indirect evidence in behaving animals supports this conjecture (e.g., Fig. 11 in (39)). However, larger tip motion may not translate into larger or faster angular motion at the follicle. The paucity of physiology data following stimuli in "follicle-centered" coordinates makes this question difficult to address.

There is a general view that rats pool information across multiple whiskers, such that sensory performance degrades smoothly with number of removed whiskers (17). In some cases rats might make up for lost whiskers with greater numbers of contacts (49), that is, compensate a change in embodiment with a change in active sensing. In one report whisker removal resulted in an increase of performance (19). It should be noted, however, that vibrissae are constantly growing and experiencing wear, providing a fluctuating embodiment even if the distribution of mechanical properties is fixed on average.

1.5. Videographic Whisker Tracking

The previous sections outline key questions relying on further understanding of whisker motions. By analogy to eye tracking, an essential tool for studying active vision, it has been necessary to develop robust high resolution whisker monitoring. Below I review methods for behavioral videography, with particular attention to design choices that benefit capture of whisker motions. I briefly touch on analysis possibilities, but development of robust automated trackers is ongoing and beyond the scope of this chapter.

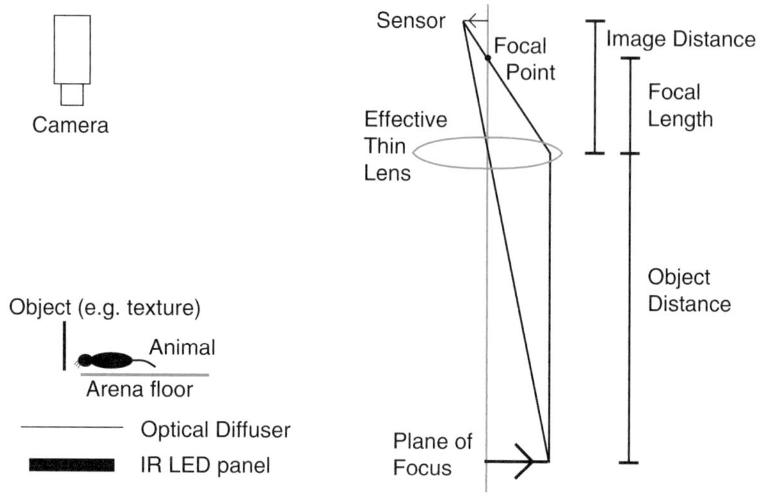

Fig. 1. *Left* Schematic of the behavioral arena and placement of camera and illumination. *Right* Geometry underlying thin lens approximation. Light rays entering parallel to the lens axis are refracted through the focal point, while rays entering towards the lens center leave in the same direction. The thin lens equations follow from these assumptions and rules for similar triangles. The thin lens is "effective" as an idealized approximation to the real compound lens (see Note 1).

2. Materials

- Behavioral arena. The arena will vary widely with experimental goals. This chapter considers a simple clear polycarbonate box (15 cm W × 20 cm L × 30 cm H), with a behavioral region of interest (bROI) containing a vertically oriented wall or a textured surface (e.g., sandpaper) for tactile exploration. The bROI alternatively could be adjacent to a raised platform (see Fig. 1).
- High-speed digital camera. The cameras considered here have megapixel resolution at rates greater than 500 frames per second (FPS); an adjustable acquisition region of interest on the imaging chip (cROI) with increasing maximal FPS for smaller cROIs; sensitivity into the infrared (>800 nm wavelengths); around 8 bit pixel depth (not necessarily high sensitivity or depth); shutter speed separately adjustable from frame rate; on-camera RAM (e.g., 4 GB) to store images prior to permanent storage; and hardware (TTL) interfaces for timing signals (an input for starting acquisition and an output indicating a frame was acquired). Monochrome cameras are usually adequate. An example meeting these criteria is the Cooke Corporation pco.1200hs CMOS camera; below we suggest other factors to guide camera selection.

- Computer, frame grabber and software for data acquisition. A PCIe-1429 (National Instruments) frame grabber connects to the pco.1200hs through a CameraLink interface, using a PCI express ×4 slot on the computer. For simplicity, I assume use of vendor supplied acquisition software (CamWare for the pco.1200hs), although custom acquisition applications could be created, e.g., in environments such as LabView or Matlab.
- Lens. A manual focus, 50-mm prime lens is likely adequate, but should be "fast," meaning it has a low minimum f-stop (e.g., 50 mm f/1.2 Nikkor #1435, Nikon Inc; a lens with minimum f/1.4 may be easier to find). A mounting adapter may be necessary (e.g., Nikon F-mount to C-mount, VA304 Bower/ General Brand).
- Infrared illumination. The wavelength should be outside the visual range of the animal (>650 nm for rodents) to eliminate visual cues and minimize subject discomfort, but within the range of the camera. High-power IR (~880 nm) LED panels are low cost and widely available. A choice compatible with the arena and bROI is a flat LED panel (IR-14, Super Circuits, with its internal light-activated switch disabled by covering the sensor).
- Optical diffuser. Glass diffusers are a common choice (e.g., DG100X100-600, ThorLabs, Inc). A simple alternative is a sheet of translucent Mylar A (DuPont, available from most art supply stores), slightly larger than the ROI, tightly held to a rigid frame.
- Mounting hardware for arena, camera and illuminators. Modular systems such as 8020 hardware (8020, Inc) provide convenient configurability and on-the-fly coarse adjustments, while use of a geared camera head (410 Geared Head, Manfrotto & Co) allows fine adjustment of the imaged area. Any hardware is adequate as long as all components are held rigidly in place, with sufficient vibration isolation.
- Test objects. I use a set of whiskers collected from earlier experiments, attached to a small stand, to use when adjusting camera parameters. Any object with fine features (e.g., a small screw) could also work.
- Calibration object. This should be a small rigid object of known spatial dimensions, used to relate distance in the image to true distances. For planar imaging, a plastic translucent ~10 cm ruler with millimeter markings is often sufficient.

3. Methods

I consider the videography process in three phases: design, acquisition and analysis. The greatest effort is devoted to design, after which acquisition should be simple. I briefly touch on some analysis possibilities. Manual analysis is straightforward, but labor intensive and slow, while the extensive literature on automated tracking provides many special-purpose but no universal solutions. However, careful consideration of the analysis phase during design can dramatically simplify the requirements for an automated tracker.

3.1. Design and Setup

High-speed videography inherently involves tradeoffs that are specific to each application. For concreteness, I consider tracking of whiskers of a freely moving animal, within a single horizontal plane of observation. Other applications require suitable modifications. The primary design specifications are the bROI, spatial resolution (pixel scale), temporal resolution, image contrast, depth of field, clip duration, image size, download time (from camera head to hard drive), and file size. Temporal resolution has two components: the frame rate (FPS), and the degree of motion blur within each image. In general most of these specifications compete with each other to determine data quality; I consider them in groups most closely related in practical terms. Although I describe the process sequentially, one typically pursues an iterative selection of tradeoffs to achieve satisfactory performance within the confines of available hardware. To implement the design, one chooses the lens, aperture setting, and placement of illumination and camera in physical space, and sets the camera FPS, shutter speed, and cROI in software.

3.1.1. Spatial Resolution

We must first select the spatial extent (bROI) and spatial resolution (PixelScale) we want to capture. For a fixed cROI, increasing the bROI decreases the spatial resolution, according to

$$\text{bROI length} = \text{cROI length} \times \text{Pixel Scale}.$$

Pixel Scale is the desired width of a region at the object location that projects onto a single pixel, e.g., in units of mm per pixel. Arrangement of the smallest acceptable bROI is usually desirable. For whisker tracking, the relevant scales are imposed by whisker diameters on the order of 0.1 mm and lengths around tens of millimeters. To capture the spread of multiple whiskers and/or allow for head motion may require a bROI slightly larger than the maximal whisker length. We thus need to cover an area of several centimeters with submillimeter resolution.

I take as example a bROI of 8 cm by 2 cm, and a PixelScale of 0.1 mm per pixel, for an image size (cROI) of 800 by 200 pixels. I orient the bROI with the long axis along a vertical wall, concentrating on whisker contacts at the expense of precontact whisks away from the surface (and out of the image frame). I plan to exploit that whiskers appear as continuous curves in the image, so appropriate smoothing along the length of the whiskers yields subpixel resolution during analysis. Actually resolving the true diameter of the whisker or some other smaller feature requires decreasing the PixelScale, leading to a reduction in the size of the bROI and/or FPS (since FPS usually decreases with increasing cROI, see below).

3.1.2. Placement and Optics

We next try to achieve the above spatial specifications via lens selection and camera placement. For most applications, a useful if not strictly true geometric guide is the thin lens approximation (50), for which

$$1/\text{FocalLength} = 1/\text{ImageDistance} + 1/\text{ObjectDistance}$$

$$\begin{aligned}\text{Magnification} &= \text{ImageSize}/\text{ObjectSize}\\ &= \text{ObjectDistance}/\text{ImageDistance}\end{aligned}$$

See Figure 1. The FocalLength is a fixed lens property. The ImageDistance is between the lens and the camera's sensor, and the ObjectDistance is from the lens to the plane at which an object is in focus (see Note 1 for additional optics information). As example, a 50-mm FocalLength lens in a camera with an ImageDistance of 60 mm will focus at an ObjectDistance of 300 mm. By varying the ImageDistance (effectively what happens when turning the focus wheel on the lens), common lens–camera combinations provide an ObjectDistance range from practically infinite to a minimum of several cm. In practice, ImageDistance is not used explicitly (nor with real lenses is it the actual distance from lens center to sensor, see Note 1), and instead one can use the equations in the rearranged form

$$\text{Magnification} = \text{PixelSize}/\text{PixelScale}$$

$$\text{ObjectDistance} = \text{FocalLength} \times (1/\text{Magnification} + 1)$$

PixelSize is the physical pixel size on the sensor, and the Magnification (in this context less than one) is determined by choice of PixelScale, which then determines the desired ObjectDistance. Continuing the example, the sensor in the pco.1200hs has square pixels of size 0.012 mm (listed on the specification sheet), requiring a Magnification of 0.12 to get the desired 0.1 mm per pixel scale. By the second equation, with a 50-mm lens the desired scaling results by placing the camera approximately 47 cm from the object. See additional lens guidance in Note 2.

To image the horizontal plane of whisking, the camera sits above the animal, facing downward, at this approximate distance (Figs. 1 and 2). The camera is held by a geared tripod head, rigidly

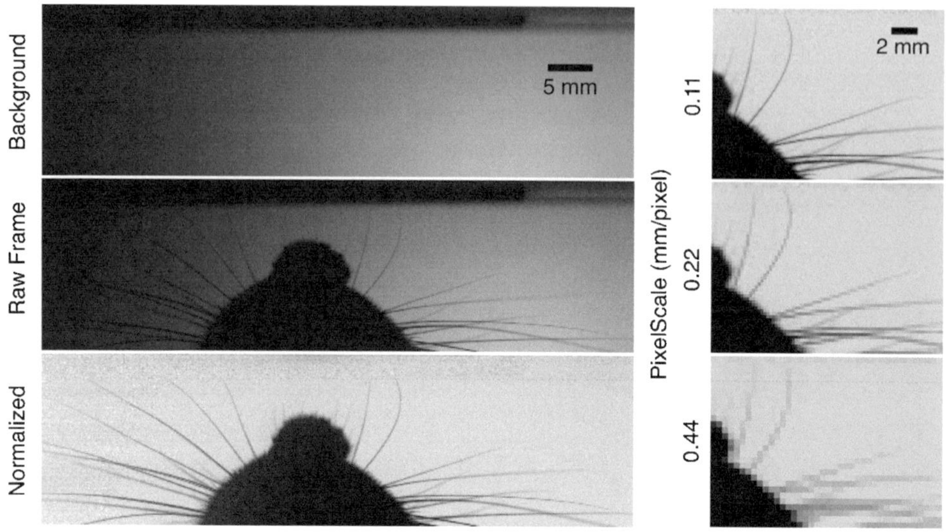

Fig. 2. *Left* Image preprocessing. Background frame consisting of an average of 100 frames before the animal is in view (*top*), single raw frame during exploration (*middle*), and same frame with each pixel divided by the corresponding pixel in the background (*bottom*). Uneven illumination is reduced, at the expense of some additional noise where the background is darker. *Right* Comparison of three PixelScales (simulated by averaging neighboring pixels). By 0.44 mm/pixel, isolated anterior whiskers could still be tracked, but posterior whiskers become blurred together.

attached to the arena frame, allowing small adjustments to image alignment just prior to acquisition. See also Notes 3–5.

3.1.3. Temporal Resolution: Frame Rate

Having determined the spatial arrangement within a single image, we now turn to parameters for the sequence of images that comprises a video clip. Our choice of frame rate FPS should balance resolving faster motions against keeping the data rate and data size manageable. We will choose FPS to bound the largest jump in pixels between two frames of the fastest moving object we want to observe. A potential difficulty with larger jumps is correctly associating the same object (whisker) across frames.

Since 1/FPS is the time between frames and 1/PixelScale relates object to image distances, an object moving with true ObjectVelocity (in mm/s) will move

$$\text{PixelDistance} = (\text{ObjectVelocity}/\text{PixelScale})/\text{FPS}$$

between consecutive frames. Solving for FPS yields

$$\text{minFPS} = \text{maxObjectVelocity}/(\text{PixelScale} \times \text{maxPixelDistance})$$

Rats and mice whisk at angular speeds up to ~1,500 degree/s, leading to linear velocities of ~500 mm/s at points on the whisker ~2 cm away from the face; again, one might exploit that whiskers appear as continuous curves in the image, so temporal resolution of whisker tips can be sacrificed given favorable tracking of the slower moving whisker base. In our example with PixelScale = 0.1 mm per pixel, if we can tolerate jumps up to 2 pixels between frames for object velocities up to 100 mm/s, the suggested FPS = 2,000 Hz. Frame rates for tracking of whisking behaviors or "simple" contacts (e.g., whisker bending against a flat object) are often slower, at 1,000 Hz or 500 Hz, due to the downward influence of data management needs and use of algorithms that track the whole whisker at once (so larger jumps can be tolerated). However, frame rates may need to be higher (>3,000 Hz) to resolve fast transient micromotions such as occur during textured surface contact (25). See also Note 6.

3.1.4. Temporal Resolution: Data Rate and Size

Data management for HSV takes planning to handle the fast generation of large files. For the types of images considered here, large lossless compression ratios are usually achievable offline. However, online data compression is not as common as in consumer video, probably due to the desire to maintain high image quality (i.e., no lossy compression) and minimize processing time. Without compression, calculation of data rates and file sizes is straightforward:

$$\text{DataRate} = \text{FrameSize} \times \text{FPS}$$

$$\begin{aligned}\text{DataSize} &= \text{FrameSize} \times \text{NumberOfFrames} \\ &= \text{DataRate} \times \text{AcquisitionDuration}\end{aligned}$$

DataSize is the total camera memory required to store a clip, or the size of the file on disk (but see Note 8). FrameSize is the total number of bytes required to store one frame, which is the number of pixels times the number of bytes ("depth") per pixel. In our example, the pco.1200hs is a 10-bit camera, requiring 2 bytes per pixel (6 unused bits). Thus a 200 × 800 cROI takes up FrameSize = 320 kB of storage. At 1,000 FPS, the DataRate is 320 MB/s, and a 4 GB on-board memory could store up to 12 s of video. At a higher 3,200 FPS, as might be used to study micromotions, the data rate is 1,024 MB/s, and the max AcquisitionDuration drops to 4 s. Once the camera memory is full, no further acquisition can occur until the data is off-loaded to permanent storage. A Camera-Link connection (or GigE or USB2.0) could in principle transmit the full 4 GB memory within a few seconds. However, a standard PC and hard drive is unlikely to save at true continuous data rates much higher than ~40 MB/s, requiring several minutes to save the data. Faster save times are possible with specialized hardware, e.g., use of RAID arrays, and frame grabber cards that bypass the PC's internal communication bus and send data directly to dedicated hard drives. Systems that stream data directly to disk in real time (without the use of onboard memory) are becoming more widely available and less expensive, but remain preferred for especially demanding applications. The design must include a hardware choice that meets acceptable data save times given the spatiotemporal acquisition requirements. See Notes 9 and 10 concerning additional data issues.

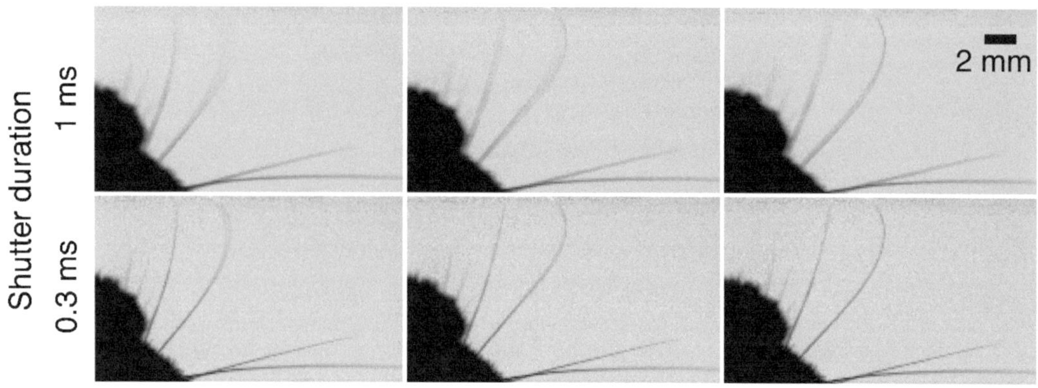

Fig. 3. Demonstration of motion blur. Both rows show three consecutive frames at ~1,000 FPS, during a whisker slip terminating in a stick. At the maximal 1 ms exposure time (*top*; simulated via averaging three frames at 3,202 FPS), the tip is lost in blur.

3.1.5. Image Intensity

Our remaining design task is to choose "intensity" parameters, for which the most prominent factors are aperture width, shutter speed, and form of illumination. The aperture is an adjustable opening in the lens, typically specified as the ratio of the focal length to the aperture diameter and written as f/FNum (e.g., at f/2 or FNum = 2 the aperture diameter is half the focal length). Lower FNums mean wider apertures, which let in more light, and tradition has fixed so-called "full stops" at multiples of the square root of two, FNum $= 2^{n/2}$ (1, 1.4, 2, 2.8, ...). Each higher full stop (usually indicated by a click in the lens when turning the aperture dial) thus halves the area of the aperture. A source of confusion in common usage is that sometimes a "stop" refers to the implied integer n, and sometimes to the actual FNum; moreover, many lenses now have settings at "third stops" or other values.

There are two common motivations for decreasing the aperture diameter: first, although uncommon in typically light-starved HSV settings, a smaller aperture can reduce pixel saturation in the presence of bright light or specular reflections. A more likely second motivation is that smaller apertures increase the range of object distances for which objects appear in acceptable focus, called the depth of field. For our 50-mm lens example, to use the widest aperture (f/1.2) at the calculated ObjectDistance requires that events of interest occur in a horizontal slice only ~1 cm thick; by f/8 most objects in the arena are in focus. See also Note 11.

The shutter speed controls how long pixels gather light in each frame, up to the maximum time of 1/FPS. Although again one typically tries to gather as much light as possible, one might increase the shutter speed (decrease exposure time) to reduce motion blur of high velocity objects (see Fig. 3). Since the sensor reports only the total light hitting it while the shutter is open, any significant motion appears spread out over multiple pixels. In whisker tracking a common issue is blur of the whisker tips. Similar to PixelDistance above, for an expected maximal velocity and acceptable amount of motion blur within each frame (expressed in pixels) one calculates

$$\text{Exposure Time} = (\text{Pixel Scale} \times \text{max Pixel Blur})/\text{max Object Velocity}$$

Exposure time and aperture width have an inverse relationship: decreasing a full stop (doubling the aperture area) and halving the ExposureTime collects the same amount of light in each frame, but in half the time. In practice, it is often easier to start with the ExposureTime at its maximum of 1/FPS and set the aperture for the desired depth of field, making later adjustments only as necessary. See also Notes 12 and 13.

3.1.6. Illumination and Backlighting

Given the aperture and shutter settings, one must provide adequate illumination to get sufficient contrast in each frame. HSV requires bright illumination to compensate for short exposure times. Moreover, freely moving subjects must be covered throughout the bROI, and for later data analysis it is best to have as uniform illumination as possible, so that location within the frame does not substantially alter appearance, and all pixels are within the camera's dynamic range. The major limits to adding more illumination are the geometry of the arena, and in some cases the need to avoid excessive heating (high-powered IR illuminators can get very hot even though not visible).

Whisker tracking does not require aesthetically rich images. A common class of generic trackers looks for "features" that remain visually invariant from frame to frame, such as a sharp corner on an object. However, whiskers exhibit few robust visual features along their length (see Note 14 about adding fiducial marks). One concentrates instead on creating as sharply defined a whisker curve as possible. Reflected light off the whiskers is generally poor (low contrast of object vs. background), and creates "complicated" objects and backgrounds with intricate, hard to parse textures that usually change as they move. These difficulties motivate use of backlighting, such that whiskers appear as shadows; this is the most common whisker HSV configuration (23, 25, 43, 51, 52). One might thus arrange the LED panel, the diffuser, and the floor of the arena as in Fig. 1. Flat panel illuminators and a diffuser create uniform background illumination. One focuses on the object, with any confounding features in the illuminator blurred by being outside the depth of field. See also Note 15.

3.2. Acquisition

3.2.1. Adjusting Optics and Calibrating Image

Acquisition proceeds by first making fine adjustments and calibrations just prior to an acquisition session, and then capturing and saving appropriate clips. It is useful to make a test object of plucked whiskers attached to stand. Test clips taken of this object in various locations throughout the bROI and at different depths assess the signal level, contrast, and effective spatial resolution. If these are judged unacceptable, the design is adjusted.

With the design in place, one acquires three clips just prior to the behavioral session. A "dark field clip," with the lens cap on (or all illumination off), allows later determination of sensor noise. It is common for sensor chips to have "hot pixels" or other biases in pixel values independent of illumination, which can introduce spurious shifts in tracked whisker positions.

In order to place true units on motion tracking, one needs a "calibration clip" (possibly a single frame) in which an object of known dimension is used to directly measure PixelScale. As a simple example, a clip of a translucent ruler allows one to determine the pixel distance between points a given distance in millimeters apart on the ruler. Dividing the pixel distance by the true distance gives the true PixelScale in millimeter per pixel. By varying the distance of the calibration object (e.g., on a micromanipulator) one can also assess depth of field, and bound the PixelScale error as the object moves out of the central plane of focus. See also Note 16.

Lastly, a "background clip" captures baseline intensity in the absence of the animal. Due to sensor noise, it is better to make an average background image from a clip of, say, 100 frames. No image movement or illumination change should occur after background clip acquisition (recall Note 5).

3.2.2. Acquiring and Saving Clips

Since many HSV applications acquire data at a rate faster than can be written to permanent storage, video acquisition is broken up in a sequence of short clips covering events of interest (e.g., a single exploration of a textured surface) stored in camera memory, interspersed with periods up to minutes where data is being written to hard drive and no further acquisition can take place. In cases where the trial duration is fixed, one simply acquires throughout the trial and then saves the clip to disk, either waiting to begin the next trial until saving is completed, or allowing for some number of trials without videography. In cases where the timing of events of interest is partially determined by the animal, as in free exploration, "ring buffering" is typically employed. In this mode, the camera continuously acquires to memory such that the currently acquired frame overwrites the oldest frame still in memory. Continuing our example, at a DataRate of 1,024 MB/s in a 4 GB camera head, the most recent 4 s of video remain available at any time, with all older frames overwritten. Once the event of interest has occurred, acquisition stops, and the relevant frames (possibly determined by online review of the clip) are saved to permanent storage. Tight prediction of the animal's behavior is not required, as the timing is determined after the fact. See also Notes 17–19.

3.3. Analysis

I touch only briefly on analysis possibilities. Raw frames are normalized by dividing each pixel by its background value (see Fig. 2), to suppress confounding background features, cancel out hot pixels, and detrend any remaining nonuniformities in background illumination. The preprocessed frame is then passed to either manual or automated analysis.

3.3.1. Manual Frame by Frame Analysis

In some applications, only minimal information is required. For example, it may be enough to note the frames in which whiskers contacted an object (23), or times of maximal whisker protraction. A primitive but functional GUI for manual tracking consists of having users click at relevant points, while stepping through frames in a loop. The output is a set of user identified whisker locations throughout the clip. To place whisker points in head-centered coordinates, some selection of facial points should also be chosen (e.g., tip of the nose and point between the eyes (20)). Such common operations are available in commercial video analysis software such as Proanalyst (Xcitex), or can be custom designed. If manual tracking is employed, it is best to randomly (and blindly) assign clips to a group of trained observers, with some degree of redundancy to assess observer variance.

3.3.2. Automated Processing

Automated tracking generally involves looking for intensity minima (assuming backlit whiskers). Best results are usually obtained by designs producing the simplest images. If interest is limited to locations of contact with fixed objects, a simple strategy is to look for minima along a predetermined line in the image. For example, one could pick a horizontal line just in front of the surface in Fig. 2. This approach is simple to implement (e.g., finding minima in the output of the improfile function in Matlab, see also Note 20), but does not estimate motions in head centered (angular) coordinates, or provide information about whisker shape.

Because whiskers become dark curves, it is possible to fit short line segments along their length, e.g., by minimizing a cost function that gives the average value of pixels under each segment. In general, edge (or better, ridge) detection is a thoroughly studied problem in computer vision. However, usually what one wants is a parametric representation of whisker locations as a function of distance from the face along the shaft. Thus in addition to finding whisker points, one must solve the problem of assembling these points into a metrically accurate curve within each image, and correctly associating the same whisker across frames. This is the major source of error with automated trackers, and an area of active work (25, 43, 51, 52, 55).

4. Notes

1. Real lenses are rarely thin, single glass elements, and the important distances are not from the physical lens center but from "principal planes" determined by the lens design; these planes may not even lie within the physical lens. However, at the scales in this application, the distinction introduces minimal error. In particular, when estimating the camera distance using FocalLength and Magnification, any difference between lens front and principal plane is usually compensated by small online adjustments. Also, there are inherent resolution limits in all lenses, e.g., due to diffraction. For the range of parameters (e.g., wavelength, aperture, and magnification) in this chapter, commonly available lenses should resolve below the pixel scale, and hence, I concentrate exclusively on sensor pixilation limits. Under some conditions (e.g., high magnification) the fundamental optical limits may become significant. Note HSV cameras tend to have larger pixels than commercial cameras.
2. Utilities to assist lens selection and use, known as *lens calculators*, are widely available online, e.g., from most lens manufacturers. Some versions include specifications for popular cameras such as sensor and pixel sizes, e.g., the free utility from Xcitex, http://www.xcitex.com/html/midas_lenscalc.php. Shorter focal length lenses (e.g., 12.5 mm) can cover a larger arena at small ObjectDistances, but with greater spatial distortions (spurious apparent curvature). Zoom lenses are usually unnecessary and may degrade image quality or stability. A too small diameter lens can produce "vignetting," in which the corners of the sensor lie outside the imaged area and are wasted. It may help to rent a range of lenses for trials.
3. Using a mirror in the optical path allows more flexible camera positioning, for example when imaging from below. The mirror should be 45 deg to the camera axis. If required, reflect the image in software to undo the orientation reversal. Cheaper mirrors may introduce "ghosts" in the image due to extra reflections off the front glass surface.
4. If a mirror is placed near the object, covering half of the image, and oriented at 45 deg, two views of the bROI are captured by a single camera. The two views can be used for 3D tracking (53). The direct and reflected optical paths must be nearly the same length for both to be in focus. See, e.g., (21, 26).
5. Drifts in intensity (e.g., due to temperature or power fluctuations in the illuminators or camera), focus drift due to lens movement, or mechanical drifts can easily occur across sessions; it is useful to try a dry run with a test object, storing clips

throughout an interval at least as long as the normal behavioral session. With even small drifts, background normalization (see below) becomes problematic, and more sophisticated preprocessing is required.

6. Frequency domain estimates are not as straightforward in video as in 1D signal processing. However, the well-known Nyquist criterion for discretely sampled data suggests FPS must be also at least twice the highest expected frequency of motion, and a rule of thumb is to oversample by a factor of five for better signal to noise.
7. DataSize can differ between camera memory and stored file. For example, the pco.1200hs efficiently packs four 10 bit pixels into five bytes in internal memory, but storage in a standard file type unpacks each pixel to two bytes (with 6 out of 16 bits wasted). A 4 GB clip in memory thus becomes 7 GB when saved.
8. Some cameras use partitioned memory to work around the saving downtime. Each partition keeps a distinct clip, and only after all partitions are full are frames saved to disk.
9. Two common ways to store HSV clips are as a numbered series of image files each containing one frame (e.g., clip1_0001.tif, clip1_0002.tif, ...) or as a single movie file (e.g., clip1.avi). Using standard file types makes later viewing easier, but both large (>4 GB) single files and directories with >10,000 files can be inconvenient to manage. It is less common, if more efficient, to break into groups of fixed-sized raw pixel files. Preference should be given to formats best suited to the analysis application.
10. DataSize is linear in FPS but quadratic in image dimensions, so where possible, reduction of cROI is more effective in reducing data requirements.
11. Depth of field also depends on focal length, ObjectDistance, PixelSize, and other factors. It is usually sufficient to determine empirically and make necessary adjustments, but some lens calculators can provide advance estimates.
12. Blur can be useful. For example, if only the amplitude of a high frequency periodic motion is needed, one simply measures the width of the blur in a single long exposure image.
13. Some cameras allow online remapping of pixel intensity values, including control of "gamma." For example, logarithmic scaling of pixel values compresses high intensities while expanding low intensities, and can help when tracking a dark object in an image with bright but experimentally meaningless hot spots. Software rescaling can put the data in a more convenient form, and decrease file sizes if dropping bits, but will not improve signal to noise.

14. One can attach a small piece of reflective material to a whisker and easily track the resulting bright spot in each image. While highly useful in some settings (e.g., (54)), the material likely perturbs the whisker, and is limited in application to a small number of points on a small number of whiskers. Fluorescent dyes are another possibility, but it can be difficult to get enough emitted light, especially in the IR, and the dyes need to be reapplied regularly.
15. Knutsen et al. (19) used backlighting for whisker tracking, but added two overhead IR LEDs to create specular reflections in the eyes, used to track head position. Often head position is determined by shape of the face.
16. This simple calibration suffices for many applications. More sophisticated calibration methods are possible, and become essential when combining multiple views for 3D tracking. See for example (53), and the Matlab toolbox developed by Jean-Yves Bouguet, at http://www.vision.caltech.edu/bouguetj/calib_doc/.
17. A ring-buffering approach can introduce selection bias into the data, prompting either fully automated acquisition, or clear, explicit rules for frame selection.
18. To synchronize HSV to other acquisition systems, e.g., for physiological recordings, reliance on multiple hardware clocks is usually too crude. Most cameras can output shutter triggers, which can be sent to a spare recording channel for later alignment; if not all frames are saved to disk, careful record keeping is required to know which triggers correspond to frames in the saved clip. Another option is to introduce "synch objects," such as a flashing IR LED with recorded driving signal; it is more accurate to interpolate between two flashes at the beginning and ending of a clip than to extrapolate forward from a single flash at the beginning. Some cameras allow acquisition of each frame to be externally driven (e.g., from an output of the physiology system), although the maximal FPS in this mode is usually below the maximal "free running" FPS.
19. LED illuminators can introduce significant noise into high gain electrophysiology recordings, often due to cheap "wall wart" power supplies; replacing them with a higher quality (e.g., benchtop) power supply may help. As always, careful grounding and shielding also may be required.
20. Objects around PixelScale size are observable as a darkening of pixels proportional to the fraction of occlusion they provide, sometimes called partial voluming. It is usually possible to localize with subpixel resolution by finding the minimum of a smooth (quadratic) function fit to intensities in a small neighborhood, rather than simply select the pixel with minimum

value. Paradoxically, slight defocusing can help by spreading object information over multiple pixels. Our eyes perform this function naturally; the point is to use smooth variation in pixel intensities to compensate for the coarse discretization of space.

References

1. Reinagel P, Zador AM (1999) Natural scene statistics at the centre of gaze. Network 10:341–350
2. Rucci M, Iovin R, Poletti M, Santini F (2007) Miniature eye movements enhance fine spatial detail. Nature 447:851–854
3. Olveczky BP, Baccus SA, Meister M (2007) Retinal adaptation to object motion. Neuron 56:689–700
4. Schumann F, Einhauser-Treyer W, Vockeroth J, Bartl K, Schneider E, Konig P (2008) Salient features in gaze-aligned recordings of human visual input during free exploration of natural environments. J Vis 8(12):11–17
5. Berg DJ, Boehnke SE, Marino RA, Munoz DP, Itti L (2009) Free viewing of dynamic stimuli by humans and monkeys. J Vis 9(19):11–15
6. Tucker VA (2000) The deep fovea, sideways vision and spiral flight paths in raptors. J Exp Biol 203:3745–3754
7. Zeil J, Hemmi JM (2006) The visual ecology of fiddler crabs. J Comp Physiol A Neuroethol Sens Neural Behav Physiol 192:1–25
8. Knutsen PM, Ahissar E (2009) Orthogonal coding of object location. Trends Neurosci 32:101–109
9. Hyson RL (2005) The analysis of interaural time differences in the chick brain stem. Physiol Behav 86:297–305
10. Christensen-Dalsgaard J, Manley GA (2005) Directionality of the lizard ear. J Exp Biol 208:1209–1217
11. Lederman SJ, Klatzky RL (1990) Haptic classification of common objects: knowledge-driven exploration. Cogn Psychol 22:421–459
12. Klatzky RL, Pellegrino J, McCloskey BP, Lederman SJ (1993) Cognitive representations of functional interactions with objects. Mem Cognit 21:294–303
13. Connor CE, Hsiao SS, Phillips JR, Johnson KO (1990) Tactile roughness: neural codes that account for psychophysical magnitude estimates. J Neurosci 10:3823–3836
14. DiCarlo JJ, Johnson KO (1999) Velocity invariance of receptive field structure in somatosensory cortical area 3b of the alert monkey. J Neurosci 19:401–419
15. Petersen CC (2007) The functional organization of the barrel cortex. Neuron 56:339–355
16. Guic-Robles E, Valdivieso C, Guajardo G (1989) Rats can learn a roughness discrimination using only their vibrissal system. Behav Brain Res 31:285–289
17. Carvell GE, Simons DJ (1995) Task- and subject-related differences in sensorimotor behavior during active touch. Somatosens Mot Res 12:1–9
18. Krupa DJ, Matell MS, Brisben AJ, Oliveira LM, Nicolelis MA (2001) Behavioral properties of the trigeminal somatosensory system in rats performing whisker-dependent tactile discriminations. J Neurosci 21:5752–5763
19. Knutsen PM, Pietr M, Ahissar E (2006) Haptic object localization in the vibrissal system: behavior and performance. J Neurosci 26:8451–8464
20. Towal RB, Hartmann MJ (2006) Right-left asymmetries in the whisking behavior of rats anticipate head movements. J Neurosci 26:8838–8846
21. Mitchinson B, Martin CJ, Grant RA, Prescott TJ (2007) Feedback control in active sensing: rat exploratory whisking is modulated by environmental contact. Proc Biol Sci 274:1035–1041
22. Stuttgen MC, Ruter J, Schwarz C (2006) Two psychophysical channels of whisker deflection in rats align with two neuronal classes of primary afferents. J Neurosci 26:7933–7941
23. von Heimendahl M, Itskov PM, Arabzadeh E, Diamond ME (2007) Neuronal activity in rat barrel cortex underlying texture discrimination. PLoS Biol 5:e305
24. Towal RB, Hartmann MJ (2008) Variability in velocity profiles during free-air whisking behavior of unrestrained rats. J Neurophysiol 100:740–752
25. Ritt JT, Andermann ML, Moore CI (2008) Embodied information processing: vibrissa mechanics and texture features shape micromotions in actively sensing rats. Neuron 57:599–613
26. Grant RA, Mitchinson B, Fox CW, Prescott TJ (2009) Active touch sensing in the rat: anticipatory and regulatory control of whisker

movements during surface exploration. J Neurophysiol 101:862–874

27. Hill DN, Bermejo R, Zeigler HP, Kleinfeld D (2008) Biomechanics of the vibrissa motor plant in rat: rhythmic whisking consists of triphasic neuromuscular activity. J Neurosci 28:3438–3455
28. O'Connor DH, Huber D, Svoboda K (2009) Reverse engineering the mouse brain. Nature 461:923–929
29. Cardin JA, Carlen M, Meletis K, Knoblich U, Zhang F, Deisseroth K, Tsai LH, Moore CI (2009) Driving fast-spiking cells induces gamma rhythm and controls sensory responses. Nature 459:663–667
30. Carvell GE, Simons DJ (1990) Biometric analyses of vibrissal tactile discrimination in the rat. J Neurosci 10:2638–2648
31. Ebara S, Kumamoto K, Matsuura T, Mazurkiewicz JE, Rice FL (2002) Similarities and differences in the innervation of mystacial vibrissal follicle-sinus complexes in the rat and cat: a confocal microscopic study. J Comp Neurol 449:103–119
32. Dorfl J (1982) The musculature of the mystacial vibrissae of the white mouse. J Anat 135:147–154
33. Knutsen PM, Biess A, Ahissar E (2008) Vibrissal kinematics in 3D: tight coupling of azimuth, elevation, and torsion across different whisking modes. Neuron 59:35–42
34. Neimark MA, Andermann ML, Hopfield JJ, Moore CI (2003) Vibrissa resonance as a transduction mechanism for tactile encoding. J Neurosci 23:6499–6509
35. Hartmann MJ, Johnson NJ, Towal RB, Assad C (2003) Mechanical characteristics of rat vibrissae: resonant frequencies and damping in isolated whiskers and in the awake behaving animal. J Neurosci 23:6510–6519
36. Mehta SB, Kleinfeld D (2004) Frisking the whiskers: patterned sensory input in the rat vibrissa system. Neuron 41:181–184
37. Arabzadeh E, Zorzin E, Diamond ME (2005) Neuronal encoding of texture in the whisker sensory pathway. PLoS Biol 3:e17
38. Hipp J, Arabzadeh E, Zorzin E, Conradt J, Kayser C, Diamond ME, Konig P (2006) Texture signals in whisker vibrations. J Neurophysiol 95:1792–1799
39. Wolfe J, Hill DN, Pahlavan S, Drew PJ, Kleinfeld D, Feldman DE (2008) Texture coding in the rat whisker system: slip–stick versus differential resonance. PLoS Biol 6:e215
40. Jadhav SP, Wolfe J, Feldman DE (2009) Sparse temporal coding of elementary tactile features during active whisker sensation. Nat Neurosci 12:792–800
41. Lottem E, Azouz R (2008) Dynamic translation of surface coarseness into whisker vibrations. J Neurophysiol 100:2852–2865
42. Denny M (2004) Stick–slip motion: an important example of self-excited oscillation. Eur J Phys 25:311–322
43. O'Connor DH, Clack NG, Huber D, Komiyama T, Myers EW, Svoboda K (2010) Vibrissa-based object localization in head-fixed mice. J Neurosci 30:1947–1967
44. Brecht M, Preilowski B, Merzenich MM (1997) Functional architecture of the mystacial vibrissae. Behav Brain Res 84:81–97
45. Szwed M, Bagdasarian K, Ahissar E (2003) Encoding of vibrissal active touch. Neuron 40:621–630
46. Solomon JH, Hartmann MJ (2006) Biomechanics: robotic whiskers used to sense features. Nature 443:525
47. Stuttgen MC, Kullmann S, Schwarz C (2008) Responses of rat trigeminal ganglion neurons to longitudinal whisker stimulation. J Neurophysiol 100:1879–1884
48. Gerdjikov TV, Bergner CG, Stuttgen MC, Waiblinger C, Schwarz C (2010) Discrimination of vibrotactile stimuli in the rat whisker system: behavior and neurometrics. Neuron 65:530–540
49. Celikel T, Sakmann B (2007) Sensory integration across space and in time for decision making in the somatosensory system of rodents. Proc Natl Acad Sci USA 104:1395–1400
50. Hecht E (2001) Optics. Addison Wesley
51. Knutsen PM, Derdikman D, Ahissar E (2005) Tracking whisker and head movements in unrestrained behaving rodents. J Neurophysiol 93:2294–2301
52. Voigts J, Sakmann B, Celikel T (2008) Unsupervised whisker tracking in unrestrained behaving animals. J Neurophysiol 100: 504–515
53. Hartley R, Zisserman A (2003) Multiple view geometry in computer vision, 2nd edn. Cambridge University Press
54. Brecht M, Schneider M, Sakmann B, Margrie TW (2004) Whisker movements evoked by stimulation of single pyramidal cells in rat motor cortex. Nature 427:704–710
55. Perkon I, Kosir A, Itskov PM, Tasic J, Diamond ME (2011) Unsupervised quantification of whisking and head movement in freely moving rodents. J Neurophysiol 105:1950–1962

Part IV

Optogenetics

Neuromethods (2012) 67: 305–338
DOI 10.1007/7657_2011_10

Published online: 13 December 2011

Light-Activated Ion Pumps and Channels for Temporally Precise Optical Control of Activity in Genetically Targeted Neurons[1]

Brian Y. Chow, Xue Han, Jacob G. Bernstein, Patrick E. Monahan, and Edward S. Boyden

Abstract

The ability to turn on and off specific cell types and neural pathways in the brain, in a temporally precise fashion, has begun to enable the ability to test the sufficiency and necessity of particular neural activity patterns, and particular neural circuits, in the generation of normal and abnormal neural computations and behaviors by the brain. Over the last 5 years, a number of naturally occurring light-activated ion pumps and light-activated ion channels have been shown, upon genetic expression in specific neuron classes, to enable the voltage (and internal ionic composition) of those neurons to be controlled by light in a temporally precise fashion, without the need for chemical cofactors. In this chapter, we review three major classes of such genetically encoded "optogenetic" microbial opsins—light-gated ion channels such as channelrhodopsins, light-driven chloride pumps such as halorhodopsins, and light-driven proton pumps such as archaerhodopsins—that are in widespread use for mediating optical activation and silencing of neurons in species from *Caenorhabditis elegans* to nonhuman primates. We discuss the properties of these molecules—including their membrane expression, conductances, photocycle properties, ion selectivity, and action spectra—as well as genetic strategies for delivering these genes to neurons in different species, and hardware for performing light delivery in a diversity of settings. In the future, these molecules not only will continue to enable cutting-edge science but may also support a new generation of optical prosthetics for treating brain disorders.

Key words: Channelrhodopsin, Optogenetics, Photosensitive proteins, Retinal, Halorhodopsin, Archaerhodopsin, Light-sensitive cation channel, Light-sensitive chloride pump, Light-sensitive proton pump, Photocontrol of behavior

1. Introduction

The ability to turn on and off specific cell types and neural pathways in the brain, in a temporally precise fashion, has begun to enable the ability to test the sufficiency and necessity of particular neural activity patterns, and particular neural circuits, in the generation of normal

[1] This chapter is an updated version of reference (141).

and abnormal neural computations and behaviors by the brain. Most electrophysiological and imaging experiments in neuroscience are correlative—comparing a neural signal observed in the brain to a behavior or pathology. In contrast, the power to manipulate specific cells and circuits is opening up the ability to understand their causal roles in brain functions. Over the last 5 years, a number of naturally occurring light-activated ion pumps and light-activated ion channels have been shown, upon genetic expression in specific neuron classes, to enable the voltage (and internal ionic composition) of those neurons to be controlled by light in a temporally precise fashion. These molecules are microbial (type I) opsins, seven-transmembrane proteins naturally found in archaea, algae, fungi, and other species, and which possess light-driven electrogenic activity or contain light-gated ion pores. These molecules, when heterologously expressed in neurons or other cells, translocate ions across cell membranes in response to pulses of light of irradiances that are easily achievable with common laboratory microscopes, LEDs, and lasers. These molecules have begun to find widespread use in neuroscience due to their ease of handling and use (each is a single gene, under 1 kb long, encoding for a monolithic protein), their lack of need for chemical supplementation in many species (they utilize the naturally occurring chromophore all-*trans* retinal, which appears to occur at sufficient quantities in the mammalian nervous system), and their high speed of operation (they can respond within tens of microseconds to milliseconds, upon delivery of light, and shut off rapidly upon cessation of light, as needed for neuroscience experiments).

Three major classes of such "optogenetic" microbial opsins have been described to date. The first class, channelrhodopsins, is exemplified by the light-gated inwardly rectifying nonspecific cation channel channelrhodopsin-2 (ChR2) from the green algae *Chlamydomonas reinhardtii* (1), which, when expressed in neurons, can mediate sizeable currents up to 1,000 pA in response to millisecond-timescale pulses of blue light (2–5), thus enabling reliable spike trains to be elicited in ChR2-expressing neurons by blue light pulse trains (Fig. 1b). Several additional channelrhodopsins useful to biologists and bioengineers have been discovered or engineered, with faster or slower kinetics, red-shifted activation, and cell-region specific targeting, and explored in detail below (6–10). The channelrhodopsins have been used to activate neurons in neural circuits in animals from worms to monkeys, and have proven powerful and easy to use. The second class of microbial opsins utilized for biological control to date, halorhodopsins, is exemplified by the light-driven inwardly directed chloride pump halorhodopsin, from the archaeal species *Natronomas pharaonis* (Halo/NpHR/pHR; (11)), which, when expressed in neurons, can mediate modest inhibitory currents on the order of 40–100 pA in response to yellow light illumination (12, 13), enabling moderate silencing of neural activity (Fig. 1c). Halorhodopsins have some intrinsic kinetic limitations,

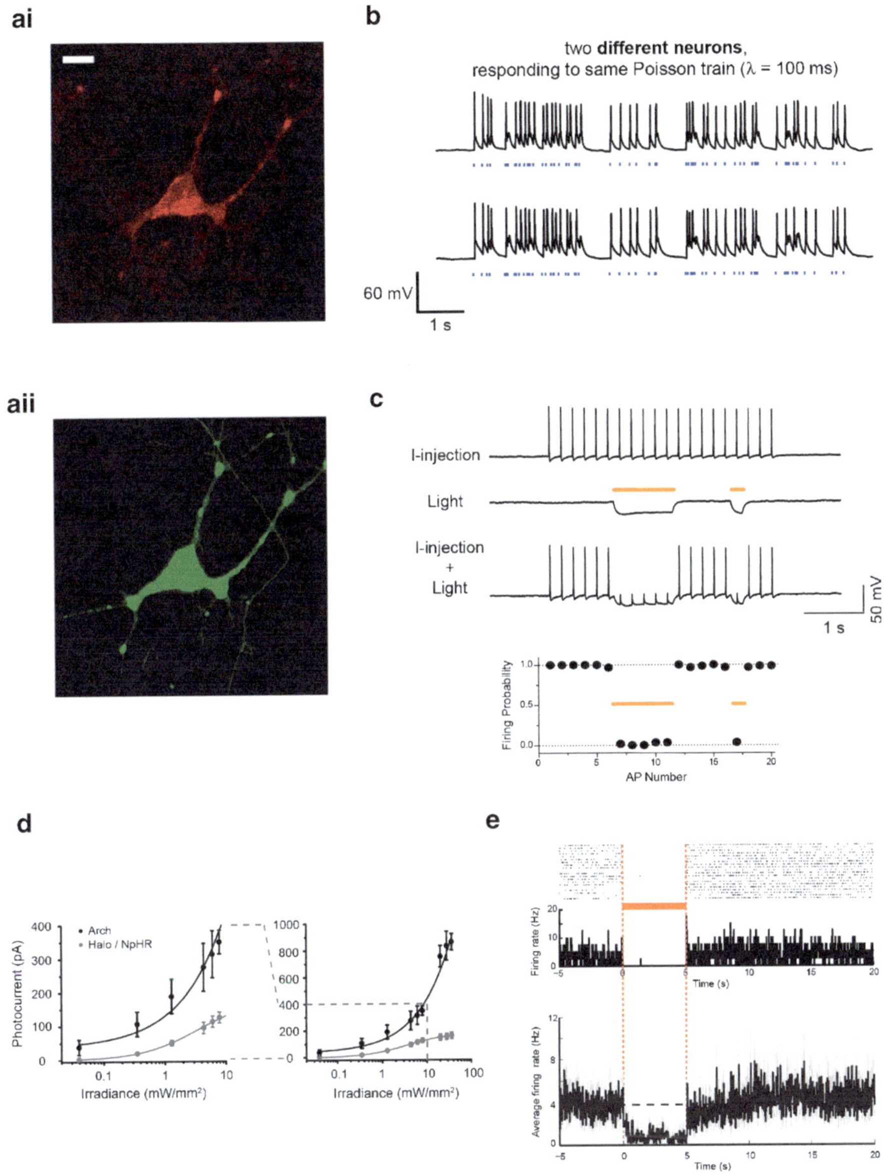

Fig. 1. Three classes of microbial opsins that enable optical neural activation and silencing tools. (**a**) Neuron expressing hChr2-mCherry (**ai**; bar, 20 μm) and Halo-GFP (**aii**). (**b**) Poisson trains of spikes elicited by pulses of blue light (*blue dashes*), in two different neurons. (**c**) Light-driven spike blockade, demonstrated (*Top*) for a representative hippocampal neuron, and (*Bottom*) for a population of neurons ($N = 7$). *I-injection*, neuronal firing induced by pulsed somatic current injection (300 pA, 4 ms). *Light*, hyperpolarization induced by periods of yellow light (*yellow dashes*). *I-injection + Light*, yellow light drives Halo to block neuron spiking, leaving spikes elicited during periods of darkness intact. *Panels* **a–c** adapted from references Boyden et al. (3) and Han and Boyden, (12). (**d**) Photocurrents of Arch vs. Halo measured as a function of 575 ± 25 nm light irradiance (or effective light irradiance), in patch-clamped cultured neurons ($N = 4$–16 neurons for each point), for low (i) and high (ii) light powers. The line is a single Hill fit to the data. (**e**) *Top*, Neural activity in a representative neuron before, during, and after 5 s of yellow light illumination, shown as a spike raster plot (*top*), and as a histogram of instantaneous firing rate averaged across trials (*bottom*; bin size, 20 ms). *Bottom*, population average of instantaneous firing rate before, during and after yellow light illumination (*black line*, mean; *gray lines*, mean ± SE; $n = 13$ units). *Panels* **d–e** adapted from reference Chow et al. (15).

with some photocycles taking tens of minutes to complete (e.g., Fig. 4a, b; (12, 14)), and halorhodopsins also require improved trafficking for expression at high levels (15–17). A third class of microbial opsin, the bacteriorhodopsins, is exemplified by the light-driven outward proton pump archaerhodopsin-3 (Arch/aR-3), from the archaeal species *Halorubrum sodomense*. Arch can mediate strong inhibitory currents of up to 900 pA in vitro (Fig. 1d), and in vivo is capable of mediating near-100% silencing of neurons in the awake behaving mouse or monkey brain in response to yellow-green light (Fig. 1e, (15, 18)). Protons are extremely effective as a charge carrier for mediating neural silencing, and proton pumps have greatly improved kinetics with respect to halorhodopsins (Fig. 4c, d), as well as a fast photocycle, and efficient trafficking to membranes. Furthermore, outward proton pumps, perhaps surprisingly, do not alter pH to a greater extent than do other opsins (such as ChR2) or than does normal neural activity. The broad and ecologically diverse class of outward proton pumps, which includes blue-green light drivable outward proton pumps such as the *Leptosphaeria maculans* opsin Mac (Fig. 5), enables, alongside the yellow-red drivable Halo, multicolor silencing of independent populations of neurons (15). Also, because the neural silencers Halo and Arch are activated by yellow or yellow-green light, and the neural depolarizer ChR2 is driven by blue light, expression of both a silencer and a depolarizer in the same cell (either by using two viruses, or by using the 2A linker to combine two opsins into a single open reading frame (12, 19)) enables bidirectional control of neural activity in a set of cells, useful for testing necessity and sufficiency of a given set of neurons in the same animal, or disruption of neural synchrony and coordination through "informational lesioning" (19).

In the sections below we describe the properties of these three opsin classes, as well as genetic (e.g., viral, transgenic) and hardware (e.g., lasers, LEDs) infrastructures for using these opsins to parse out the function of neural circuits in a wide variety of animal nervous systems. A theme of this field is that extremely rapid progress and adoption of these technologies has been driven by technology development curves in other fields such as gene therapy and optical imaging. Our hope is to convey a snapshot of this rapidly moving field as of 2011, summarizing the first half-decade of its existence, to teach both neuroengineers hoping to innovate by inventing new tools, as well as neuroscientists hoping to utilize these tools to answer new scientific questions. We will first survey the general properties of these opsins (Sect. 2), then go into the channelrhodopsins (Sect. 3), followed by the neural silencing pumps (halorhodopsins and bacteriorhodopsins, Sect. 4), the molecular strategies for delivering these genes to cells for appropriate expression in vitro and in vivo (Sect. 5), and the hardware for illumination of these opsin-expressing neurons in vitro and in vivo (Sect. 6).

2. Properties of "Optogenetic" Microbial (Type I) Opsins

The three classes of molecule described to date are from organisms such as unicellular algae, fungi, and archaea, whose native environments and membrane lipid composition are very different from those of mammalian neurons. Thus, the performance of these molecular tools in neurons can be difficult to predict based solely upon their properties in other species, and must be assessed empirically for assurance of efficacy and safety. Nevertheless, there are several molecular properties that contribute to efficacious, temporally precise optical control of neurons, which can be explored in a unified and logical fashion:

- *Initial protein expression levels.* The efficiency of ribosomal translation of a molecule is largely affected by codon optimization. It is recommended that genes be used that are codon-optimized for the target species.
- *Membrane insertion properties, protein folding, and interactions with local environment.* Increased membrane localization will result in more functional molecules and thus increased photocurrents. This property may also be inversely associated with the potential property of *toxicity*, since poorly trafficked or folded molecules may aggregate in the cytosol and endoplasmic reticulum. On the other hand, if a molecule has adverse intrinsic side effects, enhanced trafficking may exaggerate them. Furthermore, any given channel or pump will best operate under defined conditions (e.g., chloride conductance, pH, lipid environment, etc.), which may not exist in a given target cell type.
- *Innate conductance and permeability.* Channels translocate more ions per photocycle than pumps, since they open up a pore in the membrane. On the other hand, pumps can move ions against concentration gradients, unlike channels. Each opsin furthermore passes a precise set of ions in a specific cellular context, and not others.
- *Photocycle kinetics.* Both light-driven channels and pumps are described by a photocycle, the list of states that a molecule goes through after light exposure, including ion-translocating or ion pore-forming steps. The faster the photocycle, the more temporally precise the molecular function might be, and for a pump, the more ions will be translocated. (For a channel, a faster photocycle may result in the channel entering the ion pore-forming state more often, but may also reduce the time spent in the open state.) If a molecule enters an inactive

photocycle state for an enduring period of time, it may be effectively nonfunctional.

- *Photosensitivity.* Molecules may require different amounts of light to begin moving through their photocycle, based on the chromophore absorption efficiency. Furthermore, from a end-user standpoint, effective photosensitivity will appear to be a function of the overall photocycle; for example, a pump that has a slow photocycle may appear to be light insensitive (because incident photons may have no effect on the molecule during the photocycle), whereas a channel that inactivates extremely slowly may appear to be light sensitive (because each photon will result in large charge transfers).
- *Action spectrum.* Different molecules are driven by different colors of light. Multiple cell types can be orthogonally addressed with different colors of light, if they express opsins whose action spectra minimally overlap.
- *Ion selectivity.* Unlike traditional electrodes, microbial opsins can generate ion specific currents, since they will pass specific ions such as chloride (Cl^-) or calcium (Ca^{+2}). This opens up novel kinds of experimental capability, such as the ability to test the sufficiency of a given ion, in a given location, for a given biological function.

We will, in the following sections, frame current knowledge about cell-type specific optical control of neurons, in the context of decades of research in structure–function relationships of microbial (type I, or archaeal) opsins. In many ways, these molecules are similar in tertiary structure to mammalian (type II) rhodopsins (20), the pigments that confer photosensitivity to the rods and cones of the human retina. Both types are composed of seven transmembrane (7-TM) α-helices, linked by six loop segments, and their photosensitivity is enabled by a retinal bound to a specific lysine residue near the C-terminus, forming a Schiff base that undergoes a *trans–cis* or *cis–trans* isomerization upon illumination, that then induces conformational changes in the protein. However, they are evolutionarily unrelated, and their differences have important implications for their use in perturbing neuronal activity. Mammalian rhodopsins (21) are very sensitive photon detectors, optimized for sensitivity rather than speed. They utilize 11-*cis* retinal as the primary chromophore, which isomerizes to all-*trans* retinal upon absorbing a photon. The resultant structural change activates an associated G-protein, transducin, which then initiates a cascade of secondary messengers. The all-*trans* retinal dissociates from the opsin, is converted back to its 11-*cis* form, and then reassociates with the apoprotein to reconstitute a functional molecule—a process that typically takes hundreds of milliseconds, too slow to enable fast control of neurons in the central nervous system.

On the other hand, a microbial opsin utilizes all-*trans* retinal as its chromophore, which isomerizes to 13-*cis* retinal upon absorbing a photon. The chromophore does not undergo a quasi-irreversible dissociation event, but rather thermally relaxes to its active all-*trans* form in the dark (although this process can be facilitated by light). The *trans–cis* isomerization sets off several coupled structural rearrangements within the molecule that accommodate the passive conduction or active pumping of ions (22–24). Ultimately, this means that at the expense of light sensitivity, archaeal opsins can deflect the membrane potential of a cell on the scale hundreds of microseconds to a few milliseconds. However, it should be noted that genetically targetable, optical neural silencing has also been demonstrated using mammalian G-protein coupled receptors, which can couple to potassium channels (4), and genetically targetable optical neural activation has been demonstrated using melanopsins and invertebrate-style rhodopsins, at the price of temporal precision (25, 26).

As an exemplar of a well-characterized microbial opsin reagent, with crystal structure and photocycle both well-described, Fig. 2 shows the crystal structure of the light-driven chloride-pump halorhodopsin (Fig. 2a; (27)), as well as schematized structural rearrangements that are hypothesized to occur as the molecule pumps a chloride ion across the membrane and into the cytoplasm (Fig. 2b; (24)). While it is convenient to consider the molecular tools discussed here as toggle switches for turning neurons on and off, it is critical for use of these opsins to realize that the translocation of ions by microbial opsins is not as simple as a two-state toggle switch. The structural rearrangements constitute an active advancement through a complex photocycle with various intermediate states beyond the initial phototransition (Fig. 2c). There exists a very rich literature on type I microbial opsins from an evolutionary and protein structure–function perspective. The canonical molecules include the proton-pumping bacteriorhodopsin (BR), the chloride-pumping halorhodopsin (HR/sHR/HsHR) from *Halobacterium salinarum (halobium)*, and the halorhodopsin from *N. pharaonis* (Halo/NpHR/pHR). These were some of the first membrane proteins crystallized, and a myriad of structure–function studies have been performed on these molecules (11, 14, 20, 22–24, 27–45). These studies are not reviewed here, but it is important to point out their existence because much of what we know with respect to the photocycle and structure of opsins comes from these studies and from sequence homology of novel opsins to these canonical molecules. As we show later, for example, a deep understanding of the literature has enabled some researchers to derive powerful new variants of channelrhodopsin (5, 6, 8, 10), even though no crystal structure exists for this molecule.

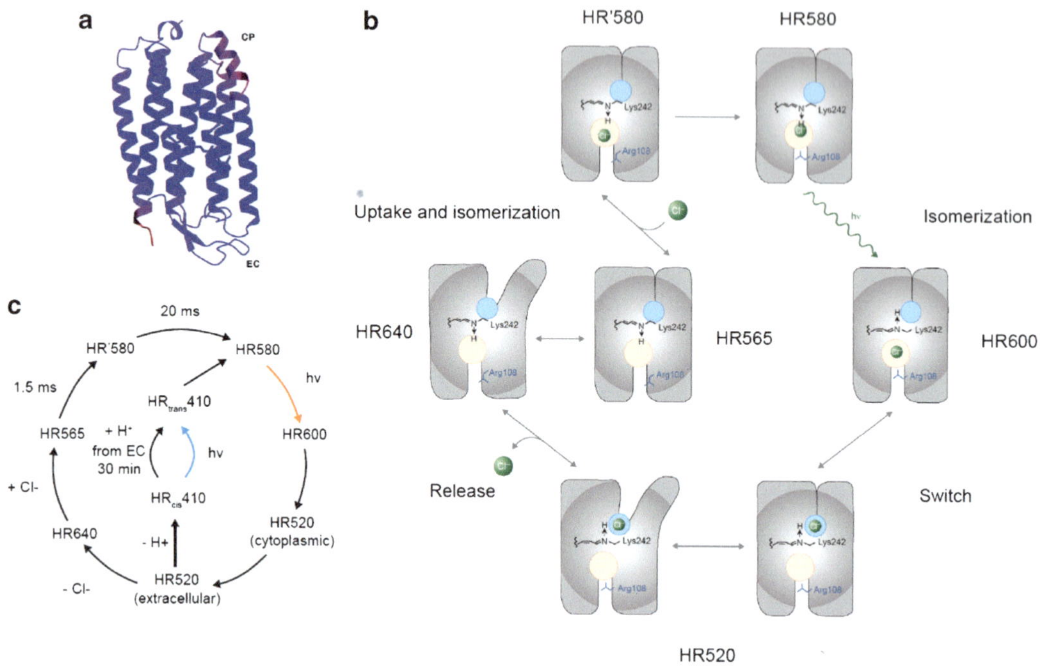

Fig. 2. Structure and function of halorhodopsin. (**a**) Crystal structure of halorhodopsin, which is composed of 7-transmembrane α-helices (7-TM) and a retinal chromophore that forms a Schiff base to a lysine near the C-terminus (from Kolbe et al. (27)). (**b**) Schematic of the halorhodopsin structural rearrangements and their relation to pumping activity at various points in the photocycle. Image modified from Essen (24). (**c**) The halorhodopsin photocycle at high-power continuous illumination on the timescale of a typical photocycle (i.e., conditions used for neural silencing, >few milliseconds). The HR410 intermediate is the origin of the long-lived inactivation in neural silencing (e.g., Bamberg et al. (14); Han and Boyden (12)).

3. Optical Neural Stimulation: Channel-rhodopsins

Channelrhodopsins are the primary photoreceptors in the eyespot of the unicellular algae that are responsible for phototactic and photophobic responses (46–48). Their name is derived from the fact that despite the sensory function, the 7-TM segment is in itself a light-activated ion channel. While the channel pore and properties remain poorly understood, it has recently been realized that channelrhodopsins are likely proton pumps like many other microbial opsins, but with a leaky step in the photocycle during which the opsin lets positive charge into cells (49). In *C. reinhardtii*, two separate channelrhodopsins were originally identified (48), one with fast kinetics but poor light sensitivity, channelrhodopsin-1 (ChR1) (50), and another with slower kinetics but improved sensitivity, channelrhodopsin-2 (ChR2) (1). Two more channelrhodopsins from *Volvox carteri* (VChR1, VChR2) have also been identified (7, 51), and as we discuss later, many more ChRs are expected to exist. ChR1-style channelrhodopsins have red-shifted action spectra (peak $\lambda_{ChR1} = 500$ nm, $\lambda_{VChR1} = 535$ nm) relative to ChR2 (peak $\lambda_{ChR2} = 470$ nm), and thus in principle ChR1-style and ChR2-style opsins could be used together to drive separate sets of neurons with two different colors of light, if suitably spectrally separated opsin pairs could be found.

3.1. Conductance, Permeability, and Context

Channelrhodopsins (abbreviated as ChRs or chops) are light-activated, inwardly rectifying cation channels that are, at neutral pH, permeable to physiologically relevant cations such as H^+, Na^+, K^+, and Ca^{2+}, with permeabilties (relative to sodium) of 1×10^6, 1, 0.5, and 0.1 respectively (1, 6, 46, 50). It is of particular note that the proton conductance (G_{H+}) is 10^6-fold larger than the sodium conductance (G_{Na+}), and thus near physiological pH, perhaps half the photocurrent is carried by protons (46); thus, ChRs may rapidly equilibrate the intracellular pH with its environment (10). Kinetic selectivity analysis has shown that the mechanism of ion selectivity is likely to be due to differential binding affinity of channelrhodopsin channel residues for different ions, not differential ion transport rates (46).

It was originally believed that ChR1 was a selective proton channel (50); however, it was later discovered that the poor photocurrents at mammalian pH were likely attributable to poor membrane localization (6), and the apparent lack of sodium currents in the original report were due to the low pH used to perform experiments in that study; the sodium conductance of ChR1 lessens at low pH (6, 46), unlike that of ChR2 (6). This highlights what is a recurrent theme throughout this chapter, that effective conductance in a heterologous system is determined not only by the innate kinetic and transport properties of the molecule but also by its trafficking and performance in the environment of the heterologous system.

The single ion channel conductance of ChR2 has been estimated at 50 fS (1), which corresponds to approximately 3×10^4 ions per second, or 300 ions per photocycle event, assuming a 10 ms turnover. This is considerably less than a typical voltage dependent sodium channel that may have a conductance on the order of ~10 pS. It has been estimated from electrophysiological data that 10^5–10^6 membrane embedded ChR2 molecules are required to cause reliable spiking in cultured rodent hippocampal neurons (52), with saturation blue light densities of several milliwatts per square millimeter both in vitro (10) and in vivo (53).

3.2. Basic View of Kinetics and Wavelength Selectivity

Figure 3a shows a typical photocurrent trace from a voltage-clamped neuron expressing ChR2 (top), and the spiking pattern that would result in current-clamp mode (bottom). There is a large transient peak with an opening time constant near 1 ms (1, 6, 10), although photocurrent onset can be measured at <200 μs (1, 3,

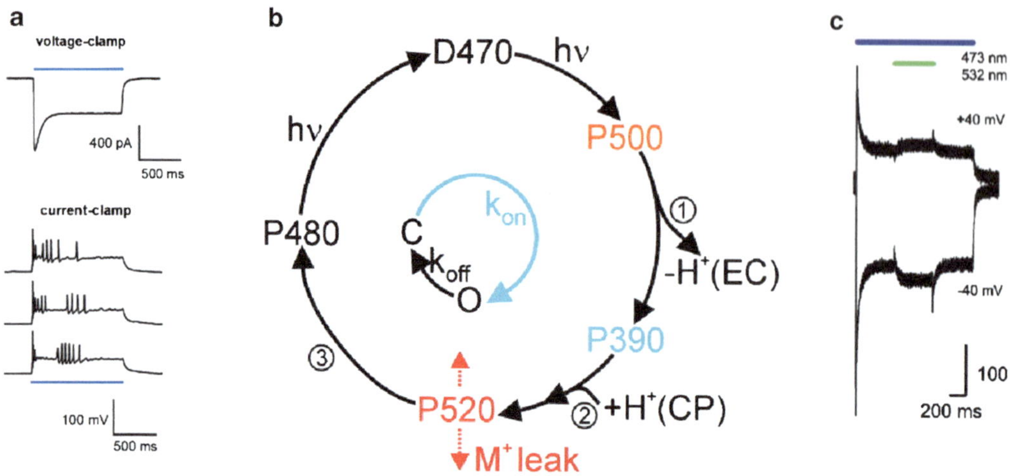

Fig. 3. Channelrhodopsin kinetic and photocycle properties, and impact on neural activity. (**a**) ChR2 currents elicited in a voltage-clamped hippocampal neuron expressing ChR2 and illuminated by blue light (*top*), and ChR2-driven spikes elicited in a current-clamped hippocampal neuron (three repetitions of the same blue light pulse in the same neuron) (*bottom*), under 1 s of blue light illumination. Adapted from Boyden et al. (3). (**b**) The photocycle of ChR2 determined by a combination of spectroscopy, site-directed mutagenesis, and electrophysiology, adapted from Feldbauer et al. (49). The *inner circle* summarizes the effective appearance of the photocycle, an approximation to the outer photocycle. (**c**) The interplay between photocycle, wavelength, and electrophysiological activity. ChR2 is excited with blue light for a brief period, and then a green light is turned on. The photocurrent initially diminishes because the channel is forced to close, but then increases because the green light also pumps the molecule into its most highly efficient state. Image modified from Bamann et al. (54).

54); this transient peak quickly decays to a stationary component that is typically <20–50% of the initial peak photocurrent (1, 3, 6, 10). Upon removing the light, ChR2 closes with a time constant of 10–20 ms (1, 6, 10). The transient photocurrent peak is highly dependent on the illumination intensity (51, 55) and history (1, 3, 10); the history dependence results from a desensitization of the transient component that takes ~5 s to recover from in the dark (3). The stationary component on the other hand, is less photosensitive and effectively history-independent (55). ChR2 absorbs maximally at 460 nm (1, 10), and the action spectra of both temporal components are nearly identical in ChR2.

The large and fast-onset peak enables ChR2-expressing neurons to spike with exquisite temporal precision on the millisecond timescale (Fig. 1b), the timescale of an action potential. However, the large inactivation (or alternatively, the small stationary component) and its slow recovery in the dark, as well as the slow closing rate of ~10–15 ms, ultimately limit the ability to drive reliable spike rates >25 Hz (3, 10) because (1) the stationary photocurrent may be too small to sufficiently depolarize a neuron to spike threshold, and (2) the channel cannot physically close quickly enough to enable de-inactivation of sodium channels. It should be noted, though, that many neurons, such as pyramidal cells, seldom fire action potentials at this rate on the individual neuron level (vs. population synchrony or rhythmogenesis).

ChR1-style channelrhodopsins (VChR1, ChR1) (7, 50) on the other hand demonstrate dramatically faster kinetics than ChR2-style channelrhodopsins (VChR2, ChR2). The stationary photocurrents of ChR1s are >70% of the peak photocurrents, and the channels open and close approximately two- to threefold faster than does ChR2. Therefore, one would expect that given comparable expression, protein folding, membrane localization, and photosensitivity (i.e., factors contributing to effective conductance), ChR1s would be capable of driving spike rates with greater fidelity than ChR2s. However, poor membrane expression limits the performance of natural ChR1-style channelrhodopsins (7, 50). Chimeras composed of the first five helices of ChR1 and last two helices of ChR2 have been constructed (6, 10, 56), and these new variants exhibit the small inactivation and action spectrum of ChR1, but the overall effective conductance of ChR2. These structure–function studies are discussed in detail later in this chapter. A point mutant of this chimera dubbed "ChIEF" (based on its composition as a [Ch] annelrhodopsin chimera with an [I]190 V substitution with domains swapped between ChR1 helix-[E] and ChR2 helix-[F]), developed by Tsien and coworkers (10), appears to be a highly improved tool for stimulating neurons. Its large stationary photocurrent and very fast channel closing, the latter conferred by the I190 V mutation, contribute to far more reliable spiking (up to 100 Hz) than ChR2.

Based on the available characterization of the channelrhodopsins from *V. carteri* (7, 51), the general characteristics are similar to those of the analogous molecules in *C. reinhardtii*. VChR2 and ChR2 have nearly identical photocycles and action spectrum (51). VChR1 and ChR1 exhibit the similar reduced inactivation, and are both red-shifted from their respective VChR2/ChR2 counterparts. It has been proposed that VChR1 could be used for multicolor optical stimulation in conjunction with ChR2, which is blue-shifted by ~70 nm, but further improvements are likely required for reliable spiking because the VChR1 photocurrents are unfortunately approximately four- to fivefold smaller (7), and also there is significant spectral overlap between VChR1 and ChR2.

3.3. Detailed Models of Kinetics and Wavelength Selectivity

As previously mentioned, it is critical to realize that the translocation of ions by opsins is not as simple as the operation of an on/off switch, but rather these opsins traverse a complex photocycle with various intermediate states beyond the initial opening of the channel. Figure 3b shows the photocycle of ChRs based on photophysical studies performed primarily by laser flash spectroscopy, physiology, and site-directed mutagenesis (8, 49, 51, 54, 57). Importantly, the intermediates of the photocycles themselves can also undergo photoreactions, and thus they may be optically driven or "short-circuited" (54) between photointermediates at much faster rates (Fig. 3c). The ChR2 photocycle initially begins in its closed dark-adapted state D470 (where the number in the state name corresponds to the peak light absorption, in nm, of the molecule in that state). The channel opens when D470 absorbs a photon, after which the molecule will become a green absorbing photoproduct or P-intermediate, P520, via thermal relaxation from shorter lived photoproducts. This initial cascade of events takes 0.2–1.5 ms, depending on the transmembrane potential. The open ChR2 can be closed by either optically pumping P520 → D470 with green light, or by decaying to P480 (via a yet to be determined intermediate), a process that takes ~6 s. The inactivation toward the stationary photocurrent may be due to molecules making the P520 → P480 transition, rather than the optically induced P520 → D470 transition that would allow the molecule to quickly open again. Assuming that ChR1 and ChR2 photocycles are topologically similar, e.g., the ChR2 D470 and P480 equate to the ChR1 peaks at 464 and 505 nm, this interpretation of the transient and stationary photocurrents is consistent with the finding that, for ChR1, the stationary photocurrent is red-shifted from the transient photocurrent (6).

The various wavelengths of absorption of rhodopsins and their intermediates throughout the photocycle arises from the different conformations of the chromophore and its local environment, which influences the chromophore charge distribution and the protonation of the retinylidene Schiff base. Figure 3c demonstrates this complex interplay in an experiment by Bamberg and coworkers (54). After blue light-excited ChR2 has reached its steady state, a green light costimulation is introduced. The photocurrent briefly diminishes because the open channel is forced to close, but the stationary photocurrent quickly improves because many molecules have been pumped back into their highly efficient, peak producing state. Thus, it is possible that slightly red-shifted or broadband illumination of ChR2 may strike a balance between optimally exciting the dark state (transient component) and repriming the dark state (driving the red-shifted intermediate photoproduct). As we discuss, optimal silencing with *N. pharaonis* halorhodopsin is analogously achieved by using both yellow light to hyperpolarize the neuron and blue light to drive the molecule out of its inactive state (12, 14).

3.4. Mutants and Variants

As previously mentioned, even though no ChR crystal structure exists at the time of this writing, useful structure–function studies have been performed based largely on sequence homology to *H. salinarum* bacteriorhodopsin. The E90Q mutation (57) has increased sodium selectivity (with respect to G_{H+}) vs. wild-type ChR2, and the H134R mutant (5) demonstrates increased conductance by approximately twofold. Various mutations to C128 (8) corresponding to bacteriorhodopsin T90, drastically slow down the rate of ChR2 closure from the open state, thus effectively creating a bistable open P520 state until illuminated with green light. By lengthening the time that ChR2 spends open on a per-photon basis, this mutation effectively decreases the amount of light needed to activate the channel, at the expense of temporal precision. In contrast, the E123T mutant, combined with the H134R mutant, speeds channel closure and increases the precision of neural action potential firing at the expense of photocurrent and light sensitivity (58), resulting in a reagent nicknamed ChETA.

Chimeras of ChR1 and ChR2 have been constructed by several researchers (6, 10, 56), one of which was that composed of ChR1 helices A–E and ChR2 helices F–G (called abcdeFG, ChEF, or ChR1/$2_{5/2}$ by various investigators). These chimeras displayed the small inactivation of ChR1, but the large photocurrents of ChR2 on account of improved membrane localization and light sensitivity (based on quantitative confocal fluorescence microscopy, (6)). An I190 V substitution to ChEF led to the molecule, "ChIEF," capable of driving more reliable fast spiking due to the much larger stationary current and faster channel closing kinetics after light offset (10). During these studies, it was also discovered that a single point mutation to wild-type ChR1, E87Q, eradicates its pH-dependent spectral shifts, and increases inactivation during illumination (56).

The fact that the poor effective conductance of ChR1 can be largely attributed to membrane localization rather than its photophysical properties highlights the importance of considering and improving the trafficking of heterologously expressed molecules. In particular, ChRs are not localized to the outer membrane, but rather the eyespot, in *C. reinhardtii*, and the membrane composition of the organism is less than 20% phosphoglyceride (59), a primary lipid type in mammalian neurons. As we later discuss in the context of halorhodopsins, the use of signaling peptides can improve outer membrane localization and reduce aggregation in the cytosol, endoplasmic reticulum, and Golgi apparatus.

Along similar lines of using signal peptides to alter trafficking, the myosin binding domain (MBD) peptide promotes subcellular localization of opsins to neuronal dendrites (9). This subcellular localization strategy may prove to be helpful for enabling driving of electrical activity in specific neural compartments, or for high-resolution connectomic mapping in vivo. Two-photon excitation is a powerful laser excitation technique that enables submicron resolution in 3-D (60) relatively deep into the brain (~750 μm, or a significant fraction of the thickness of the mouse cortex), but its ability to induce action potentials in a neuron expressing ChR2 is limited by the interplay between molecule density and the extent of optical depolarization with respect to time (61, 62). The probability of inducing an action potential, at low powers that are not destructive to tissue, is relatively low using a traditional raster scan because the fraction of molecules excited at any point in time is small, and most photons that do hit the membrane are wasted (since the open time of ChR2 is long relative to a femtosecond laser photon delivery rate). Thus, with most conventional two-photon laser scanning methods, the aggregate contributions of the serially excited molecules never sufficiently depolarize the whole neuron to spike threshold. However, Rickgauer and Tank have demonstrated that neurons expressing ChR2 can be reliably excited by two-photon microscopy by optimizing the scan pattern to deliver light optimally to the cell membrane, in a fashion that reaches the maximum surface area while minimizing wastage of photons on already-light-driven channelrhodopsin molecules (62).

3.5. Diversity

Unlike microbial rhodopsins from archaea, significantly less is known about the photoelectrogenic molecules of unicellular algae, the only organisms known to date to have naturally occurring light-activated channels. Photoelectric responses have been measured in several green flagellates, as well as phylogenetically distant cryptophytes (47, 63–65). Interestingly, the two-component phototaxis strategy employed by *C. reinhardtii*, in which the response is mediated by a fast (ChR1) and slow (ChR2) rhodopsin, appears to be general (47), which begs the question whether chimeras of their respective rhodopsins will also result in kinetic improvements and variants with interesting properties. Thus, as more phototaxis-mediating rhodopsins are isolated and sequenced, or as perhaps new depolarizing rhodopsin types are discovered, new molecular tools for controlling neurons will surely emerge.

4. Optical Neural Silencing: Halorhodopsins and Bacteriorhodopsins

Whereas traditional electrodes can stimulate neurons with temporal precision (albeit without cell type specificity), they are incapable of silencing neurons in order to assess their necessity for given neural computations, behaviors, and pathologies. Therefore, there is a large need for spatio-temporally precise methods for optical inhibition of neurons. Inwardly rectifying chloride pumps and outwardly rectifying proton pumps, halorhodopsins (HRs, hops) and bacteriorhodopsins (BRs, bops), respectively, are electrogenic pumps that when heterologously expressed are capable of sufficiently hyperpolarizing a neuron to silence its activity (Fig. 1c–e; (12, 15, 66)). They are thus far known to exist in every kingdom except for animals: archaea (22, 23, 67–69), bacteria (70–76), fungi (77, 78) and algae (79). In addition to their opposite electrophysiological effect, HRs and BRs differ primarily from channelrhodopsins in that their physiological functions is chiefly due to their role as pumps as opposed to operating as passive channels, and thus can translocate ions against concentration gradients (but, typically only one ion per photocycle). Much is known about the photocycles and structure–function relationships of HRs and BRs because they have been crystallized (24, 27, 28, 80, 81) and heavily characterized via spectroscopy, mutagenesis, and physiology for decades.

This section focuses on two molecules in particular: *N. pharaonis* halorhodopsin (Halo/NpHR) and *H. sodomense* bacteriorhodopsin (Arch/AR-3), also known as an archaerhodopsin (that is, a bacteriorhodopsin from the *halorubrum* genus). Halorhodopsins were shown in 2007 to be capable of mediating modest optical neural hyperpolarizations, and since have been improved in trafficking to boost their currents (12, 13, 16); bacteriorhodopsins were shown in 2009 to be able to mediate very powerful and kinetically versatile silencing of multiple neural populations with different colors of light (18). We discuss in the following sections "Conductance, permeability, and context" and "Kinetics and wavelength selectivity" of halorhodopsins and bacteriorhodopsins for these two classes separately, followed by a joint discussion of the "Mutants and variants" and genomic "Diversity" in a unified section.

4.1. Halorhodopsins: Conductance, Permeability, and Context

N. pharaonis halorhodopsin (NpHR, Halo) is a highly selective, inwardly rectifying, chloride pump, which can also conduct larger monovalent anions (82). It has a reversal potential of approximately −400 mV (82), and its chloride-dependence of pumping activity (full- and half-saturating chloride concentrations: $[Cl^-]_{saturation} = 20$ mM, $[Cl^-]_{1/2} = 2.5$ mM) (83) is appropriate for operation in mammalian cells, in contrast to *H. salinarum* halorhodopsin, which is not capable of effective operation in mammalian neurons (15), presumably because of its large chloride dependency: $[Cl^-]_{saturation} = 5$ M and $[Cl^-]_{1/2} = 200$ mM (83, 84). In the absence of any signal peptide sequences to improve trafficking and membrane localization, Halo has been reported to generate 40–100 pA of hyperpolarizing current (12, 18, 66), with the differences in measured photocurrents between studies largely attributable to the power and wavelength of excitation used. This photocurrent is approximately 10–25-fold less than typical peak depolarizing currents generated by ChR2, highlighting one potential disadvantage inherent to a pump that translocates one ion per photocycle (e.g., Halo) vs. a channel that conducts 300 ions per photocycle (e.g., ChR2). To mediate these currents, there are an estimated ten million membrane-embedded Halo molecules per neuron (as assessed in hippocampal neuron culture). Because the expression levels are so high, Halo is known to form puncta or intracellular blebs, aggregating in the endoplasmic reticulum (ER) and Golgi apparatus (16, 17). These issues are somewhat addressed by attaching trafficking enhancement sequences to the molecule, e.g., sequences from the KiR2.1 protein (eNpHR, eNpHR3.0), which increases the effective conductance several fold by increasing membrane expression (16, 85).

Recently, we have discovered that the crux-halorhodopsin (HR from the *haloarcula* genus) from *Haloarcula marismortui*, canonically known as cHR-5 (69, 86), produces similar photocurrents to Halo with more uniform expression; even when highly overexpressed under high copy number transfection conditions, no puncta or intracellular blebbing is observed (15). This molecule may better express than Halo in vivo, but it is unknown at this moment whether it will ultimately be more efficacious at altering mammalian behavior, given their statistically insignificant difference in photocurrent. However, the prolactin (Prl) ER-location sequence in conjunction with a signal sequence from a MHC class I antigen triples the Halo photocurrent (15, 18); we are now trying out multiple trafficking sequences in combination to see if they boost current further. However, it is important to note that if halorhodopsins have other side effects that are due to the protein's intrinsic properties—as an example, one paper quantitates the significantly altered neuronal capacitance that results from expressing halorhodopsin in neurons in vivo (87)—then boosting expression may only make such side effects worse.

4.2. Halorhodopsins: Kinetics and Wavelength Selectivity

N. pharaonis halorhodopsin is capable of silencing weakly firing neurons on the millisecond timescale with its ~100 pA-scale currents, with rapid onset and offset (12), but during long periods of illumination, all halorhodopsins that we have tested so far and that have current (from *N. pharaonis*, *H. sodomense*, *Haloarcula vallismortis*, *H. marismortui*, and *Salinibacter ruber*) inactivate by approximately 30% every 15 s of illumination at 1–10 mW/mm^2 yellow (593 nm) light (Fig. 4a, b; (12, 18)). This slow inactivation stands in contrast to ChR2, which responds to light with a large transient peak that decays within seconds, followed by a stable stationary photocurrent. For all of the halorhodopsins named above, recovery in the dark from light-induced inactivation is slow, with a time constant of tens of minutes, as has been described for some halorhodopsins earlier (12, 14, 39) (Fig. 4a, b). This long-lasting inactivation property may hinder the use of halorhodopsins for silencing for prolonged periods, e.g., during repeated behavioral trials. Importantly, for all halorhodopsins investigated, the inactive photoproduct can be driven back into its active pumping state with a short (e.g., subsecond duration) pulse of blue or UV light (12, 14); thus, optimal use of Halo for neural silencing requires both yellow and blue light to be delivered to the same set of neurons, which is possible (88) but can complicate optics setups.

The Halo photocycle is shown in schematic form in Fig. 2c. The time constants listed are the limiting ones, with all other transitions <100 μs. It should be noted that the names for

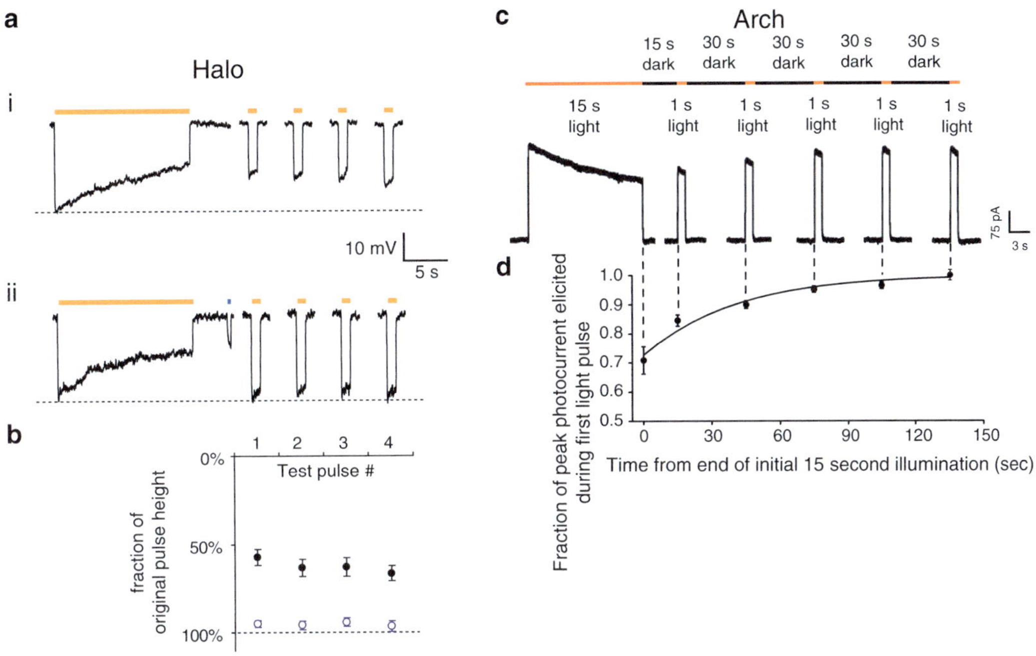

Fig. 4. Kinetic comparisons between halorhodopsins and archaerhodopsins. (**ai**) Time course of Halo-mediated hyperpolarizations in a representative current-clamped hippocampal neuron during 15 s of continuous yellow light, followed by four 1-s test pulses of yellow light (one every 30 s, starting 10 s after the end of the first 15-s period of yellow light). (**aii**) Time course of Halo-mediated hyperpolarization for the same cell exhibited in (**ai**), but when Halo function is facilitated by a 400-ms pulse of blue light in between the 15-s period of yellow light and the first 1-s test pulse. (**b**) Population data for blue-light facilitation of Halo recovery ($N = 8$ neurons). Plotted are the hyperpolarizations elicited by the four 1-s test pulses of yellow light, normalized to the peak hyperpolarization induced by the original 15-s yellow light pulse. *Dots* represent mean ± S.E.M. *Black dots* represent experiments when no blue light pulse was delivered (as in Fig. 5ai.). *Open blue dots* represent experiments when 400 ms of blue light was delivered to facilitate recovery (as in Fig. 5aii.). (**c**) Raw current trace of a neuron lentivirally infected with Arch, illuminated by a 15 s light pulse (575 ± 25 nm, irradiance 7.8 mW/mm^2), followed by 1 s test pulses delivered starting 15, 45, 75, 105, and 135 s after the end of the 15 s light pulse. (**d**) Population data of averaged Arch photocurrents ($N = 11$ neurons) sampled at the times indicated by the *vertical dotted lines* that extend into Fig. 4c.

canonical spectroscopic states of *H. salinarum* halorhodopsin—the K, L, N, and O photointermediates—have not been used here because the order of the N- and O-states in *N. pharaonis* halorhodopsin is still somewhat debated (42, 89, 90). In the dominant photocycle, Halo absorbs a photon and then within tens of microseconds, quickly releases a chloride ion into the cytoplasm during the HR520 → H640 transition, via short-lived intermediates that "switch" the chloride location within the molecule from the extracellular loading domain to the cytoplasmic release domain. The molecule then reisomerizes from the HR640 state and takes up a chloride ion from the extracellular side, a process that takes ~1.5 ms; it then forms the HR' state, which finally relaxes to the active ground state with a time constant of ~20 ms. However, halorhodopsins can enter an alternate photocycle (middle trajectory within Fig. 2c), most notably under prolonged or bright illumination, as might occur during in vivo neural silencing. The 13-*cis* retinalydene Schiff base becomes deprotonated (releasing a proton into the cytoplasm and thus introducing a small depolarizing proton current) and forms a long-lived intermediate HR410 (14, 39). In the dark, the halorhodopsin will remain in this inactive state for a duration on the order of 30 min (39). This formation of HR410 is the origin of the long inactivation observed in neurons expressing halorhodopsin and the ability to recover the active state using the short blue light pulse (12, 14). In contrast, as we discuss in detail in the next section, archaerhodopsins (bacteriorhodopsins from the *Halorubrum* genus) spontaneously recover in the dark under physiological conditions.

4.3. Bacteriorhodopsins: Conductance, Permeability, and Context

Arch, canonically known as archaerhodopsin-3 (AR-3) from *H. sodomense*, is a yellow-green-light sensitive outwardly rectifying proton pump with nearly an order of magnitude increase in hyperpolarizing current over any characterized natural halorhodopsin (15, 18), attaining neuronal currents up to 900 pA in vitro in response to light powers easily achievable in vitro or in vivo. The efficacy of these proton pumps is surprising, given that protons occur, in mammalian tissue, at a millionfold-lower concentration than the ions carried by the optical control molecules described above. This high efficacy may be due to the fast photocycle of Arch (see also (91, 92)), but it may also be due to the ability of high-pKa residues in proton pumps to mediate proton uptake (91, 93).

Arch is a highly efficacious tool in vivo, with cortical neurons in the awake behaving mouse undergoing a median of 97.1% reductions in firing rate for periods of seconds to minutes (Fig. 1e) (18), and safely expresses for months in both mice and monkeys when virally delivered in vivo. Due to the larger currents, Arch enables very large (e.g., order of magnitude-scale) increases in addressable volume of tissue silenceable, over earlier reagents. We thoroughly investigated the safety of Arch function. To date, blebbing issues that have affected the usage of halorhodopsins have not been observed in vitro or in vivo for Arch, but membrane trafficking sequences may still prove effective at boosting expression beyond the natural state (the Prl sequence, which greatly magnifies Halo current, slightly increases Arch current in neurons). Furthermore, we have not observed changes in cell membrane capacitance or other passive neural properties, as has been reported with halorhodopsin expression (see above). From a end-user standpoint, illumination of Arch neurons was safe: spike rates measured in vivo were not significantly different before vs. after periods of optical neural silencing. Biophysically, pH changes in neurons expressing Arch and undergoing illumination were minimal, plateauing rapidly at alkalinizations of ~0.1–0.15 pH units; the fast stabilization of pH_i

may reflect a self-limiting influence that rapidly limits proton concentration swings, and may contribute to the safe operation of Arch in neurons, as observed in mice and monkeys. Indeed, the changes in pH observed in cells expressing Arch and being illuminated are comparable in magnitude to those observed during illumination of ChR2-expressing cells (10) (due to the proton currents carried by ChR2 (1, 46)) and are also within the magnitudes of changes observed during normal neural activity (94–97). We have observed that other archaerhodopsins from other *Halorubrum* strains are also particularly powerful molecular reagents (work in progress). In contrast, the canonical *H. salinarum* bacteriorhodopsin, well known to poorly function in *Escherichia coli*, successfully produced modest photocurrents in mammalian neurons, which highlights the importance of not assuming all molecules will express and traffic the same in different organisms. (In contrast, *E. coli* does not support any detectable expression of ChR2, which expresses well in neurons; target species influences in modulation of opsin function should not be underestimated).

4.4. Bacteriorhodopsins: Kinetics and Wavelength Selectivity

Unlike all of the halorhodopsins we have screened to date (including not only the natural halorhodopsins described above, but also products of halorhodopsin site-directed mutagenesis aimed at improving kinetics), which after illumination remained inactivated for tens of minutes, Arch spontaneously recovers function in seconds in the dark (Fig. 4c, d), more like the light-gated cation channel channelrhodopsin-2 (ChR2) than like halorhodopsins. This feature is particularly useful for in vivo behavior work because it dramatically simplifies the necessary optical hardware; the need to use only one wavelength of light also increases the available bandwidth for multicolor silencing in multiple cell-types. This spontaneous recovery has also been observed by us with other archaerhodopsins, and thus may be a general feature of archaerhodopsins as a whole (work in progress).

Arch is maximally excited with green-yellow light ($\lambda = 561$ nm), a fairly common peak wavelength for proton pumps. Thus, it is backward compatible with halorhodopsin-driving equipment. Proton pumps naturally exist that are activated by many colors of light, in contrast to chloride pumps, which are primarily driven by yellow-orange light (even with significant mutagenesis of retinal-flanking residues, (15)). The light-driven proton pump from *L. maculans*, here abbreviated Mac, has an action spectrum strongly blue-shifted relative to that of the light-driven chloride pump Halo (Fig. 5a). We found that Mac-expressing neurons could undergo 4.1-fold larger hyperpolarizations with blue light than with red light, and Halo-expressing neurons could undergo 3.3-fold larger hyperpolarizations with red light than with blue light, when illuminated with appropriate powers and filters (Fig. 5b). Accordingly, we could demonstrate selective silencing of spike firing in Mac-expressing

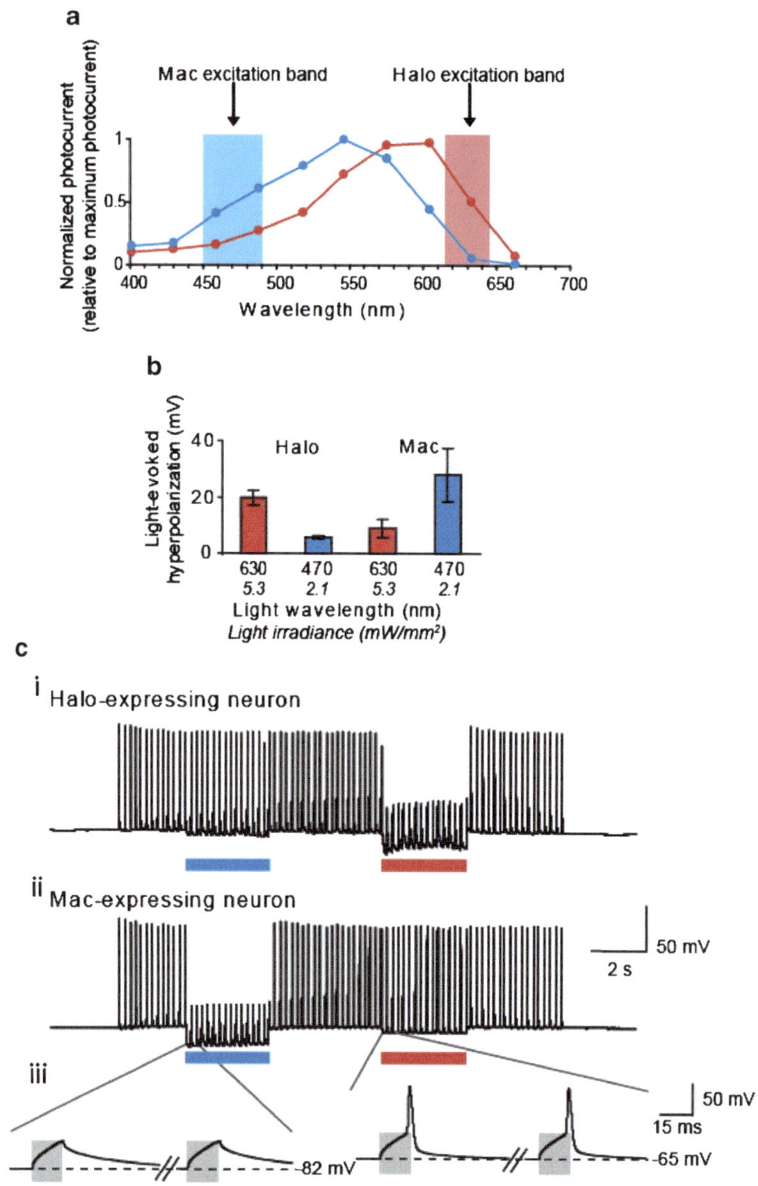

Fig. 5. Multicolor silencing of two neural populations, enabled by blue- and red-light drivable ion pumps of different classes. (**a**) Action spectra of Mac vs. Halo; *rectangles* indicate filter bandwidths used for multicolor silencing in vitro. Blue light power is via a 470 ± 20 nm filter at 5.3 mW/mm^2, and red light power is via a 630 ± 15 nm filter at 2.1 mW/mm^2. (**b**) Membrane hyperpolarizations elicited by blue vs. red light, in cells expressing Halo or Mac ($N = 5$ Mac-expressing neurons, $N = 6$ Halo-expressing neurons). (**c**) Action potentials evoked by current injection into patch clamped cultured neurons transfected with Halo (**ci**) were selectively silenced by the red light but not by the blue light, and vice versa in neurons expressing Mac (**cii**). *Gray boxes* in the *inset* (**ciii**) indicate periods of patch clamp current injection.

neurons in response to blue light, and selective silencing of spike firing in Halo-expressing neurons in response to red light (Fig. 5c). Thus, the spectral diversity of proton pumps points the way toward independent multicolor silencing of separate neural populations.

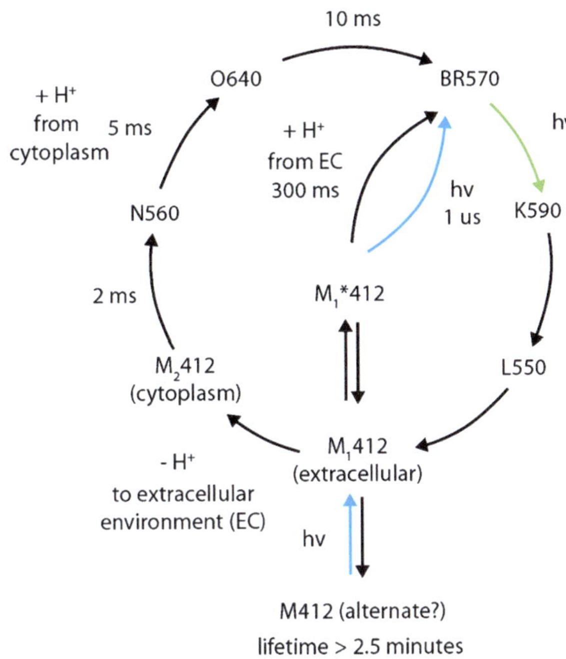

Fig. 6. The photocycle of the *H. salinarum* bacteriorhodopsin. As in Figs. 2c and 3b, the photocycle has been simplified to reflect the dominant photocycle at large continuous illumination on the timescale of a typical photocycle (i.e., conditions used for neural silencing, >few milliseconds). The M412 alternate intermediate is the origin of long-lived inactivation with bacteriorhodopsin. In contrast, Arch spontaneously quickly recovers from this state in the dark.

This result opens up novel kinds of experiment, in which, for example, two neuron classes, or two sets of neural projections from a single site, can be independently silenced during a behavioral task.

Figure 6 shows the photocycle of the canonical *H. salinarum* bacteriorhodopsin, as representative of the photocycle of the class of proton pumps (the photocycles of archaerhodopsins are not as well characterized, and may well be different) (22, 93). The data that has led to the synthesis of the modern model of the bacteriorhodopsin photocycle provide many of the insights that have led to solid investigations into this field as a whole. As in Figs. 2c and 3b, it has been simplified to represent the dominant photocycle expected at large and continuous illumination on the timescale of a typical photocycle (e.g., many milliseconds, as would be used for neural silencing). Upon absorbing a photon, bacteriorhodopsin forms its L550 intermediate (L = lumi) within microseconds, after quickly transferring a proton from the retinylidene Schiff base (at the Lys-216 position) to the Asp-85 proton acceptor. This transfer triggers the proton releasing group (PRG), containing Glu-204 and Glu-194, to release its own proton (98, 99), during the L → M transition. After a "switching" step during which the Schiff base reorients itself on the cytoplasmic side, it is then reprotonated during the M → N transition via the Asp-96 donor residue, which in turn picks up a proton from the cytoplasm in the next transition (N → O). The chromophore reisomerizes to the all-*trans* state during this transition as well. Finally, the bacteriorhodopsin relaxes back to its ground state as the proton release group is reloaded from the Asp-85 residue. Like halorhodopsin, bacteriorhodopsin can become trapped in a long-lasting light-unresponsive state (Fig. 6, "M412 alternate") that requires blue light to reenter the normal photocycle; this could partly explain the lower currents observed with BR when compared to Arch.

4.5. Halorhodopsins and Bacteriorhodopsins: Mutants and Variants

For decades, researchers have been making mutants of bacteriorhodopsins and halorhodopsins for structure–function studies reviewed in many earlier publications (e.g., (22, 24, 93)). However, point mutations that improve these molecules as neural silencing tools have yet to be reported. Given that the conductance of a pump is limited by the fact that they move only one ion per photocycle, it would be highly desirable to mutate halorhodopsin into a channel, or channelrhodopsin into an anion channel. Tuning the action spectrum of both bacteriorhodopsins and halorhodopsins may be achieved by mutating the retinal-flanking residues (75, 76, 100, 101). Mutations that alter ion selectivity, such as the most well-known example of this is the bacteriorhodopsin D85T mutation that converts it into a chloride pump (102), could allow ion-specific currents to be mimicked. A Ca^{2+} selective pump, for example, would have powerful impact on enabling powerful studies of plasticity, synaptic transmission, and cellular signaling. Given that crystal structures for many bacteriorhodopsins and halorhodopsins exist, we anticipate that functionally oriented and applied site-directed mutagenesis will be a highly active field of research in neuroengineering. The fact that archaerhodopsins on the whole represent a class of molecules that express particularly well in mammalian cells possibly indicates properties of their lipid-interacting residues, or perhaps "signal sequence"-like activity by their loop regions. The known crystal structures of these molecules (80, 81) may provide key insights into the trafficking of heterologously expressed molecules and their membrane insertion.

4.6. Diversity

Light-activated proton and chloride pumps are known to exist in far more organisms than do light-gated cation channels, and proton pumps are particularly prevalent, as all opsins described to date likely have at least some proton pumping capability (22, 23, 67–79, 103, 104). Even though many proton pumps maximally absorb blue-green to green wavelengths (which opens up, as shown in Fig. 5, the possibility, alongside yellow-light driven chloride pumps, for multiple-color silencing of distinct neural populations), light-activated hyperpolarizing currents carried by protons have been observed across the whole visible spectrum, from deep blue (via a sensitizer) (105) to far red (106) (>650 nm), although the red-light sensitive current likely originates from a receptor that triggers an H^+-ATPase, as opposed to direct light-mediated ion translocation.

The discovery or creation of a purely genetically encoded light-activated inhibitory channel would be highly desirable. In addition to seeking natural molecules and site-directed mutants, linking a non-light-gated ion-channel to a type I archaeal rhodopsin may be another promising approach to creating such a molecular tool (107). In this way, a light-activated shunt could be created, which would more closely mimic natural mechanisms of neural inhibition in the brain.

5. Molecular Targeting of Microbial Opsins to Different Cell Types

The number of papers using optical neural control in species ranging from *C. elegans* to mouse to nonhuman primate is increasing exponentially each year, and so we do not attempt to review the literature comprehensively. In each species, opsins have been used to test the necessity and sufficiency of neurons, cell types, muscles, neural pathways, brain regions, and other entities in behaviors, pathologies, and neural computations. Opsins have proven valuable in exploring neural dynamics in multiple mammalian brain structures as well. We focus on highlighting principles that govern how to best use these opsins in various settings, from a molecular biology standpoint (Sect. 5) and from a physical–optical standpoint (Sect. 6). From a molecular biology standpoint, these opsins can be delivered to neurons in almost any conventional way that genes are delivered into cells or into organisms. Transgenic mice have been made with ChR2, for example (108), mice and monkeys have been injected with lentiviruses, adeno-associated viruses (AAV), and other viruses encoding for light-gated proteins (2, 52, 109–111), and rodents and chicks have been electroporated in utero with plasmids encoding for light-gated proteins (4, 53, 112). In each case, different parameters of the technique can be selected so as to enable specific cell types, pathways, or regions to be selectively labeled or to selectively express the opsin. Transgenic *C. elegans* have been made expressing ChR2 using conventional methods (5), as have transgenic Drosophila (113) and zebrafish (114). For these latter species, supplementation with all-*trans*-retinal may be beneficial, whereas mammalian brains seem to operate microbial opsins without need for supplementation. (It is possible that in the future, genetically engineering retinal lacking organisms, such as invertebrates, to produce retinal within their nervous systems may be of use, e.g., by expressing within them enzymes that can produce retinal from vitamin precursors (115)).

For mammalian nervous systems, a large variety of possible strategies exist for conveying opsin genes into specific cell types. For example, transgenic mice can be made through BAC transgenic, knock-in, or other methods, but such strategies are not common for other species yet. For viral delivery, cell type specific promoters can be inserted upstream of the opsin to target various excitatory, inhibitory, and modulatory neurons (e.g., (19, 116–119)); the size of the promoter is limited by the virus type (AAV viruses hold typically 4–4.5 kb total, whereas lentiviruses hold typically 8–10 kb maximum). The surface or coat proteins that a virus bears can also modulate which cell types will take up the virus; for example, lentiviruses may favor excitatory neurons of the cortex, whereas certain AAV serotypes may favor inhibitory cells (120). Lentiviruses can be pseudotyped—fabricated with a coat protein of desired targeting capacity, e.g., with rabies glycoprotein that leads to lentiviruses that travel retrogradely (121)—whereas AAVs can be engineered with biotinylation sites that enable, upon streptavidin conjugation, targeting of potentially arbitrary substrates (122). Retroviruses, which preferentially label dividing cells, have been used to deliver ChR2 to newborn neurons of the dentate gyrus of the hippocampus (123). Other viruses such as rabies virus and pseudorabies virus can be used, with unique tracing capabilities including the ability to go retrogradely across multiple synapses (124, 125). Viral particles typically have to be injected directly into the brain, often through stereotactic targeting to a specific brain area, since the potent blood–brain barrier typically precludes systemic delivery of large viral particles (although see (126)). We have recently described a parallel injector array that can deliver viruses into complex three-dimensional configurations (127). Finally, in utero electroporation of ChR2-GFP into embryonic mice at embryonic day ~15.5 has been found to selectively label pyramidal cells in layers 2/3 (53, 112).

Of potential interest is the possibility of a new generation of neural prosthetics, which can accomplish the synthetic neurobiology mission of repairing the nervous system by enabling optical input of information to sculpt neural dynamics and overcome pathology. To deliver these genes in a safe, efficacious, and enduring way, viruses such as AAV may be valuable; AAV has been used in over 600 human patients in gene therapy clinical trials without a single serious adverse event (128) and has been successfully used with opsin delivery. The ability to optically control specific targets within the brain may enable more potent and side-effect free therapies than possible with existing electrical and magnetic neuromodulation therapies, or with drugs which often are nonspecific and have side effects. Already several groups have prototyped blindness therapies that may enable new approaches to a

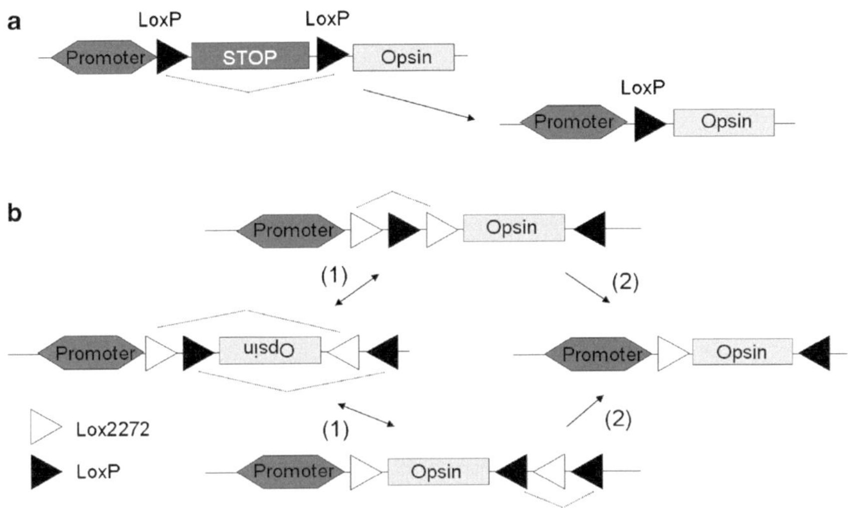

Fig. 7. Transgenic mouse expressing Cre within specific cells, coupled to lox-containing AAV viruses, enable cell-type specific opsin expression. (**a**) A pair of loxP recombination sequences meditate the removal of the transcriptional and translational stop cassette containing multiple polyadenylation signals (STOP), in the presence of Cre (provided in transgenic mice within specific cell types), to initiate gene expression. (**b**) Two pairs of loxP-type recombination sequences (FLEX) for stable inversion proceeds in two steps: (1) inversion followed by (2) excision. loxP and lox2272 are orthogonal recombination sites. (Adapted from Kuhlman and Huang (111) and Atasoy et al. (110)).

currently intractable set of disorders, those in which photoreceptors degenerate within the retina (129, 130). To that end, the recent assessment of brain and immune function in nonhuman primates after ChR2 expression and activation, which showed a lack of harmful effects in a preliminary study (109), may pave the way toward new ideas for neural prosthetics for humans.

One strategy that has become widely popular, is to inject the brain of a mouse expressing Cre recombinase in a specific cell type, with a virus that either bears an opsin preceded by a lox-flanked stop cassette, or a virus that bears an opsin reversed and flanked by pairs of lox sites in a specific configuration (Fig. 7; (110, 111)). Given the very large number of Cre transgenic mice in existence, and that are being generated, this strategy is likely to be very useful, at least for mice. Transgenic mice can be made that express Cre in extremely cell specific ways (e.g., through 3′ UTR knockins at the ends of cell type-specific genes), and then viruses can be rapidly made as new opsins are created, thus enabling cell type-specific expression without requiring the difficult process of trimming cell-specific promoters to fit within the small viral payload, or the difficult process of making transgenic mice for each new opsin tool that is developed (given the rapid pace of innovation, as shown in Sects. 1–4).

6. Hardware for Optical Neural Control

For in vitro use, xenon lamps (e.g., Sutter DG-4, Till Photonics Polychrome) equipped with fast-moving mirrors or monochromators can be used for flexible delivery of fast (millisecond-timescale), brigh light pulses to biological samples on microscopes. Fluorescence filters can be used to deliver light of the appropriate wavelength (e.g., GFP excitation filter for ChR2, rhodamine or Texas Red excitation filter for Arch, Texas Red excitation filter for Halo, and GFP or YFP excitation filter for Mac). Recently, many companies such as Thorlabs have begun to sell LEDs or LED arrays compatible with microscope fluorescence illuminators, which sell for a small fraction of the price of a full lamp setup. Or fiber-coupled LEDs can simply be placed nearby to the sample (131). Confocal and two-photon microscopes, or more generally scanning laser methodologies, can be used to drive opsins, as have been described in a variety of papers (112, 132, 133). Recently, digital micromirror displays (DMDs) have come forth as potentially useful for photostimulating in complex patterns, comprising millions of individual pixels that can be toggled either on or off (134, 135).

In vivo, the brain scatters light starting within a few hundred microns of an optical source, and absorbs light starting within a few millimeters of distance. Thus, most efforts for in vivo neuromodulation focus on delivering light to a volume of tissue, ranging from a very small volume containing a few hundred cells to a large volume (say, a few cubic millimeters) containing many thousands of cells. One widespread method is to use a laser coupled to an optical fiber (Fig. 8a shows a versatile setup that couples multiple colors of laser light into a single fiber; simpler commercially available single-color laser-coupled fibers with TTL control are also available), and to insert the fiber into a cannula implanted in the brain (Fig. 8b shows a hand-built one for mice; commercial versions from companies such as Plastics One can also be built) or directly into the brain (Fig. 8c shows a setup for monkey), or to couple the fiber to an implanted fiber via a ferrule. An optical commutator (e.g., from Doric Lenses) can be placed between the fiber that inserts into the brain and the fiber that connects to the laser, to allow free rotation (66). Arrays of custom-targetable optical fibers, each coupled to a miniaturized light source (e.g., a raw die LED) and targeted to a unique target, will open up the ability to drive activity in complex 3-D patterns, enabling the perturbation of complexly shaped structures as well as the ability to perturb targets in a patterned fashion (136, 137).

Aside from the key advantage of being able to manipulate a specific cell type, another key advantage of optical stimulation is the lack of electrical artifact as compared to conventional electrical

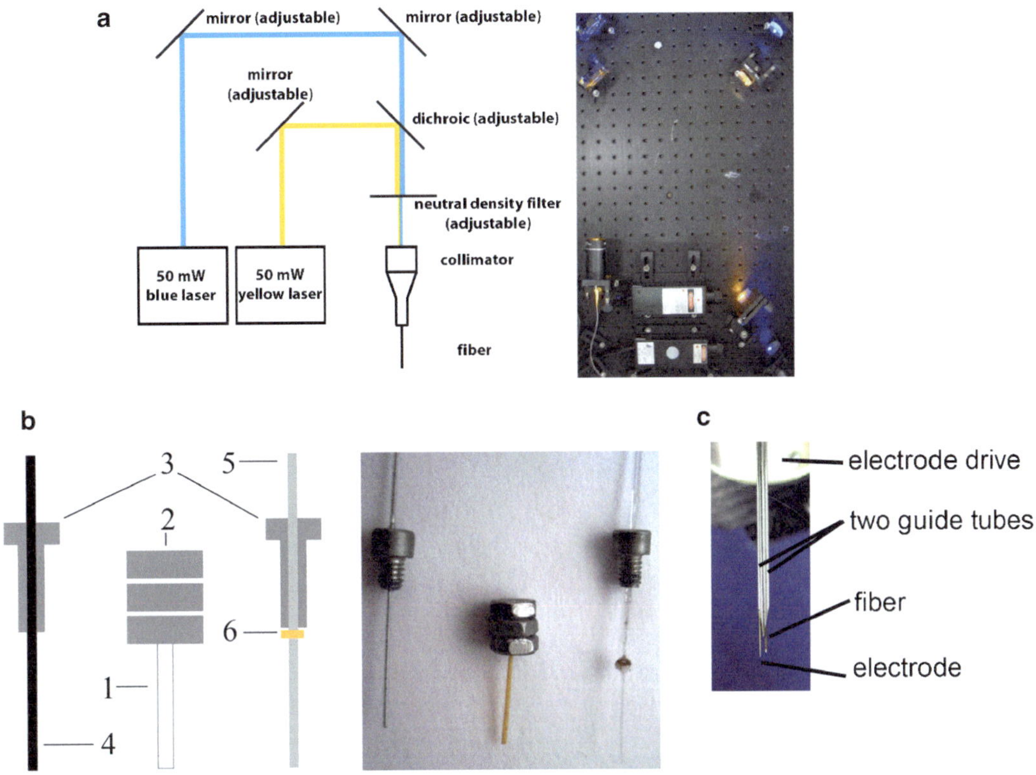

Fig. 8. (**a**) Schematic design (*left*) and picture (*right*) of an optics assembly used to couple blue and yellow laser light into a single optical fiber for in vivo neural modulation. A pictured assembly, lacking a neutral density filter, shows the hardware laid out on a standard optics breadboard. (**b**) Schematic design (*left*) and picture (*right*) of a system for targeting and securing optical fibers within the brain. A polyimide cannula (1, 250 μm ID), designed to terminate at the locus of optical modulation, is epoxied to a stack of hex nuts (2, sized 2–56) which will be secured to the skull with dental cement. Vented screws (3, sized 2–56), which have holes in their centers, screw into the nuts while leaving a path open to the cannula. A dummy wire (4, 230 μm stainless steel wire) may be epoxied to the screw to seal the craniotomy when the optics are not in use. An optical fiber (5, 230 μm OD silica fiber) is allowed free rotation without vertical displacement by a plastic washer (6, homemade) which is epoxied to the fiber and sandwiched in between the vented screw above and the cannula below. (**c**) Apparatus for optical activation and electrical recording. Photograph, showing optical fiber (200 μm diameter) and electrode (200 μm shank diameter) in guide tubes. Adapted from Han et al. (19).

stimulation methods. However, despite the lack of electrical artifact, light does produce a voltage deflection when the electrode tip was illuminated (19, 138). For example, Fig. 9 shows traces recorded on a tungsten electrode in saline being illuminated by a pulsed laser beam, as adapted from (19). This voltage deflection slowly evolved over many tens of milliseconds, and accordingly was only recorded when the electrode voltage filtered at 0.7–170 Hz to examine local field potentials, (Fig. 9, top traces of each panel). This voltage deflection was not recorded when the electrode voltage was filtered at 250–8,000 Hz to detect spike signals; (Fig. 9, bottom traces of each panel). When light illuminated parts of the electrode other than the tip, no artifact was recorded; only illumination of the

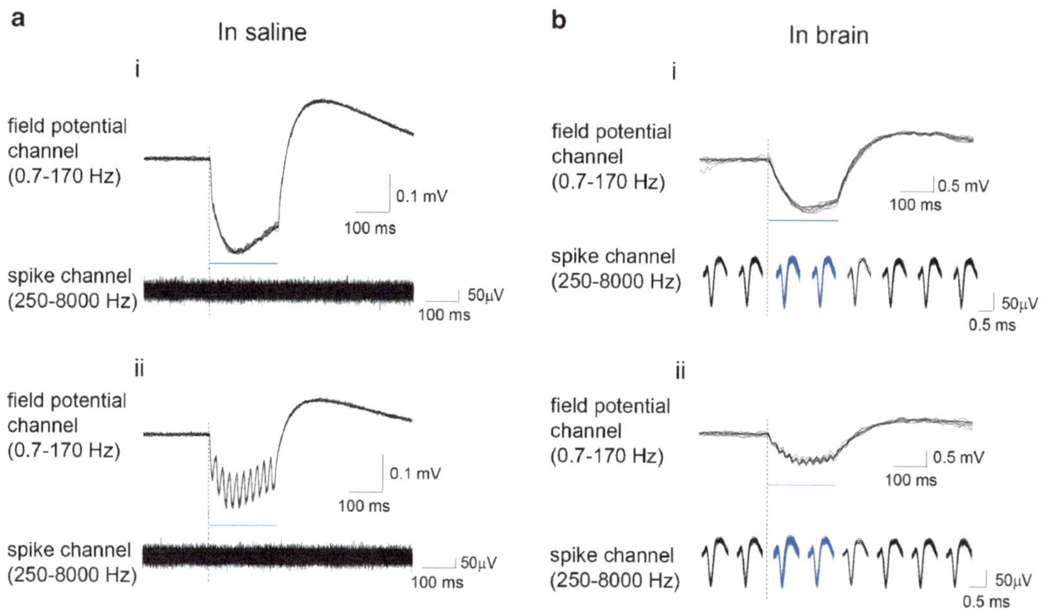

Fig. 9. Voltage deflections observed on tungsten electrodes immersed in saline (**a**) or brain (**b**), upon tip exposure to 200 ms blue light pulses (**bi**) or trains of 10 ms blue light pulses delivered at 50 Hz (**bii**). Light pulses are indicated by *blue dashes*. Electrode data was hardware filtered using two data acquisition channels operating in parallel, yielding a low-frequency component ("field potential channel") and a high-frequency component ("spike channel"). For the "spike channel" traces taken in brain (**b**), spikes were grouped into 100 ms bins, and then the binned spikes were displayed beneath corresponding parts of the simultaneously acquired "field potential channel" signal (59 and 53 repeated light exposures for **bi** and **bii** respectively). (Shown are the spikes in eight such bins—the two bins before light onset, the two bins during the light delivery period, and the four bins after light cessation.) For all other signals shown, ten overlaid traces are plotted. From Han et al. (19).

tip–saline interface resulted in the voltage transient. This phenomenon is consistent with a classical photoelectrochemical finding, the Becquerel effect, in which illumination of an electrode placed in saline can produce a voltage on the electrode (139, 140). Consistent with the generality of the Becquerel effect as a property of electrode-electrolyte interfaces, this artifact is observed on various electrode materials, such as stainless steel, platinum–iridium, silver/silver chloride, gold, nichrome, or copper. Similar slowly evolving voltage deflections were observed when tungsten electrodes were used to record neural activity in the brain, during optical stimulation. Because the optical artifact was slowly evolving over many tens of milliseconds, spike waveforms were detected without corruption by the artifact. However, local field potentials and field oscillations, which reflect coherent neural dynamics in the range of Hertz to tens of Hertz, may be difficult to isolate from this Becquerel artifact using the electrodes here tested. Notably, we have not seen the artifact with pulled glass micropipettes (such as previously used in (3) and (12), or in the mouse recordings with pulled glass pipettes in (109)). Thus, for recordings of local field potentials and other slow signals of importance for neuroscience, hollow glass electrodes may prove useful.

References

1. Nagel G et al (2003) Channelrhodopsin-2, a directly light-gated cation-selective membrane channel. Proc Natl Acad Sci USA 100:13940–13945
2. Ishizuka T, Kakuda M, Araki R, Yawo H (2006) Kinetic evaluation of photosensitivity in genetically engineered neurons expressing green algae light-gated channels. Neurosci Res 54:85–94
3. Boyden ES, Zhang F, Bamberg E, Nagel G, Deisseroth K (2005) Millisecond-timescale, genetically targeted optical control of neural activity. Nat Neurosci 8:1263–1268
4. Li X et al (2005) Fast noninvasive activation and inhibition of neural and network activity by vertebrate rhodopsin and green algae channelrhodopsin. Proc Natl Acad Sci USA 102:17816–17821
5. Nagel G et al (2005) Light activation of channelrhodopsin-2 in excitable cells of *Caenorhabditis elegans* triggers rapid behavioral responses. Curr Biol 15:2279–2284
6. Wang H et al (2009) Molecular determinants differentiating photocurrent properties of two channelrhodopsins from chlamydomonas. J Biol Chem 284:5685–5696
7. Zhang F et al (2008) Red-shifted optogenetic excitation: a tool for fast neural control derived from *Volvox carteri*. Nat Neurosci 11:631–633
8. Berndt A, Yizhar O, Gunaydin LA, Hegemann P, Deisseroth K (2009) Bi-stable neural state switches. Nat Neurosci 12:229–234
9. Lewis TL Jr, Mao T, Svoboda K, Arnold DB (2009) Myosin-dependent targeting of transmembrane proteins to neuronal dendrites. Nat Neurosci 12:568–576
10. Lin JY, Lin MZ, Steinbach P, Tsien RY (2009) Characterization of engineered channelrhodopsin variants with improved properties and kinetics. Biophys J 96:1803–1814
11. Lanyi JK, Duschl A, Hatfield GW, May K, Oesterhelt D (1990) The primary structure of a halorhodopsin from *Natronobacterium pharaonis*. Structural, functional and evolutionary implications for bacterial rhodopsins and halorhodopsins. J Biol Chem 265:1253–1260
12. Han X, Boyden ES (2007) Multiple-color optical activation, silencing, and desynchronization of neural activity, with single-spike temporal resolution. PLoS One 2:e299
13. Zhang F et al (2007) Multimodal fast optical interrogation of neural circuitry. Nature 446:633–639
14. Bamberg E, Tittor J, Oesterhelt D (1993) Light-driven proton or chloride pumping by halorhodopsin. Proc Natl Acad Sci USA 90:639–643
15. Chow B et al (2010) High-performance genetically targetable optical neural silencing by light-driven proton pumps. Nature 463 (7277):98–102
16. Gradinaru V, Thompson KR, Deisseroth K (2008) eNpHR: a Natronomonas halorhodopsin enhanced for optogenetic applications. Brain Cell Biol 36:129–139
17. Zhao S et al (2008) Improved expression of halorhodopsin for light-induced silencing of neuronal activity. Brain Cell Biol 36:141–154
18. Chow B, Han X, Qian X, Boyden ES (2009) High-performance halorhodopsin variants for improved genetically-targetable optical neural silencing. Frontiers in Systems Neuroscience. Conference Abstract: Computational and systems neuroscience. doi:10.3389/conf.neuro.10.2009.03.347
19. Han X, Qian X, Stern P, Chuong AS, Boyden ES (2009) Informational lesions: optical perturbation of spike timing and neural synchrony via microbial opsin gene fusions. Front Mol Neurosci. doi:10.3389/neuro.02.012.2009
20. Henderson R, Schertler GF (1990) The structure of bacteriorhodopsin and its relevance to the visual opsins and other seven-helix G-protein coupled receptors. Philos Trans R Soc Lond B Biol Sci 326:379–389
21. Palczewski K (2006) G protein-coupled receptorrhodopsin. Annu Rev Biochem 75:743–767. doi:10.1146/annurev.biochem.75.103004.142743
22. Lanyi JK (2004) Bacteriorhodopsin. Annu Rev Physiol 66:665–688. doi:10.1146/annurev.physiol.66.032102.150049
23. Lanyi JK (1986) Halorhodopsin: a light-driven chloride ion pump. Annu Rev Biophys Biophys Chem 15:11–28. doi:10.1146/annurev.bb.15.060186.000303
24. Essen LO (2002) Halorhodopsin: light-driven ion pumping made simple? Curr Opin Struct Biol 12:516–522
25. Zemelman BV, Lee GA, Ng M, Miesenbock G (2002) Selective photostimulation of genetically chARGed neurons. Neuron 33:15–22
26. Lin B, Koizumi A, Tanaka N, Panda S, Masland RH (2008) Restoration of visual function in retinal degeneration mice by ectopic

expression of melanopsin. Proc Natl Acad Sci USA 105:16009–16014
27. Kolbe M, Besir H, Essen LO, Oesterhelt D (2000) Structure of the light-driven chloride pump halorhodopsin at 1.8 A resolution. Science 288:1390–1396
28. Luecke H, Schobert B, Richter HT, Cartailler JP, Lanyi JK (1999) Structure of bacteriorhodopsin at 1.55 A resolution. J Mol Biol 291:899–911
29. Braiman MS, Stern LJ, Chao BH, Khorana HG (1987) Structure-function studies on bacteriorhodopsin. IV. Purification and renaturation of bacterio-opsin polypeptide expressed in *Escherichia coli*. J Biol Chem 262:9271–9276
30. Gilles-Gonzalez MA, Engelman DM, Khorana HG (1991) Structure-function studies of bacteriorhodopsin XV. Effects of deletions in loops B-C and E-F on bacteriorhodopsin chromophore and structure. J Biol Chem 266:8545–8550
31. Mogi T, Stern LJ, Chao BH, Khorana HG (1989) Structure-function studies on bacteriorhodopsin. VIII. Substitutions of the membrane-embedded prolines 50, 91, and 186: the effects are determined by the substituting amino acids. J Biol Chem 264:14192–14196
32. Mogi T, Marti T, Khorana HG (1989) Structure-function studies on bacteriorhodopsin. IX. Substitutions of tryptophan residues affect protein-retinal interactions in bacteriorhodopsin. J Biol Chem 264:14197–14201
33. Mogi T, Stern LJ, Hackett NR, Khorana HG (1987) Bacteriorhodopsin mutants containing single tyrosine to phenylalanine substitutions are all active in proton translocation. Proc Natl Acad Sci USA 84:5595–5599
34. Marti T et al (1991) Bacteriorhodopsin mutants containing single substitutions of serine or threonine residues are all active in proton translocation. J Biol Chem 266:6919–6927
35. Marinetti T, Subramaniam S, Mogi T, Marti T, Khorana HG (1989) Replacement of aspartic residues 85, 96, 115, or 212 affects the quantum yield and kinetics of proton release and uptake by bacteriorhodopsin. Proc Natl Acad Sci USA 86:529–533
36. Mogi T, Stern LJ, Marti T, Chao BH, Khorana HG (1988) Aspartic acid substitutions affect proton translocation by bacteriorhodopsin. Proc Natl Acad Sci USA 85:4148–4152
37. Subramaniam S, Greenhalgh DA, Khorana HG (1992) Aspartic acid 85 in bacteriorhodopsin functions both as proton acceptor and negative counterion to the Schiff base. J Biol Chem 267:25730–25733
38. Brown LS, Needleman R, Lanyi JK (1996) Interaction of proton and chloride transfer pathways in recombinant bacteriorhodopsin with chloride transport activity: implications for the chloride translocation mechanism. Biochemistry 35:16048–16054
39. Hegemann P, Oesterhelt D, Steiner M (1985) The photocycle of the chloride pump halorhodopsin. I: Azide-catalyzed deprotonation of the chromophore is a side reaction of photocycle intermediates inactivating the pump. EMBO J 4:2347–2350
40. Blanck A, Oesterhelt D (1987) The halo-opsin gene. II. Sequence, primary structure of halorhodopsin and comparison with bacteriorhodopsin. EMBO J 6:265–273
41. Rudiger M, Oesterhelt D (1997) Specific arginine and threonine residues control anion binding and transport in the light-driven chloride pump halorhodopsin. EMBO J 16:3813–3821
42. Varo G et al (1995) Light-driven chloride ion transport by halorhodopsin from *Natronobacterium pharaonis*. 1. The photochemical cycle. Biochemistry 34:14490–14499
43. Tittor J et al (1997) Chloride and proton transport in bacteriorhodopsin mutant D85T: different modes of ion translocation in a retinal protein. J Mol Biol 271:405–416
44. Tittor J, Oesterhelt D, Bamberg E (1995) Bacteriorhodopsin mutants D85N, D85T and D85,96N as proton pumps. Biophys Chem 56:153–157
45. Varo G, Needleman R, Lanyi JK (1995) Light-driven chloride ion transport by halorhodopsin from *Natronobacterium pharaonis*. 2. Chloride release and uptake, protein conformation change, and thermodynamics. Biochemistry 34:14500–14507
46. Berthold P et al (2008) Channelrhodopsin-1 initiates phototaxis and photophobic responses in Chlamydomonas by immediate light-induced depolarization. Plant Cell 20:1665–1677. doi:10.1105/tpc.108.057919
47. Sineshchekov OA, Govorunova EG, Spudich JL (2009) Photosensory functions of channelrhodopsins in native algal cells. Photochem Photobiol 85:556–563
48. Sineshchekov OA, Jung KH, Spudich JL (2002) Two rhodopsins mediate phototaxis to low- and high-intensity light in *Chlamydomonas reinhardtii*. Proc Natl Acad Sci USA 99:8689–8694

49. Feldbauer K et al (2009) Channelrhodopsin-2 is a leaky proton pump. Proc Natl Acad Sci USA 106:12317–12322
50. Nagel G et al (2002) Channelrhodopsin-1: a light-gated proton channel in green algae. Science 296:2395–2398
51. Ernst OP et al (2008) Photoactivation of channelrhodopsin. J Biol Chem 283:1637–1643
52. Zhang F, Wang LP, Boyden ES, Deisseroth K (2006) Channelrhodopsin-2 and optical control of excitable cells. Nat Methods 3:785–792
53. Huber D et al (2008) Sparse optical microstimulation in barrel cortex drives learned behaviour in freely moving mice. Nature 451:61–64
54. Bamann C, Kirsch T, Nagel G, Bamberg E (2008) Spectral characteristics of the photocycle of channelrhodopsin-2 and its implication for channel function. J Mol Biol 375:686–694
55. Nikolic K, Degenaar P, Toumazou C (2006) Modeling and engineering aspects of channelrhodopsin2 system for neural photostimulation. Conf Proc IEEE Eng Med Biol Soc 1:1626–1629
56. Tsunoda SP, Hegemann P (2009) Glu 87 of channelrhodopsin-1 causes pH-dependent color tuning and fast photocurrent inactivation. Photochem Photobiol 85:564–569
57. Ritter E, Stehfest K, Berndt A, Hegemann P, Bartl FJ (2008) Monitoring light-induced structural changes of channelrhodopsin-2 by UV-visible and Fourier transform infrared spectroscopy. J Biol Chem 283:35033–35041. doi:10.1074/jbc.M806353200
58. Gunaydin LA et al (2010) Ultrafast optogenetic control. Nat Neurosci 13(3):387–92
59. Harwood JL, Guschina IA (2009) The versatility of algae and their lipid metabolism. Biochimie 91:679–684
60. Zipfel WR, Williams RM, Webb WW (2003) Nonlinear magic: multiphoton microscopy in the biosciences. Nat Biotechnol 21:1369–1377
61. Mohanty SK et al (2008) In-depth activation of channelrhodopsin 2-sensitized excitable cells with high spatial resolution using two-photon excitation with a near-infrared laser microbeam. Biophys J 95:3916–3926
62. Rickgauer JP, Tank DW (2009) Two-photon excitation of channelrhodopsin-2 at saturation. Proc Natl Acad Sci USA 106:15025–15030
63. Sineshchekov OA et al (2005) Rhodopsin-mediated photoreception in cryptophyte flagellates. Biophys J 89:4310–4319
64. Sineshchekov OA, Litvin FF, Keszthelyi L (1990) Two components of photoreceptor potential in phototaxis of the flagellated green alga *Haematococcus pluvialis*. Biophys J 57:33–39
65. Litvin FF, Sineshchekov OA, Sineshchekov VA (1978) Photoreceptor electric potential in the phototaxis of the alga *Haematococcus pluvialis*. Nature 271:476–478
66. Gradinaru V et al (2007) Targeting and readout strategies for fast optical neural control in vitro and in vivo. J Neurosci 27:14231–14238
67. Ihara K et al (1999) Evolution of the archaeal rhodopsins: evolution rate changes by gene duplication and functional differentiation. J Mol Biol 285:163–174
68. Mukohata Y, Ihara K, Tamura T, Sugiyama Y (1999) Halobacterial rhodopsins. J Biochem 125:649–657
69. Klare JP, Chizhov I, Engelhard M (2008) Microbial rhodopsins: scaffolds for ion pumps, channels, and sensors. Results Probl Cell Differ 45:73–122
70. Antón J et al (2005) *Salinibacter ruber*: genomics and biogeography. In: Gunde-Cimerman N, Plemenitas A, Oren A (eds) Adaptation to life in high salt concentrations in archaea, bacteria and eukarya. Kluwer Academic Publishers, Dordrecht, Netherlands, pp 257–266
71. Balashov SP et al (2005) Xanthorhodopsin: a proton pump with a light-harvesting carotenoid antenna. Science 309:2061–2064
72. Beja O, Spudich EN, Spudich JL, Leclerc M, DeLong EF (2001) Proteorhodopsin phototrophy in the ocean. Nature 411:786–789
73. Beja O et al (2000) Bacterial rhodopsin: evidence for a new type of phototrophy in the sea. Science 289:1902–1906
74. Friedrich T et al (2002) Proteorhodopsin is a light-driven proton pump with variable vectoriality. J Mol Biol 321:821–838
75. Kelemen BR, Du M, Jensen RB (2003) Proteorhodopsin in living color: diversity of spectral properties within living bacterial cells. Biochim Biophys Acta 1618:25–32
76. Kim SY, Waschuk SA, Brown LS, Jung KH (2008) Screening and characterization of proteorhodopsin color-tuning mutations

in *Escherichia coli* with endogenous retinal synthesis. Biochim Biophys Acta 1777:504–513

77. Brown LS (2004) Fungal rhodopsins and opsin-related proteins: eukaryotic homologues of bacteriorhodopsin with unknown functions. Photochem Photobiol Sci 3:555–565
78. Waschuk SA, Bezerra AG, Shi L, Brown LS (2005) Leptosphaeria rhodopsin: bacteriorhodopsin-like proton pump from a eukaryote. Proc Natl Acad Sci USA 102:6879–6883. doi:10.1073/pnas.0409659102
79. Tsunoda SP et al (2006) H^{+}-pumping rhodopsin from the marine alga Acetabularia. Biophys J 91:1471–1479
80. Yoshimura K, Kouyama T (2008) Structural role of bacterioruberin in the trimeric structure of archaerhodopsin-2. J Mol Biol 375:1267–1281
81. Enami N et al (2006) Crystal structures of archaerhodopsin-1 and -2: common structural motif in archaeal light-driven proton pumps. J Mol Biol 358:675–685
82. Seki A et al (2007) Heterologous expression of Pharaonis halorhodopsin in *Xenopus laevis* oocytes and electrophysiological characterization of its light-driven Cl^{-} pump activity. Biophys J 92:2559–2569
83. Okuno D, Asaumi M, Muneyuki E (1999) Chloride concentration dependency of the electrogenic activity of halorhodopsin. Biochemistry 38:5422–5429
84. Muneyuki E, Shibazaki C, Wada Y, Yakushizin M, Ohtani H (2002) Cl(−) concentration dependence of photovoltage generation by halorhodopsin from *Halobacterium salinarum*. Biophys J 83:1749–1759
85. Gradinaru V et al (2010) Molecular and cellular approaches for diversifying and extending optogenetics. Cell 141(1):154–65
86. Baliga NS et al (2004) Genome sequence of *Haloarcula marismortui*: a halophilic archaeon from the Dead sea. Genome Res 14:2221–2234
87. Tonnesen J, Sorensen AT, Deisseroth K, Lundberg C, Kokaia M (2009) Optogenetic control of epileptiform activity. Proc Natl Acad Sci USA 106:12162–12167
88. Bernstein JG et al (2008) Prosthetic systems for therapeutic optical activation and silencing of genetically-targeted neurons. Proc Soc Photo Opt Instrum Eng 6854:68540H
89. Ludmann K, Ibron G, Lanyi JK, Varo G (2000) Charge motions during the photocycle of pharaonis halorhodopsin. Biophys J 78:959–966
90. Chizhov I, Engelhard M (2001) Temperature and halide dependence of the photocycle of halorhodopsin from *Natronobacterium pharaonis*. Biophys J 81:1600–1612
91. Ming M et al (2006) pH dependence of light-driven proton pumping by an archaerhodopsin from Tibet: comparison with bacteriorhodopsin. Biophys J 90:3322–3332
92. Lukashev EP et al (1994) pH dependence of the absorption spectra and photochemical transformations of the archaerhodopsins. Photochem Photobiol 60:69–75
93. Lanyi JK (2006) Proton transfers in the bacteriorhodopsin photocycle. Biochimica et Biophysica Acta (BBA) – Bioenergetics 1757:1012–1018
94. Bevensee MO, Cummins TR, Haddad GG, Boron WF, Boyarsky G (1996) pH regulation in single CA1 neurons acutely isolated from the hippocampi of immature and mature rats. J Physiol 494(Pt 2):315–328
95. Chesler M (2003) Regulation and modulation of pH in the brain. Physiol Rev 83:1183–1221
96. Meyer TM, Munsch T, Pape HC (2000) Activity-related changes in intracellular pH in rat thalamic relay neurons. Neuroreport 11:33–37
97. Trapp S, Luckermann M, Brooks PA, Ballanyi K (1996) Acidosis of rat dorsal vagal neurons in situ during spontaneous and evoked activity. J Physiol 496(Pt 3):695–710
98. Brown LS et al (1995) Glutamic acid 204 is the terminal proton release group at the extracellular surface of bacteriorhodopsin. J Biol Chem 270:27122–27126
99. Phatak P, Ghosh N, Yu H, Cui Q, Elstner M (2008) Amino acids with an intermolecular proton bond as proton storage site in bacteriorhodopsin. Proc Natl Acad Sci USA 105:19672–19677
100. Henderson R et al (1990) Model for the structure of bacteriorhodopsin based on high-resolution electron cryo-microscopy. J Mol Biol 213:899–929
101. Man-Aharonovich D et al (2004) Characterization of RS29, a blue-green proteorhodopsin variant from the Red Sea. Photochem Photobiol Sci 3:459–462
102. Sasaki J et al (1995) Conversion of bacteriorhodopsin into a chloride ion pump. Science 269:73–75
103. Iwamoto M et al (2004) Proton release and uptake of pharaonis phoborhodopsin (sensory rhodopsin II) reconstituted into phospholipids. Biochemistry 43:3195–3203

104. Sudo Y, Iwamoto M, Shimono K, Sumi M, Kamo N (2001) Photo-induced proton transport of pharaonis phoborhodopsin (sensory rhodopsin II) is ceased by association with the transducer. Biophys J 80:916–922
105. Boichenko VA, Wang JM, Antón J, Lanyi JK, Balashov SP (2006) Functions of carotenoids in xanthorhodopsin and archaerhodopsin, from action spectra of photoinhibition of cell respiration. Biochimica et Biophysica Acta (BBA) – Bioenergetics 1757:1649–1656
106. Serrano EE, Zeiger E, Hagiwara S (1988) Red light stimulates an electrogenic proton pump in Vicia guard cell protoplasts. Proc Natl Acad Sci USA 85:436–440
107. Moreau CJ, Dupuis JP, Revilloud J, Arumugam K, Vivaudou M (2008) Coupling ion channels to receptors for biomolecule sensing. Nat Nanotechnol 3:620–625
108. Wang H et al (2007) High-speed mapping of synaptic connectivity using photostimulation in channelrhodopsin-2 transgenic mice. Proc Natl Acad Sci USA 104:8143–8148
109. Han X et al (2009) Millisecond-timescale optical control of neural dynamics in the nonhuman primate brain. Neuron 62:191–198
110. Atasoy D, Aponte Y, Su HH, Sternson SM (2008) A FLEX switch targets channelrhodopsin-2 to multiple cell types for imaging and long-range circuit mapping. J Neurosci 28:7025–7030
111. Kuhlman SJ, Huang ZJ (2008) High-resolution labeling and functional manipulation of specific neuron types in mouse brain by Cre-activated viral gene expression. PLoS One 3: e2005
112. Petreanu L, Huber D, Sobczyk A, Svoboda K (2007) Channelrhodopsin-2-assisted circuit mapping of long-range callosal projections. Nat Neurosci 10:663–668
113. Schroll C et al (2006) Light-induced activation of distinct modulatory neurons triggers appetitive or aversive learning in Drosophila larvae. Curr Biol 16:1741–1747
114. Douglass AD, Kraves S, Deisseroth K, Schier AF, Engert F (2008) Escape behavior elicited by single, channelrhodopsin-2-evoked spikes in zebrafish somatosensory neurons. Curr Biol 18:1133–1137
115. Yan W et al (2001) Cloning and characterization of a human beta, beta-carotene-15,15′-dioxygenase that is highly expressed in the retinal pigment epithelium. Genomics 72:193–202
116. Dittgen T et al (2004) Lentivirus-based genetic manipulations of cortical neurons and their optical and electrophysiological monitoring in vivo. Proc Natl Acad Sci USA 101:18206–18211
117. Chhatwal JP, Hammack SE, Jasnow AM, Rainnie DG, Ressler KJ (2007) Identification of cell-type-specific promoters within the brain using lentiviral vectors. Gene Ther 14:575–583
118. Adamantidis AR, Zhang F, Aravanis AM, Deisseroth K, de Lecea L (2007) Neural substrates of awakening probed with optogenetic control of hypocretin neurons. Nature 450:420–424
119. Tan W et al (2008) Silencing preBotzinger complex somatostatin-expressing neurons induces persistent apnea in awake rat. Nat Neurosci 11:538–540
120. Nathanson JL, Yanagawa Y, Obata K, Callaway EM (2009) Preferential labeling of inhibitory and excitatory cortical neurons by endogenous tropism of adeno-associated virus and lentivirus vectors. Neuroscience 161:441–450
121. Wickersham IR, Finke S, Conzelmann KK, Callaway EM (2007) Retrograde neuronal tracing with a deletion-mutant rabies virus. Nat Methods 4:47–49
122. Stachler MD, Chen I, Ting AY, Bartlett JS (2008) Site-specific modification of AAV vector particles with biophysical probes and targeting ligands using biotin ligase. Mol Ther 16:1467–1473
123. Toni N et al (2008) Neurons born in the adult dentate gyrus form functional synapses with target cells. Nat Neurosci 11:901–907
124. Wickersham IR et al (2007) Monosynaptic restriction of transsynaptic tracing from single, genetically targeted neurons. Neuron 53:639–647
125. Banfield BW, Kaufman JD, Randall JA, Pickard GE (2003) Development of pseudorabies virus strains expressing red fluorescent proteins: new tools for multisynaptic labeling applications. J Virol 77:10106–10112
126. Foust KD et al (2009) Intravascular AAV9 preferentially targets neonatal neurons and adult astrocytes. Nat Biotechnol 27:59–65
127. Chan SC, Bernstein JG, Boyden ES (2010) Scalable fluidic injector arrays for viral targeting of intact 3-D brain circuits. J Vis Exp 35:1489
128. (2007) Retracing events. Nat Biotechnol 25:949
129. Bi A et al (2006) Ectopic expression of a microbial-type rhodopsin restores visual responses in mice with photoreceptor degeneration. Neuron 50:23–33

130. Lagali PS et al (2008) Light-activated channels targeted to ON bipolar cells restore visual function in retinal degeneration. Nat Neurosci 11:667–675
131. Campagnola L, Wang H, Zylka MJ (2008) Fiber-coupled light-emitting diode for localized photostimulation of neurons expressing channelrhodopsin-2. J Neurosci Methods 169:27–33
132. Rickgauer JP, Tank DW. In: Neuroscience 2008. Society for Neuroscience
133. Petreanu L, Mao T, Sternson SM, Svoboda K (2009) The subcellular organization of neocortical excitatory connections. Nature 457:1142–1145
134. Farah N, Reutsky I, Shoham S (2007) Patterned optical activation of retinal ganglion cells. Conf Proc IEEE Eng Med Biol Soc 2007:6369–6371
135. Guo ZV, Hart AC, Ramanathan S (2009) Optical interrogation of neural circuits in *Caenorhabditis elegans*. Nat Methods 6:891–896
136. Bernstein J et al (2008) A scalable toolbox for systematic, cell-specific optical control of entire 3-D neural circuits in the intact mammalian brain. Society for Neuroscience
137. Bernstein JG et al (2009) Modulation of fear behavior via optical fiber arrays targeted to bilateral prefrontal cortex. Society for Neuroscience
138. Ayling OG, Harrison TC, Boyd JD, Goroshkov A, Murphy TH (2009) Automated light-based mapping of motor cortex by photoactivation of channelrhodopsin-2 transgenic mice. Nat Methods 6:219–224
139. Honda K (2004) Dawn of the evolution of photoelectrochemistry. J Photochem Photobiol A Chem 166:63–68
140. Gratzel M (2001) Photoelectrochemical cells. Nature 414:338–344
141. Chow YC, Han X, Bernstein JG, Monahan PE, Boyden ES (2011) Light-activated ion pumps and channels for temporally precise optical control of activity in genetically targeted neurons. In: Chambers JJ, Kramer RH (eds) Photosensitive molecules for controlling biological function. Neuromethods, vol 55. Springer, New York, doi: 10.1007/978-1-61779-031-7

Neuromethods (2012) 67: 339–355
DOI 10.1007/7657_2011_23

Published online: 1 December 2011

Integrated Optogenetic and Electrophysiological Dissection of Local Cortical Circuits In Vivo

Jessica A. Cardin

Abstract

The recent development of optical methods for controlling the activity of specific populations of neurons holds tremendous promise for understanding the roles of neural subtypes in local circuits, a major goal of systems neuroscience. Optogenetic tools, in combination with electrophysiology, allow identification and manipulation of specific groups of excitatory and inhibitory neurons in active neural networks in vivo in a spatially and temporally precise manner. Here, we provide a detailed description of methods for integrating optogenetics with more traditional electrophysiological approaches as a means to probe these complex interactions.

Key words: Optogenetic, Interneuron, Oscillation, Channelrhodopsin-2, Halorhodopsin, Artifact, Light-evoked activity, Cortex, Electrophysiology, In vivo

1. Introduction

Neural circuits in the cortex comprise interconnected sets of excitatory and inhibitory neurons. Within these two broad classes, cells can be distinguished as belonging to a number of distinct subpopulations. This is especially true of inhibitory interneurons, which can be divided into multiple categories based on molecular expression profiles, morphology, synaptic connectivity, and intrinsic firing properties (1–3). This diversity of neuronal types generates a complex series of cellular interactions and provides a significant degree of functional flexibility to the local circuit. Under active conditions in vivo, each neuron receives a continuous barrage of excitatory and inhibitory synaptic inputs from the surrounding network and, in turn, contributes spike activity back to the network (4). This continuous reciprocal exchange makes it challenging to determine the roles of specific classes of neurons in generating ongoing patterns of network activity, such as oscillations. The development of technologies that regulate the activity of specific cells is thus critical to exploring their contribution to local network computations and ultimately to cognition. Optogenetic tools, which allow optical control of neuronal activity by incorporation of artificially generated light-sensitive proteins into the cell membrane, are a powerful method for dissecting neural circuits. This recently developed technology provides a reliable and physiologically relevant method for stimulating and suppressing the activity of specific neural populations in vitro (5–7) and in vivo (8–15), and can potentially be applied to both experimental and therapeutic ends.

Recent studies have emphasized the use of cell type-specific expression of the light-sensitive nonspecific cation channel channelrhodopsin-2 (ChR2), which is activated by ~470 nm light, as a tool for determining the role of inhibitory interneurons in gamma oscillations and other structured cortical activity patterns in vivo and in information processing in vitro (7, 8, 15). This kind of experiment requires integration of a number of techniques, including efficient viral transduction, cell-type specific expression of light-activated channels, optical stimulation, and intra- and extracellular recording techniques in the intact brain. This combined approach, detailed below, provides a richly detailed view of functional cellular interactions.

1.1. Strategies for Expression of Optogenetic Tools

A number of approaches have been successfully used to insert light-activated proteins into neurons that are otherwise not responsive to optical signals. These include in utero electroporation (11), constitutive and inducible expression in transgenic mice (16), and viral approaches either alone or in association with the well-characterized Cre-loxP system (7, 8, 14). Each method has advantages and disadvantages for achieving cell type-specific expression at functional levels. Generation of transgenic mice expressing cell type-specific ChR2 under the control of a specific promoter is effective, but relatively expensive and time-intensive. In addition, little is known about the potential toxicity of extremely long-term expression of optogenetic tools, such as ChR2, the light-activated chloride pump Halorhodopsin (eNpHR) (17, 18), or the light-activated proton pumps Arch and Mac (19). In utero electroporation can be targeted to general cell classes by restricting the treatment to particular sites and specific prenatal days (20–25). One strength of this approach is that it provides whole-brain expression with potentially high levels of cell-type specificity. A limitation is that the insertion of probes into the fetal brain may potentially have a negative impact on some experiments. In addition, cell-type specificity using this approach is somewhat limited to the resolution provided by cell birth date and broad localization within the germinal zones (26). However, combining this approach with appropriate cell type-specific promoters can yield a very high degree of specificity.

One common approach for expression of optogenetic tools in brain tissues is via viral transduction. Viral vectors are relatively inexpensive to produce and can be administered directly to target

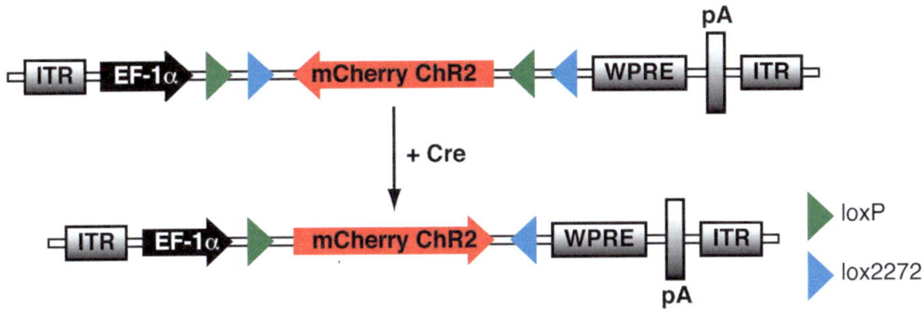

Fig. 1. AAV DIO *ChR2-mCherry* construct for Cre-dependent, cell type-specific expression of ChR2 in vivo. In this construct, ChR2-mCherry starts in the antisense direction, flanked by two sets of incompatible lox sites. In the presence of Cre recombinase, ChR2-mCherry is flipped into the sense direction and is then expressed under the control of the strong promoter EF-1α. *ITR* inverted terminal repeat; *pA* poly(A); *WPRE* woodchuck hepatitis B virus posttranscriptional element.

brain regions with limited tissue damage and a high degree of resulting transduction efficacy. Lentiviral vectors expressing optogenetic tools have been used in cortex, but do not yield a substantial degree of spatial spread around the initial site of delivery (5, 9–11, 27). Recent evidence suggests that lentiviruses exhibit significant tropism for excitatory neurons (28) and thus may be useful for targeting this population but undesirable for targeting inhibitory interneurons. Adeno-associated viruses (AAVs), which have been used in gene therapy approaches (29–31), provide extensive spatial spread and high expression levels in cortex when used with appropriate promoters (Fig. 1).

1.2. Cell-Type Specificity

A robust method for targeting genes of interest to specific cell types is the well-characterized Cre-loxP system. A more recent innovation, the "double-floxed" inverted (DIO) viral vector, has been used in recent studies to produce enhanced specificity and high expression levels (7, 8, 14, 32). The viral vector AAV *D*ouble-floxed *I*nverted *O*pen reading frame-ChR2-mCherry (AAV DIO ChR2-mCherry) (Fig. 1) has recently been used in knock-in and transgenic Cre mice to achieve expression of ChR2 specifically in neuronal populations of interest (8). In the AAV DIO ChR2-mCherry vector, two incompatible loxP variants flank an inverted version of ChR2 fused to the fluorescent marker mCherry. In the presence of Cre recombinase, a stochastic recombination of either loxP variant occurs, resulting in inversion of ChR2-mCherry into the sense direction and expression of light-activated channels under the control of the strong promoter EF1α (Elongation Factor 1α) (33). Cre-dependent expression of light-activated proteins, such as ChR2 and eNpHR, is particularly well suited to targeting expression to cell types that lack comprehensively identified promoter sequences. In many instances, cellular promoters are not fully known or exceed the size limitation imposed by using a viral vector. Cellular promoters alone may also lack the high degree of specificity or expression levels necessary for physiologically useful optogenetic manipulations. Effective use of optogenetic tools in vitro and in vivo depends on a minimal effective level of expression of the light-activated proteins, thus limiting the use of small promoter elements (7). This can be overcome by inserting a strong general promoter into the viral vector and utilizing the high degree of cell-type specificity conferred by the Cre expression pattern of a transgenic line.

Most often, Cre-dependent viral vectors are constructed with a loxP-flanked transcriptional stop cassette preceding the gene of interest. However, the intrinsic size limitations of viral vectors preclude the use of large stop cassettes. In consequence, only small stop cassettes may be used, leading to a low but easily detectable rate of expression of the gene of interest in the absence of Cre (6, 7). The use of the inverted ChR2-mCherry sequence in a double-floxed vector eliminates the need for a polyadenylation stop cassette in the viral construct, as expression is dependent on both Cre-dependent recombination and inversion into the sense direction. For a complete explanation of the double-floxed inverted strategy, see Cardin et al. (8, 34) or Sohal et al. (7).

1.3. Probing Neural Dynamics

Once expression of light-activated channels has been achieved in the cell type of interest, light activation or suppression of neural activity can be used to probe neural network dynamics in vivo in combination with recording techniques. ChR2 can be used to evoke action potentials with a high degree of precision and reliability (Fig. 2a). Repetitive stimulation can be used to evoke patterns of activity that elucidate network architecture, as highlighted by the sustained gamma oscillations observed in response to repetitive stimulation of fast-spiking interneurons (Fig. 2b). Alternatively, spontaneous or ongoing activity can be suppressed with a similar degree of precision (Fig. 2c).

One challenge raised by the combination of extrinsic network modulation and traditional methods of electrophysiology is that optical stimulation can induce a variety of electrical artifacts during such recordings (34). These artifacts likely derive from multiple causes, including photoelectric and temperature-dependent effects. While they are relatively easy to avoid in vitro, they have proven more troubling in vivo and have prevented straightforward recording of field potentials, the most commonly used measure of cortical and hippocampal network state, in several recent studies (10, 35). However, these issues can be resolved using several techniques detailed here (see Note 4.5). The steps outlined here are for expression and stimulation using ChR2; modifications specific to eNpHR expression are described in Note 4.2.

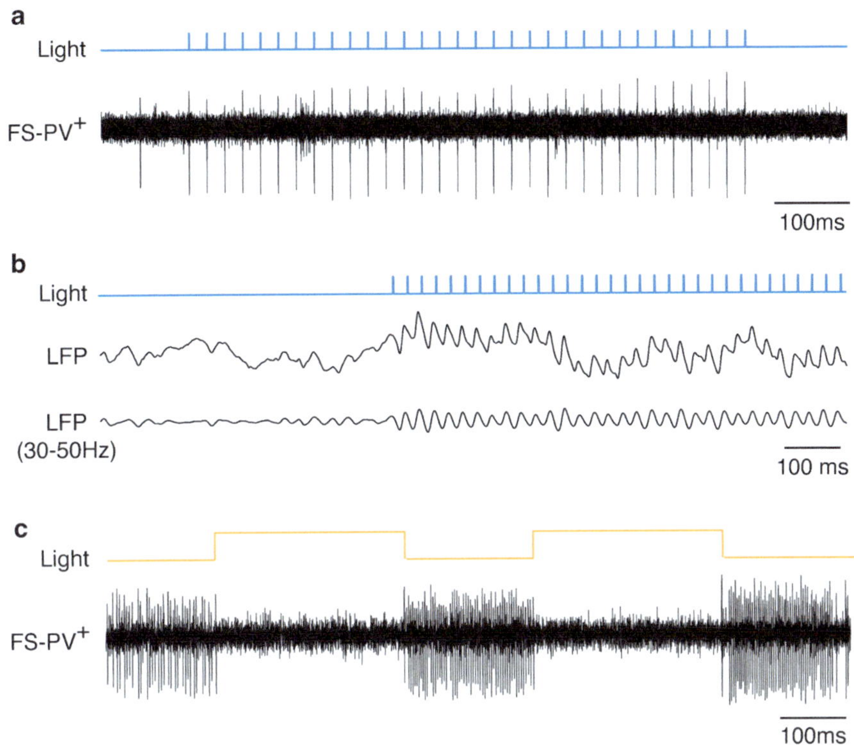

Fig. 2. Optogenetic manipulation of neural activity in cortex. (**a**) Pulses of blue light (473 nm, 1 ms/pulse) applied to the cortical surface in a PV-Cre mouse 7 days after injection of AAV-DIO-ChR2-mCherry. The isolated neuron, a parvalbumin-expressing, fast-spiking inhibitory interneuron (FS-PV^+) in cortical layer 2/3, produced a single action potential in response to each light pulse. (**b**) Stimulation of FS-PV^+ interneurons in cortex at 40 Hz generated a sustained gamma (30–50 Hz) oscillation in the local field potential (LFP). (**c**) Suppression of spontaneous FS-PV^+ firing in a PV-Cre mouse by pulses of yellow light (593.5 nm) applied to the cortical surface 22 days after injection of AAV-DIO-eNpHR3.0-eYFP, a double-floxed construct carrying eNpHR v3.0 under the control of the EF-1 α promoter.

2. Materials

2.1. Optics

- ~473 nm laser (e.g., Optoengine LLC or IkeCool Corp.)
- Breadboard (1/2″ solid aluminum)
- Posts (1/2″)
- FiberPort Coupler (e.g., Thorlabs PAF-SMA-5-A FiberPort and HCP L bracket Mount)
- Optical Fiber (e.g., Thorlabs BFH22-200 or BFH37-200)
- SMA connectors (e.g., Thorlabs Fiber SMA Connectorization kit, SMA 905 connectors)
- Optical power meter (e.g., Thorlabs PM100D power meter, S120C power sensor)

- Beamsplitter (e.g., Thorlabs BP-108 on mount BP-107)
- Neutral density filter and mount (e.g., Thorlabs NE03A on mount KM100T)
- Photodiode and mount (e.g., SM05SPD1A on mount LM05XY)
- BNC adapter (e.g., Thorlabs T4290)

2.2. Stereotaxic and Surgical Apparatus

- Experimental animals (mice)
- Stereotaxic frame with anesthesia head holder and arms for electrode holders (e.g., David Kopf, Inc. Model #930, Model #1760-61, Model #1460, Model #923-B)
- Stereotaxic injector system (e.g., Stoelting QSI; #53311)
- Heating blanket
- Surgical tools (fine forceps and scissors, scalpel)
- High-speed dental drill
- Drill bits
- Sterile saline
- Anesthetic (Isoflurane)
- Isoflurane vaporizer
- Induction chamber (Glass jar or plastic container containing cotton with Isoflurane)
- Tank of compressed O_2
- Local anesthetic (lidocaine 1%)
- Analgesic (buprenorphine or an NSAID)
- Iodine-based wash (betadine)
- Cyanoacrylate glue
- Eye lubricant
- Sutures (6–0 gauge)
- Hair trimmer

2.3. Virus and Pipette Preparation

- AAV of desired serotype at titer of ~10^{12} (can be stored for approximately 1 year at −80°)
- Dye marker (e.g., Fast Green 2.5%)
- Pipette puller (e.g., Sutter Instrument Co. Sutter Model P-97)
- Graduated borosilicate glass capillaries (e.g., Drumond Wiretrol I Calibrated Micropipet, 10 μl size)
- Microfils (26 g)
- Mineral oil (White, light)

2.4. Optogenetic Stimulation and Electrophysiological Recordings

- Electrophysiology equipment for extracellular or intracellular recordings
- Extracellular electrodes (e.g., single tungsten electrodes, FHC Inc.; prefabricated tetrode arrays, NeuroNexus Technologies; or twisted wire stereotrodes or tetrodes)

3. Methods

3.1. Optics

The laser and optics components should be mounted on posts, attached firmly to the breadboard, and aligned so that the laser beam hits the center of the fiberport, which will focus the light at the proximal end of the optical fiber (Fig. 3). The SMA connector can be attached to the optical fiber with epoxy, and the fiber can then be attached to the output connector on the fiber port. The fiber end should be polished with polishing papers to ensure even light transmission. Before beginning any stimulation, it is critical to calculate the input–output function of the laser with the optical power meter (Fig. 4a). Most small lasers produce variable light output depending on voltage input, and some light will be lost in the fiber. It is therefore important to calibrate the exact light intensity at the tip of the optical fiber (mW/mm^2) at each input voltage level used during the experiment. Splitting off a small portion of the laser beam to a photodiode provides a temporally accurate measure of laser output that can be recorded via BNC cable to the recording equipment throughout an experiment.

A low output intensity is usually sufficient to activate ChR2 or eNpHR (see Note 4.2). A light power level of 2 mW at the distal tip

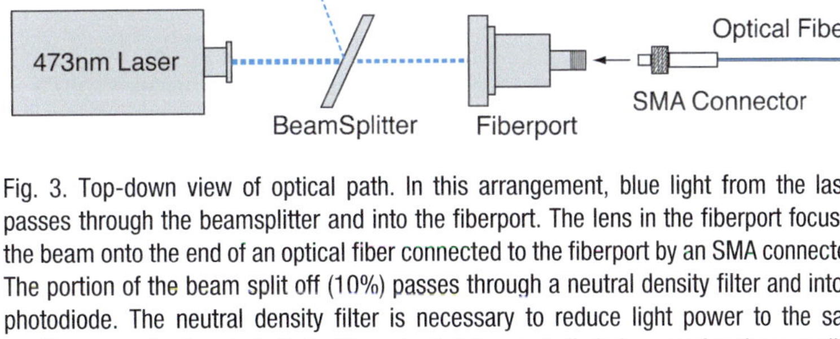

Fig. 3. Top-down view of optical path. In this arrangement, blue light from the laser passes through the beamsplitter and into the fiberport. The lens in the fiberport focuses the beam onto the end of an optical fiber connected to the fiberport by an SMA connector. The portion of the beam split off (10%) passes through a neutral density filter and into a photodiode. The neutral density filter is necessary to reduce light power to the safe working range for the photodiode. The output of the photodiode is passed to the recording equipment via a BNC cable. When calibrated using an optical power meter, this arrangement allows for an online readout of laser power with ms-level accuracy.

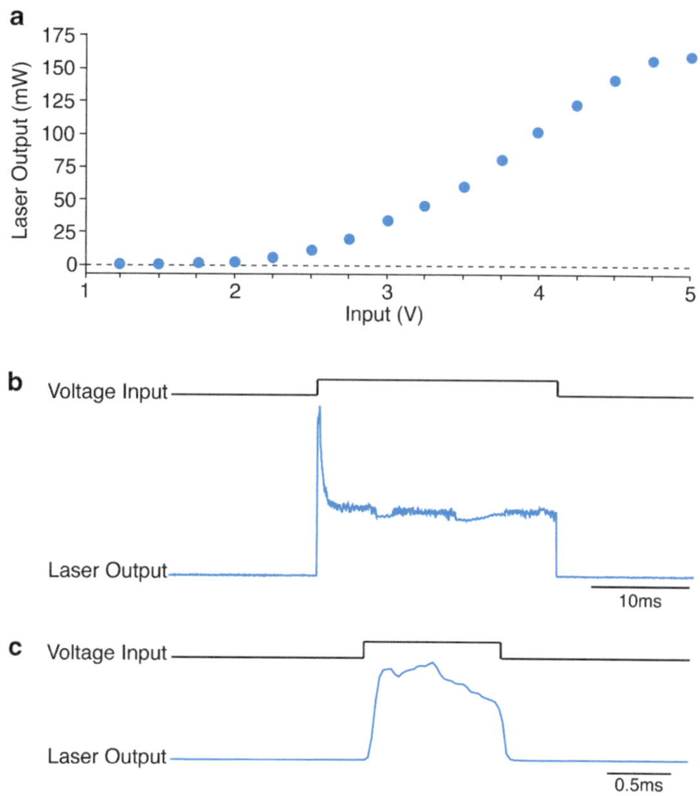

Fig. 4. Nonlinear aspects of laser function. (**a**) Small lasers have a generally sigmoidal input–output function, as shown by this example from a nominally 150 mW 473 nm laser. Below the voltage threshold for activation (<2 V), no light is emitted. Above the threshold (>2 V, light power increases rapidly before saturating at 169 mW. (**b**) Pulsed activation of the laser anywhere in the input–output range generates a time-varying response, with an initial peak in power followed by a plateau. (**c**) Short pulses (<2 ms) take advantage of the relative uniformity of the initial peak and produce a light power profile that approximates a step function.

of a 200 μm core diameter fiber produces ~64 mW/mm^2, considerably above the minimal level for in vivo stimulation. Most optical fibers can be mounted on a standard electrode manipulator and positioned with micron accuracy in the craniotomy. For fibers that are lowered into the tissue, use an appropriate fiber-stripping tool to remove the cladding from the end of the fiber to minimize total diameter.

3.2. Stereotaxic Apparatus

The animal should be placed stably in the stereotaxic setup with the head appropriately positioned to use landmarks for injections and recordings. Connect the isoflurane vaporizer output to the input on the anesthesia head holder. Initial isoflurane levels immediately after induction may be as high as 2%, but can be reduced to ~0.5% as the animal's anesthesia state becomes stable.

3.3. Virus and Pipette Preparation

Thaw the selected viral vector (see Note 4.1). Add a small amount (~0.05–0.1 μl) of Fast Green to 1 μl of the virus solution. Pull glass pipettes for injection. These should have a shank of several mm and a sharp tip. Using a syringe and a Microfil, fill the pipette completely to the tip with mineral oil. Break the tip back slightly with forceps or scissors to allow free flow of the oil. Final tip diameter should be ~40–80 μm.

3.4. Virus Injection

1. Anesthetize the animal briefly in the induction chamber and shave the scalp above the area of interest.
2. Place the animal securely in the stereotaxic apparatus on a heating blanket (37°C) in a sterile surgical environment or as required by institutional policy. Make sure the isoflurane vaporizer feed is at an appropriate level. Test the animal's withdrawal reflex by pinching a toe to ensure that a surgical level of anesthesia has been achieved prior to proceeding with surgery.
3. Lift the scalp with forceps and administer several small subcutaneous injections of lidocaine or a similar local anesthetic. Lidocaine will take ~2–3 min to be effective.
4. Using a sterile scalpel blade, made a single incision through the scalp. The incision should run along the anterior–posterior axis to expose stereotaxic markers on the skull. Push aside the underlying connective tissue to expose the skull.
5. Using appropriate stereotaxic coordinates, mark the site of intended virus injection.
6. Make a single burr hole in the skull with the drill at the marked injection site. The drill bit should just reach the bottom of the skull, and a thin layer of skull may remain after drilling. The overall diameter of the burr hole should not exceed 100 μm. Using a 30 g syringe needle, remove a small piece of any remaining skull tissue so that the dura can be seen. The open area of dura should be no more than several tens of microns in diameter. Flush the drill site with sterile saline during and after drilling to remove bone dust and reduce heat generated by the drill bit. Excessive pressure can cause the drill bit to penetrate to the dura or underlying tissue, causing damage and tissue inflammation post-injection.
7. Fill the tip of the prepared injection pipette with ~1 μl of the virus previously tinted with Fast Green. Position the pipette over the injection site and lower until the tip touches the exposed dura. This point is the vertical "zero."
8. Lower the pipette through the dura to the desired tissue depth.
9. Inject virus at a rate of not more than 0.1 μl/min. When the injection is completed, allow a minimum of 10 min additional rest time before retracting the pipette. Raise the pipette slowly

in two stages, allowing another 5 min of rest at the halfway point. Rapid virus injection or pipette withdrawal will cause the virus to flow up the pipette track, reducing the actual injected volume.

10. Pull the edges of the skin incision together and clean lightly with Betadine. Holding the edges together with forceps, make two sutures using standard surgical techniques. Apply small amounts of cyanoacrylate glue to seal the skin and cover the sutures. The burr hole does not require additional filling or sealing.
11. Allow the animal to recover from the anesthetic and treat with an analgesic (Buprenorphine or an NSAID) for 3 days following the surgery.
12. For cortical and hippocampal injections of AAV DIO ChR2-mCherry, 6 days post-injection recovery time is sufficient to generate functional levels of ChR2 expression in both inhibitory and excitatory neurons. Longer wait times will allow higher expression levels of ChR2 and may be necessary for other brain regions. Halorhodopsin expression requires longer wait times (see Note 4.2). Lentivirus expression may require 10–14 days to peak.

3.5. Optogenetic Stimulation and Electrophysiological Recordings

1. After recovery and expression time, anesthetize the animal and apply local anesthetic as described above. Using forceps, remove any remaining glue and sutures from the scalp. Reapply local anesthetic at intervals throughout the experiment.
2. Using the burr hole as a marker, make a new craniotomy next to the injection site. One side of the craniotomy may adjoin the injection site, but the electrode site should be about 200 μm away from the injection to reduce the possibility of recording from damaged or inflamed tissue. The craniotomy should be large enough to accommodate positioning both the optical fiber and the electrodes. As before, the drill should not touch the dura, or the underlying brain tissue may become damaged. Bleeding will significantly reduce transmission of laser light and will contribute to edema.
3. Using a fine syringe tip (30 g), size 5 forceps, or very fine dissecting scissors, make a small incision in the dura to allow access for electrodes. Avoid compressing the tissue while pulling on the dura.
4. Place the extracellular electrodes in the brain tissue at the desired angle and depth by lowering them with a manipulator. If more than one electrode is being used, or if a single large electrode is used, allow approximately 30 min for recovery after electrode placement before commencing recordings. Note that both local field potential and spiking signals can be disrupted by damage

from electrode movement. If intracellular electrodes are being used, complete step 5 before placing intracellular pipettes.

5. Place the optical fiber close to the electrodes so that the cut surface of the fiber is flat against the exposed dura. Ensure that no blood remains on the dura, as this will block the transmission of light. If the optical fiber is being placed in the brain tissue, lower the exposed fiber section in a manner similar to that used for electrodes. For intracellular recordings, lower the pipette and apply warm agar after positioning the optical fiber.
6. Apply brief light pulses at a moderate intensity (~0.5–10 ms duration at 20–70 mW/mm^2) to evoke or suppress activity from cells expressing light-activated channels (see Note 4.3). If no change in spiking activity is observed in response to light stimuli, continue moving electrodes in small steps (<5 μm) until evoked spiking or suppression occurs. If using eNpHR, it may be necessary to give longer light pulses to observe an impact on spontaneous spiking (see Note 4.2). If optoelectric artifacts are observed, adjust the angle between the electrodes and optical fiber (see Note 4.5).

4. Notes

4.1. Cell-Type Specificity: Choosing a Vector and Mouse Line

Because different viruses and serotypes exhibit distinct tropisms for particular cell types (27, 29, 31, 36), it is critical to select an appropriate viral vector. Recent work has used both lentiviruses (5, 9–11) and AAVs (7, 13–15, 34) to express light-activated channels in hippocampus and cortex. The degree of viral spread from the injection site varies with virus and tissue type. One advantage of using AAV is its broader spatial spread, which allows experimental manipulations some distance from the initial injection site. We routinely observe expression over a distance of up to 2 mm of cortical tissue (34). Because this spread is useful for many in vivo experiments, the protocol outlined here specifies the use of an AAV. However, the same procedures may be used for a lentivirus as long as additional safety precautions (BSL2) are taken.

The use of a double-floxed (DIO) construct allows expression of light-activated channels specifically in the subpopulation of neurons that express Cre recombinase. A large number of commercially available transgenic, BAC, and knock-in mouse lines express Cre in specific populations of excitatory neurons (e.g., CW2, T29-1, T29-2) or inhibitory neurons (e.g., PV-Cre). The DIO construct confers a high degree of specificity of ChR2 expression (7, 14, 34). However, individual Cre lines can exhibit varied degrees of specificity of Cre expression. A nonspecific Cre line will result in expression of light-activated channels in a varied population of cells, despite the DIO construct. A fluorescent tag fused to the expressed channel, such as mCherry or eYFP, allows for post-hoc identification and mapping of the ChR2-expressing population and confirmation of the specificity of that expression. It is important to also make control injections of the virus into mice that do not express Cre recombinase to confirm that expression of the light-activated channels is Cre-dependent.

4.2. Channelrhodopsin vs. Halorhodopsin

The protocol outlined here can be used for experiments involving ChR2 or eNpHR. Halorhodopsin routinely requires longer expression times to reach a functionally effective level. Whereas ChR2 expression levels may be high enough to recruit neuronal firing after 6–10 days, eNpHR may not reach effective levels for 18–24 days. Halorhodopsin is activated by ~593 nm light, which requires a separate laser. Once activated, eNpHR rapidly adapts to the light stimulus and becomes somewhat less effective (17, 18), suggesting that long stimulus times may not produce optimal effects. An additional light-activated silencer, Arch, has recently become available (19) and may offer an alternative method for hyperpolarizing target cells.

Detecting the impact of eNpHR activation in vivo may be challenging. Because cells in the intact brain receive a constant barrage of synaptic inputs, the 5–10 mV hyperpolarization evoked by eNpHR activation is unlikely to completely eliminate firing. One potential method of identifying cells expressing light-activated channels is to co-express ChR2 and eNpHR. However, the activation spectra for ChR2 and eNpHR overlap slightly, (17) raising the possibility that light stimulation may evoke a small amount of opposing depolarization.

Early versions of Halorhodopsin exhibited a high degree of cellular toxicity due to trafficking error in the endoplasmic reticulum. Halorhodopsin version 3.0 has eliminated much of this issue and appears to be correctly inserted into the cell membrane (18). However, to date there has not been a long-term study of the impact of ChR2 or eNpHR3.0 expression on cell health.

4.3. Electrophysiology

Single extracellular electrodes or multielectrode arrays may be placed several hundred microns away from the injection site, ensuring that recordings are made from undamaged tissue. Optogenetic stimulation also works well in conjunction with intracellular recording techniques. Latency to light-evoked spiking will vary significantly with cell type, stimulation intensity, tissue depth, region of the cell stimulated, and anesthesia state. For anesthetized experiments, the EEG or LFP should be continuously monitored to confirm the stability of the animal's state. Control recordings should be made from untransduced tissue to confirm that light-evoked signals are the result of the optogenetic manipulation.

If no light-evoked activity is observed, several steps may help identify the underlying problem. First, histology should always be performed to confirm expression of the light-activated protein. Second, viral transduction does not always occur uniformly around the injection site. Moving electrodes in small steps around the injection site can help locate areas of high expression. Third, light levels may not be sufficient to evoke activity or the light may not be transmitted to the brain. It is critical to measure the light power at the tip of the optical fiber and to make sure no blood obscures the tip. Fourth, the targeted cell type may be sparsely present in the tissue, requiring a careful search with electrodes. Finally, the cells expressing the light-activated channel may be too far from the fiber to be activated by light at the surface. Lowering a small diameter fiber into the tissue may be more effective.

4.4. Light Stimulation Parameters

Stimulation via an optical fiber, outlined in the protocol here, is one of the simplest and more effective methods for light stimulation. Light from the point source at the tip of the optical fiber spreads in an approximate cone through neural tissue. Light power from a fiber at the surface of the cortex decreases with tissue depth, resulting in variability of the effective light drive with distance from the light source (37). Surface positioning of an optical fiber provides effective stimulation of cells in the supragranular and granular layers of cortex. Light stimulation can be delivered through mouse dura or thinned skull. Alternatively, optical fibers of small diameter (25–100 μm) can be placed in the brain tissue to stimulate deeper cells. Light intensity is measured in mW/mm^2, and a safe stimulation range for in vivo stimulation is up to ~75 mW/mm^2 for short pulses (0.5–50 ms). We have observed tissue damage in cortex immediately under the site of optical stimulation from sustained stimulation (>500 ms) at levels above 100 mW/mm^2. Damage caused by very high levels of light power can be observed as a compressed area of tissue at the site of stimulation.

Very long or continuous stimulation can cause abnormal activity patterns and excitotoxicity. The pulsed output of a small laser is usually nonlinear, with an initial peak followed by a drop to a steady level (Fig. 4b). Short pulses take advantage of the reliability of the initial peak and provide a stimulation waveform approximating a step function (Fig. 4c). For experiments requiring long pulses of stimulation, it may be advantageous to use a steady-state light source and either a mechanical shutter or a Pockels cell to deliver timed pulses of light stimulation. In turn, power transmitted by the optical fiber will vary with fiber type (multimode or single mode) and numerical aperture (NA), with a small NA giving a narrow angle of light spread from the fiber and a high NA giving a broader angle of spread. Power will also vary as a function of the radius (r) of the fiber, as the area of the initial cross-section of the light at the tip of the fiber increases as πr^2.

4.5. Optoelectric Artifacts

Direct and indirect illumination of metal by a laser beam will result in electrical artifacts (10, 34, 35). These effects can be elicited by laser light hitting the tip of a metal electrode, a ground electrode, or a silver wire in a glass pipette. Optoelectric artifacts are proportional in amplitude to the total laser power and the amount of exposed metal and have been demonstrated in recordings from tungsten, platinum/iridium, nichrome, steel, gold, and silver electrodes. The onset of the artifact signal is precisely aligned with the onset of the light stimulus (Fig. 5a, b), rather than showing the several millisecond delay of evoked neural activity. Light-evoked artifacts typically exhibit both an onset ramp that continues for the duration of the light if the pulse is brief (~1 ms) and a slower offset time constant (~2–10 ms decay time for a 1 ms pulse) that can appear as an overshoot followed by a return to baseline (Fig. 5a). Optoelectric artifacts can contaminate both LFP and intracellular recordings and do not depend on the presence of ChR2 or eNpHR. These artifacts have little effect on extracellular recordings of action potentials and are a less significant problem if neither intracellular nor field potential recordings are necessary.

Several strategies have been successfully used to minimize the contribution of light-evoked artifacts to electrophysiological recordings in vivo. The most effective method is to reduce the total exposure of metal to the light by limiting the duration of the light pulse and decreasing the exposed metal surface. For field potential or intracellular recordings, one effective approach is to use glass electrodes with a distal metal wire insert that is not contacted by direct or reflected light. For field potential recordings made with glass electrodes, adding a nonreflective, opaque coating to the glass (e.g., acrylic paint or lacquer) will further reduce light transmission to the silver wire.

While glass electrodes are a useful solution, many in vivo experimental applications require metal electrodes, either for chronic recordings or for multiple parallel recordings of neurons in different sites. We have found that tungsten electrodes with small exposed tip surfaces, such as twisted wire stereotrodes or tetrodes with wire diameters <20 μm, exhibit smaller optoelectric artifacts than do electrodes with larger exposed surfaces, such as standard prefabricated tungsten electrodes (FHC, Inc.) or laminar silicon probes (Neuronexus, Inc.). In simultaneous local field recordings in cortex in vivo, optically induced artifacts on twisted wire stereotrodes were ~85% smaller than those on nearby commercially produced single tungsten electrodes (34).

An important additional step to reduce optoelectric artifacts is to adjust the relative angle between the light source and the recording electrodes (Fig. 5c). Directly attaching an optical fiber to a parallel metal electrode or laminar electrode is useful for experiments requiring optogenetic manipulations in deep cortical layers or subcortical structures. This combined configuration gives

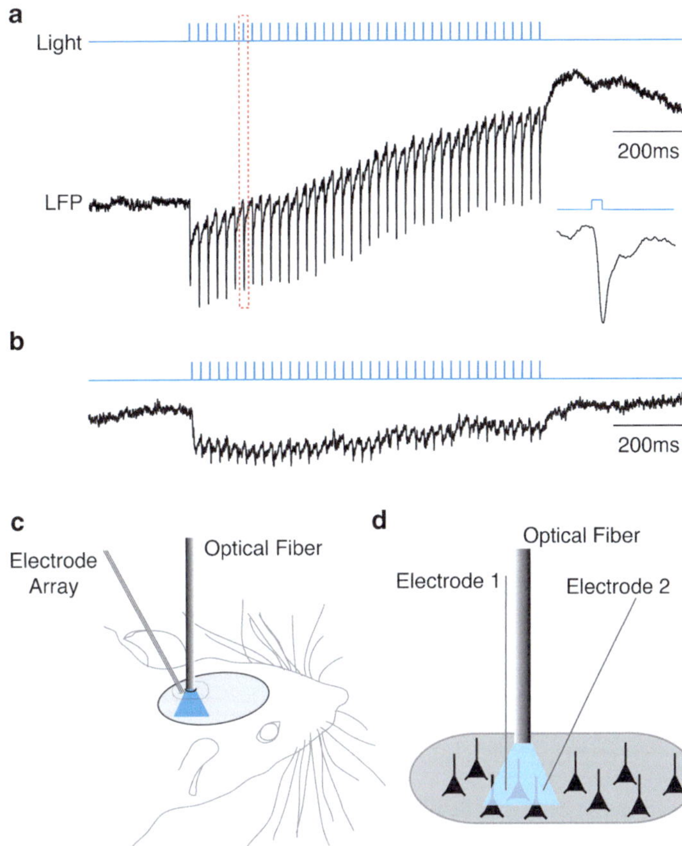

Fig. 5. Optoelectric artifacts. Laser light striking metal electrodes generates an optoelectric artifact with both fast and long time constants in the local field potential. (**a**) Blue light stimulation (40 Hz, 1 ms pulses) of an untransduced cortical area with no ChR2 expressed. In this example, the light is directly striking the electrode (12 μm tungsten wire stereotrode). Each light pulse elicits a fast transient (*inset panel*), and the continuing series of pulses also generates a large deflection of the LFP that lasts for several hundred milliseconds after the light has turned off. (**b**) Similar artifacts generated by indirect light striking the electrode in an area expressing ChR2. (**c**) One strategy for reducing these artifacts is to adjust the angle between the electrode and the light source. (**d**) Increasing the relative angle reduces the amount of metal exposed to the light cone from the optical fiber.

precise placement of the optical stimulus near the recorded neural population and allows the stiffer electrode to provide mechanical stability to the optical fiber. However, this approach increases the undesired light exposure to the portion of the electrode shank that is recording neural signals and thus increases optoelectric artifacts. Near-complete elimination of light-evoked artifacts can be achieved by using thin wire electrodes and adjusting the angle of the electrodes so that minimal exposed metal lies directly in the cone of laser light beneath the tip of the fiber (Fig. 5d). Bipolar electrode recordings eliminate small amplitude optical artifacts by subtracting locally generated signals, but may not completely eradicate larger artifacts.

Acknowledgments

This work was funded by NIH/NEI R00 EY018407, the Whitehall Foundation, and the Esther A. and Joseph Klingenstein Foundation.

References

1. Ascoli GA, Alonso-Nanclares L, Anderson SA et al (2008) Petilla terminology: nomenclature of features of gabaergic interneurons of the cerebral cortex. Nat Rev Neurosci 9 (7):557–568
2. Markram H, Toledo-Rodriguez M, Wang Y et al (2004) Interneurons of the neocortical inhibitory system. Nat Rev Neurosci 5 (10):793–807
3. Moore CI, Carlen M, Knoblich U et al (2010) Neocortical interneurons: from diversity, strength. Cell 142(2):189–193. doi:S0092-8674(10) 00774-9 [pii]. 10.1016/j.cell.2010.07.005
4. Pare D, Shink E, Gaudreau H et al (1998) Impact of spontaneous synaptic activity on the resting properties of cat neocortical pyramidal neurons in vivo. J Neurophysiol 79 (3):1450–1460
5. Boyden ES, Zhang F, Bamberg E et al (2005) Millisecond-timescale, genetically targeted optical control of neural activity. Nat Neurosci 8(9):1263–1268
6. Kuhlman SJ, Huang ZJ (2008) High-resolution labeling and functional manipulation of specific neuron types in mouse brain by cre-activated viral gene expression. PLoS One 3 (4):e2005
7. Sohal VS, Zhang F, Yizhar O et al (2009) Parvalbumin neurons and gamma rhythms enhance cortical circuit performance. Nature 459(7247):698–702
8. Cardin JA, Carlen M, Meletis K et al (2009) Driving fast-spiking cells induces gamma rhythm and controls sensory responses. Nature 459(7247):663–667
9. Gradinaru V, Mogri M, Thompson KR et al (2009) Optical deconstruction of parkinsonian neural circuitry. Science 324(5925):354–359
10. Han X, Qian X, Bernstein JG et al (2009) Millisecond-timescale optical control of neural dynamics in the nonhuman primate brain. Neuron 62(2):191–198
11. Huber D, Petreanu L, Ghitani N et al (2008) Sparse optical microstimulation in barrel cortex drives learned behaviour in freely moving mice. Nature 451(7174):61–64
12. Petreanu L, Huber D, Sobczyk A et al (2007) Channelrhodopsin-2-assisted circuit mapping of long-range callosal projections. Nat Neurosci 10(5):663–668
13. Petreanu L, Mao T, Sternson SM et al (2009) The subcellular organization of neocortical excitatory connections. Nature 457 (7233):1142–1145
14. Tsai HC, Zhang F, Adamantidis A et al (2009) Phasic firing in dopaminergic neurons is sufficient for behavioral conditioning. Science 324 (5930):1080–1084
15. Adesnik H, Scanziani M (2010) Lateral competition for cortical space by layer-specific horizontal circuits. Nature 464(7292):1155–1160. doi: nature08935 [pii]. 10.1038/nature08935
16. Arenkiel BR, Peca J, Davison IG et al (2007) In vivo light-induced activation of neural circuitry in transgenic mice expressing channelrhodopsin-2. Neuron 54(2):205–218
17. Gradinaru V, Zhang F, Ramakrishnan C etal (2010) Molecular and cellular approaches for diversifying and extending optogenetics. Cell 141(1):154–165. doi:S0092-8674(10) 00190-X [pii]. 10.1016/j.cell.2010.02.037
18. Zhang F, Wang LP, Brauner M et al (2007) Multimodal fast optical interrogation of neural circuitry. Nature 446(7136):633–639. doi: nature05744 [pii]. 10.1038/nature05744
19. Chow BY, Han X, Dobry AS et al (2010) High-performance genetically targetable optical neural silencing by light-driven proton pumps. Nature 463(7277):98–102. doi: nature08652 [pii]. 10.1038/nature08652
20. Marin O, Rubenstein JL (2003) Cell migration in the forebrain. Annu Rev Neurosci 26:441–483
21. Saito T, Nakatsuji N (2001) Efficient gene transfer into the embryonic mouse brain using in vivo electroporation. Dev Biol 240(1): 237–246
22. Tabata H, Nakajima K (2001) Efficient in utero gene transfer system to the developing mouse brain using electroporation: visualization of neuronal migration in the developing cortex. Neuroscience 103(4):865–872

23. Tabata H, Nakajima K (2002) Neurons tend to stop migration and differentiate along the cortical internal plexiform zones in the reelin signal-deficient mice. J Neurosci Res 69 (6):723–730
24. Tabata H, Nakajima K (2003) Multipolar migration: the third mode of radial neuronal migration in the developing cerebral cortex. J Neurosci 23(31):9996–10001
25. Wonders C, Anderson SA (2005) Cortical interneurons and their origins. Neuroscientist 11(3):199–205
26. Borrell V, Yoshimura Y, Callaway EM (2005) Targeted gene delivery to telencephalic inhibitory neurons by directional in utero electroporation. J Neurosci Methods 143(2):151–158
27. Jasnow AM, Rainnie DG, Maguschak KA et al (2009) Construction of cell-type specific promoter lentiviruses for optically guiding electrophysiological recordings and for targeted gene delivery. Methods Mol Biol 515:199–213
28. Nathanson JL, Yanagawa Y, Obata K et al (2009) Preferential labeling of inhibitory and excitatory cortical neurons by endogenous tropism of adeno-associated virus and lentivirus vectors. Neuroscience 161(2):441–450
29. Fu H, Muenzer J, Samulski RJ et al (2003) Self-complementary adeno-associated virus serotype 2 vector: global distribution and broad dispersion of aav-mediated transgene expression in mouse brain. Mol Ther 8 (6):911–917
30. Taymans JM, Vandenberghe LH, Haute CV et al (2007) Comparative analysis of adeno-associated viral vector serotypes 1, 2, 5, 7, and 8 in mouse brain. Hum Gene Ther 18 (3):195–206
31. Van Vliet KM, Blouin V, Brument N et al (2008) The role of the adeno-associated virus capsid in gene transfer. Methods Mol Biol 437:51–91
32. Atasoy D, Aponte Y, Su HH et al (2008) A flex switch targets channelrhodopsin-2 to multiple cell types for imaging and long-range circuit mapping. J Neurosci 28(28):7025–7030
33. Livet J, Weissman TA, Kang H et al (2007) Transgenic strategies for combinatorial expression of fluorescent proteins in the nervous system. Nature 450(7166):56–62
34. Cardin JA, Carlen M, Meletis K et al (2010) Targeted optogenetic stimulation and recording of neurons in vivo using cell-type-specific expression of channelrhodopsin-2. Nat Protoc 5(2):247–254. doi:nprot.2009.228 [pii]. 10.1038/nprot.2009.228
35. Ayling OG, Harrison TC, Boyd JD et al (2009) Automated light-based mapping of motor cortex by photoactivation of channelrhodopsin-2 transgenic mice. Nat Methods 6(3):219–224
36. Tenenbaum L, Chtarto A, Lehtonen E et al (2004) Recombinant aav-mediated gene delivery to the central nervous system. J Gene Med 6(Suppl 1):S212–S222
37. Aravanis AM, Wang LP, Zhang F et al (2007) An optical neural interface: in vivo control of rodent motor cortex with integrated fiberoptic and optogenetic technology. J Neural Eng 4 (3):S143–S156

Neuromethods (2012) 67: 357–372
DOI 10.1007/7657_2011_11

Published online: 3 December 2011

Combining Optical Stimulation with Extracellular Electrophysiology in Behaving Mice

Joshua H. Siegle

Abstract

Optical manipulation allows researchers to alter the activity of genetically defined neural populations with millisecond-scale temporal accuracy. This facilitates the activation and inactivation of functionally distinct subnetworks at precise times relative to behavior, something that was not previously possible. This chapter describes the construction of an implant for simultaneous light delivery and extracellular electrophysiology in unanesthetized, unrestrained mice. The implant contains eight stereotrodes surrounding a central fiber optic cable, is compact and lightweight, and can be modified for other recording/stimulation configurations.

Key words: Optogenetics, Channelrhodopsin, Stereotrodes, Chronic recording, Behavior, Mice

1. Introduction

Perturbing cellular activity with light, a subset of the field of optogenetics, is on its way to becoming a standard technique in neuroscience. The range of such interventions that have become available over the last five years is astounding, and there are undoubtedly more to come. Optical manipulation has already been applied in species ranging from *Caenorhabditis elegans* (1) to primates (2). In mice, light-based activation of genetically defined cell populations in deep brain nuclei (3), subregions of cortical columns (4), and the retinal surface (5) are a few of many examples that showcase the flexibility of this technique.

The popularity of optical manipulation is likely to endure because this method fills a critical void in systems neuroscience: the ability to perturb specific cell types with flexible spatial resolution *on the time-scale of behavior*. Previously available methods, such as pharmacology and conditional genetics, can target specific cell types, or even specific receptors, but their effects last orders of magnitude longer than most behavioral events. Electrical stimulation, on the other hand, can activate essentially random groups of cells on submillisecond timescales (6), but more functionally specific interventions on the network scale remain untenable. Optical manipulation combines the functional specificity of conditional genetics and pharmacology with the temporal precision of electrical stimulation, making it a powerful tool.

To find links between network activity and organism-level behaviors, it is desirable to simultaneously perturb and monitor network activity in unanesthetized animals. At the very least, this involves delivering light through the skull, reading out brain states (via electrical, optical, or magnetic means), and ensuring that the animals involved are treated humanely. To carry out experiments that involve stimulation and recording over the course of many days, or require that the animals be able to move freely, part of the preparation must be packaged into a compact implant that can remain on the animal's head indefinitely. For mice, this presents considerable challenges.

This chapter contains instructions for building a chronic implant that enables simultaneous optical stimulation and extracellular electrophysiology in awake mice. The implant is easy to construct, customizable, and inexpensive. The final implant is 19 mm high, 24 mm in diameter, and weighs around 1.5 g, small enough for mice to carry without strain. It has been used to stimulate and record from light-sensitive cells in cortex, thalamus, and hippocampus (Fig. 1), and will be adaptable for use in other regions. The chronic nature of the implant means that recording and stimulation can continue in the same animal for many months.

1.1. Issues to Consider

Moving from acute to chronic preparations requires overcoming a number of technical hurdles. (For a description of experiments in anesthetized animals involving optogenetics and electrophysiology, see (4, 7, 8); for detailed methods, see (9) and Cardin et al. (this volume).) All of the electrodes, light guides, and connectors must fit into a container that is lightweight, compact, and durable enough to remain on the animal for weeks to months. The implant must have interfaces for light delivery that allow the animal to be plugged-in and unplugged many times without the need for anesthesia, and without any discomfort.

These requirements impose limits on the number of electrodes that can be used, and the area of brain tissue that can be exposed to light. Both experimental and theoretical studies of light penetration through the brain have estimated the range of effective illumination for a single light source to be on the order of 1 mm^3 (10–12). Reaching a larger region requires more light sources.

Combining optical stimulation and electrophysiology also involves contending with light artifacts, which occur when photons strike the interface between a metal electrode and a conductive

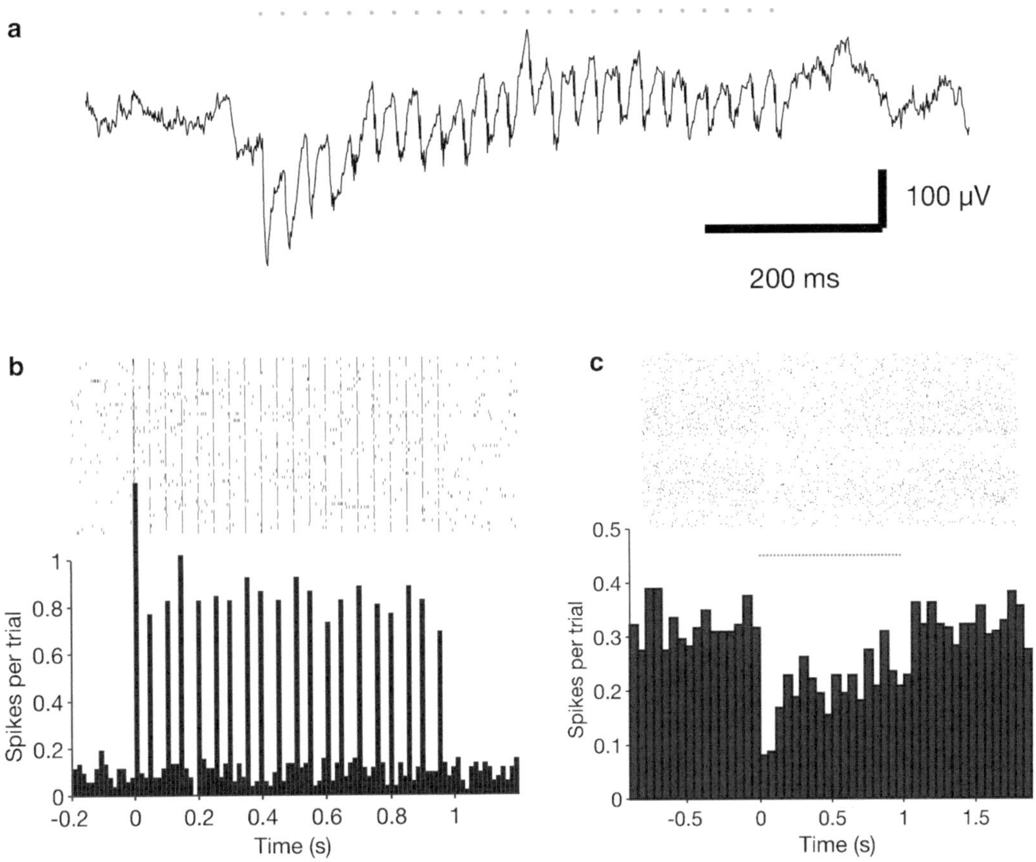

Fig. 1. Example recordings with the implant described in this chapter (or a variant with movable electrodes). All recordings were obtained in awake animals expressing light-sensitive ion channels. Recordings took place between 2 weeks and 2 months after implantation. (**a**) Mean local field potential in cortex in response to repeated trains of 1 ms light pulses at 40 Hz (indicated by *gray dots*). (**b**) Raster plot and histogram for a hippocampal single-unit driven by 1 ms light pulses at 20 Hz beginning at 0 s. (**c**) Suppression of a single unit in thalamus in response to 2 ms light pulses at 40 Hz (indicated by *gray dots*).

solution. Light artifacts have been characterized in previous optogenetic experiments (2, 9). If one does not take precautions to avoid them, artifacts from long light pulses can interfere with local field potential recordings, sometimes saturating the signal. In the most insidious cases, short pulses of light can actually cause fast deflections of the electrical signal that may be detected as spikes. Special coatings have been developed for reducing artifact amplitude (13). Another effective solution (implemented here) is to move the electrodes far enough away from the light source that artifacts do not impair the recordings.

1.2. The Range of Options Available

Once these limitations are considered, the approach must be refined to suit one's experimental needs. Previously published methods for combining optogenetics and electrophysiology in behaving mice can be divided into three categories.

In the first, an electrode is attached to a fiber optic and inserted through a guide cannula into the region of interest (14, 15). The main advantage of this technique is that it is simple to build and easy to implant. However, it has significant drawbacks, including the difficulty of inserting and removing the fiber/electrode combination, the inability to record from the same cell populations over the course of multiple sessions, the limited region of recording, and the additional damage done by inserting a cannula. There is also a risk of light artifact when the light source is close to the recording electrode.

To increase the number of recording sites, others have combined fiber optics with silicon probes (16). There have also been efforts to design customizable probes that can target arbitrary 3D locations (17). This is a tenable option for chronic recording, although the probes can be delicate and expensive.

The third technique involves modifying existing designs for movable tetrode wire arrays with a fiber optic cable that can deliver light to the tissue surrounding the electrodes (18, 19). Movable tetrodes increase the yield and stability of single units. Although building these implants is labor-intensive, they can be easily customized and are inexpensive in comparison to commercially available silicon probes.

1.3. The Current Design

The implant design described in this chapter is most similar to the third technique, but differs in one crucial regard. Instead of using movable electrodes, this implant contains an array of stationary electrodes surrounding a central fiber optic cable. This dramatically reduces the build time, and the final implant weighs a small fraction of what it otherwise would. Stationary tetrodes will obviously make it more difficult to isolate single units in the recordings, but for many applications, the yield will be sufficient. In practice, each stereotrode yields one to three single units in cortex and thalamus. For other areas, yield might be higher or lower, depending on the density and activity of cells. Stationary tetrodes are also not optimal for targeting small nuclei or subregions or larger structures, as fine adjustments of electrode location may be necessary to hit the desired target.

Many aspects of this design had to be customized, as there is currently no commercially available solution for combining optical stimulation and electrophysiology in chronic preparations. The design described in this chapter includes a custom plastic body, which was printed from a digital model using stereolithography. There is also a custom electrode interface board (EIB), designed with freely available software, which was necessary to accommodate the centrally placed fiber optic cable. All files for ordering and modifying these parts are available online, which will allow readers to have everything manufactured as is, or to customize them as they see fit (see Note 1 for more information on altering the design).

Another important aspect of this design is the type of fiber optic connector used. Standardized connectors (such as LC, FC/PC, and SMA) are too heavy and bulky for use with mice. Instead, the current design uses a fiber optic encased in a metal ferrule. An identical metal ferrule within a zirconia sleeve is used to attach the fiber on the animal to a second fiber connected to a light source. These parts are available from Doric Lenses (Québec, Canada), an optical components supplier that now specializes in optogenetics equipment.

These implants can be used to determine how the effects of optical manipulation performed in anesthetized animals differ in awake animals. Many recent optogenetic experiments have characterized physiological effects of optical stimulation in anesthetized animals, while using optical stimulation alone in behaving animal. Confirming that the manipulations are similar in both scenarios will be a crucial control.

Most importantly, the technique described in this chapter will allow experimenters to monitor how changes in optogenetic drive change with both neural *and* behavioral states. It can also be used to explore the effects of activating or inactivating certain regions at precise times relative to behavior. Combining optogenetics and electrophysiology in awake animals will allow us to learn more about the function of specific cell types, or to co-opt cells with known functions to achieve a particular physiological or behavioral outcome. In conjunction with digital signal processing, the implant described here can be used in experiments involving online feedback. It will also be a valuable tool in applying optogenetics for understanding and treating mouse models of disease, which are likely to have both behavioral and neurophysiological phenotypes.

Extensions of this technique allow simultaneous recording and stimulation across more regions. In addition, the use of a wireless headstage, combined with head-mounted LEDs, could alleviate the need to use tethers during the experiment.

In the following sections, detailed methods for constructing an implant for simultaneous chronic optical manipulation and electrophysiology are presented.

2. Materials

Label	Item	Quantity per implant	Supplier	Notes
A	Body	1	Stereolithography company, such as APProto	STL and SolidWorks files available online; printed using Accura 55 (ABS-like) plastic
B	Electrode interface board	1	Sunstone Circuits	PCB123 files available online
C	Omnetics connector	1	Neuralynx	18-Pin male connector
D	Large gold pins	18	Neuralynx	Available in quantities of 500 or 1,000
E	Implantable fiber optic patch cord	1	Doric Lenses	Item # MFC_200/260/-0.22_100mm_1.25_MFR
F	17-Gauge thin-wall stainless steel cannula	1 (3.5 mm)	Small Parts	Item # HTX-17T
G	20-Gauge stainless steel cannula	8 (2.0 mm each)	Small Parts	Item # HTX-20T
H	29-Gauge stainless steel cannula	1 (25 mm)	Small Parts	Item # HTXX-29R
I	30-Gauge polyimide tubing	8 (20 mm each)	Small Parts	Item # TWPT-0071
J	Stainless steel ground wire	30 mm	A-M Systems	Item #791400
	Transparency film	1 Sheet		Template for protective cone available online
	Aluminum foil	Approximately 50 mm × 50 mm		
K	0-80 screws, 1/4″	2	Small Parts	Item # B000FN42TC
	Electrode wire	8	Various	See Note 2 for information

See Figs. 2 and 3 for illustrations of each item

General tools:

Stereomicroscope

Fiber cleaver (Thorlabs item S90W) and polishing film (Thorlabs item LFG5P)

Drill and drill bits (#54, #61), available from Small Parts

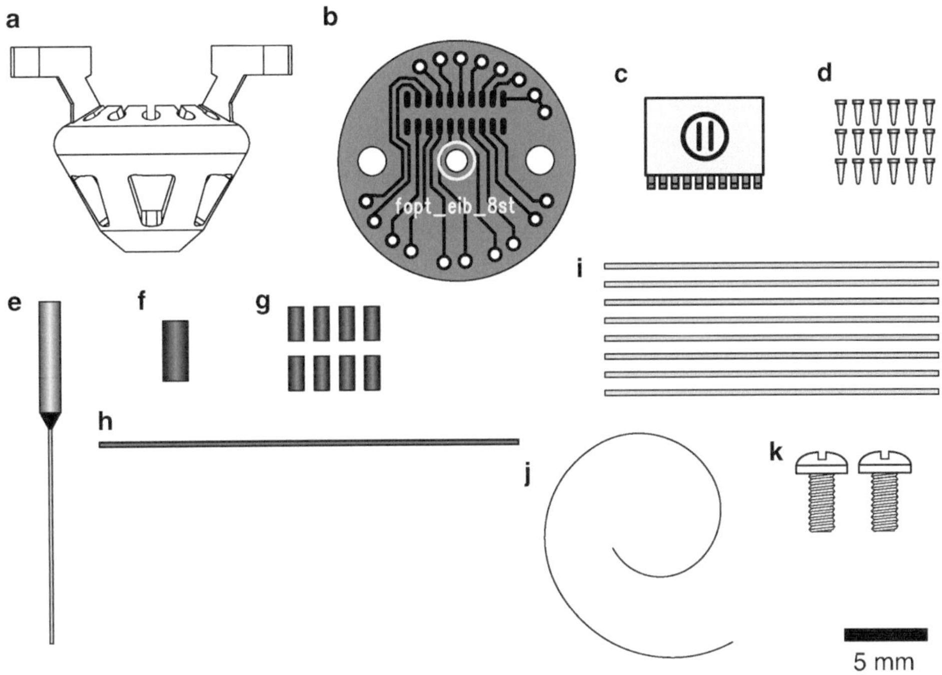

Fig. 2. The raw materials for implant construction. Implant body (**a**), electrode interface board (**b**), connector (**c**), gold pins (**d**), fiber optic (**e**), 17-gauge cannula (**f**), 20-gauge cannulae (**g**), 29-gauge cannula (**h**), 30-gauge polyimide tubes (**i**), ground wire (**j**), 0–80 screws (**k**). All parts are drawn to scale.

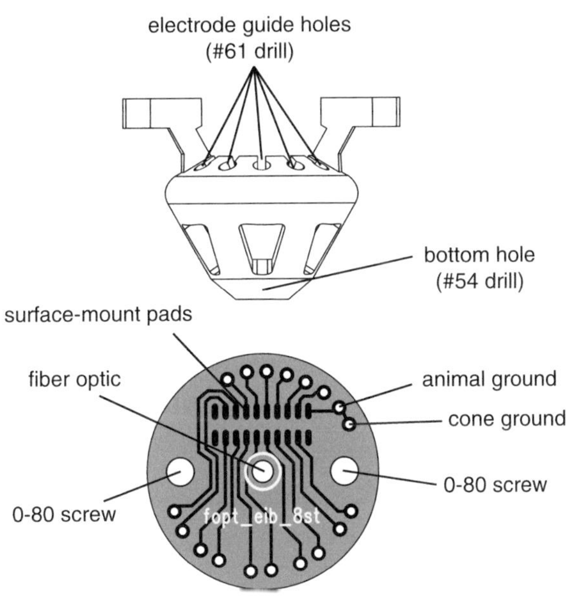

Fig. 3. Important attributes of the implant body (*top*) and the electrode interface board (*bottom*).

Soldering iron, solder, solder wick

Pliers

Scissors

Forceps

Electrode plating station (gold solution, impedance tester, current generator)

Epoxy (Z-Poxy recommended)

Cyanoacrylate glue (Great Planes "Medium" recommended)

Razor blades

Silver Print or a conductive silver pen

Electrical tape

Dial calipers

Black paint marker

3. Methods

The completed implant consists of a computer-designed plastic body for structure, polyimide tubes serving as electrode guides, electrodes connected to an electrode interface board, and a central fiber optic cable to illuminate brain tissue. The whole implant is surrounded by a cone made of foil and plastic, which adds protection as well as electrical shielding. The customized nature of this design stems from the need to accommodate the fiber optic cable. Although almost all aspects of this design can be changed to suit one's experimental needs, following the instructions in this chapter results in a finished product that can be immediately implanted and used for simultaneous electrophysiology and optical manipulation in mice. Refer to Fig. 4 for illustrations of the major steps in the construction process.

3.1. Part 1: Preparing the Implant Body

1. Drill out the bottom hole of one implant body with a #54 drill bit. To ensure that the hole is straight, the use of a drill press is recommended. If used carefully, a handheld drill will also work.
2. Using a #61 drill bit, manually drill out the holes that will serve as electrode guides.
3. Place eight 50-mm polyimide tubes into the 17-gauge stainless steel cannula. About 10% of each polyimide should protrude from one end.
4. Insert the 29-gauge cannula in the center of the polyimides. This cannula will serve as a placeholder for the fiber optic cable, and will eventually be removed. If there is extra space in the

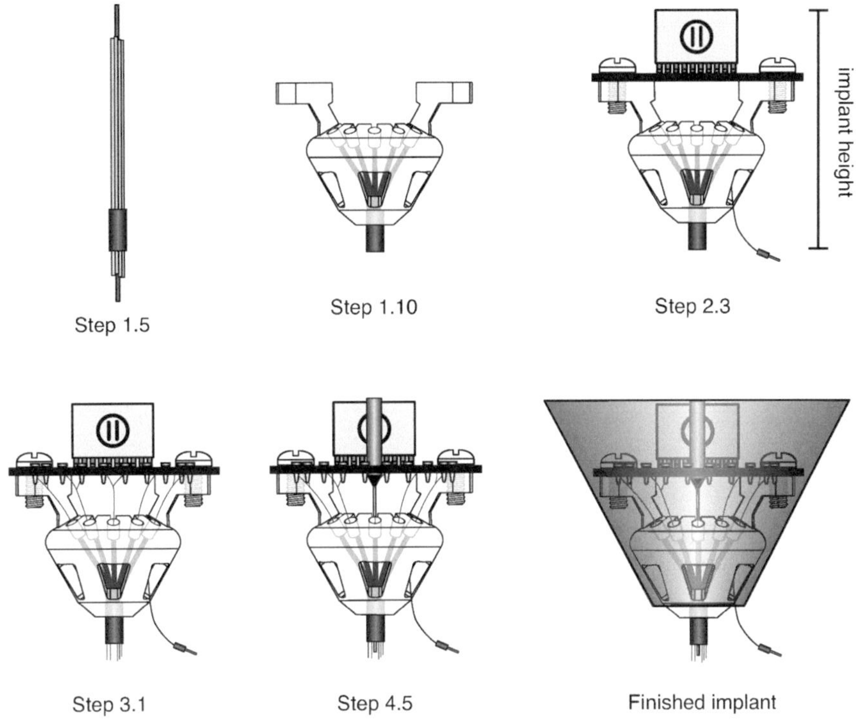

Fig. 4. Illustrations of the major steps involved in constructing the implant. *Step 1.5*: Eight polyimide tubes and fiber placeholder (29-gauge cannula) are placed in the 17-gauge cannula and glued with cyanoacrylate. *Step 1.10*: Each polyimide tube is placed in its respective guide hole and is held in place by a 20-gauge cannula. *Step 2.3*: The electrode interface board is screwed onto the implant body, and the ground wires are put in place. *Step 3.1*: Electrodes are attached to the electrode interface board and cut to the desired length. *Step 4.5*: The fiber optic is placed within the implant and glued in place. *Finished implant*: The protective cone is added as the final step.

cannula, it can be filled with additional polyimide tubes (see Note 1 for information on customizing this configuration).

5. Apply a ring of gap-filling cyanoacrylate glue around the top and bottom of the 17-gauge cannula. Enough glue should be applied to hold the polyimides in place, but too much glue will cause them to adhere to the 29-gauge cannula in the center. Allow glue to dry for at least 1 h.
6. Carefully remove the cannula from the center of the polyimide tubes. If it is stuck in place, try to extract it with a twisting motion. If the cannula does not move, or if some of the polyimides come out along with it, repeat steps 3–6 with a new set of materials.
7. Using a razor blade, cut the shorter end of the polyimides flush with the edge of the cannula.

8. Slide the 17-gauge cannula, polyimide-end first, into the bottom of the implant body, so approximately 2 mm protrudes out the-bottom.
9. Slide a 20-gauge cannula over each polyimide to secure it in its respective hole. Try to keep the polyimides from twisting around one another. While performing this step, you may want to reinsert the 29-gauge cannula to ensure that the hole for the fiber optic remains open. To align the electrodes with the electrode interface board, there should be 5 in one half, and 3 in the other.
10. Once everything is in place, cut the polyimides so they protrude about 1 mm above the cannulae. The 20-gauge cannulae should sit snugly in their holes, but they can also be glued in place with epoxy. Finally, a ring of epoxy should be applied to the point at which the 17-gauge cannula contacts the implant body.

3.2. Part 2: Electrode Interface Board

1. Surface mounting the connector:
   (a) Under a microscope, place a small amount of solder on each of the surface-mount pads on the EIB.
   (b) Align the feet of the connector on the pre-soldered pads. The Omnetics logo must be facing toward the center of the EIB.
   (c) Touch each foot individually with the tip of the soldering iron and press down until it contacts the solder. Repeat until all the feet are connected. A fine-tipped soldering iron is recommended.
   (d) Test the connections between the pins on the connector and the holes in the EIB using a multimeter or impedance meter. Any pins that are not connected should be resoldered. If two adjacent feet are connected by a solder bridge, remove the excess with a solder wick.
   (e) Apply epoxy around the base of the connector to seal everything in place. Be careful not to fill any of the holes in the EIB with epoxy.
2. Attach the EIB to the drive body with 0–80 screws, ensuring that the pin holes are properly aligned. The holes do not need to be tapped. The connector should be placed on the half of the base with the three polyimides.
3. Using gold pins, attach ground wires to the EIB. Strip the insulation off both ends for approximately 3 mm. The first wire (approximately 20 mm) will serve as the animal ground while the second (approximately 10 mm) will be attached to the cone in step 5.4.

4. Measure the distance from the top of the connector to the tip of the cannula. This distance will be crucial for determining the length of your fiber optic.
5. Under a microscope, look at the bottom of the implant. To map each opening to its corresponding channel on the EIB, place a small wire through each polyimide until it sticks out the other end. Make a note (either on a drawing or photograph of the polyimide configuration) of the nearest channel for each opening.

3.3. Part 3: Loading the Electrodes

1. Under a microscope, carefully load the electrodes of your choice into the top of the guide tubes, and connect them to the EIB using gold pins (see Note 2 for information on electrodes). Cutting each electrode to the right length immediately after insertion is recommended. The length will depend on the depth of the targeted brain structure from the cortical surface.
2. Test the connections for each electrode by placing the tips in a saline bath and touching one lead of an impedance meter to the gold pins and the other lead to the saline. If any electrodes are not connected, they should be removed and reloaded.
3. Plate the electrode tips with gold to enhance biocompatibility and reduce impedance. Target impedance should be between 200 and 300 kΩ.

3.4. Part 4: Inserting the Fiber Optic

1. Using a fiber cleaving kit, cut the fiber to the desired length. The correct length will be the distance measured in step 2.4, plus the distance that the fiber will be inserted into the brain. For example, if the height of the implant was 19 mm, and you need the fiber to penetrate 1 mm below the cortical surface, the overall length of the fiber (from the top of the metal ferrule to the cleaved tip of the fiber) should be 20 mm.
2. Polish the fiber tip with 5 μm silicon carbide sheet to remove rough edges.
3. Use a black paint marker (or black paint) to coat the portion of the fiber that will sit within the implant. Do not paint any portion of the fiber that will enter the brain.
4. Measure light transmission through the fiber using an optical power meter. Make a note of the difference in power between the end of the fiber connected to the light source and the cleaved tip of the fiber that will illuminate the brain. Typical light loss through the connector is 10–50%.
5. Carefully insert the fiber through the hole in the EIB, and feed it through the hole made by the 29 gauge cannula in step 1.4. Stop when the top of the metal ferrule is at the same height as the top of the connector on the EIB.

6. Glue the fiber in place with epoxy. Make sure to apply epoxy on both the top and bottom of the EIB. Be liberal with glue so that the fiber will not break free of the EIB.

3.5. Part 5: Adding the Cone

1. Construct the cone:
   (a) Print the template (available online) onto a sheet of transparency paper.
   (b) Coat the transparency film with a thin layer of epoxy, and lay a sheet of aluminum foil on top. Apply pressure to remove air bubbles. Allow the epoxy to dry for at least 1 h.
   (c) Use scissors to cut around the edge of one of the templates. Fold it into a cone shape by aligning the flat edge with the dashed guide marks on the opposite side.
   (d) Apply cyanoacrylate glue to hold the cone in place, and allow it to dry for at least 3 h.
2. Flip the drive upside down and place the cone onto it, taking care not to contact the electrodes. Make sure the ground wire is protruding from the cone on the electrode side. You will need at least 12 mm of wire to ensure that the ground wire can reach the ground screw during surgery.
3. Apply a ring of epoxy where the cone meets the implant body, and allow it to dry.
4. Use silver print or a conductive silver pen to connect the cone wire to the aluminum foil and allow it to dry. If the silver print is not completely dry before the next step, it can result in a bad connection between the ground wire and the cone.
5. Epoxy the cone wire in place. Next, place a thick ring of epoxy between the edge of EIB and the cone to seal in electrodes. Make sure you do not get any epoxy on the connector or the fiber optic connector.
6. Make a cap for the implant by cutting a circle out of transparency paper and attaching with electrical tape.

Once the epoxy is dry, the implant is ready to go. See Note 3 for advice on surgery and Note 4 for recording tips.

4. Notes

1. *Notes on Customization*: Certain aspects of this design are easy to alter, such as the length of the fiber optic and the length of the electrodes. The precise parameters used will depend on the brain structure the experimenter wishes to target. The implant is small enough that it can fit on the

animal's head along with other components, such as a head post, EEG screws, cannulae for drug delivery, etc. Nevertheless, this design has been optimized for recording and stimulating the same structure, and must be modified for applications in which stimulating and recording in different areas is required.

With basic knowledge of 3D modeling software (SolidWorks, Dassault Systèmes, Concord, MA) and printed circuit board software (PCB123, Sunstone Circuits, Mulino, OR), it is possible to add additional holes in the implant body based on stereotactic coordinates of the target structures. It is also possible to increase the number of electrode channels by adding more electrode guides. When adding new holes to the design, keep in mind that there is some inaccuracy in the stereolithography printing process; it is therefore best to make holes slightly smaller than desired and drill them out manually. For more information on custom-printed implants, see ref. (20).

The current design includes an 18-pin Omnetics connector, which is compatible with headstage preamplifiers made by Neuralynx, Plexon, Blackrock Microsystems, and others. With the right adapters, it can be made compatible with almost any commercially available headstage. However, there are certain circumstances in which a different connector is desirable. By modifying the design for the EIB, other types of connectors can be accommodated. Care should be taken to ensure that all the holes on the EIB (for screws and fiber optic) align with those on the implant body.

The digital files for producing custom parts are available as Supplementary Online Material at The online version of this article (doi: 10.1007/7657_2011_11) contains supplementary material, which is available to authorized users.

The EIB file (opto_ephys_EIB_8st.123) can be opened and edited using PCB123 software from Sunstone Circuits (http://www.sunstone.com/pcb123.aspx). Printed circuits boards must be ordered from within the software.

The STL file for printing the implant body (opto_ephys_implant_body.stl) can be submitted directly to a stereolithography service, such as American Precision Protoyping (http://www.approto.com/). Printing with an ABS-like material (such as Accura-55) is recommended. Alternatively, the original SolidWorks file (opto_ephys_implant_body.sldprt) can be edited and exported as an STL file in order to further customize the design.

The cone template (cone_template.pdf) can be opened with any PDF viewer and printed onto transparency paper using a standard laser printer.

2. *Notes on Electrode Choice*: This implant design has been used primarily with nichrome or tungsten stereotrodes, but it is also compatible with tetrodes or single-wire electrodes. When choosing electrodes, it is important to consider their stiffness. Flimsier electrodes will be more stable in brain tissue, but they might not penetrate as easily. If only stereotrodes are desired,

but dual-wire electrodes are not stiff enough, one can twist four-wire electrodes, but only connect two wires to the EIB. This will add stiffness, while keeping the electrode density the same. For instructions on tetrode fabrication, see ref. (21).

With the 16-channel EIB provided in this design, it is possible to accommodate 4 tetrodes, 8 stereotrodes, 16 single-wire electrodes, or a combination of all three. To increase or decrease the number of electrodes, one must either change the size of the bottom cannula or change the size of the polyimide guide tubes.

Please note that a dense array of electrodes can severely damage brain tissue when inserted as a single bundle. If each electrode is housed in its own 30-gauge polyimide tube and distributed around a central fiber optic cable, as described above, the potential for damage is minimized. Any new configuration of electrodes should be implanted in a test animal to ensure that no tissue injury (beyond that expected from the individual electrode tracks) has occurred.

3. *Notes on Surgery*: Implantation should take place in accordance with standard sterile surgical practice. For detailed protocols, see ref. (22).

   Some advice specific to this application:

   - To hold the implant in place during surgery, it is useful to have a female Omentics connector attached to a stereotaxic probe. Another option is to use a probe holder specific for cannulae (such as MyNeuroLab.com product #39463017), which also happens to be the perfect size to hold the Omnetics connector.
   - If you are using a mouse line that requires a viral injection to express light-sensitive proteins, the injection can either be performed concurrently with electrode/fiber implantation, or at an earlier date. For viruses that take a longer time to express, an earlier injection may be desirable. However, the buildup of connective tissue on the skull following the original surgery may make the second surgery more difficult.
   - A durotomy is recommended, especially if the tip of the fiber optic will penetrate into the brain. However, the mouse dura is very thin, and electrodes may have no trouble punching through it. For stimulation and recording in cortex, a durotomy may not be necessary.
   - At some point in the surgery (usually after the implant has been cemented in place), the ground wire from the implant must be connected to the ground screw. This can be done with silver print, but it is often easier to create a system with two mating pins.

4. *Notes on Recording*: Once the animal has recovered from surgery, recording can begin. See ref. (23) for more information on chronic recordings in mice.

Some issues that may arise in combining optogenetics and electrophysiology:

- The fiber attached to the animal must be connected to a second fiber connected to a light source, usually a laser or an LED. A cable with the appropriate connector on the other end can be purchased from Doric Lenses along with the implantable fibers.
- For most applications, the fiber is flexible enough that commutation is not necessary. However, a fiber optic rotary joint can be used to allow the fiber to rotate freely.
- Be sure to check if any apparent effects of light stimulation are actually light artifacts. Any deflections in the local field potential that are temporally locked to light onset or offset with submillisecond latency may be artifactual. Waveforms collected around the time of light presentation should be checked by eye to ensure they look like real action potentials. Light-evoked deflections in the local field potential should be treated with suspicion until light artifact can be ruled out as a possible cause.
- The duration and intensity of light pulses should be carefully considered in all optogenetic experiments, but these parameters are especially important when dealing with awake animals. Prolonged excitation or prolonged disinhibition can lead to seizures or excitotoxic damage. Begin your experiment with the shortest and weakest stimulus possible to avoid harming the animal. A good starting intensity is 1 mW.

Acknowledgments

I would like to thank members of the laboratories of Matthew Wilson, Christopher Moore, and Susumu Tonegawa for advice on in vivo electrophysiology and optogenetics. This work was supported by an NDSEG Fellowship.

References

1. Nagel G, Brauner M, Liewald JF et al (2005) Light activation of channelrhodopsin-2 in excitable cells of *Caenorhabditis elegans* triggers rapid behavioral responses. Curr Biol 15:2279–2284
2. Han X, Qian X, Bernstein JG et al (2009) Millisecond-timescale optical control of neural dynamics in the nonhuman primate brain. Neuron 62:191–198
3. Gradinaru V, Mogri M, Thompson KR et al (2009) Optical deconstruction of Parkinsonian neural circuitry. Science 324:354–359
4. Adesnik H, Scanziani M (2010) Lateral competition for cortical space by layer-specific horizontal circuits. Nature 464:1155–1160
5. Thyagarajan S, van Wyk M, Lehmann K et al (2010) Visual function in mice with photoreceptor degeneration and transgenic expression

of channelrhodopsin 2 in ganglion cells. J Neurosci 30:8745–8758
6. Histed MH, Bonin V, Reid RC (2009) Direct activation of sparse, distributed populations of cortical neurons by electrical microstimulation. Neuron 63:508–522
7. Sohal VS, Zhang F, Yizhar O et al (2009) Parvalbumin neurons and gamma rhythms enhance cortical circuit performance. Nature 459:698–702
8. Cardin JA, Carlén M, Meletis K et al (2009) Driving fast-spiking cells induces gamma rhythm and controls sensory responses. Nature 459:663–667
9. Cardin JA, Carlén M, Meletis K et al (2010) Targeted optogenetic stimulation and recording of neurons in vivo using cell-type-specific expression of Channelrhodopsin-2. Nat Protoc 5:247–254
10. Aravanis AM, Wang LP, Zhang F et al (2007) An optical neural interface: in vivo control of rodent motor cortex with integrated fiberoptic and optogenetic technology. J Neural Eng 4: S143–S156
11. Bernstein JG, Han X, Henninger MA et al (2008) Prosthetic systems for therapeutic optical activation and silencing of genetically-targeted neurons. Proc Soc Photo Opt Instrum Eng 6854:68540H
12. Huber D, Petreanu L, Ghitani N et al (2008) Sparse optical microstimulation in barrel cortex drives learned behaviour in freely moving mice. Nature 451:61–64
13. Zorzos AN, Dietrich A et al (2009) Light-proof neural recording electrodes. Soc Neurosci (Abstract)
14. Adamantidis AR, Zhang F, Aravanis AM et al (2007) Neural substrates of awakening probed with optogenetic control of hypocretin neurons. Nature 450:420–424
15. Tsai HC, Zhang F, Adamantidis A et al (2009) Phasic firing in dopaminergic neurons is sufficient for behavioral conditioning. Science 324:1080–1084
16. Royer S, Zemelman BV, Barbic M et al (2010) Multi-array silicon probes with integrated optical fibers: light-assisted perturbation and recording of local neural circuits in the behaving animal. Eur J Neurosci 31 (12):2279–91
17. Zorzos AN, Bernstein JG et al (2009) Integrated microstructure lightguides for ultradense optical neural control of 3-dimensional neural circuits. Soc Neurosci (Abstract)
18. Ritt JT, Siegle JH et al (2009) Optical manipulation of cortical subnetworks during active touch in awake mice using channelrhodopsin-2. Soc Neurosci (Abstract)
19. Kvitsiani D, Ranade S et al (2009) Optical tagging of genetically defined cell types for extracellular recording in behaving rodents. Soc Neurosci (Abstract)
20. Kloosterman F, Davidson TJ, Gomperts SN et al (2009) Micro-drive array for chronic in vivo recording: drive fabrication. J Vis Exp 26(pii):1094
21. Nguyen DP, Layton SP, Hale G et al (2009) Micro-drive array for chronic in vivo recording: tetrode assembly. J Vis Exp 26(pii):1098
22. Oliveira LMO, Dimitrov D (2008) Surgical techniques for chronic implantation of microwire arrays in rodents and primates. In: Nicolelis MAL (ed) Methods for neural ensemble recording, 2nd edn. CRC, Boca Raton, FL
23. Dzirasa K (2008) Chronic recordings in transgenic mice. In: Nicolelis MAL (ed) Methods for neural ensemble recording, 2nd edn. CRC, Boca Raton, FL

Part V

Morpho-Functional Approaches

Neuromethods (2012) 67: 375–389
DOI 10.1007/7657_2011_12

Published online: 25 November 2011

Morphological Approaches to the Anatomical Dissection of Neuronal Circuits

David C. Lyon

Abstract

The purpose of this introductory chapter is to provide an overview of a number of techniques used to study the brain's anatomical organization, with particular focus on the investigation of the mammalian neocortex, the large outer surface of the brain and the key structure responsible for complex perceptual and cognitive abilities of mammals. I briefly examine how traditional and more recently developed methods involving histochemistry, neuronal tracing, cell morphology and photostimulation are being used to reveal cortical anatomy at broad and very fine levels of specificity. In addition, in the last section, I review recent advances in neuronal tracing technology that capitalize on the neurotropic properties of rabies virus to enable retrograde labeling of neural circuits at single cell resolution. In the chapters that follow, detailed descriptions of new methodologies for characterizing neuronal cell populations (Herculano-Houzel), as well as high-resolution techniques for mapping local cortical networks (Radkinow et al) are discussed. These approaches are proving to be invaluable in unraveling the many neuroanatomical issues that remain unresolved.

Key words: Architecture, Cell morphology, Cell type specificity, Immunohistochemistry, Neuronal tracers, Fluorescent labeling, Glycoprotein, Photostimulation, Rabies virus, Retrograde transport

1. Introduction

Determining the anatomical organization of the nervous system is considered the fundamental first step in understanding how the brain operates (1). Neuroanatomical investigation ultimately involves the visualization of both the broader and finer specificity of neuronal architecture and interconnecting circuits. In many ways, it is most desirable to directly link the anatomical structure to its function and this can now be done at a fairly high level of detail, especially in combination with advanced optogenetic techniques (see Part IV of this volume). Yet, despite extensive anatomical study dating back to early last century (2, 3), there remain many gaps in our knowledge of the anatomical structures that make up the mammalian brain. This is especially true for the mammalian neocortex (cortex), the large region covering the surface of the brain that is primarily responsible for complex perception and cognition.

The cortical anatomy can be characterized at several levels of specificity, each offering a high degree of complexity. At the broadest level, the cortex is subdivided into areas, or organs of the brain (2), many of which can be defined through distinct architectonics and patterns of connections (4–7). At a finer level of anatomical resolution, within a cortical area neuronal cell bodies are arranged into distinct layers and sublayers (2, 8–10). In addition, in the same area of the cortex, neurons can also be clustered vertically across layers into columns which represent repeated and orderly spaced functional modules (7, 11, 12). Many of these layers and modules can be defined through stains for neurochemical markers, such as for cytochrome oxidase, and they can exhibit distinct patterns of connections within and between cortical areas (13–16). At perhaps the finest resolution of neuroanatomical specificity, it is now clear that neighboring neurons in the same column or layer can have very different connection patterns and are thus likely to give rise to distinct functional outcomes (17). This chapter provides a brief look at some of the traditional and more recently developed methods involving histochemistry, neuronal tracing, and photostimulation that are being used to reveal cortical anatomy at multiple levels of specificity. In addition, the last section of this chapter highlights a new technique involving a genetically modified rabies virus that allows for the highest, to date, specificity in neuroanatomy, i.e., the anatomical labeling of the neural circuitry impinging onto a single neuron (18–20).

2. Architectonic Techniques for Cortical Characterization

For sensory and motor cortex, which occupy a great deal of the cortex, traditional histochemical stains have been especially useful for defining primary, and some secondary areas. For example, silver stains for myelin using the Gallyas (21) technique and stains for the metabolic enzyme cytochrome oxidase (CO) (22) have been extensively used to reveal several sensory and motor areas in a whole host of mammalian species, ranging from monotremes such as the duck-billed platypus to several species of monkey (6, 7, 23). A key approach used by these researchers is a flattened preparation, where the cortex is carefully removed from the underlying cortical structures and any sulci that are present are unfolded by severing the connections made by the underlying white matter. An excellent description of the flattening procedure for the entire macaque monkey cortex is provided in Sincich et al. (24).

By using flattened preparations, several cortical areas can be observed at once, a perspective that can make it easier to interpret the full extent of each identified area. This can be seen quite nicely in a whole flat mount preparation of marmoset monkey cortex from Lyon and Kaas (25) as shown in Fig. 1.

In mammals with greatly expanded cortical territory there are many more than the handful of primary and secondary areas and these additional areas are not reliably identified through traditional architectonic techniques. In monkeys for example, well over a dozen areas just for visual cortex have been proposed primarily based on connectivity patterns and retinotopic maps (4, 7, 26, 27), yet only three to four of these areas can be reliably identified through CO or myeloarchitecture (7, 25, 28). See for example, areas V1, V2, V3, and MT in Fig. 1. However a number of other stains are now being applied to cortical investigation and are helping to define the remaining regions. For example, histochemical stains such as for free ionic zinc, and immunohistochemical stains for neurofilament protein epitopes (SMI-32), for the calcium-binding proteins calbindin and parvalbumin, and for the vesicle glutamate transporter protein (VGlut2) have helped to substantiate the borders of several areas in a prosimian primate (16) and close primate relative, the tree shrew (29).

More detailed architectonic features include differences in laminar thickness, cell density, and gross morphological characteristics such as cell soma size, as well as various histochemical stains. These characteristics are better examined in coronal or parasagittal sections and are also useful in distinguishing between cortical areas. For example, within the mammalian brain, primary motor cortex (M1, or Brodmann's area 4) is characterized by a large *layer 5*, which is comprised of pyramidal neurons with relatively large soma sizes. On the other hand, primary visual cortex (V1, or Brodmann's area 17) is characterized by a large *layer 4* containing densely packed granular, or spiny stellate cells. Species specific laminar characteristics are also common. For example, *layer 2/3* of primary and secondary somatosensory cortex (S1 and S2) in rodents is distinguished by a barreloid pattern revealed through CO stains (23). On the other hand, in primates, barrel fields are not present in somatosensory cortex, but blobs of high cytochrome activity are a hallmark of *layer 2/3* in V1 (7, 30, 31) as shown in Fig. 1d. Stains are now being targeted to newer markers for more detailed laminar study (10). These include the combination of procedures recently employed by Wong and Kaas (16, 29), which have revealed clearer boundaries of sublamina in a number of cortical areas. Using these newer procedures, an example of the very high level of detail that can be achieved through architectonic stains was demonstrated by Preuss and Coleman (32) where they used immunohistochemical stains for calbindin and the Cat-301 antibody to reveal a repeating pattern of modules within *layer 4A* that is unique to human V1.

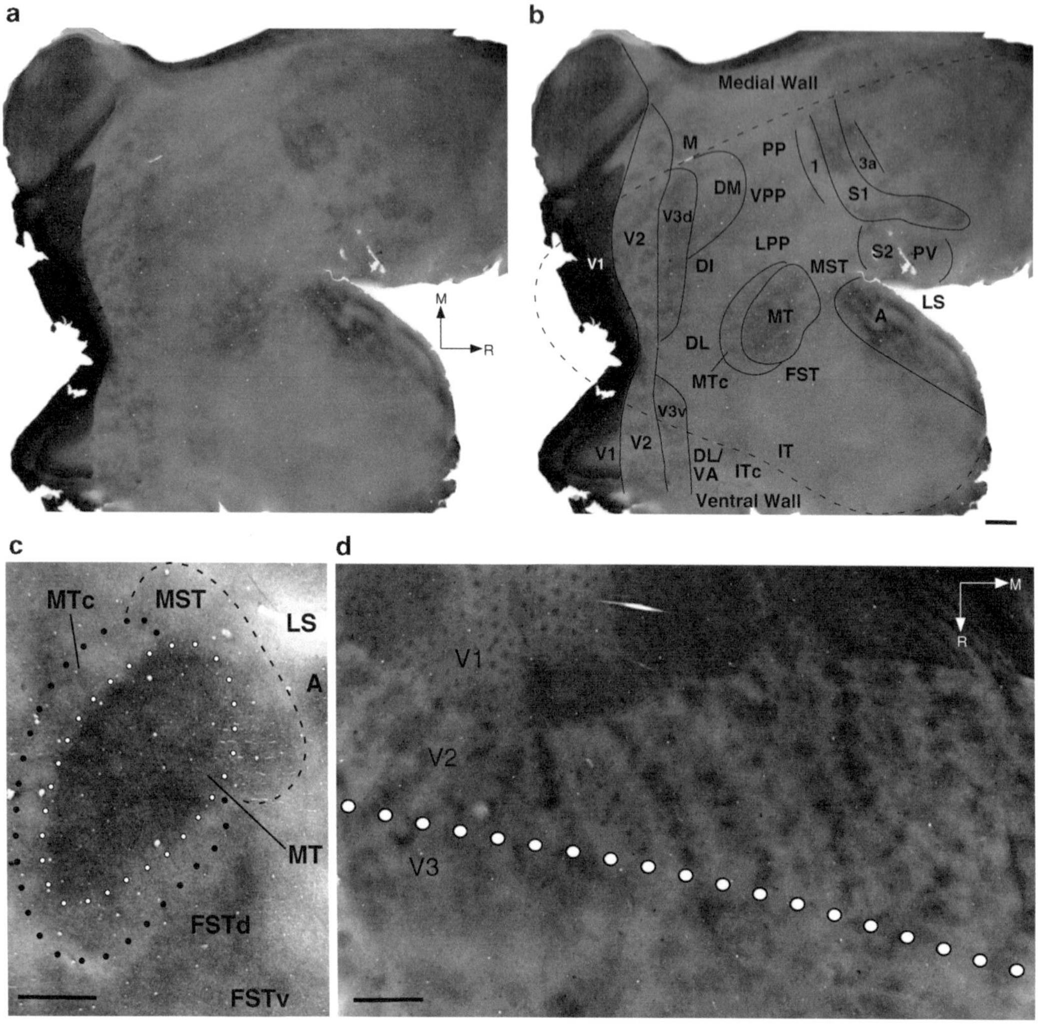

Fig. 1. Cytochrome (CO) and myeloarchitecture of flattened New World monkey cortex. Nearly an entire marmoset monkey hemisphere processed for CO is shown in (**a**), and is duplicated and labeled in (**b**). As in all primates, the lateral border of V1 is most easily recognized due to a sharp decrease in CO density from V1 to V2. Areas V2 and MT show typical modular CO staining patterns of primates, the *narrow dark and light bands* in V2 and *dark and light patches* in MT, and their borders are clearly identified. Just lateral to V2 we identify dorsal (d) and ventral (v) halves of V3 as two 1.5–2 mm narrow strips of cortex roughly half the width of V2. Visual areas DM, DI, FST, and MST are better identified through myelin staining (see (**c**)). Locations of other visual areas such as DL, VPP, LPP, and ITc are based on connection patterns and are indicated in (**b**) for reference. The *dashed black line* represents the estimated boundary of the dorsal and lateral surface of the cortex prior to flattening. Areas S1, 3a, and A stain dark for CO, and are also shown for reference (for all abbreviations see Lyon and Kaas (25)). (**c**) A more highly magnified image of flattened marmoset cortex processed for myelin showing MT and surrounding visual areas. MT (outlined by *white dots*) stains the darkest, compared to surrounding cortex, and consists of modular-like dark and light patches of myelin. MTc (outlined by *black dots*) is seen as a thin crescent of lighter myelin surrounding the majority of caudal and lateral MT. MST (*dashed line*) located at the rostral border of MT is less uniformly stained, probably due to uneven flattening near the lateral sulcus (LS; see (**b**)). Areas FSTd and FSTv are visible as moderately myelinated regions of cortex located successively lateral to ventral MTc. (**d**) CO modules in V1 and V2 shown in the flattened cortex of a squirrel monkey. In V1, the dark CO patches form small, discreet blobs, whereas in V2 the blobs are larger and strung together to form stripes, or bands. All scale bars = 2 mm. The images of marmoset and squirrel monkey visual cortex are reproduced from Lyon and Kaas (25, 39) with permission.

3. Neuronal Tracers

Knowledge about morphological connectivity has been primarily gleaned from the use of anterograde and retrograde neuronal tracers. These substances are passively taken up at the soma or synaptic terminals and then, transported back to either the axon terminals or cell soma, respectively. More traditional tracers, that are still used today (11), include the bidirectional tracers (they transport both retrograde and anterograde), horseradish peroxidase (HRP (33)), and wheat germ agglutinin conjugated to HRP (WGA-HRP), as well as the anterograde tracer *Phaseolus vulgaris* Leukoagglutinin (PhaL (34)). Also widely used are dextrans, which can be conjugated with various fluorescent dyes (35, 36) allowing for multiple distinguishable retrograde tracers in the same animal (25, 37–39) (see Fig. 2). The various fluorescent conjugates of cholera toxin subunit-b (CTB), the nontoxic subunit, are also very popular as bidirectional tracers (40, 41). CTB transports long distances relatively fast (~3 days) and provides different levels of sensitivity depending on the amount of tracer injected extracellularly. Larger injections of CTB (~0.2 μl) are more effective at revealing interareal connections over several millimeters of cortex (25) (see Fig. 2), whereas very small injections (~0.02 μl) can be used to examine relatively local connectivity, ≤250 μm (42).

The types of questions that can be addressed through retrograde or anterograde tracers depend on where the tracers are injected. The most effective way to use a retrograde tracer is to inject it into a well characterized cortical area. The resulting retrograde labeled cell bodies can then be used to identify new morphologically connected cortical areas. For example, multiple retrograde tracer injections in V2 of the rarely studied tree shrew revealed several distinct clusters of labeled neurons from each of these tracers (37). Coupled with myelin staining patterns of the entire flattened cortex to identify V1 and V2, the clustered pattern of retrograde label beyond V2 was used to demonstrate the existence of six additional extrastriate visual areas. Even in one of the most commonly studied species for visual cortex connectivity, the macaque monkey (4), new discoveries, of a ventral area V3 for example, have been made using neuronal tracer injections into V1 (7, 25, 38, 39) (see Fig. 2).

The connectivity patterns of neuronal tracers are also useful in identifying modular organization within cortical areas. For example, functionally defined orientation domains which are now commonly observed and studied in V1 of felines, tree shrews, and primates through intrinsic signal optical imaging (7), were shown to be highly interconnected using neuronal tracer injections into V1 of the tree shrew (43, 44). These specific networks that are formed between functionally defined modules of V1 demonstrate

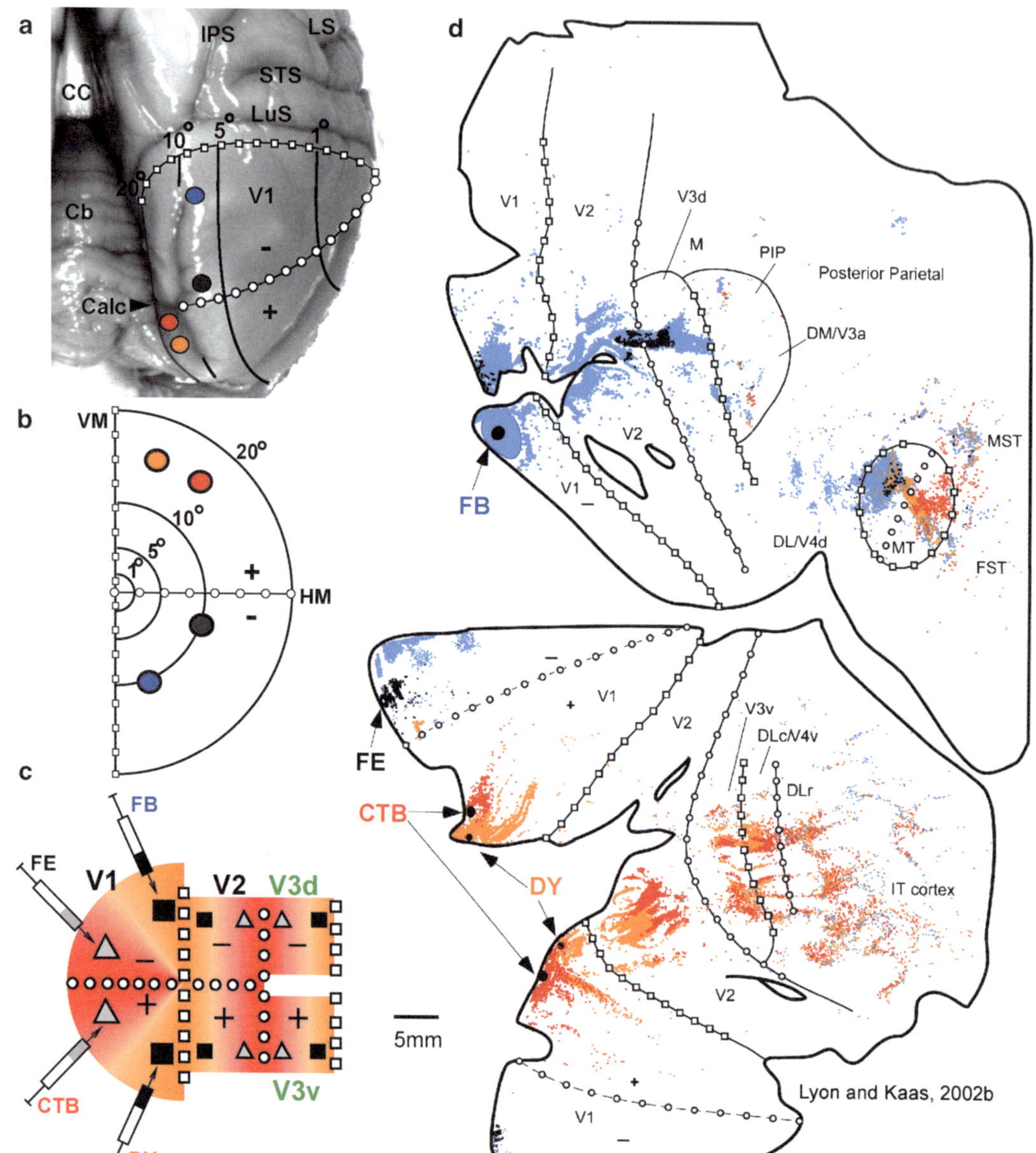

Fig. 2. The retinotopic organization of macaque V1 is well established and its connection patterns can be used to identify higher order visual areas. (**a**) A caudal posterior view of a macaque monkey brain showing the locations of four retrograde tracer injection sites (*colored circles*) along the banks of the calcarine sulcus (Calc). The four injections, two into upper (+) and two into lower (−) field representations were located closer to the horizontal (*dotted line*) or vertical (*squared line*) meridians. (**b**) The relative positions of the four injection sites shown on the contralateral visual hemifield. (**c**) A scheme showing the predicted pattern of label from the four injections. The key issue for this study was whether ventral V3 (V3v) connections with V1 existed. If so, they should represent a mirror reversal of the pattern of label in V2. (**d**) The actual pattern of labeled cells shown on flattened macaque visual cortex. CTB (*red*) and diamidino yellow (DY; *orange*) injections into ventral V1 labeled neurons in the ventral V3 region and the pattern of *red to orange* is a mirror reversal of the pattern in ventral V2 (*orange to red*) as predicted in (**c**). A similar mirror reversal pattern from V2 to V3 is seen in dorsal V3 (V3d) for the *black* (fluoremerald; FE) and *blue* (fast blue; FB) tracers. Retinotopic patterns of connections are also seen for MT and to some extent in V4. This figure is adapted from Lyon and Kaas (38) with permission.

a more detailed level of anatomy that would not be revealed through architectonic methods. On the other hand, through the use of neuronal tracers the architectonically defined CO blobs in monkey V1 have also been shown to preferentially interconnect (45) and to have architectonically distinct targets in higher visual areas (15, 46).

Additional retrograde tracer strategies can also be useful for identifying connectivity at an even finer resolution. For example, neuronal tracer injections into two distinct higher level visual areas, V2 and MT (V5) of the macaque monkey, demonstrated that separate cell types in layer 4B of V1 project to two separate areas. Specifically, spiny stellate cells were labeled through retrograde tracer injections into MT, whereas pyramidal neurons were labeled by retrograde tracer injections into V2 (47). This study used a modified rabies virus that not only acted as a retrograde tracer but also filled the neuron with green fluorescent protein (GFP) to allow for complete morphological reconstruction (48). For more details on the rabies virus see the last section of this chapter.

Cell type specific tracing of connections can be achieved even more directly through novel tools developed through genetic manipulation of neurons and viral technology (18, 49). For example, using a transgenic mouse line that expresses Cre recombinase in only one phenotypically defined population of neurons (i.e., serotonergic or neuropeptide Y neurons) Enquist and colleagues have made extracellular injections of conditional pseudorabies virus recombinants (BA2000 or BA2001) that can only replicate in these Cre-expressing neurons (49–51). The pseudorabies viruses act as either a transsynaptic (BA2001) or monosynaptic (BA2000) retrograde tracer (51) transported from the initially infected host cells. Because the viruses have been engineered to express GFP, neurons can be easily identified.

4. Photostimulation

Traditional neuronal tracers described above typically leave a dense amount of extracellular label surrounding the immediate region of an injection site and therefore tend to totally obscure the local connection patterns. This effect can be compensated for by making unusually small injections 0.02 μl (42) instead of the typical 0.2 μl (i.e., (25, 38, 39)). Nevertheless, analysis of neuronal tracers is typically limited to longer-range (>250 μm) intrinsic connections, connections with other cortical areas, and connections with subcortical structures.

To enable a detailed examination of the functional connectivity of local circuits (<250 μm), Callaway and Katz (52) developed the method of photostimulation using caged glutamate. The process involves very local photostimulation of a living brain slice bathed in the caged glutamate while a neuron is being recorded intracellularly. Single photon photostimulation uncages glutamate within a small region (50 μm) causing only neurons at this location to fire. In addition, this approach can be extended by stimulating sequentially the tissue in a grid-like pattern, allowing the mapping of the functional connectivity of an entire column (53). Intracellular recording of a single neuron is used to reveal excitatory or inhibitory postsynaptic potentials, the occurrence of which can be correlated to the site of the photostimulation to provide a detailed map of the location of inputs to the recorded neuron. Not only are distinct inputs and outputs determined with regard to cortical layer through histochemical stains of the slice, but because the individual recorded neurons are filled with dye and characterized morphologically, distinct patterns of laminar inputs to specific cell types can also be revealed. Various forms of this photostimulation technique have been used as a precise tool for dissecting cell type specific intracortical circuits in investigations ranging from macaque monkey V1 to rat barrel cortex (8, 54). Furthermore, recent use of the photostimulation method has taken advantage of transgenic mouse lines that selectively express fluorescent markers in specific cell types to characterize the laminar specificity of inputs to nine distinct inhibitory neurons in mouse somatosensory cortex (55).

Perhaps the finest scale of cortical morphofunctional connectivity is achieved by using photostimulation in conjunction with paired recordings. These experiments demonstrated that neurons of the same cell type located right next to each other within the same cortical layer can have very different connections. For example, it was found that only 20% of neighboring pyramidal neurons in layer 3 receive common inputs from layer 4 (17). This is consistent with electron microscopy based anatomical results showing that local circuitry is highly refined, and suggests that widespread, seemingly diffuse axonal projections actually have very specific postsynaptic targets (56–59).

5. Genetically Modified Rabies Virus as a Fine Scale Retrograde Tracer

While the photostimulation technique described above allows the detailed description of the laminar location of inputs to a single morphologically characterized neuron, the approach is limited to local connections that remain present and viable in a 400 μm thick slice of cortex. Therefore, the fine scale specificity of longer range neuronal circuits remains unexplored with this approach and the development of new tools is needed. The use of rabies virus to anatomically label the local and long range inputs to a single cortical neuron provides a novel solution to this issue (19, 20, 48). The virus has been modified to prevent its spreading transsynaptically (see Sect. 5.2) and has been pseudotyped to allow the selective infection of neurons (see Sect. 5.3). Before discussing these two modifications in detail, the characteristics of wild type rabies are described first.

5.1. Wild Type Rabies

Wild type rabies virus acts as a highly effective transsynaptic neuronal tracer (18, 60–62). The rabies glycoprotein, which is embedded in the viral envelope (see Fig. 3a), serves as the ligand necessary for infection and transneuronal spread (63–65). The receptor for rabies glycoprotein appears to be located only at presynaptic nerve terminals (66). Therefore, injection of rabies directly into the brain results in the infection of neurons exclusively through axon terminals. The virus then spreads in a retrograde direction to replicate in the soma, invades postsynaptic terminals and spread to additional neurons only through presynaptic contacts (18, 60, 62).

Unlike traditional nonviral tracers described earlier, only a small amount of rabies virus is needed for labeling circuits as only a single viral particle is sufficient to spread throughout the brain (67). Neurons are thus very well labeled by virions, allowing for detailed morphological reconstruction following histochemical staining for the rabies nucleocapsid protein (see description of procedure in Nassi et al. (68)). However, to extract meaningful data the spread of the virus needs to be spatially restricted to prevent the infection of the entire central nervous system. One approach for controlling the spread of rabies is to limit the postinjection survival time. For example, 72 h allows the CVS-11 strain of rabies to spread disynaptically, but not to a third synapse (see (18)). Such an approach has been used to reveal several, previously unidentified disynaptic pathways of the primate visual cortex (13, 68, 69). Therefore, controlled use of wild type rabies can be an effective tool.

5.2. Glycoprotein-Deleted Rabies

The spread of rabies virus can also be limited by controlling the genetic expression of the rabies glycoprotein (64). Wickersham and colleagues (48) removed the rabies glycoprotein genome from SAD-B19 vaccine strain of rabies and replaced it with the gene for enhanced GFP (Fig. 3a). Importantly, during production of this glycoprotein-deleted (ΔG) rabies virus (SADΔG-GFP) glycoprotein was supplied so that it could be incorporated into the viral envelope (but not into the genome). Without any glycoprotein in the viral envelope the virus would not be able to infect any neurons (see above). Thus, by including the glycoprotein in the envelope but excluding it from the genome, this virus acts as

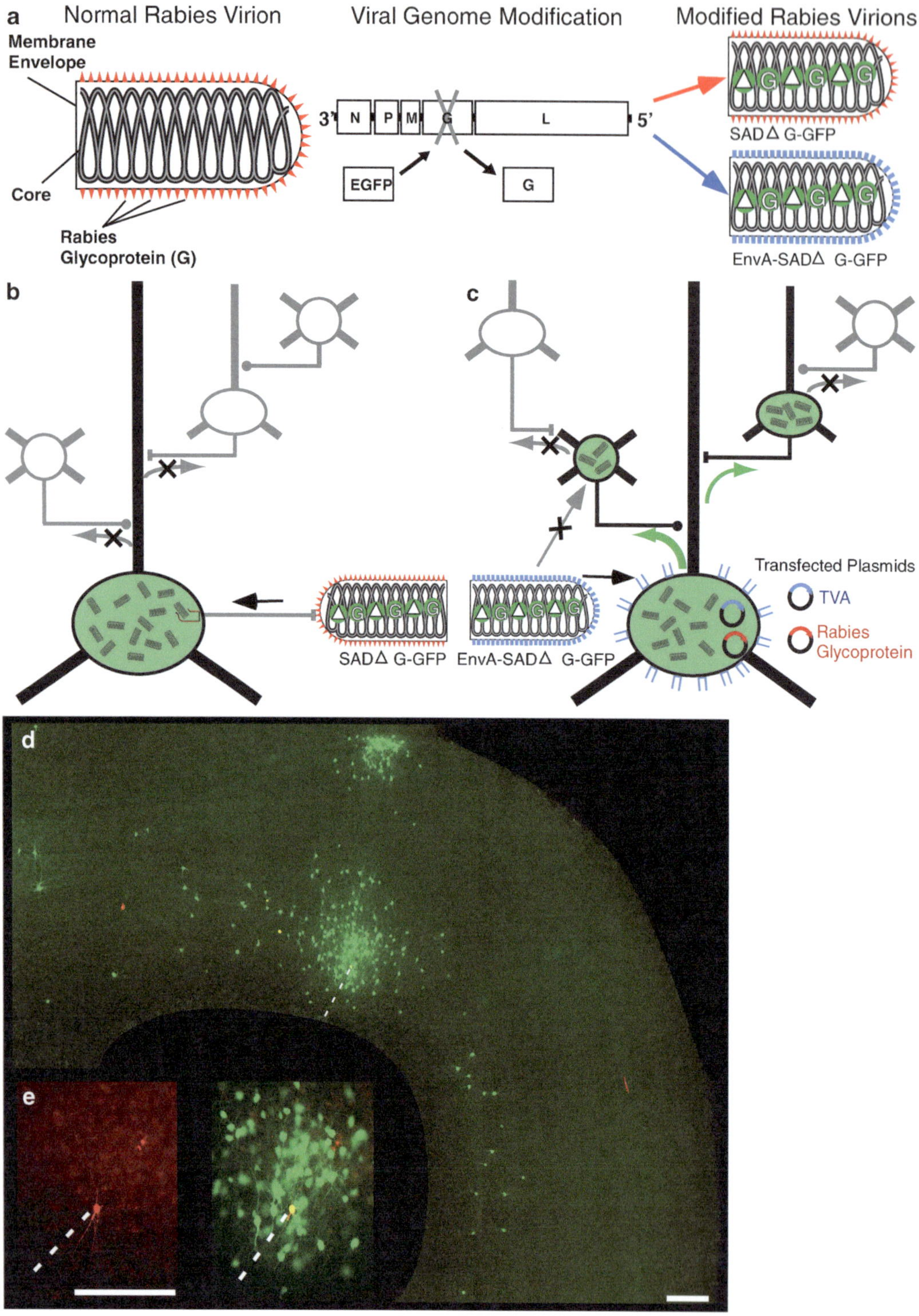

Fig. 3. Using a modified and pseudotyped rabies virus to retrogradely trace the inputs to a single neuron. (**a**) *Scheme of the rabies virion*. The viral core consists of the RNA genome and associated proteins surrounded by the membrane envelope in which the rabies glycoprotein (G) is embedded. The gene for the rabies glycoprotein was removed (ΔG) and replaced with

a monosynaptic retrograde tracer that infects axon terminals at the site of brain injection (Fig. 3b). The virus is then transported retrogradely to the soma where it replicates and fills the soma and dendrites completely with GFP (Fig. 3f; see examples in (47, 48, 70)). Importantly, the virus cannot spread across another synapse.

5.3. EnvA Pseudotyped Rabies

The advantages of the modified SADΔG-GFP rabies described above is that it will not spread transsynaptically and the first order neurons that it infects become completely filled with GFP allowing for detailed morphological characterization. In essence, the SADΔG-GFP variant acts as a nonspecific retrograde tracer, infecting all terminal endings present at the site of injection.

An additional modification to the SADΔG-GFP was made to enable highly specific, genetically targeted infection of a single neuron. Wickersham and colleagues (20) replaced the rabies glycoprotein in the membrane envelope with the envelope A glycoprotein (EnvA) from the avian sarcoma and leukosis virus, a process known as pseudotyping. EnvA limits infection to cells that express the TVA receptor, a protein found in birds but not mammals (71). Thus, the resulting pseudotyped virus, EnvA-SADΔG-GFP, can be targeted specifically to mammalian neurons that are made to express the TVA receptor (20) (Fig. 3c). Since the viral genome of EnvA-SADΔG-GFP is the same as SADΔG-GFP, once inside the cell, it codes for replication of the SADΔG-GFP variant. Because the glycoprotein has been deleted from the genome of this variant (see above), complementation through the introduction of a rabies glycoprotein plasmid is required to enable the virus to spread to presynaptically connected neurons.

Fig. 3. (continued) one for GFP (center). Two versions of this virus have been made, one that incorporates its native glycoprotein (designated SADΔG-GFP, *right, top*), and one that is pseudotyped with the envelope A protein (EnvA) from the avian sarcoma and leukosis virus (designated EnvA-SADΔG-GFP, *right, bottom*). (**b**) *Retrograde infection with SADΔG-GFP.* The SADΔG-GFP virus acts as a nonspecific retrograde tracer, infecting all terminal endings present at the site of injection. While the virus replicates in the soma of infected neurons, filling it with high levels of GFP, the virions produced lack the rabies glycoprotein in their envelope membrane. Therefore, the virus can not spread to presynaptically connected neurons. (**c**) *Targeting infection of a single neuron and labeling of presynaptic neurons with EnvA-SADΔG-GFP.* Because this virus has been pseudotyped with EnvA it can only infect neurons expressing TVA, the receptor for EnvA that is naturally expressed in birds but not mammals. Furthermore, because the EnvA-SADΔG-GFP virus produces new SADΔG-GFP in the host neuron the virus is unable to spread to presynaptic neurons (as described in **b**). For this single cell tracing system to work, a plasmid containing three genes needs to be inserted into the target cell: (1) the gene for TVA, so the virus can enter the soma, (2) the gene for rabies glycoprotein, so the virus can spread to presynaptically connected neurons, and (3) a marker protein such as mCherry, so the host cell can be identified. Following a 24-h period to allow for expression of these three proteins the EnvA-SADΔG-GFP virus is injected near the transfected neuron. The transfected neuron, which is already expressing mCherry, will become infected and then over 24 h express GFP. Over an additional 3 days, presynaptic neurons will also become infected and begin to express GFP as well. The virus cannot spread beyond these directly connected presynaptic cells because they were not transfected with the rabies glycoprotein plasmid. *Panels* (**a**) and (**c–f**) are modified from Wickersham et al. (20) with permission.

In summary, for the single cell tracing system to work three genes need to be inserted into the target neuron prior to delivering the EnvA-SADΔG-GFP virus: (1) the gene for TVA, so that the virus can enter the soma, (2) the gene for rabies glycoprotein, to allow the virus spread to presynaptically connected neurons, (3) the gene for a marker protein such as mCherry for cell identification. Following a 24 h period to allow for expression of these three proteins the EnvA-SADΔG-GFP virus is injected near the transfected neuron. The transfected neuron which is already expressing mCherry will become infected and then over the next day begin to express GFP. Over an additional 3 days presynaptic neurons will also become infected and start expressing GFP (Fig. 3d). Importantly, the virus cannot spread beyond these directly connected presynaptic cells because they are not producing rabies glycoprotein plasmid (no plasmid was delivered to these cells; see Fig. 3c).

The pseudotyping of the SADΔG-GFP with EnvA is a key factor in the selective genetic targeting of a neuron. The remaining limitation for tracing the inputs to single neurons is delivery of the plasmids containing the TVA and rabies glycoprotein genes. Initially, this was done in cultured slices of neonatal rat cortex, where plasmids were delivered into a small random sample of neurons using biolistics (20). These in vitro experiments were used to demonstrate that dozens of presynaptically connected neurons could be well labeled locally within a cortical column and at a distance of several millimeters horizontally from the host cell (Fig. 3d). The in vivo delivery of TVA and rabies glycoprotein plasmids has just recently been published. This exciting work by Marshel and colleagues (19) used 2-photon microscope guided single-cell electroporation (72) to successfully transfect a single neuron with the three genes, thereby allowing for targeted EnvA-psuedotyped rabies infection and complementation; each transfected cell was successfully infected, and a number of presynaptically connected neurons were labeled.

This newly developed ability to genetically target and label neural circuits of individual neurons provides a powerful new tool for understanding the fine scale structure of the brain. Additional modifications to the approach can be made to allow for functional investigation of the single-cell circuit as well. For example, a blind juxtacellular patch procedure can be used for functional characterization of the host neuron prior to delivering genes through SCE, as can be done already for delivery of smaller molecules (i.e., (73)). In addition, genes other than for the expression of GFP can be inserted into the rabies genome for delivery of genetic activity indicators or light sensitive ion channels that can be observed and manipulated optically, as suggested by others (18, 74). Perhaps in the very near future, through the combination of genetic manipulation of the rabies virus and advanced methods for gene delivery, new experiments will be used to identify and relate neuronal circuits to neuronal function at this unprecedentedly precise level of specificity. Lastly, this technique will be most powerful when used in combination with traditional and emerging histological techniques described above that allow for anatomical identification of functional modules and layers.

References

1. Parker D (2010) Neuronal network analyses: premises, promises and uncertainties. Philos Trans R Soc Lond B Biol Sci 365:2315–2328
2. Brodmann K (1909) Vergleichende Lokalisationslehre der Grosshirnrinde. Barth, Leipzig
3. Jones EG (2007) Neuroanatomy: Cajal and after Cajal. Brain Res Rev 55:248–255
4. Felleman DJ, Van Essen DC (1991) Distributed hierarchical processing in the primate cerebral cortex. Cereb Cortex 1:1–47
5. Kaas JH (1987) The organization of neocortex in mammals: implications for theories of brain function. Annu Rev Psychol 38:129–151
6. Krubitzer L (1995) The organization of neocortex in mammals: are species differences really so different? Trends Neurosci 18:408–417
7. Lyon DC (2007) The evolution of visual cortex and visual systems. In: Krubitzer LA, Kaas JH (eds) Evolution of nervous systems. Academic, Oxford
8. Callaway EM (2002) Cell type specificity of local cortical connections. J Neurocytol 31:231–237
9. Hassler R (1967) Comparative anatomy in the central visual systems in day- and night-active primates. In: Hassler R, Stephan H (eds) Evolution of the forebrain. Thieme, Stuttgart
10. Yamamori T, Rockland KS (2006) Neocortical areas, layers, connections, and gene expression. Neurosci Res 55:11–27
11. Catania KC, Henry EC (2006) Touching on somatosensory specializations in mammals. Curr Opin Neurobiol 16:467–473
12. Livingstone M, Hubel D (1988) Segregation of form, color, movement, and depth: anatomy, physiology, and perception. Science 240:740–749
13. Nassi JJ, Callaway EM (2009) Parallel processing strategies of the primate visual system. Nat Rev Neurosci 10:360–372
14. Rockland KS (1997) Elements of cortical architecture: hierarchy revisited. In: Rockland KS, Kaas JH, Peters A (eds) Cerebral cortex. Plenum Publishing Co., New York
15. Sincich LC, Horton JC (2005) The circuitry of V1 and V2: integration of color, form, and motion. Annu Rev Neurosci 28:303–326
16. Wong P, Kaas JH (2010) Architectonic subdivisions of neocortex in the Galago (*Otolemur garnetti*). Anat Rec (Hoboken) 293:1033–1069
17. Yoshimura Y, Dantzker JL, Callaway EM (2005) Excitatory cortical neurons form fine-scale functional networks. Nature 433:868–873
18. Callaway EM (2008) Transneuronal circuit tracing with neurotropic viruses. Curr Opin Neurobiol 18:617–623
19. Marshel JH, Mori T, Nielsen KJ et al (2010) Targeting single neuronal networks for gene expression and cell labeling in vivo. Neuron 67:562–574
20. Wickersham IR, Lyon DC, Barnard RJ et al (2007) Monosynaptic restriction of transsynaptic tracing from single, genetically targeted neurons. Neuron 53:639–647
21. Gallyas F (1979) Silver staining of myelin by means of physical development. Neurol Res 1:203–209
22. Wong-Riley M (1979) Changes in the visual system of monocularly sutured or enucleated cats demonstrable with cytochrome oxidase histochemistry. Brain Res 171:11–28
23. Catania KC (2002) Barrels, stripes, and fingerprints in the brain – implications for theories of cortical organization. J Neurocytol 31:347–358
24. Sincich LC, Adams DL, Horton JC (2003) Complete flatmounting of the macaque cerebral cortex. Vis Neurosci 20:663–686
25. Lyon DC, Kaas JH (2001) Connectional and architectonic evidence for dorsal and ventral V3, and dorsomedial area in marmoset monkeys. J Neurosci 21:249–261
26. Rosa MG, Tweedale R (2005) Brain maps, great and small: lessons from comparative studies of primate visual cortical organization. Philos Trans R Soc Lond B Biol Sci 360:665–691
27. Van Essen DC (2004) Organization of visual areas in macaque and human cerebral cortex. In: Chalupa LM, Werner JS (eds) The visual neurosciences. MIT, Cambridge

28. Kaas JH, Lyon DC (2001) Visual cortex organization in primates: theories of V3 and adjoining visual areas. Prog Brain Res 134:285–295
29. Wong P, Kaas JH (2009) Architectonic subdivisions of neocortex in the tree shrew (*Tupaia belangeri*). Anat Rec (Hoboken) 292:994–1027
30. Casagrande VA, Kaas JH (1994) The afferent, intrinsic, and efferent connections of primary visual cortex in primates. In: Peters A, Rockland KS (eds) Cerebral cortex. Plenum Publishing Co., New York
31. Horton JC, Hubel DH (1981) Regular patchy distribution of cytochrome oxidase staining in primary visual cortex of macaque monkey. Nature 292:762–764
32. Preuss TM, Coleman GQ (2002) Human-specific organization of primary visual cortex: alternating compartments of dense Cat-301 and calbindin immunoreactivity in layer 4A. Cereb Cortex 12:671–691
33. Mesulam MM (1982) Tracing neural connections with horseradish peroxidase. Wiley, New York
34. Gerfen CR, Sawchenko PE (1984) An anterograde neuroanatomical tracing method that shows the detailed morphology of neurons, their axons and terminals: immunohistochemical localization of an axonally transported plant lectin, Phaseolus vulgaris leucoagglutinin (PHA-L). Brain Res 290:219–238
35. Novikova L, Novikov L, Kellerth JO (1997) Persistent neuronal labeling by retrograde fluorescent tracers: a comparison between Fast Blue, Fluoro-Gold and various dextran conjugates. J Neurosci Methods 74:9–15
36. Schmued L, Kyriakidis K, Heimer L (1990) In vivo anterograde and retrograde axonal transport of the fluorescent rhodamine-dextran-amine, Fluoro-Ruby, within the CNS. Brain Res 526:127–134
37. Lyon DC, Jain N, Kaas JH (1998) Cortical connections of striate and extrastriate visual areas in tree shrews. J Comp Neurol 401:109–128
38. Lyon DC, Kaas JH (2002) Evidence for a modified V3 with dorsal and ventral halves in macaque monkeys. Neuron 33:453–461
39. Lyon DC, Kaas JH (2002) Evidence from V1 connections for both dorsal and ventral subdivisions of V3 in three species of New World monkeys. J Comp Neurol 449:281–297
40. Angelucci A, Clasca F, Sur M (1996) Anterograde axonal tracing with the subunit B of cholera toxin: a highly sensitive immunohistochemical protocol for revealing fine axonal morphology in adult and neonatal brains. J Neurosci Methods 65:101–112
41. Ericson H, Blomqvist A (1988) Tracing of neuronal connections with cholera toxin subunit B: light and electron microscopic immunohistochemistry using monoclonal antibodies. J Neurosci Methods 24:225–235
42. Marino J, Schummers J, Lyon DC et al (2005) Invariant computations in local cortical networks with balanced excitation and inhibition. Nat Neurosci 8:194–201
43. Bosking WH, Zhang Y, Schofield B et al (1997) Orientation selectivity and the arrangement of horizontal connections in tree shrew striate cortex. J Neurosci 17:2112–2127
44. Rockland KS, Lund JS (1982) Widespread periodic intrinsic connections in the tree shrew visual cortex. Science 215:1532–1534
45. Yabuta NH, Callaway EM (1998) Cytochrome-oxidase blobs and intrinsic horizontal connections of layer 2/3 pyramidal neurons in primate V1. Vis Neurosci 15:1007–1027
46. Federer F, Ichida JM, Jeffs J et al (2009) Four projection streams from primate V1 to the cytochrome oxidase stripes of V2. J Neurosci 29:15455–15471
47. Nassi JJ, Callaway EM (2007) Specialized circuits from primary visual cortex to V2 and area MT. Neuron 55:799–808
48. Wickersham IR, Finke S, Conzelmann KK et al (2007) Retrograde neuronal tracing with a deletion-mutant rabies virus. Nat Methods 4:47–49
49. Enquist LW, Card JP (2003) Recent advances in the use of neurotropic viruses for circuit analysis. Curr Opin Neurobiol 13:603–606
50. Braz JM, Enquist LW, Basbaum AI (2009) Inputs to serotonergic neurons revealed by conditional viral transneuronal tracing. J Comp Neurol 514:145–160
51. DeFalco J, Tomishima M, Liu H et al (2001) Virus-assisted mapping of neural inputs to a feeding center in the hypothalamus. Science 291:2608–2613
52. Callaway EM, Katz LC (1993) Photostimulation using caged glutamate reveals functional circuitry in living brain slices. Proc Natl Acad Sci USA 90:7661–7665
53. Katz LC, Dalva MB (1994) Scanning laser photostimulation: a new approach for analyzing brain circuits. J Neurosci Methods 54:205–218
54. Shepherd GM, Svoboda K (2005) Laminar and columnar organization of ascending excitatory projections to layer 2/3 pyramidal

neurons in rat barrel cortex. J Neurosci 25:5670–5679

55. Xu X, Callaway EM (2009) Laminar specificity of functional input to distinct types of inhibitory cortical neurons. J Neurosci 29:70–85
56. Callaway EM (2004) Close encounters: how cortical neurons find and connect to their correct synaptic partners depends on the cell type. Neuron 43:156–158
57. da Costa NM, Martin KA (2009) Selective targeting of the dendrites of corticothalamic cells by thalamic afferents in area 17 of the cat. J Neurosci 29:13919–13928
58. Stepanyants A, Tamas G, Chklovskii DB (2004) Class-specific features of neuronal wiring. Neuron 43:251–259
59. Thomson AM, Morris OT (2002) Selectivity in the inter-laminar connections made by neocortical neurones. J Neurocytol 31:239–246
60. Kelly RM, Strick PL (2000) Rabies as a transneuronal tracer of circuits in the central nervous system. J Neurosci Methods 103:63–71
61. Ugolini G (1995) Specificity of rabies virus as a transneuronal tracer of motor networks: transfer from hypoglossal motoneurons to connected second-order and higher order central nervous system cell groups. J Comp Neurol 356:457–480
62. Ugolini G (2008) Use of rabies virus as a transneuronal tracer of neuronal connections: implications for the understanding of rabies pathogenesis. Dev Biol (Basel) 131:493–506
63. Conzelmann KK, Cox JH, Schneider LG et al (1990) Molecular cloning and complete nucleotide sequence of the attenuated rabies virus SAD B19. Virology 175:485–499
64. Etessami R, Conzelmann KK, Fadai-Ghotbi B et al (2000) Spread and pathogenic characteristics of a G-deficient rabies virus recombinant: an in vitro and in vivo study. J Gen Virol 81:2147–2153
65. Gaudin Y, Ruigrok RW, Knossow M et al (1993) Low-pH conformational changes of rabies virus glycoprotein and their role in membrane fusion. J Virol 67:1365–1372
66. Lafon M (2005) Rabies virus receptors. J Neurovirol 11:82–87
67. Coulon P, Rollin P, Aubert M et al (1982) Molecular basis of rabies virus virulence. I. Selection of avirulent mutants of the CVS strain with anti-G monoclonal antibodies. J Gen Virol 61(Pt l):97–100
68. Nassi JJ, Lyon DC, Callaway EM (2006) The parvocellular LGN provides a robust disynaptic input to the visual motion area MT. Neuron 50:319–327
69. Lyon DC, Nassi JJ, Callaway EM (2010) A disynaptic relay from superior colliculus to dorsal stream visual cortex in macaque monkey. Neuron 65:270–279
70. Larsen DD, Wickersham IR, Callaway EM (2007) Retrograde tracing with recombinant rabies virus reveals correlations between projection targets and dendritic architecture in layer 5 of mouse barrel cortex. Front Neural Circuits 1:5
71. Barnard RJ, Elleder D, Young JA (2006) Avian sarcoma and leukosis virus-receptor interactions: from classical genetics to novel insights into virus-cell membrane fusion. Virology 344:25–29
72. Kitamura K, Judkewitz B, Kano M et al (2008) Targeted patch-clamp recordings and single-cell electroporation of unlabeled neurons in vivo. Nat Methods 5:61–67
73. Joshi S, Hawken MJ (2006) Loose-patch-juxtacellular recording in vivo – a method for functional characterization and labeling of neurons in macaque V1. J Neurosci Methods 156:37–49
74. Luo L, Callaway EM, Svoboda K (2008) Genetic dissection of neural circuits. Neuron 57:634–660

Neuromethods (2012) 67: 391–403
DOI 10.1007/7657_2011_13

Published online: 1 December 2011

The Isotropic Fractionator: A Fast, Reliable Method to Determine Numbers of Cells in the Brain or Other Tissues

Suzana Herculano-Houzel

Abstract

Stereological techniques that estimate cell numbers require specific training and elaborate sampling strategies to infer total numbers of cells in well-defined structures of measurable volume. The isotropic fractionator is a fast and inexpensive method that requires little specific training and few materials before it can be used to quantify total numbers of neuronal and nonneuronal cells in the whole brain or any dissectible regions thereof. It consists in transforming highly anisotropic (paraformaldehyde fixed and dissected) brain structures into homogeneous, isotropic suspensions of cell nuclei which can be counted and identified morphologically and immunocytochemically as neuronal or nonneuronal. Estimates of total cell, neuronal and nonneuronal, numbers can be obtained within 24 h, and vary by less than 10% among samples of the same structure. Since the estimates obtained are independent of brain volume, they can be used in comparative studies of brain volume variation among species and in studies of phylogenesis, development, adult neurogenesis, and pathology.

Key words: Numbers of neurons, Brain size, Isotropic fractionator, Allometry, NeuN, Stereology

1. Introduction

How do numbers of neurons and glial cells compare across structures, species, ages, conditions, or experimental manipulations? Traditionally, stereological methods, such as the optical dissector and fractionator, are used to estimate numbers of cells in discrete brain regions (1). These methods, however, are very time consuming, require familiarity with stereological techniques, depend on accurate measurement of structure volume, and either must be restricted to well-defined structures of isotropic architecture and measurable volume or require elaborate sampling strategies to ensure that the counted samples are representative.

To circumvent these drawbacks, we created an alternative method, the isotropic fractionator (2), which estimates the total number of neuronal and nonneuronal cells independently from tissue volume and anisotropy, and can be applied to the whole brain or any dissectible structure. It does not require familiarity with stereological techniques, and yields reproducible estimates of total numbers of cells and neurons within a single day. We have already employed this method in comparative studies of the cellular composition of the brain of different mammalian species (3–5), including the entire human brain (6), and also in a study of the changes in the cellular composition of the developing rat brain (7). Other possible uses include the analysis of pathological alterations in the cellular composition of the brain and of experimental manipulations expected to affect it.

The isotropic fractionator relies on the single assumption that every cell in the brain contains one and only one nucleus. The method consists in processing fixed brains, as a whole or dissected into subregions, into an isotropic suspension of isolated nuclei in which cytoarchitectural heterogeneities have been literally dissolved. Since this suspension has a known, defined volume and can be made homogeneous by agitation, the total number of nuclei therein—and therefore the total number of cells in the original tissue—can be estimated by determining the density of nuclei in small aliquots of the suspension. Once the total cell number is known, the proportion of neurons is determined by immunocytochemical detection of Neuronal Nuclear antigen (NeuN), expressed in all nuclei of most neuronal cell types (notable exceptions are Purkinje cells, inferior olive neurons, mitral cells, and photoreceptors) and not in nonneuronal cells (8, 9), and the number of nonneuronal cells derived by subtraction. Alternatively, morphological criteria can be used to determine numbers of readily identifiable nuclear types, such as those of Purkinje cells.

The single important limitation of the isotropic fractionator is that it relies on nuclear labels to identify specific subpopulations of cell nuclei. This eliminates several useful neuronal and glial cell markers which are restricted to the cytoplasm, such as MAP2 and GFAP. On the other hand, more and more cell type-specific transcription factors are being discovered, which expands the applications of the method.

The isotropic fractionator is a simple, fast, and reliable method for counting cells in brain (or other) tissue that consists of only a few steps.

Tissue fixation. The tissue to be analyzed is ideally perfusion fixed, and postfixation by immersion in 4% paraformaldehyde is required. Good fixation is mandatory for the technique to work; 2 weeks of postfixation suffice to render the cell nuclei extremely resistant to mechanical friction.

Tissue dissection. This step limits how reliable individual counts are, so the technique is best applied to easily dissectible structures. Whole brain can also be used, which requires little dissection.

Tissue dissociation. The dissected structures are homogenized by mechanical friction in a saline detergent solution, which dissolves the cell membranes while maintaining the nuclear membranes intact, thus turning the tissue into a suspension of free nuclei.

Counting of total cell numbers. Samples of the DAPI-stained nuclei suspension (made homogeneous by agitation) are counted in a hemocytometer. Typically, this step takes 15–20 min per structure to be counted.

Heat-induced epitope retrieval. This is a necessary step when tissue has been overfixed, that is, when it has been postfixed for over 2 months.

Immunocytochemical identification of specific nuclear types. Neuronal nuclei in most structures can be identified by the expression of the NeuN antigen. Alternatively, morphological criteria can be used. Immunocytochemical processing takes 5–6 h, after which it takes approximately 30 min to determine the proportion of specific nuclear types in each structure. Alternatively, flow cytometry can be used in this step to speed up the determination of the proportion of immunolabeled nuclei (10). However, visual identification of immunolabeled nuclei has several advantages: it ensures a quality control of the samples, since stained debris are easily discernible from labeled nuclei under the microscope; it allows the identification of specific cell types by their nuclear morphology; and it is readily accessible to any laboratory equipped with an upright fluorescence microscope.

Storage of free nuclei. Since the immunocytochemical identification of specific nuclear types requires only a small aliquot of the nuclear suspension, the remaining volume can be stored in antifreezing solution for later reanalysis, for instance when other antibodies become available, or for photographical documentation.

2. Materials

2.1. Tissue Fixation

- Saline (NaCl, 0.9%).
- Paraformaldehyde (4% solution in phosphate buffer).
- Peristaltic pump.

2.2. Tissue Dissection

- Surgical instruments.
- Stereoscopic microscope.
- Analytic balance.

2.3. Tissue Dissociation

- Tenbroeck tissue grinder (glass homogenizer, 2, 7. or 40 ml capacity, depending on the size of the tissue of interest).

- Dissociation solution: 1 l of 1% Triton X-100 in 40 mM sodium citrate.
- Pasteur pipettes.
- Centrifuge tubes (15 or 50 ml).
- PBS.
- Clinical centrifuge.

2.4. Counting of Total Cell Numbers

- Diamino-phenyl-indol (DAPI, Molecular Probes), 10 mg/l.
- Neubauer chamber (hemocytometer), improved (see Note 1).
- Micropipette, 10 μl.
- Fluorescence microscope, upright (see Note 2).

2.5. Heat-Induced Epitope Retrieval

- 0.2 M boric acid, pH 9.0 (adjusted with lentils of NaOH).
- PBS.
- 1.5-ml Eppendorf tubes.
- Microcentrifuge.
- Heating bath or oven at 70°C.

2.6. Immunocytochemical Identification of Specific Nuclear Types

- Anti-NeuN mouse IgG (Millipore ab377).
- Normal goat serum.
- Fluorescent anti-mouse secondary antibody.
- DAPI (Molecular Probes), 10 mg/l.

2.7. Storage of Free Nuclei

- Sucrose, 30% in PB.
- Antifreezing solution (30% glycerol and 30% ethylene glycol in 0.024 M phosphate buffer; use 10× dilution of 0.24 M phosphate buffer, which is 30.8 g NaOH and 117.12 g NaH_2PO_4 in 4 l of distilled water).

3. Methods

3.1. Tissue Fixation

1. Perfuse the animal of interest with saline followed by 4% paraformaldehyde. This step can be skipped for small tissues, such as embryonic mouse brains or invertebrate ganglia (which can be immersion fixed) (see Note 3).
2. Remove the brain from the skull; remove the dura mater and major superficial blood vessels from the fixed brain (for other body structures, remove any connective tissue and blood vessels).
3. Allow the tissue to postfixate by immersion in 4% paraformaldehyde. The ideal postfixation time is 2–4 weeks (see Note 4).

3.2. Tissue Dissection

4. Under a stereoscope, dissect the regions of interest.
5. Weigh the dissociated regions of interest to be counted so that cell densities can be determined later.
6. Store the dissected regions of interest in 4% paraformaldehyde if further postfixation is required (that is, if the tissue is not yet well-fixed or has not been immersion fixed for at least 2 weeks); store in PBS for immediate use or in antifreezing solution for long-term storage until dissociation (see step #45).

3.3. Tissue Dissociation

7. Place the dissociated tissue inside the homogenizer tube. Small tissues (up to about 50 mg or 2 mm diameter) can be dissociated directly. Large tissues, such as whole cortical areas (up to 1 g of tissue), need to be diced in a Petri dish prior to homogenizing; wash any leftovers in the dish and in your scalpel into the homogenizer tube using the dissociation solution (it is fundamental that ALL tissue is homogenized). Much larger tissues, such as very large cortical areas, whole cerebella, spinal cords, or whole brains, must be divided into smaller portions of about 1 g to be dissociated separately and combined in a graduated cylinder (see note #10).
8. Add dissociation solution to the homogenizer to a final volume that is about ten times larger than the tissue to be dissociated, and large enough for the glass tube to be filled with fluid when the piston is fully inserted into the homogenizer. Use this volume to wash down into the tube any tissue fragments that stick to the walls of the homogenizer (see Note 5).
9. Insert the piston and homogenize the tissue by making simultaneous up-and-down and rotating movements with the piston. Hold the tube upright at all times to avoid any spills (see Note 6). Use care not to let air into the tube, or foam will result (see Note 7).
10. Homogenization is over when no more tissue fragments are visible, which should take 10–30 min of grinding (see Note 8). To check for remaining tissue fragments, raise the homogenizer against the light, with the piston fully inserted, and look for small specks of tissue while gently rotating the piston. When no more specks are visible, proceed to the next step.
11. Wash the walls of the homogenizer. First, place the tube in a stand, remove the piston carefully, and use a fresh volume of dissociation solution to wash the walls of the piston into the tube. Once no material is left on the piston, place it on the bench and use another fresh volume of dissociation solution to wash the material left on the walls of the homogenizer into the tube (see Note 9). This collects into the main suspension any nuclei left behind on the glass walls.

12. Collect the nuclei suspension into a 15- or 50-ml graduated centrifuge tube. To do this, do not pour the contents of the homogenizer into the graduated tube (see Note 10). Rather, use a long Pasteur pipette to pick up the suspension, starting from the bottom of the tube where the nuclei are densest, and transfer it to the graduated tube. Once all of the suspension is transferred, use a fresh volume of dissociation solution to wash the walls of the homogenizer, collecting into the bottom of the tube any nuclei left on the glass walls. Transfer this volume to the same graduated tube. Repeat the wash about two more times or as needed using small volumes in each wash, until the fluid in the transfer pipette is perfectly clear (see Note 11).
13. Now that the entire suspension containing all cell nuclei in the original structure is in the graduated tube, add DAPI to it, diluting 20–50× from a 10 mg/l stock solution. The dilution required depends on how dense the nuclei suspension is; for a same suspension volume, cell-denser structures, such as the cerebellum, require more DAPI (for instance, a 20× dilution only) to achieve sufficient labeling that is stable under UV illumination.
14. Use PBS to complete the suspension volume to a defined value in the graduated tube. The precise value is not important as long as it can be read with precision on the tube (see Note 12).

3.4. Counting of Total Cell Numbers

15. Make the nuclei suspension homogeneous by inverting the graduated tube 10–20 times, taking care to avoid forming foam.
16. Immediately after making the suspension homogeneous, collect four 10 μl samples and place them in different chambers of two hemocytometers. Allow 1–2 min for nuclei to sink.
17. Under low magnification in the fluorescence microscope, verify the quality of the preparation. Check if the nuclei are free and well-distributed, without clumps (Fig. 1, left) (see Note 13). Under high magnification (400×), verify that the nuclei are well-labeled with DAPI, appearing brightly blue under UV illumination and that the vast majority are intact, with preserved contours (see Note 14).
18. Count the number of nuclei in a same, known volume for each chamber. For instance, if nuclei in all 25 fields of the center grid are counted, that amounts to the number of cell nuclei in a volume of 0.0001 ml of the suspension (see Note 15). If there are at least 50–60 nuclei in the ensemble of the 25 central fields, then proceed to step 19. If not, then go to step 18a.
    (a) Increase the density of nuclei in the suspension to be counted. This step is necessary to ensure that the counts are reproducible and not subject to Poisson variation.

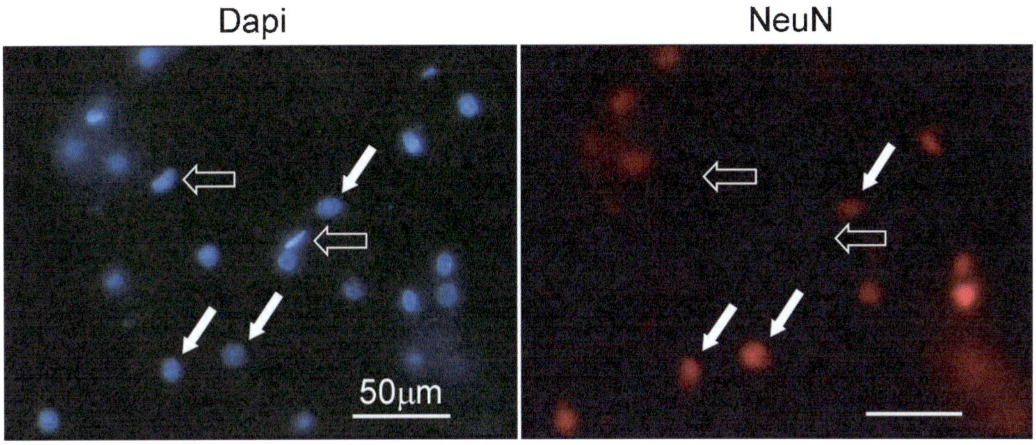

Fig. 1. Appearance of isolated, DAPI- and NeuN-stained nuclei. Isolated nuclei have a clearly delimited contour, indicating that the nuclear membrane is intact, and are brightly stained with DAPI (*left*). Notice that the nuclear morphology is preserved by the fixation. Nuclei that are also stained by labeling with anti-NeuN antibody are readily identifiable against a dark background of virtually inexistent, inspecific staining (*right*). *Filled arrows*, examples of NeuN-positive nuclei; *open arrows*, examples of NeuN-negative nuclei. Scale bar, 50 μm.

To do this, spin down the graduated tube (see Note 16) with the nuclei suspension in a clinical centrifuge (the exact time required depends on the diameter of the rotor; typically, 8 min at 3,200 rpm suffice to ascertain that all nuclei are collected in a pellet). Remove the supernatant with a Pasteur pipette, without disturbing the pellet; and add PBS to a precise final volume that is small enough to ensure that there are more than 60 nuclei per 0.0001 ml of the suspension. Repeat step 18.

19. Average the number of nuclei counted in the same volume across the four samples, and verify if the coefficient of variation (CV, which amounts to the standard deviation of the values divided by the average value) is smaller than 0.15. Typically, the CV is smaller than 0.10. If it is not, then count additional samples until CV < 0.15 (see Note 17), and calculate the average value.
20. Calculate the total number of nuclei in the suspension by simply multiplying the number of nuclei per milliliter by the total suspension volume. If you determine the average number of nuclei in a volume of 0.0001 ml (that is, the 25 central fields of the chamber), then multiply that average by 10,000 to find the number of nuclei per ml of the suspension; next, multiply that value by the total volume of the suspension.
21. Make the suspension homogeneous again by inverting the graduated tube and collect 1 ml into an Eppendorf tube.

22. Spin down nuclei for 5 min in a microcentrifuge. Remove supernatant with a micropipette, without disturbing the pellet.
23. If the tissue of interest has been in fixative for less than 6 weeks, then proceed to step 28. If not, proceed to step 24.

3.5. Heat-Induced Epitope Retrieval

24. Add 1 ml of PBS to the pellet and dissolve it completely using the micropipette if necessary. Do not vortex. Spin down nuclei again and remove supernatant.
25. Repeat step 24 for a total of three times (that is, wash nuclei three times in PBS).
26. Add 1 ml of 0.2 boric acid, pH 9.0, to the pellet (to breakdown the excess of aldehydes and the autofluorescence that they cause) and dissolve the pellet completely using the micropipette if necessary. Do not vortex.
27. Incubate at 70°C for 45 min, and then spin down nuclei to remove supernatant.

3.6. Immunocytochemical Identification of Specific Nuclear Types (see Note 18)

28. Add 1 ml of PBS to the pellet and dissolve it completely using the micropipette if necessary. Do not vortex. Spin down nuclei again and remove supernatant.
29. Repeat step 28 for a total of three times (that is, wash nuclei three times in PBS).
30. Add 199 μl of PBS to the pellet and 1 μl of anti-NeuN primary antibody (1:200 dilution). Alternatively, use the appropriate dilution of the desired antibody. Dissolve the pellet completely using the micropipette. Do not vortex.
31. Incubate at room temperature for 2 h, preferably under agitation.
32. Spin down the nuclei for 5 min in a microcentrifuge. Remove supernatant with a micropipette without disturbing the pellet.
33. Add 1 ml of PBS to the pellet and dissolve it completely using the micropipette if necessary. Do not vortex. Spin down nuclei again and remove supernatant.
34. Repeat step 33 for a total of three times (that is, wash nuclei three times in PBS).
35. Add the secondary antibody in the required dilution in the presence of 10% normal goat serum and 10% DAPI. Dissolve the pellet completely using the micropipette. Do not vortex.
36. Incubate at room temperature for 2 h, preferably under agitation.
37. Spin down the nuclei for 5 min in a microcentrifuge. Remove supernatant with a micropipette without disturbing the pellet.
38. Add 1 ml of PBS to the pellet and dissolve it completely using the micropipette if necessary. Do not vortex. Spin down nuclei again and remove supernatant.

39. Repeat step 38 for a total of three times (that is, wash nuclei three times in PBS).
40. Add 1 ml of PBS to the pellet and dissolve it completely using the micropipette if necessary. Do not vortex.
41. After agitating the tube, remove a 4 μl sample and place it in the Neubauer chamber. Allow 1–2 min for the nuclei to set.
42. Under 400× magnification, determine the percentage of DAPI-labeled nuclei that also show labeling with the antibody (Fig. 1). The total number of nuclei that need to be counted depends on the frequency of antibody-stained nuclei; the smaller the percentage of stained nuclei, the larger the total number of nuclei that must be counted. Given that usually 30–90% of nuclei in any brain structure are NeuN positive, a minimum of 500 nuclei should be counted.
43. Determine the total number of neurons (that is, NeuN-positive nuclei) in the structure as (% $NeuN^+$ nuclei × total number of nuclei)/100.
44. Determine the total number of nonneuronal cells (that is, NeuN-negative nuclei) in the structure as (total number of cells − $NeuN^+$ cells).

3.7. Storage of Free Nuclei

45. For later reanalysis of nuclei (for instance, with another antibody or morphological criterion), collect the remaining nuclei for storage by spinning them down in a centrifuge.
46. Discard the supernatant and resuspend the nuclei in 30% sucrose in PB. Allow for at least a couple of hours for the nuclei to equilibrate and start to descend in the tube (see Note 19).
47. Spin down the nuclei, discard the supernatant, and resuspend the nuclei in antifreezing solution. Allow for at least a couple of hours for the nuclei to equilibrate and start to descend in the tube (see Note 20).
48. Store at −20°C until the next use. Frozen nuclear suspensions can be utilized for immunocytochemistry. In that case, agitate the frozen suspension before taking a 1 ml sample for processing, spin down the nuclei, and proceed to step 24 (if epitope retrieval is required) or to step 28.

4. Notes

1. The "improved" Neubauer chamber has a reflexive coating that is necessary for visualizing the markings in the chamber under the fluorescence microscope with the very little transmitted light required for discerning stained from nonstained nuclei.

2. An inverted microscope does not do as the Neubauer chamber is too thick to allow for the light to reach the sample, and cannot be placed upside down on the chariot (or the coverslip will slide).
3. The only drawback of immersion fixation, supposing that the tissue is fixed well-enough by immersion, is that blood cells are not removed. Considering that most of them are red blood cells, which do not have nuclei in mammalian species, and that blood occupies at most 5% of brain volume (11), the total cell count in the tissue should not be affected significantly. Beware of red blood cells, however, if you count avian or reptile tissue that has not been perfused. However, these cells are easily distinguished from their slim, elongated nuclear morphology, and it suffices not to count them.
4. Shorter postfixation times or no postfixation yields nuclei that are not resistant enough to the dissociation step and break open or clump together after centrifuging. The result is visible under the microscope as DAPI-stained broken nuclei and free nucleoli or as nuclear "grape bunches." On the other hand, postfixation for longer than 6 weeks requires heat-induced epitope retrieval for antibody staining to be achieved. This can be done as described in the text, but is best avoided if possible.
5. The exact volume of dissociation solution used in the homogenization is not important as long as it is large enough to fill the tube and allow all cell membranes to be dissolved and small enough not to make spills likely. If dissociation is difficult (that is, if tissue fragments resist your efforts), add more dissociation solution to the homogenizer—or, next time, dissociate smaller fragments of tissue at a time (the homogenates can be combined into a single graduated cylinder for counting).
6. This method is based on collecting ALL the cell nuclei in the structure of interest into a suspension of free nuclei. If the nuclei-dense preparation is spilled at this step, the structure should be discarded, as there is no telling what percentage of all nuclei were lost in the spill.
7. This is a quite physical part of the method, and can require a good amount of force. Do resist the urge to rest the homogenizer on the bench as you apply force to the piston: this spares the already strained glass bottom of the tube from further pressure from the bench, which otherwise greatly increases the chances of having the homogenizer break, spilling your sample and hurting you in the process.
8. The amount of time required to homogenize the tissue depends on the amount of tissue in the homogenizer, but it should never take longer than about 40 min. The problem with too long dissociations is the increasing risk of disrupting

nuclei, resistant as they may be. Visual inspection under the microscope, at the cell counting step, lets you know whether the homogenization took too long, as you are able to see many broken nuclei. If that is the case, the structure should be discarded. The next time that you homogenize a similarly large structure, try dicing it into smaller pieces first. Cutting the tissue into thin slices that can easily slide between the glass walls of the homogenizer (rather than small cubes) is the best way to decrease homogenization time.

9. This volume, like all the other washing volumes, should be as small as possible because it has to fit inside the tube of the homogenizer, and ideally the total volume should be either smaller than 15 ml (when using a 7-ml homogenizer) or 50 ml (when using a 40-ml homogenizer). For the large, 40-ml homogenizer, I usually use 4–6 ml of solution to wash the piston and 8–15 ml to wash the walls of the tube.
10. When homogenizing large structures, such as the entire cortex of a macaque monkey, it is necessary to go through steps 7–12 several times, once for each small portion of the tissue to be counted. In this case, collect all of the homogenates into the same graduated cylinder. Until you have enough experience with the method, expect to collect a total of at least 50 ml of nuclei suspension per gram of tissue.
11. Again, do this transfer step with great care not to spill any drop of the nuclei-rich suspension. To avoid having to discard your sample due to spills, hold the graduated tube as close to the mouth of the homogenizer as possible.
12. To read this volume with precision, there should be no foam or bubbles in the suspension. If there are, let the suspension rest for a few minutes after adding DAPI before adjusting its final volume. This volume must be precise because it is the value used to multiply the cell density determined in the hemocytometer to obtain the final number of cell nuclei in the suspension.
13. The occasional clump of nuclei does not disturb the counts. When an isolated clump of nuclei is encountered, it should be counted as one nucleus only. What should be checked here is if the homogenization was complete, in which case the vast majority of nuclei should be intact and free.
14. If more than a small minority of nuclei are broken, discard the structure. Typically, fewer than 1–2% of the nuclei are damaged. If many clumps of nuclei are visible, return the suspension to the homogenizer for further homogenization (from step 7 above). If DAPI staining is weak or fades rapidly under UV illumination, add more DAPI to the suspension, taking care to complete the final volume to a new, precise value (step 14).

15. The exact volume of the Neubauer chamber that is counted is irrelevant as long as the same volume is counted for all samples and the density of nuclei per milliliter of the suspension is calculated accordingly. In our lab, we count four different sets of 0.000016 ml (that is, four of the 25 central squares) for each sample (drop of the suspension) and use the sum of the four sets (that is, 0.000064 ml) to calculate CVs. This allows us to discard samples that are internally anisotropic, which may happen due to debris stuck to the coverslip or air bubbles, for example.
16. For large structures, the nuclei suspension is typically in a graduated cylinder. In our experience, there is never a need to concentrate the suspension when dealing with large structures, such as whole cerebella or cerebral cortex.
17. Values of CV above 0.15 are seldom obtained, and if they occur this is an indication that the nuclei suspension was not well-homogenized. In that case, consider starting over: make the suspension homogeneous again by inverting the tube and take four new samples for counting.
18. Alternatively, morphological criteria can be used already at the step of "counting of total cell numbers."
19. The nuclei immediately float in the sucrose solution. Use a Pasteur pipette to make sure that the pellet is dissolved into the sucrose solution.
20. The nuclei immediately float in the antifreezing solution. Use a Pasteur pipette to make sure that the pellet is dissolved into the antifreezing solution.

Acknowledgments

Thanks to Roberto Lent for encouraging and supporting the creation of this method, to Jon Kaas for continued support, to Christine Collins for improving the method with the FACS variation, and to Paul Manger for insights on tissue storage. The development and improvement of this method were made possible by grants from CNPq and FAPERJ.

References

1. West MJ (1999) Stereological methods for estimating the total number of neurons and synapses: issues of precision and bias. Trends Neurosci 22:51–61
2. Herculano-Houzel S, Lent R (2005) Isotropic fractionator: a simple, rapid method for the quantification of total cell and neuron numbers in the brain. J Neurosci 25:2518–2521

3. Herculano-Houzel S, Mota B, Lent R (2006) Cellular scaling rules for rodent brains. Proc Natl Acad Sci USA 103:12138–12143
4. Herculano-Houzel S, Collins C, Wong P et al (2007) Cellular scaling rules for primate brains. Proc Natl Acad Sci USA 104:3562–3567
5. Sarko DK, Catania KC, Leitch DB et al (2009) Cellular scaling rules of insectivore brains. Front Neuroanat 3:8
6. Azevedo FAC, Carvalho LRB, Grinberg LT et al (2009) Equal numbers of neuronal and non-neuronal cells make the human brain an isometrically scaled-up primate brain. J Comp Neurol 513:532–541
7. Bandeira FC, Lent R, Herculano-Houzel S (2009) Changing numbers of neuronal and non-neuronal cells underlie postnatal brain growth in the rat. Proc Natl Acad Sci USA 106:14108–14113
8. Mullen RJ, Buck CR, Smith AM (1992) NeuN, a neuronal specific nuclear protein in vertebrates. Development 116:201–211
9. Gittins R, Herrison PJ (2004) Neuronal density, size and shape in the human anterior cingulated cortex: a comparison of Nissl and NeuN staining. Brain Res Bull 63:155–160
10. Collins CE, Young NA, Flaherty DK, Airey DC, Kaas JH (2009) A rapid and reliable method of counting neurons and other cells in brain tissue: a comparison of flow cytometry and manual counting methods. Front Neuroanat 9:5
11. Lauwers F, Cassot F, Lauwers-Cances V, Puwanarajah P, Duvernoy H (2008) Morphometry of the human cerebral cortex microcirculation: general characteristics and space-related profiles. Neuroimage 39:936–948

Neuromethods (2012) 67: 405–431
DOI 10.1007/7657_2011_14

Published online: 25 November 2011

Morpho-Functional Mapping of Cortical Networks in Brain Slice Preparations Using Paired Electrophysiological Recordings

Gabriele Radnikow, Robert Heinz Günter, Manuel Marx, and Dirk Feldmeyer

Abstract

The study of neuronal microcircuits with paired electrophysiological recordings from synaptically coupled neurons in brain slices has revealed a large variety of neuronal cell types with highly distinct and connection-specific characteristics of synaptic transmission. In combination with simultaneous biocytin fillings paired recordings permit correlated structural and functional analyses of pre- and post-synaptic neurons, including technically challenging approaches such as a quantal analysis of identified unitary synaptic connections. Here, we present the technical procedures for successful paired recordings, methods to obtain an optimal neuronal morphology, the working principle for neuronal reconstructions, and the basis and caveats in estimating neuronal connectivity from paired recording data. The paired recording technique will remain an important approach in the analysis of neuronal connectivity in the brain, in particular with respect to the finer details of synaptic transmission, since it allows the morphological and functional characterisation of both pre- and post-synaptic neuronal cell types which are not possible using other methods to study neuronal connectivity.

Key words: Neuronal connectivity, Paired recordings, Neuronal reconstructions, Axonal domain, Dendritic domain, Innervation domain, Synaptic contacts, Structure–function correlations

1. Introduction

The way information is encoded in the brain is largely determined by the neuronal microcircuits involved in this process. The rules that govern the neuronal connectivity and thus the make-up of such microcircuits have therefore gained considerable attention and the past 20 years saw the development of a number of new experimental approaches. However, studying connectivity is a multifaceted problem. To describe a neuronal microcircuit, it is necessary to identify the structural and functional characteristics of both the pre- and post-synaptic neurons, the type of synapse (electrical and/or chemical), the properties of the synaptic response (time course,

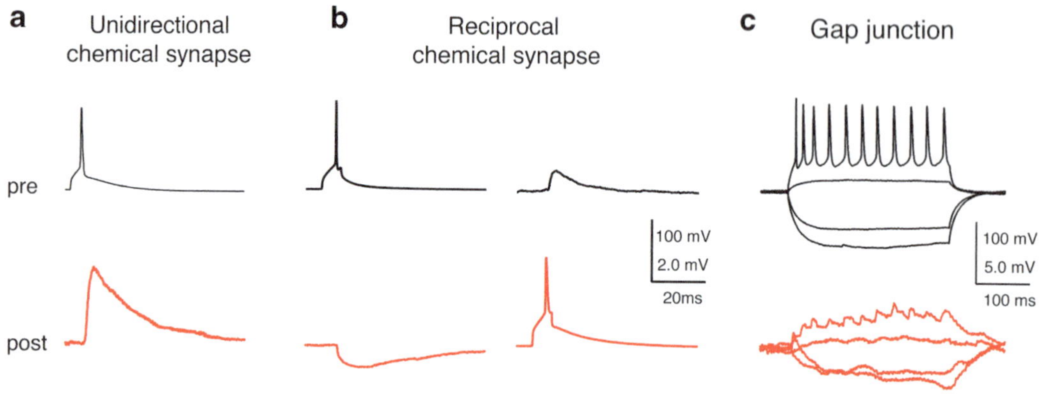

Fig. 1. Paired recordings from synaptically coupled neurons. The figure shows paired recordings from neurons that display different modes of synaptic coupling. (**a**) A pre-synaptic action potential (*black trace*) in a glutamatergic cortical neuron elicits a monosynaptic EPSP in another glutamatergic neuron (*red trace*). (**b**) An action potential (*black trace*) in a pre-synaptic inhibitory interneuron evokes a post-synaptic hyperpolarising IPSP (*red trace*) in an excitatory neuron. In turn, an action potential in the excitatory neuron (*red trace*) elicits an EPSP in the inhibitory interneuron (*black trace*). (**c**) A family of hyper- and depolarising voltage steps elicited by current injections into a neuron (*black traces*) is reflected in a gap-junction coupled second neuron (*red traces*). Note the slower rise and decay of the voltage response in the coupled neuron and the strong amplitude attenuation.

post-synaptic receptors etc.), as well as the number and location of synaptic contacts. For this, paired recordings from synaptically coupled neurons ("paired or multiple recordings") are therefore the method of choice as no other available technique is presently able to investigate all these parameters simultaneously. Figure 1a–c illustrates that paired recordings permit the analysis of identified neuronal connections at unidirectional or reciprocal chemical as well as electrical synapses. Using a paired recording approach it is possible to record from reciprocal synaptic connections in which one neuron is inhibitory and the other excitatory (Fig. 1b) or to record from gap junction-coupled pairs of neurons (Fig. 1c). Even recordings from neurons that are connected via both gap junctions and chemical synapses are feasible (e.g. (1)). So far, optical stimulation techniques are unable to achieve this.

Paired recordings from synaptically coupled neurons have been performed in a variety of tissues (2–7), initially with sharp microelectrodes and later with patch pipettes to investigate the properties of unitary monosynaptic connections. Infrared video microscopy combined with patch-clamp recording techniques further facilitated this approach (8–10). Thus, it was possible to record from visually identified neurons in defined cortical microcircuits with an improved temporal resolution. Paired recordings from synaptically coupled neurons are therefore particularly well suited for an in-depth investigation of synaptic transmission at identified neuronal connections (11–13). A few studies also demonstrated that paired recordings are feasible under in vivo conditions (14–16).

While a number of optical stimulation approaches have been used successfully to study the large-scale synaptic connectivity (17–29), simultaneous recordings from synaptically coupled neurons are particularly well suited to characterise the properties of both post- and pre-synaptic neurons in unitary synaptic connections. To date, optical stimulation (using photo-release of caged glutamate compounds or photo-activation of channelrhodopsins with LEDs or LASERs) does not allow the identification of the pre-synaptic neuron in a synaptic connection and/or the exact localisation of synaptic contacts for monosynaptic connections. However, studies in recent years have demonstrated that not only GABAergic interneurons but also excitatory neurons are functionally and morphologically highly diverse (30–33) which emphasises the need for a morphological identification of both the pre- and post-synaptic neuron within a connection.

Another advantage of the paired recording technique is that it allows to probe for reciprocal synaptic connections (Fig. 1b, (34, 35)). It is also possible to test for the presence of functional gap junctions (Fig. 1c) and their possible coexistence with chemical synaptic contacts (1, 36–38).

When combined with simultaneous biocytin staining via the recording pipette (Fig. 2a; (34, 35, 39)), paired recording approaches allow a correlation of the functional and structural properties of a synaptic connection, both at the light and electron microscopic level. A quantitative morphological analysis can be performed after the reconstruction of the recorded neuronal cell pair. In particular, it is possible to identify the location of synaptic contacts. This can be done directly using electron microscopic techniques (11, 35, 40–44) or using calcium imaging (45, 46) of synaptic contacts, but these studies are also extremely challenging and laborious. However, pure electron microscopic or imaging approaches do not allow the unambiguous identification of particular neuronal cell types and make a detailed analysis of synaptic contacts between larger populations of pre- and post-synaptic neurons rather difficult.

It is possible to predict the degree of synaptic connectivity between two morphologically identified neurons by analysing the spatial extent of the pre-synaptic axon and its degree of overlap with the dendrite of the post-synaptic neuron from reconstructions performed at the light microscopic level (Fig. 2b). The hypothesis on which this approach is based is that the probability of finding a synaptic contact is proportional to both the density of axons (belonging to the population of putative pre-synaptic neurons) projecting into a predefined region and the density of dendrites (belonging to the population of putative post-synaptic neurons) projecting into the same region (see Sect. 3.6 and Figs. 9, 10 for a detailed description of this technique). The concept suggesting that axonal projections target all dendrites within their reach in a random fashion was

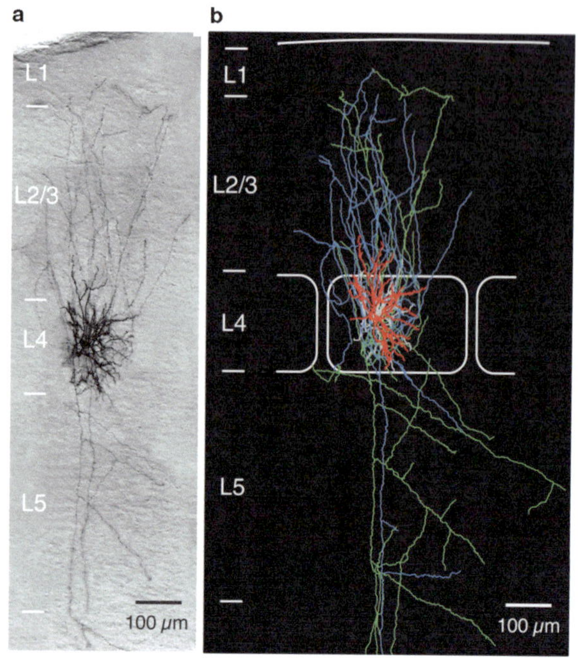

Fig. 2. Pair of synaptically coupled neurons. (**a**) Halftone photograph of a biocytin-labelled pair of synaptically coupled neurons in layer 4 of the rat somatosensory (barrel) cortex. Cortical layers are indicated on the *left*. (**b**) Computerised reconstruction of the dendritic and axonal arbours of the pre- and post-synaptic excitatory neurons in layer 4. The dendrites of the pre-synaptic neuron are in *red* and its axon in *blue*, the dendrites of the post-synaptic neuron are in *light grey* and its axon in *green*. The outlines of the barrels at the level of layer 4 and the pia are also given; cortical layers are indicated on the *left*. Such reconstructions are the basis for realistic computer models of neuronal networks.

originally proposed for the thalamocortical input to cortical layer 4 (47). It was subsequently extended as a principle to all thalamocortical afferents (48) and generalised as a "rule" for the neuronal connectivity in the neocortex ("Peter's rule", e.g. (49)). The strong assumption of a proportionality can be weakened into the following argument: the number of synapses in a given volume is a monotonous function of the amount of axons and dendrites in that volume. This somewhat weaker formulation may allow to include more specific synaptic connectivity (i.e. "exceptions to Peter's rule", e.g. summarised in (49, 50)). Another modification of the assumption of locally random connectivity is to consider potential synaptic connectivity on a smaller length scale (51–53), i.e. to consider axons not just within a whole cortical layer but in the vicinity (i.e. within few micrometres) of the post-synaptic dendrite (for a detailed discussion see (54)). In this chapter we will show that morphological reconstructions made from biocytin-labelled neurons can predict indeed the location of synaptic contacts, but we will also alert the reader to potential pitfalls of this methodology.

Paired recordings are certainly a very time-consuming and labour-intensive approach to study the neuronal connectivity. However, the information gained from such experiments is often much more detailed than that from connectivity studies using other technical approaches, in particular because it permits a correlation of the physiological and morphological characteristics of synaptic transmission. It is therefore not likely that paired recording techniques will be superseded completely by other methods in the near future.

2. Materials

2.1. Set-Up

1. Dual or multiple patch clamp amplifier (HEKA USB Double, Triple or Quadro, *HEKA Elektronik, Lambrecht/Pfalz, Germany*) including two to four preamplifiers.
2. Olympus BX50W (*Olympus, Hamburg, Germany*) or ZEISS AxioExaminer (*Zeiss, Oberkochen, Germany*) microscope equipped with Nomarski differential interference contrast (DIC) optics for infrared (IR) light or a Dodt gradient contrast (55, 56). The microscope should have two independent camera ports. Two objectives are required: a 4× objective for slice overviews and a 40× or 60× water-immersion objective (with numerical apertures $\geq$0.8) with a long working distance to allow the placement of the patch pipette required for paired recordings.
3. Infrared video cameras (TILL VX55, *Till Photonics GmbH, Gräfelfing, Germany*) to monitor the slice at low and high magnification.
4. Workstation *Infrapatch* 240 including a motorised *x–y* stage on which the microscope is mounted (which allows to move the microscope independent of the slice chamber) and a motorised focus axis for the microscope and *x–y–z* manipulators for two to four patch clamp preamplifiers (*Luigs and Neumann GmbH, Ratingen, Germany*).
5. Computer with software to control the patch clamp (Patchmaster, *HEKA Elektronik, Lambrecht/Pfalz, Germany*) and visualisation software.
6. Manometers and switchable pressure valves for multiple patch pipettes in combination with a suction pulser unit (*Sigmann Elektronik, Hüffenhardt, Germany*).
7. Peltier device for the experimental chamber and temperature control unit (*Luigs and Neumann GmbH, Ratingen, Germany*).
8. Custom-made perfusion system with perfusion pump (*Watson-Marlow, Rommerskirchen, Germany*); perfusion speed set to 4–6 ml min^{-1} for optimal oxygenation.

9. Anti-vibration table (*Newport Spectra-Physics GmbH, Darmstadt, Germany*).
10. Faraday cage (*HEKA Elektronik, Lambrecht/Pfalz, Germany*).

2.2. Slicing

1. Custom-made perspex tools for reproducible cutting angles (Fig. 3a).
2. Dissection tools such as fine and coarse forceps, normal scissors, fine spring scissors, spatulas, bone cutters, etc. (*Fine Science Instruments, Heidelberg, Germany*).

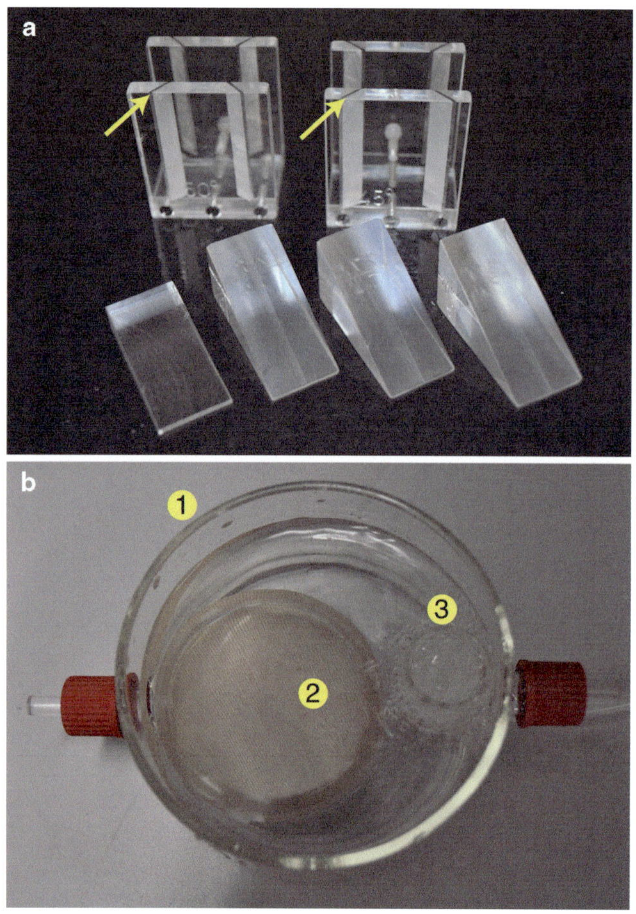

Fig. 3. Tools for slicing and slice incubation chamber. (**a**) For the preparation of brain slices with an optimal connectivity for a specific synaptic connection tools with defined cutting angles need to be used for a reproducible cutting procedure. These tools are made of perspex and allow to cut the brain on ramps with different slope angles (varying from 10° to 30°) and under different orientations (45°, 50° or 55° to the midline; see *arrows*). These cuts help to maximise the connectivity but have to be adjusted with regard to age and brain size of the animal. (**b**) After cutting, the slices are kept in a custom-made glass incubation chamber (1) in which the slices are kept on a plastic mesh that is held by a perspex ring attached to the chamber wall (2). The slices are constantly perfused with 95% O_2/5% CO_2 through a bubbler (3). In case the incubation solution needs to be heated, the chamber is transferred to a water bath.

3. Various beakers and dishes (pre-filled with well-oxygenated and cooled artificial cerebrospinal fluid (ACSF)) to keep the brain or sections of it.
4. For slicing we use a vibrating Dosaka Microslicer DTK-1000 (*Ted Pella Inc., Redding, CA, USA*) with an excentric motor or a "zero z" vibration slicer (*Microm International GmbH, Walldorf, Germany*). Brain slices in excellent condition from both immature and mature animals were obtained using either of these slicers.
5. Incubation chamber: Submerged chamber (custom-made from glass) with a perspex holder for a mesh to keep the slices and a bubbler for optimal oxygenation during the recovery period (Fig. 3b).

2.3. Equipment for Patch Pipettes

1. Patch pipettes were pulled from borosilicate glass (*Hilgenberg GmbH, Malsfeld, Germany*) with a programmable Sutter P-97 Flaming/Brown micropipette puller (*HEKA Elektronik, Lambrecht/Pfalz, Germany*).
2. Syringes with fine *Microfil* needles to back-fill patch pipettes (*World Precision Instruments Germany GmbH, Berlin, Germany*).
3. Solution filters (0.2 μm pore; *Millipore GmbH, Schwalbach/Taunus, Germany*) to filter the intracellular (patch pipette) solution.

2.4. Solutions

2.4.1. Extracellular Solutions

1. For slicing of brains from immature/juvenile animals the following solution was used (in mM): 125 NaCl, 2.5 KCl, 25 glucose, 25 $NaHCO_3$, 1.25 NaH_2PO_4, 1 $CaCl_2$, 5 $MgCl_2$, 3 Myo-inositol, 2 Na-pyruvate and 0.4 ascorbic acid; the osmolarity of the solution was ~310 mOsmol l^{-1}. The Ca^{2+} concentration was lowered to reduce potentially excitotoxic synaptic transmission during slicing.
2. For slices from brain of more mature/adult animals a sucrose containing solution was used (in mM): 206 sucrose, 2.5 KCl, 25 glucose, 25 $NaHCO_3$, 1.25 NaH_2PO_4, 1 $CaCl_2$, 3 $MgCl_2$, 3 Myo-inositol, 2 Na-pyruvate and 0.4 ascorbic acid. This solution improved the viability of mature neurons and permitted long-term recordings (>2 h) in the recording solution (see below) and has an osmolarity of ~310 mOsmol l^{-1}.
3. During recording an artificial cerebrospinal fluid (ACSF) of the following composition was used (in mM): 125 NaCl, 2.5 KCl, 25 glucose, 25 $NaHCO_3$, 1.25 NaH_2PO_4, 2 $CaCl_2$ and 1 $MgCl_2$; the osmolarity of the solution was ~310 mOsmol l^{-1}.

All extracellular solutions were constantly gassed with 95% O_2 and 5% CO_2 to maintain a physiological pH and adequate oxygenation levels.

2.4.2. Intracellular Solutions

All pipette solutions for intracellular recordings are based on K-gluconate.

1. For excitatory connections the intracellular solution had a composition as follows: 105 mM K-gluconate, 30 mM KCl, 10 mM HEPES, 10 mM phosphocreatine, 4 mM ATP-Mg^{2+}, 0.3 mM. The pH was adjusted to a value of 7.3 with KOH; the osmolarity of the solutions was ~300 mOsmol l^{-1}. The chloride reversal potential of this solution is close to −40 mV.
2. For recordings from GABAergic connections the following internal solution was used: 135 mM K-gluconate, 4 mM KCl, 10 mM HEPES, 10 mM phosphocreatine, 4 mM ATP-Mg^{2+}, 0.3 mM GTP. The pH was adjusted to a value of 7.3 with KOH; the osmolarity of the solutions was ~300 mOsmol l^{-1}. The chloride reversal potential of this solution is close to −90 mV.

   Biocytin (*Sigma-Aldrich, München, Germany*) at a concentration of 2–5 mg ml^{-1} was routinely added to these two internal solutions to label the neurons during the electrophysiological recordings.
3. For cell-attached stimulation (see below), we used a modified version of these internal solutions containing: 105 mM Na-gluconate, 30 mM NaCl, 10 mM HEPES, 10 mM phosphocreatine, 4 mM ATP-Mg^{2+}, 0.3 mM GTP (adjusted to pH 7.3 with NaOH); the osmolarity of the solutions was ~300 mOsmol l^{-1}.

2.5. Materials for Histochemistry

1. Multi-well plates (6 or 12 wells) (*BD Bioscience, Heidelberg, Germany*).
2. Gelatinised standard object slides or adhesive, silane-coated Histobond microscope slides (*Marienfeld Lab. Glassware, Lauda-Königshof, Germany*).
3. Ultra-thin cover slips (0.04–0.06 mm; *Menzel Gläser, Braunschweig, Germany)* to minimise the distance between the fixated slice and microscope objective.
4. Phosphate buffer.
5. H_2O_2 (hydrogen peroxide) solution (*Merck, Darmstadt, Germany*) to block the endogenous peroxidase activity in brain slices.
6. Triton X100 (*Sigma-Aldrich Chemie, München, Germany*) to permeabilise cell membranes. TritonX should not be used when slices are processed for electron microscopy.

7. Fixatives
   (a) for light microscopy: 4% paraformaldehyde (*Sigma-Aldrich Chemie, München, Germany*) in 0.1 M phosphate buffer.
   (b) for electron microscopy: 2.5% glutaraldehyde and 1% paraformaldehyde (both *Sigma-Aldrich Chemie, München, Germany*) in 0.1 M phosphate buffer.
8. Vectastain ABC (Avidin-Biotin Complex) elite kit (*Biozol Diagnostica Vertrieb GmbH, Eching, Germany*).
9. *Moviol* embedding medium containing 6.0 g glycerol, 2.4 g Moviol 4-88, 12 ml Tris–HCl (0.2 M, pH 8.5) (all Merck, Darmstadt, Germany) and 6 ml distilled water.
10. *Eukitt* embedding medium (*Fluka, München, Germany*; Eukitt is registered trademark of *O. Kindler GmbH, Freiburg, Germany*).
11. 3-3′-diaminobenzidine in tablet form (*Sigma-Aldrich Chemie, München, Germany*). 1 ml aliquots of this solution should be prepared in advance and stored for later use in a freezer at −18°C.

2.6. Set-Up for Computerised Reconstructions

1. Olympus BX50W (*Olympus, Hamburg, Germany*) equipped with a computerised stage for digital three-dimensional (3D) reconstructions of the dendritic and axonal morphology (*Microbrightfield Europe, Magdeburg, Germany*).
2. Neurolucida and Neuroexplorer software (*Microbrightfield Europe, Magdeburg, Germany*) running on a PC.

3. Methods

Protocols for patch clamp recordings from neurons in brain slices have been reported in numerous previous publications (see e.g. (57–59)). To avoid being overly repetitive, we refer the reader to these publications and focus here on the equipment and procedures relevant for paired recordings from synaptically coupled neurons.

3.1. Brain Slice Preparation

The preservation of axonal projections and synaptic connections is of paramount importance for paired recording studies in brain slices. In our studies, we found that we have to determine the optimal slicing conditions (i.e. the optimal cutting angles) for each neuronal connection and developmental stage under investigation in test experiments. For this, brain slices are cut using different cutting angles (which can be done using the tools shown in Fig. 3a). Subsequently, neurons are labelled with biocytin via a patch pipette (see below). By comparing brain slices cut under different conditions and with different cutting angles we determine the conditions under which dendritic and axonal truncations of pre- and post-synaptic neurons are minimal thereby maximising the probability of finding synaptic connections between these cells.

The general procedures for brain dissection and slice preparation are in accordance with the German animal welfare act and the Federation of European Laboratory Animal Science Associations (FELASA) guidelines. The animal is anaesthetised using isoflurane and killed by cervical dislocation. All dissection and slicing procedures are carried out in ice-cold ACSF (prepared by adding frozen ACSF to freshly prepared solution). Sacrificing of young animals can be done using a pair of scissors while for older animals a guillotine is required. The skull should be opened with caution to avoid damage of the underlying brain tissue. The brain is then carefully removed from the skull and trimmed according to the brain region and synaptic connection under study. Subsequently, the brain is glued on the chilled stage of a vibrating microslicer that is covered with ice-cold ACSF with high Mg^{2+} and low Ca^{2+} to minimise synaptic transmission. Slices are cut at low speed and high vibration frequencies (preferentially using a "zero z" slicer, see 2.2) at a thickness of 250–500 μm. The slices are then transferred to an incubation chamber (Fig. 3b) that is constantly gassed with 95% O_2 and 5% CO_2. The formation of gas bubbles should be tightly controlled to avoid floating of the slices, as this damages the brain tissue and impairs finding synaptic connections. The solution in the incubation chamber is at room temperature and in our experiments we found that heating (to 30–35°C) during the recovery period is not essential for obtaining healthy slices.

The age of the animal is always a highly critical factor for determining neuronal connectivity. Generally, neurons appear to be healthier in juvenile brain tissue and dendrite and axon truncations are also less common. For this reason, neuronal connectivity is frequently studied in brain slices of rats or mice aged 12–16 days. In more mature brain tissue (aged 25 days and older) the viability of neurons is often lower and dead neurons are frequently encountered. Connectivity studies in more mature animals are therefore significantly more difficult.

We cut slices from brains of older (>30 postnatal days old) animals in a sucrose containing external solution (see Sect. 2.4) to improve the survival of neurons. In addition, increasing the oxygen level (>30%) during anaesthesia may be advantageous for cell survival in mature animals although we have not tested this systematically. Since the survival of slices from mature animals is significantly shorter than from immature animals, we have reduced the incubation time to about 30 min.

Brains from very immature animals (aged 1–7 days) should be cut at slower vibration speeds than more mature brains as the tissue is still rather soft. When cutting early postnatal brains, we use a slice thickness of ~400 μm to achieve a sufficiently high slice stability and connectivity rate.

3.2. Set-Up for Electrophysiology

An illustration of a typical experimental set-up for paired recordings in our laboratory is shown in Fig. 4. The essential element of such a set-up is an upright microscope with a motorised focus axis mounted on a motorised *x–y* table. It should be equipped with infrared-differential interference contrast optics (or Dodt contrast illumination) to achieve a high visibility for neurons located deep in the slice (>60 μm). The slices are kept in glass-bottomed experimental chambers during the experiment. Thin coverslips are used to permit appropriate focusing of the microscope condenser with high numerical aperture and short working distance.

For synaptic connections that are not local (i.e. where pre- and post-synaptic somata are more than 200 μm apart) it is helpful to have a low- *and* a high-power video camera port on the microscope. This allows both an overview of the neuronal cell pair as well as a fine control of the electrode placement. High-precision micromanipulators to move the patch pipette virtually vibration-free are also indispensable for a paired recording set-up. This is particularly important when changing electrodes after finding a synaptic connection (see below). In addition, the motorised focus axis and *x–y* table helps to reliably recover positions of neurons that are not located in the same field of view (e.g. (60–63)).

For the perfusion of the recording chamber with extracellular solution we use peristaltic pumps to ensure a constant flow rate. We obtained best results for cell viability by using flow rates of 4–6 ml min^{-1}. Lower flow rates resulted in a rapid deterioration of the slice while much higher rates caused turbulence in the experimental solution. All solutions used during the experiment should be temperature-controlled, preferentially at a site close to the inflow into the experimental chamber to avoid overheating. We found that optimal temperatures for paired recordings lie between 32 and 35°C since the oxygenation can be better maintained at these than at higher temperatures.

3.3. Searching for Synaptic Connections

Since the connectivity in brain slices is finite and for some connections also very low, a special "searching" procedure has to be used to find synaptically connected pairs of neurons. For this, a "post-synaptic" neuron is patched in the whole-cell configuration of the patch clamp technique using pipettes with a long and slender shank since this reduces movement artefacts and minimises the pressure exerted on the brain tissue by introducing the electrodes into the slice.

Subsequently, putative pre-synaptic neurons are patched in the loose-seal configuration (seal resistance 100–300 MΩ) using a "searching" patch pipette of 7–10 MΩ resistance (34) filled with a modified high Na^+ internal solution for cell-attached stimulation.

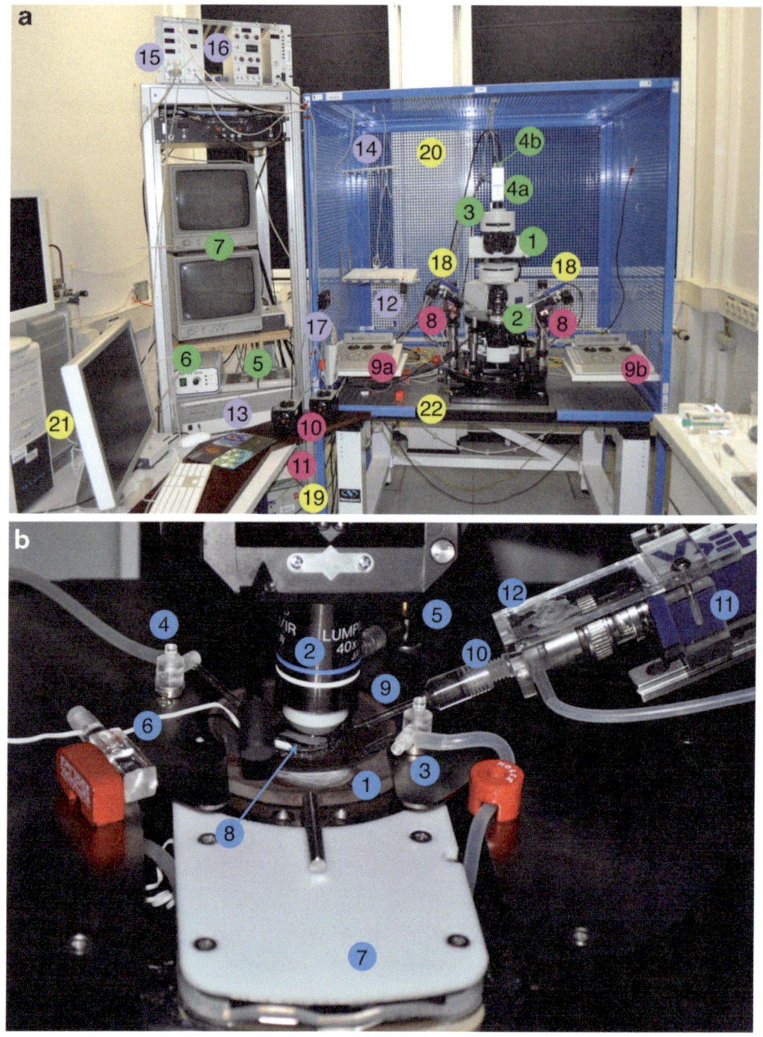

Fig. 4. Set-up for paired recording. (**a**) Photograph of a typical patch clamp setup (in a Faraday cage) used for paired recordings in brain slices. The different parts of the paired recording set-up are colour-coded: Visualisation/imaging equipment (*green*, 1–7), mechanical manipulators (*purple*, 8–11), perfusion and pressure system (*light blue*, 12–17), electrophysiology equipment and air table (*yellow*, 18–22). The different parts of the set-up are as follows: (1) Microscope for infrared-differential interference contrast video microscopy equipped with a motorised focus axis and mounted on a motorised *x*–*y* table. (2) Objective holder with a 4× objective for slice overviews and a 40× water-immersion objective with long working distance (2 mm) to visualise and patch individual cells, (3) Magnifying lenses (1×–4×) for front camera (4a). The rear camera (4b; hidden behind the front camera) is equipped with 0.5× lens to increase the field of view. (5) Camera controllers, (6) Light control unit, (7) TV screens for front and rear camera. (8) Three axis manipulators to position patch clamp electrodes, (9a) Remote control/keypad for manipulators and (9b) Remote control/keypad for the focus axis and *x*–*y* table of the microscope, (10) Controller cubes and (11) Unit containing controller electronics for micromanipulators/focus axis/*x*–*y* table. (12) Peristaltic pump and glass "dripper" for the inflowing solution. While the pump maintains a constant flow rate, the "dripper" interrupts the continuous solution flow, thereby reducing electrical noise. (13) Controller for Peltier

We did not use normal ACSF in the "searching" pipette because an accidental breakthrough (which occurs occasionally) would then result in cell death. "Searching" patch pipettes should have a higher resistance than the pipettes used during the actual recording to avoid accidental breakthrough.

The command potential in the "searching" pipette is set to about −60 mV in current clamp mode. Action potentials are evoked by applying a 3–5 ms current pulse (0.1–2.0 nA) at a stimulation frequency of 0.1 Hz. If an action potential is induced, this is in most cases visible as a small deflection on the voltage trace recorded in the putative "pre-synaptic" neuron. If no synaptic response is elicited in the post-synaptic neuron, a new neuron should be patched in the loose-patch configuration. Changing the "searching" patch pipette is not necessary because only a "loose" seal between patch pipette and cell membrane is required. A "loose" seal can be obtained also with patch pipettes that are not perfectly clean. We were able to test as many as 30 putative pre-synaptic neurons at a relatively high frequency before the "searching" pipette had to be replaced. The technique described here is therefore suitable for identifying synaptic connections even when they have a low connectivity ratio (60, 62, 64) as is the case for some long-range connections or synaptic connections in very immature tissue.

For connections with high connectivity ratios (as is the case for many GABAergic connections) a different procedure to find a connection can be used. Here, a pre-synaptic neuron is patched in the whole-cell configuration and a prospective post-synaptic partner neuron is also patched in the whole-cell mode. An action potential is elicited in the pre-synaptic neuron and the post-synaptic neuron is monitored for a possible post-synaptic potential. In case the prospective post-synaptic neuron is unresponsive, a new neuron is patched with a new patch electrode until a responsive "post-synaptic" neuron

Fig. 4. (continued) device (see *Panel* **b**, (7)), (14) Tubing for the gassing (95% O_2/5% CO_2) of experimental solutions, (15) Pressure monitor and (16) Suction pump for patch electrodes, (17) Switchable valve and control unit for suction pump (*black box*), (18) Patch clamp preamplifiers, (19) Patch clamp amplifier with AD board, (20) Faraday cage, (21) Computer to run electrophysiological and imaging recording software and (22) Anti-vibration table. (**b**) Slice chamber (1) for paired recordings from synaptically coupled neurons in brain slices. A 40× water-immersion objective for infrared light (2) is used to obtain a high-power image of the synaptically coupled neurons. Holders for solution inflow (3), solution outflow (4), the bath electrode (behind the objective, not fully visible) (5) and a temperature probe (6) are placed around the slice chamber. The inflowing ACSF is heated using a Peltier device (7). The slice is placed directly under the objective and held by a small "harp" made from U-shaped platinum wire and strings of dental floss (8). One patch electrode (9) mounted in an electrode holder (10) that is connected to a preamplifier (11) is also visible. Note that the preamplifier is equipped with a stabilising perspex device (12) to prevent movements of the electrode holder when applying suction to the pipette.

is found. For some neuronal connections with high connectivity ratios this technique works quite well, however, the technique is not feasible when the connectivity ratio falls below 10%.

Very immature tissue has very few synaptic connections which may also run-down quickly during recording. Therefore, the stimulation frequency must be low (about one to two stimuli per minute!) and exposure of immature neurons to the depolarising high K^+ internal solutions should be minimised to avoid excessive action potential firing and the concomitant run-down of the synaptic response.

It is also possible to search for gap junction connections in a fashion similar to that described for chemical synapses. For this, the putative coupled neuron is stimulated in the loose-patch configuration with hyperpolarising current pulses about 100–200 ms in length. The current amplitude should be increased gradually to avoid damage to the neuron by excessive stimulation. If the first neuron responds with a hyperpolarisation of equal length, a gap junction coupled pair is established.

3.4. Patch Clamp Recording and Biocytin Filling

In case the cell-attached stimulation resulted in synaptic responses in the post-synaptic neuron at short latency (i.e. within 5 ms), the "searching" pipette was withdrawn. The pre-synaptic cell was then re-patched with a recording pipette (6–8 MΩ; the actual tip size depends on the size of the patched neuron) filled with biocytin-containing intracellular solution. Action potentials were elicited in the whole-cell (voltage recording) mode to characterise the synaptic connection. We use biocytin concentrations of 2–5 mg ml^{-1} internal solution for good staining results; biocytin has to diffuse into the cell for at least 15 min to allow a good recovery of dendritic and axonal profiles in the subsequent histological processing (see below). The combination of electrophysiological data and the cell morphology will then allow a correlation of the structural and functional properties of an identified synaptic connection.

For a successful labelling of the neurons the size of the patch electrodes is a very critical factor. Large electrodes (with low resistance, 2–4 MΩ) provide a good electrical access which is advantageous for the electrophysiological recordings. However, it is more likely that the electrode retraction after the experiment results in a partial or a complete removal of the soma and proximal dendrites from the slice or to an incomplete resealing (seal resistance not in the GΩ but in the 100 MΩ range) and a subsequent dye leakage from the cell, especially after long-term (>2 h) recordings. Large electrodes also decrease the viability of small diameter neurons during the experiment. Therefore, patch pipettes of a smaller size and higher resistance (6–8 MΩ) are preferable; only for large diameter pyramidal neurons pipettes with lower resistance values are used.

3.5. Histochemical Processing

To study the structural basis of neuronal connectivity, a complete morphological recovery of the recorded neurons is important. Over the years, we have optimised our procedures and provide here a detailed description.

3.5.1. Fixation

After the recording, the slices are transferred into small vials and the excess extracellular solution is removed. For light microscopy, brain slices are fixated overnight in 4% paraformaldehyde dissolved in 0.1 M phosphate buffer. For electron microscopy, the slices are fixated in 1.0% paraformaldehyde and 2.5% glutaraldehyde dissolved in 0.1 M phosphate buffer for a maximum of 24 h. Thereafter, the slices should be transferred to a fixative-free solution to avoid tissue damage due to an overly long exposure to the fixative. This is particularly important for slices that are processed for electron microscopy.

3.5.2. Histochemical Processing of Brain Slices for Light Microscopy

Biocytin has a high affinity for avidin and can be labelled permanently using avidinylated visualisation reagents. Coupling to a horseradish-peroxidase-(HRP)-fluorescent dye was introduced first by Horikawa and Armstrong (65) and subsequently optimised by a biocytin-HRP complex for a better visualisation (66, 67). The histological processing to visualise the biocytin-filled neurons was done by a streptavidin-biotinylated horseradish peroxidase reaction catalysed with diaminobenzidine (68). All procedures for the histochemical processing are carried out in 0.1 M phosphate-buffer (pH 7.3–7.4). In our laboratory, we use the following procedure (69–71) to obtain optimal development of the biocytin-labelling in recorded neurons:

Following overnight fixation, brain slices are rinsed six to eight times each for 10 min to remove the fixative. Subsequently, the endogenous peroxidase activity of neurons is blocked by an incubation in a 3% H_2O_2 solution in 0.1 M phosphate buffer (about 2 ml per slice) lasting at least 20 min. Application of H_2O_2 results in the intense formation of bubbles and should be continued until no further bubble formation is observed. Brain slices of immature animals appear to have a higher endogenous peroxidase activity; therefore, blocking endogenous peroxidases will take longer than for slices from more mature brains. Following this, the slices need to be rinsed again for at least six to eight times (for 10min each) in 0.1 M phosphate buffer to remove any excess H_2O_2.

The histological processing to visualise the biocytin-filled neurons was done by a streptavidin-biotinylated horseradish peroxidase reaction catalysed with diaminobenzidine as the chromogen (68). For this, the ABC solution is prepared by mixing 150 μl of reagent A to 150 μl of reagent B; this mixture is then dissolved in a solution with 150 μl 10% Triton X100 (only for light microscopy) and 14.55 ml phosphate buffer (Reagent A:Reagent B:Triton X100: phosphate buffer = 1:1:1:97) and kept for about 30 min before use. Then, the phosphate buffer is removed entirely and the ABC reagent solution added. The slices are incubated on a shaker for 1 h in the dark at room temperature and subsequently overnight at 4°C in a fridge.

On the next morning, the slices are incubated for a further hour at room temperature and subsequently rinsed at least four times for 10 min each in phosphate buffer. Then, each slice is incubated on a shaker for about 30 min in the dark in 2 ml of a Nickel-intensified 3-3′-diaminobenzidine solution. This solution is prepared by adding 17.5 mg 3-3′-diaminobenzidine to 25 ml 0.1 M phosphate buffer; 5–10 μl 1% $(NH_4)_2Ni(SO_4)_2$ and 5 μl 1% $CoCl_2$ are added drop by drop while continuously stirring the solution in order to intensify the reaction. Diaminobenzidine (both as solid and in solution) should only be used under the fume hood and handled with utmost care since it is extremely carcinogenic even at very low concentrations. Waste diaminobenzidine solution must be inactivated with sodium peroxide for a proper disposal.

The chromogenic reaction is started by adding 6.5 μl 3% H_2O_2 to the diaminobenzidine solution. The slices are incubated for another 30–60 s under light microscopic control to monitor the development of a brown-black biocytin stain. The reaction is stopped by transferring the slices into phosphate buffer. Then, the slices are washed again several times (at least six to eight times) with phosphate buffer to remove all diaminobenzidine solution. The brain slices are then mounted on adhesive, silane-coated Histobond or gelatinised standard object slides with a fine brush and all excess phosphate buffer is removed. For subsequent embedding, we used either the hydrophilic, viscous Moviol or the hydrophobic Eukitt medium.

3.5.3. Embedding and Preparation of Permanent Specimens

An optimal visualisation of the biocytin-labelled neurons is of paramount importance for neuronal reconstructions and hence for obtaining realistic estimates of the potential synaptic connectivity. This does not only depend on the biocytin concentration in the cell and a successful histological processing but also on the embedding medium used for permanent storage of brain slices on microscopic slides. The two types of embedding media used in our laboratory for light microscopy, Moviol, a partially hydrolysed polyvinyl alcohol, and the polyacrylate Eukitt differ substantially in their optical and mechanical properties (71).

Moviol is frequently used in cytochemical and histochemical studies because of its ease of use, in particular since the time-consuming dehydration after the fixation and histochemical processing is not necessary (70, 72–75) and dehydration artefacts are therefore avoided. Moviol can be removed easily for subsequent investigations, e.g. electron microscopic studies because it does not cure completely and is therefore a viscous rather than a solid medium. For embedding, the slices are mounted on slides and

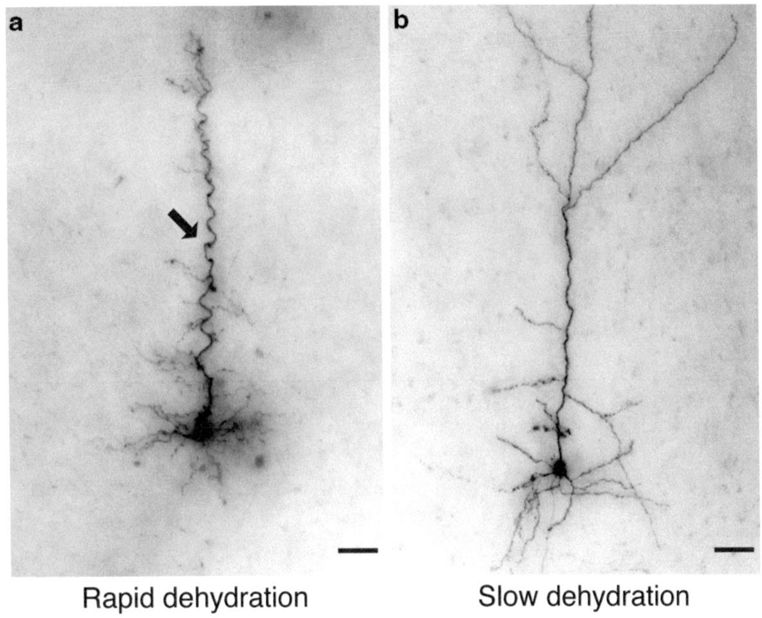

Fig. 5. "Corkscrew" appearance of dendritic profiles in rapidly dehydrated brain sections can be avoided using slow dehydration procedures. (**a**, **b**) Two biocytin-filled pyramidal neurons embedded in Eukitt are shown. Note that the apical dendrite of the pyramidal neuron shown in (**a**) reveals a tortuous, "corkscrew"-like appearance (*arrow*) which results from a rapid dehydration in ethanol/xylol. The pyramidal neuron shown in (**b**) displays no "corkscrew" distortions as a result of a careful dehydration in slowly incrementing ethanol concentrations. Scale bars: 100 μm. Taken from (71) with permission.

covered with 250 μl Moviol, air bubbles are removed and a thin coverslip is placed on top.

Before the slices can be embedded in Eukitt they are transferred to adhesive slides and dried overnight in a moist chamber with at least 80% humidity and then air-dried for a further 10 min at room temperature. The slices need to be dehydrated prior to embedding in Eukitt. For this, ten dehydration steps in increasing concentrations of ethanol (20–100%) are used. A slow dehydration is essential because the rapid dehydration procedures commonly used in histochemistry result in severe distortions of dendritic and axonal structures ("corkscrew" effect; see Fig. 5). Following the dehydration, the slices are incubated for 10 min in xylol and covered with one to two drops of Eukitt medium before embedding them with ultrathin coverslips. The slices are then air-dried for 20 min at room temperature.

While embedding in Moviol is less demanding and labour-intensive than embedding in Eukitt, the latter medium has some advantages which make the more laborious procedures worthwhile. The slices in Eukitt have a higher contrast and the cortical lamination is better discernible than in the slices in embedded Moviol

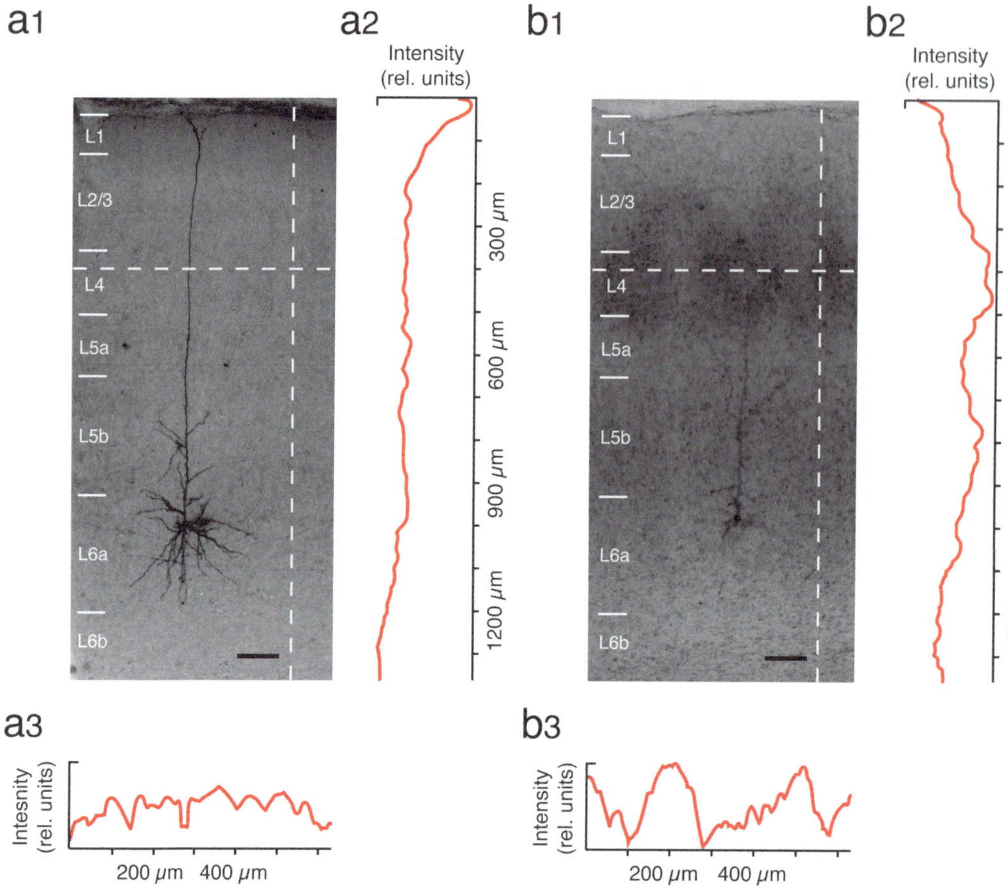

Fig. 6. Optical effects of different embedding media. (**a1**) Overview of a barrel cortex brain slice section visualised by transmission bright-field microscopy embedded in Moviol and (**b1**) in Eukitt (scale bars in **a1**, **b1** = 100 μm). Since the emphasis is on the labelling of cortical layers and cytoarchitectonic units in this figure; the neuron in (**b1**) is therefore not in the plane of focus and thus not clearly visible. Note the homogeneous shading and the low contrast in the Moviol-embedded slice; by contrast, the neocortical lamination is readily discernible in the Eukitt-embedded slice. (**a2**, **b2**) Vertical light intensity profiles through all layers and (**a3**, **b3**) horizontal light intensity profiles through layer 4 of the brain slices shown in (**a1**) and (**b1**), respectively. The regions where the intensity profiles were recorded are marked by *dashed lines*. The light intensity per cortical length (*red lines*) is given in arbitrary units; bright structures have a high light intensity. Slices embedded in Moviol have a linear light intensity so that layer borders are not clearly discernible and barrels at the level of layer 4 are not visible. For Eukitt-embedded slices the borders between cortical layers and sublaminae and between layer 4 barrels are visible as changes in light intensity.

(Fig. 6). In the somatosensory cortex, the cytoarchitectonic unit coined barrels (76) are clearly apparent when using Eukitt embedding. By contrast, for Moviol-embedded slices, an additional histochemical staining for cytochrome oxidase (71, 77, 78) is required to visualise the barrel structures.

Furthermore, in the Moviol-embedded slices biocytin-labelled axons often appear to be discontinuous and frequently only the pre-synaptic boutons can be seen; connections between dendritic

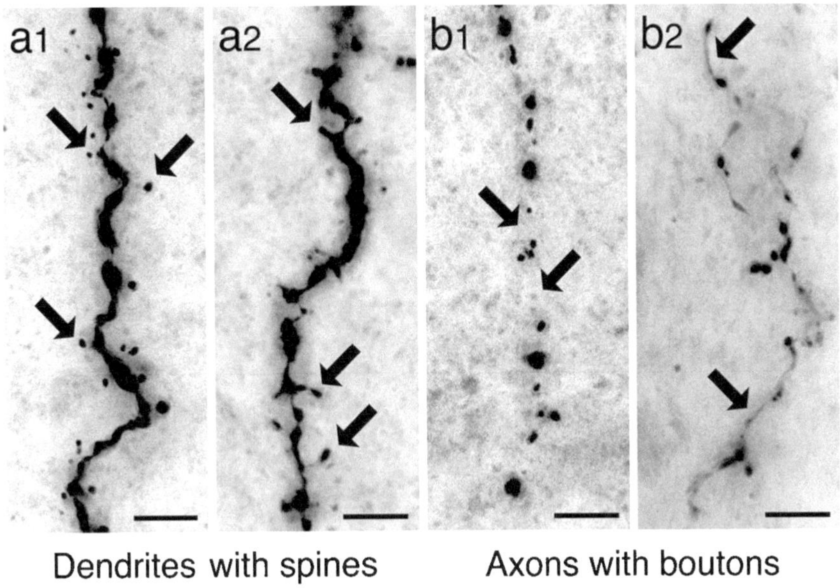

Fig. 7. High magnification images of neuronal structures in slices embedded in Moviol or Eukitt. Extended focal images of dendritic and axonal structures are shown. Biocytin-labelled dendrites in Moviol-embedded slices (**a1**) appear to be less contiguous and do not have clear transitions (*arrows*) to their dendritic spines compared to those in Eukitt-embedded slices. (**a2**) Axon collaterals in Moviol-embedded slices (**b1**) are often incompletely filled and only traceable by detached axonal boutons, in particular in more mature tissue. In Eukitt-embedded slices (**b2**) biocytin-labelling of axon collaterals generally show a continuous staining which simplifies the neuronal tracing. Scale bars: 5 μm. Taken from (71) with permission.

shafts and spine heads (i.e. spine necks) are often not clearly visible. This contrasts markedly with Eukitt-embedded slices in which discontinuities of axonal and dendritic processes are significantly less frequent and finer details can be resolved (Fig. 7). Eukitt is therefore the superior embedding medium when neuronal structures need to be analysed with high resolution or with regard to the underlying cortical lamination.

3.5.4. Neuronal Reconstruction and Shrinkage Correction

Slices with biocytin-labelled pairs of neurons are examined under the light microscope at high magnification using either a 100× or a 60× oil immersion objective with high numerical aperture ($N_A = 1.4$) for best resolution. Subsequently, computer-aided 3D reconstructions of the labelled neurons are performed using NEUROLUCIDA, a morphology analysis system for neuron tracing that is coupled to a motorised microscope. Dendritic and axonal arborisation of the pre- and post-synaptic neurons are reconstructed at the highest possible magnification to detect even small arborisations. This is particularly essential for the axonal arborisations where the more distal portions are only faintly visible. Care should be taken when dendritic or axonal processes descend into the slice-a change of the objective to a lower power magnification may be necessary to avoid destroying the coverslip and ultimately the slice. Since no microscope is perfectly stable, frequent alignments between the labelled neuron and the reconstruction should be performed for all three spatial dimensions to avoid distortions due to error propagation. The NEUROLUCIDA software is also suitable to perform bouton counts on the axon and spine counts on dendrites, respectively. A differential distribution of spines/boutons with respect to cortical layers may help to characterise a particular neuronal cell type. For a quantitative morphological analysis the NEUROEXPLORER software is used; it can provide measurements of the length, polarity and collateralisation of the dendritic and axonal branches.

Following the reconstruction, the labelled neurons are inspected at high magnification for putative synaptic contacts. Such contacts are identified as close appositions of a bouton on the pre-synaptic axon and a post-synaptic dendritic shaft or spine in the same plane of focus. These potential synaptic contacts are marked in the reconstruction and their dendritic location can be determined.

When inspecting the computer-aided reconstruction of neuronal cell pairs, it is evident that a substantial amount of deformation and shrinkage occurred, which severely distorts the axonal and dendritic geometry. While no dehydration-dependent shrinkage is observed when using the semi-liquid Moviol as embedding medium (because dehydration is not required) spatial deformations in particular in the *z*-dimension will occur when such a medium is subjected to pressure-for example when the embedded slice is covered with a coverslip. Since these deformations are not homogeneous, it is very difficult to correct for them and involves a comparison of the neuronal morphology in the living (using a fluorescent marker) and the fixated, biocytin-labelled slices (79).

On the other hand, Eukitt is a solid embedding medium. It is therefore possible to back-calculate the original size by applying shrinkage correction factors, despite the occurrence of both dehydration-dependent shrinkage and mechanical deformation. In a recent study, we have determined these shrinkage correction factor (71). While *x* and *y* shrinkage factors are roughly similar (1.05 for *x*, 1.13 for *y*) there is substantially more shrinkage in the *z*-direction (correction factor 2.09). Figure 8 illustrates how the application of shrinkage correction recovers the neuronal geometry of a layer 6B pyramidal cell. In the uncorrected reconstruction the basal dendritic tree appears as a flattened ellipsoid; when applying the appropriate shrinkage correction an almost circular basal dendritic arbour will result (Fig. 8b). Such corrected reconstructions can then be utilised to infer the actual 3D geometry of the neurons and may be used to construct computational neuron models with a realistic morphology.

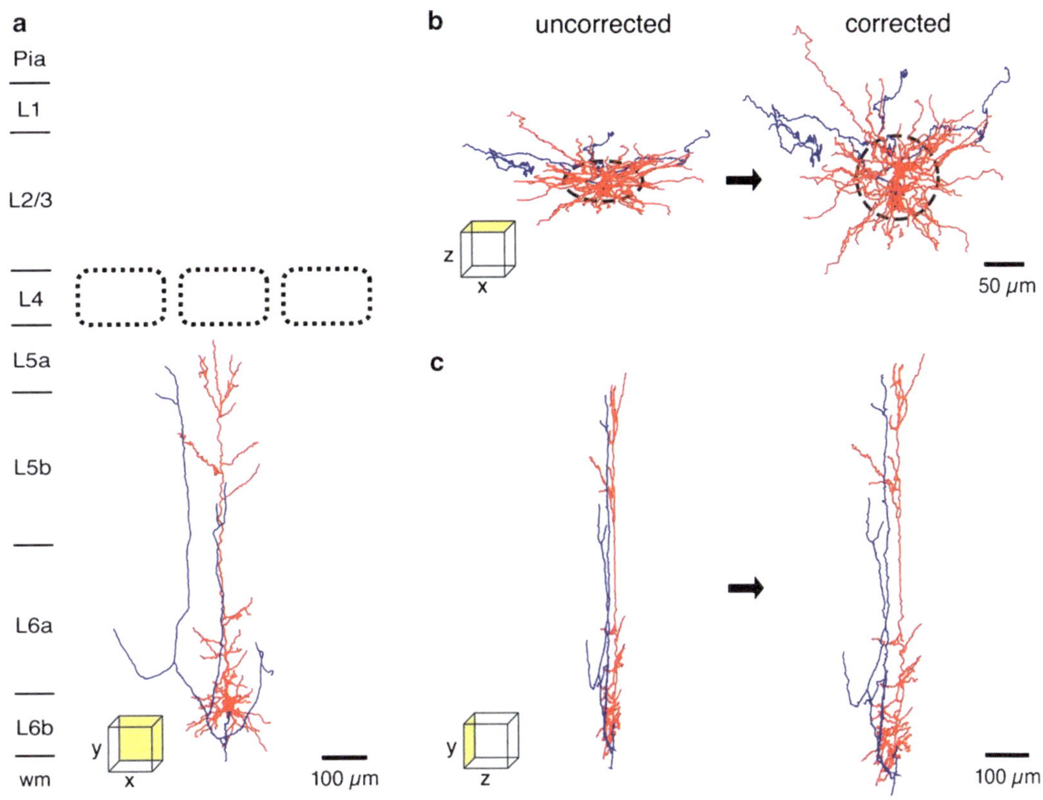

Fig. 8. Shrinkage correction of neuronal slices. (**a**) Three-dimensional (3D) reconstruction of a pyramidal neuron (layer 6B) of the barrel cortex shown in the *x–y* plane. The same neuron rotated in *y–z* direction (**b**) and *x–z* direction (**c**) before and after shrinkage correction, respectively. Dendrites are shown in *red* and axons in *blue*. The *cubes* indicate the direction of rotation. Cortical barrel fields in layer 4 (**a**) are indicated by *closed dashed lines*. Shrinkage correction was performed to recover the original 3D morphology. Note that the shrinkage is particularly prominent in the *z*-direction (**b**, **c**) and less so in the *x*- and *y*-direction. For pyramidal neurons, this results in an elipsoid basal dendritic field in the uncorrected reconstruction; applying the appropriate shrinkage correction results in an almost circular dendritic field in accordance with data from in vivo fillings (**b**). Taken from (71) with permission.

3.6. Density Maps and Quantitative Morphological Measurements

To obtain an estimate of the synaptic connectivity between two identified cell types an approach comparable to that used in thalamocortical axonal tracing studies is pursued (47). The computerised 3D reconstructions of the biocytin-labelled pre-synaptic axonal and post-synaptic dendritic domains of synaptically coupled pairs of neurons are used to construct maps of the length of the axonal and dendritic branches; these are projected in the 2D plane and measured in a 50 × 50 μm Cartesian grid, yielding a raw density map (Fig. 9; (69)). Initially, "length density" maps are constructed for individual pre-synaptic axon and post-synaptic dendrites of each synaptically coupled neuronal cell pair. Subsequently, the individual maps are superimposed to yield average axonal and dendritic "length density" maps for a given connection. Such average axonal and dendritic "length density" maps can be seen in Fig. 10a2–3, b2–3.

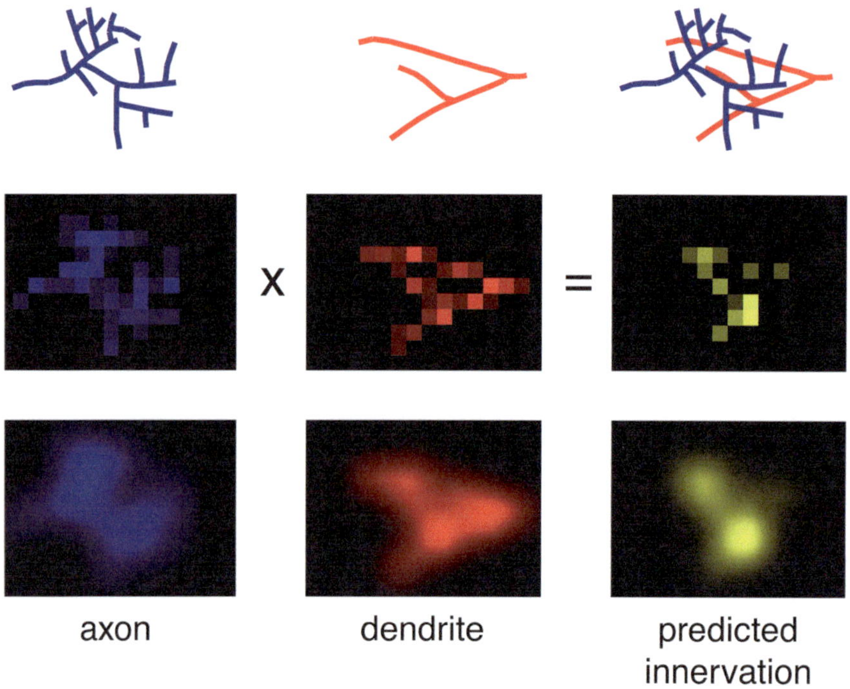

Fig. 9. Innervation domain construction. Quantification of the axo-dendritic overlap interpreted as the probability of synaptic innervation: neurite trees (*top row*) are converted to 3D neurite path length density maps (*middle row*, unit μm per μm^3), which can be low-pass filtered and interpolated (*bottom row*). The product of axonal and dendritic length density can be interpreted as the probability of establishing a synaptic contact (predicted innervation probability, for details see text; modified from (54) with permission).

Spatial low-pass filtering of these maps was performed by convolution with a Gaussian kernel ($\sigma = 50$ μm); the average projection volume thus obtained is then quantified by computing the iso-axonal-density surface that contained a substantial fraction of the total axonal path length (80% represented by the closed white contour lines in Fig. 10a2–3, b2–3). These were constructed using bicubic interpolation in Mathematica 4.1 (*Wolfram Research, Champaign, IL*).

In case an alignment of the maps with respect to the barrel centre is required, barrel borders were identified in the low power (4× objective) bright-field micrographs made from the brain slice preparation (62, 69, 80, 81).

Next, the analysis of axonal projection geometries can be used to infer synaptic connectivity with post-synaptic neurons. For this, the average pre-synaptic axonal and post-synaptic dendritic "length density" maps obtained in the way outlined above were multiplied to calculate the predicted "innervation domain" between the pre- and the post-synaptic neuron (69). Spatial low-pass filtering of the "innervation domain" maps was performed as specified for the axonal and dendritic "length density" maps. Its outline can be calculated as the iso-innervation density line comprising e.g. 80% of the total

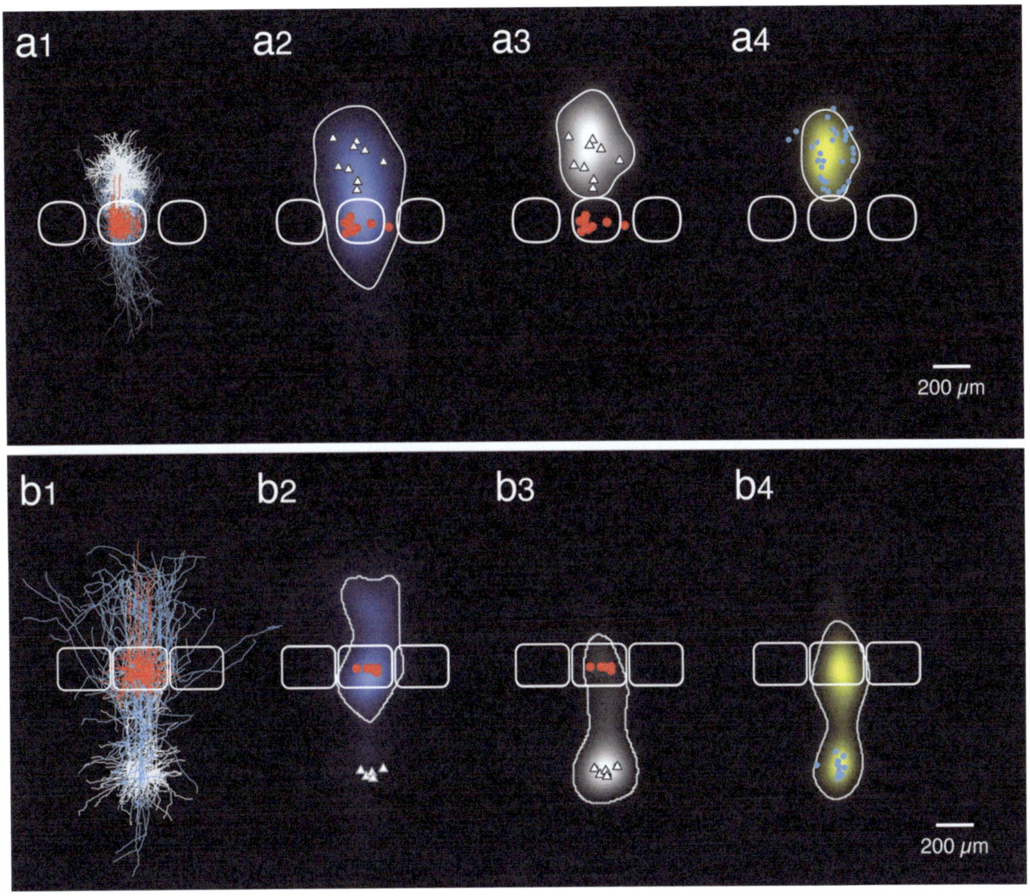

Fig. 10. Predictions of synaptic connectivity using density plots. Pre-synaptic axonal projections and post-synaptic dendritic domains allow the prediction of the degree of synaptic connectivity from the construction of "innervation domains". (**a1**) Overlay of reconstructions of synaptically connected pairs of layer 4 spiny neurons (dendrites: *red*, axons: *blue*) and layer 2/3 pyramidal neurons (dendrites: *white*) with the centre of the barrel in layer 4 as reference point. (**a2**) Average axonal length density map of the pre-synaptic layer 4 spiny neuron axonal projection shown in (**a1**). (**a3**) Average dendritic length density map of the post-synaptic layer 2/3 pyramidal neuron dendrites shown in (**a1**). The position layer 4 spiny neuron somata is marked by *red dots* in (**a2**, **a3**), that of layer 2/3 pyramidal cell somata by *white triangles*, respectively. (**a4**) Innervation density map (the product of **a2** and **a3**) for the layer 4-to-layer 2/3 pyramidal neuron connection. The *light blue dots* indicate the location of light-microscopically identified synaptic contacts. Note that for this connection the vast majority of contacts are distributed throughout the predicted innervation domain, as predicted by Peter's rule. (**a2**–**a4** modified from (69) with permission). (**b**) Similar as in (**a**): (**b1**) Overlay of reconstructions of synaptically connected pairs of layer 4 star pyramid neurons (dendrites: *red*, axons: *blue*) and layer 6A pyramidal cells (dendrites: *white*) with the centre of the barrel in layer 4 as reference point. (**b2**) Average axonal length density map of the pre-synaptic layer 4 star pyramidal neuron axonal projection shown in (**b1**). (**b3**) Average dendritic length density map of the post-synaptic layer 6A pyramidal cell dendrites shown in (**b1**). The position layer 4 star pyramidal neuron somata is marked by *red dots* in (**b2**, **b3**), that of layer 6A pyramidal cell somata by *white triangles*, respectively. (**b4**) Innervation density map (the product of **b2** and **b3**) for this layer 4-to-layer 6A pyramidal neuron connection. The *light blue dots* indicate the location of light-microscopically identified synaptic contacts. In contrast to the layer 4-to-layer 2/3 connection synaptic contacts for this connection are found exclusively in the lower part of the predicted innervation domain, i.e. on the basal dendrites off the L6A pyramidal cells (Data from (60)). The *white contour lines* superimposed on the density maps in (**a2**, **a3**, **b2**, **b3**) enclose 80% of the integrated axonal and dendritic density, respectively. *Contour lines* in (**a4**, **b4**) comprise 80% of the total innervation density.

innervation density (closed white contours in Fig. 10a4, b4). However, it is not self-evident that these predicted innervation domains correspond indeed to regions with a high probability of finding synaptic contacts between pre- and post-synaptic neurons. For pairs of neurons that had shown mono-synaptic transmission, synaptic contacts were searched as described above as close appositions of the pre-synaptic axon and the post-synaptic dendrite at the light microscopic level. However, the resolution of light microscopy is not sufficient to distinguish a close apposition from a true synaptic contact. Selected slices are further processed for electron microscopy and high magnification images are taken of such putative synaptic locations to verify the existence of synaptic contacts based on the common ultrastructural criteria of synapses. We were able to demonstrate that between 80 and 90% of light-microscopically identified contacts were indeed contacts at the ultrastructural level (11, 35, 63, 81).

An overlay of the predicted innervation domain and identified synaptic contacts (Fig. 10a4) then showed that the vast majority of the synaptic contacts were within the 80% innervation domain, as we were able to demonstrate for several neuronal connections ((61, 69, 80, 81), for a review see (54)).

In a recent study, however, we could show that this rule does not apply for all connections. For the specific synaptic connection shown in Figure 10b, a layer 4 star pyramidal neuron to layer 6A pyramidal cell connection, synaptic contacts were exclusively found on the basal dendritic tree although the outlines of the innervation domain would have predicted contacts on the apical dendrite (Fig. 10b4). It is therefore essential to identify synaptic contacts independently (using light microscopic examination and preferentially ultrastructural verification with the electron microscope).

Acknowledgement

The authors would like to thank Werner Hucko for his excellent technical assistance and the Helmholtz Alliance for Systems Biology and the DFG Research Group 'Barrel Cortex Function' for financial support.

References

1. Beierlein M, Gibson JR, Connors BW (2000) A network of electrically coupled interneurons drives synchronized inhibition in neocortex. Nat Neurosci 3:904–910
2. Hughes GM, Tauc L (1968) A direct synaptic connexion between the left and right giant cells in Aplysia. J Physiol 197:511–527
3. Korn H, Triller A, Mallet A et al (1981) Fluctuating responses at a central synapse:

n of binomial fit predicts number of stained presynaptic boutons. Science 213:898–901
4. Faber DS, Korn H (1980) Single-shot channel activation accounts for duration of inhibitory postsynaptic potentials in a central neuron. Science 208:612–615
5. Miles R (1990) Synaptic excitation of inhibitory cells by single CA3 hippocampal pyramidal cells of the guinea-pig in vitro. J Physiol 428:61–77
6. Thomson AM, West DC (1993) Fluctuations in pyramid-pyramid excitatory postsynaptic potentials modified by presynaptic firing pattern and postsynaptic membrane potential using paired intracellular recordings in rat neocortex. Neuroscience 54:329–346
7. Bolshakov VY, Siegelbaum SA (1995) Regulation of hippocampal transmitter release during development and long-term potentiation. Science 269:1730–1734
8. Stuart GJ, Dodt HU, Sakmann B (1993) Patch-clamp recordings from the soma and dendrites of neurons in brain slices using infrared video microscopy. Pflugers Arch 423:511–518
9. Dodt HU, Zieglgänsberger W (1990) Visualizing unstained neurons in living brain slices by infrared DIC-videomicroscopy. Brain Res 537:333–336
10. MacVicar BA (1984) Infrared video microscopy to visualize neurons in the in vitro brain slice preparation. J Neurosci Methods 12:133–139
11. Silver RA, Lübke J, Sakmann B, Feldmeyer D (2003) High-probability uniquantal transmission at excitatory synapses in barrel cortex. Science 302:1981–1984
12. Biro AA, Holderith NB, Nusser Z (2005) Quantal size is independent of the release probability at hippocampal excitatory synapses. J Neurosci 25:223–232
13. Gulyas AI, Miles R, Sik A, Toth K, Tamamaki N, Freund TF (1993) Hippocampal pyramidal cells excite inhibitory neurons through a single release site. Nature 366:683–687
14. Bruno RM, Sakmann B (2006) Cortex is driven by weak but synchronously active thalamocortical synapses. Science 312:1622–1627
15. Crochet S, Chauvette S, Boucetta S et al (2005) Modulation of synaptic transmission in neocortex by network activities. Eur J Neurosci 21:1030–1044
16. Matsumura M, Chen D, Sawaguchi T et al (1996) Synaptic interactions between primate precentral cortex neurons revealed by spike-triggered averaging of intracellular membrane potentials in vivo. J Neurosci 16:7757–7767
17. Petreanu L, Huber D, Sobczyk A et al (2007) Channelrhodopsin-2-assisted circuit mapping of long-range callosal projections. Nat Neurosci 10:663–668
18. Bureau I, von Saint PF, Svoboda K (2006) Interdigitated paralemniscal and lemniscal pathways in the mouse barrel cortex. PLoS Biol 4:e382
19. Shepherd GM, Svoboda K (2005) Laminar and columnar organization of ascending excitatory projections to layer 2/3 pyramidal neurons in rat barrel cortex. J Neurosci 25:5670–5679
20. Schubert D, Kötter R, Luhmann HJ et al (2006) Morphology, electrophysiology and functional input connectivity of pyramidal neurons characterizes a genuine layer Va in the primary somatosensory cortex. Cereb Cortex 16:223–236
21. Schubert D, Kötter R, Zilles K et al (2003) Cell type-specific circuits of cortical layer IV spiny neurons. J Neurosci 23:2961–2970
22. Schubert D, Staiger JF, Cho N et al (2001) Layer-specific intracolumnar and transcolumnar functional connectivity of layer V pyramidal cells in rat barrel cortex. J Neurosci 21:3580–3592
23. Yoshimura Y, Callaway EM (2005) Fine-scale specificity of cortical networks depends on inhibitory cell type and connectivity. Nat Neurosci 8:1552–1559
24. Yoshimura Y, Dantzker JL, Callaway EM (2005) Excitatory cortical neurons form fine-scale functional networks. Nature 433:868–873
25. Dantzker JL, Callaway EM (2000) Laminar sources of synaptic input to cortical inhibitory interneurons and pyramidal neurons. Nat Neurosci 3:701–707
26. Peron S, Svoboda K (2011) From cudgel to scalpel: toward precise neural control with optogenetics. Nat Methods 8:30–34
27. Petreanu L, Mao T, Sternson SM et al (2009) The subcellular organization of neocortical excitatory connections. Nature 457:1142–1145
28. Scanziani M, Häusser M (2009) Electrophysiology in the age of light. Nature 461:930–939
29. Adesnik H, Scanziani M (2010) Lateral competition for cortical space by layer-specific horizontal circuits. Nature 464:1155–1160
30. Molnár Z, Cheung AF (2006) Towards the classification of subpopulations of layer V pyramidal projection neurons. Neurosci Res 55:105–115
31. Groh A, Meyer HS, Schmidt EF et al (2010) Cell-type specific properties of pyramidal neurons in neocortex underlying a layout that is modifiable depending on the cortical area. Cereb Cortex 20:826–836
32. Doyle JP, Dougherty JD, Heiman M et al (2008) Application of a translational profiling

approach for the comparative analysis of CNS cell types. Cell 135:749–762
33. Brown SP, Hestrin S (2009) Intracortical circuits of pyramidal neurons reflect their long-range axonal targets. Nature 457:1133–1136
34. Feldmeyer D, Egger V, Lübke J, Sakmann B (1999) Reliable synaptic connections between pairs of excitatory layer 4 neurones within a single 'barrel' of developing rat somatosensory cortex. J Physiol 521(Pt 1):169–190
35. Markram H, Lübke J, Frotscher M et al (1997) Physiology and anatomy of synaptic connections between thick tufted pyramidal neurones in the developing rat neocortex. J Physiol 500 (Pt 2):409–440
36. Deans MR, Gibson JR, Sellitto C et al (2001) Synchronous activity of inhibitory networks in neocortex requires electrical synapses containing connexin36. Neuron 31:477–485
37. Galarreta M, Hestrin S (2002) Electrical and chemical synapses among parvalbumin fast-spiking GABAergic interneurons in adult mouse neocortex. Proc Natl Acad Sci U S A 99:12438–12443
38. Galarreta M, Hestrin S (2001) Electrical synapses between GABA-releasing interneurons. Nat Rev Neurosci 2:425–433
39. Deuchars J, West DC, Thomson AM (1994) Relationships between morphology and physiology of pyramid-pyramid single axon connections in rat neocortex in vitro. J Physiol 478(Pt 3):423–435
40. Gray EG (1959) Axo-somatic and axo-dendritic synapses of the cerebral cortex: an electron microscope study. J Anat 93:420–433
41. Uchizono K (1965) Characteristics of excitatory and inhibitory synapses in the central nervous system of the cat. Nature 207:642–643
42. Colonnier M (1968) Synaptic patterns on different cell types in the different laminae of the cat visual cortex. An electron microscope study. Brain Res 9:268–287
43. Tamás G, Buhl EH, Somogyi P (1997) Fast IPSPs elicited via multiple synaptic release sites by different types of GABAergic neurone in the cat visual cortex. J Physiol 500(Pt 3):715–738
44. Tamás G, Buhl EH, Lörincz A et al (2000) Proximally targeted GABAergic synapses and gap junctions synchronize cortical interneurons. Nat Neurosci 3:366–371
45. Oertner TG, Sabatini BL, Nimchinsky EA et al (2002) Facilitation at single synapses probed with optical quantal analysis. Nat Neurosci 5:657–664
46. Koester HJ, Johnston D (2005) Target cell-dependent normalization of transmitter release at neocortical synapses. Science 308:863–866
47. Peters A (1979) Thalamic input to the cerebral cortex. Trends Neurosci 2:183–185
48. White EL (1979) Thalamocortical synaptic relations: a review with emphasis on the projections of specific thalamic nuclei to the primary sensory areas of the neocortex. Brain Res 180:275–311
49. Braitenberg V, Schüz A (1998) Cortex: statistics and geometry of neuronal connectivity, 2nd edn. Springer, Berlin, Heidelberg, New York, p 249
50. White EL (2007) Reflections on the specificity of synaptic connections. Brain Res Rev 55:422–429
51. Stepanyants A, Hirsch JA, Martinez LM et al (2008) Local potential connectivity in cat primary visual cortex. Cereb Cortex 18:13–28
52. Shepherd GM, Stepanyants A, Bureau I et al (2005) Geometric and functional organization of cortical circuits. Nat Neurosci 8:782–790
53. Stepanyants A, Chklovskii DB (2005) Neurogeometry and potential synaptic connectivity. Trends Neurosci 28:387–394
54. Helmstaedter MN, Feldmeyer D (2010) Axons predict neuronal connectivity within and between cortical columns and serve as primary classifiers of interneurons in a cortical column. In: Feldmeyer D, Lübke JHR (eds) New aspects of axonal structure and function, 1st edn. Springer Science + Business Media, New York, Dordrecht, Heidelberg, London, pp 141–155
55. Eder M, Zieglgänsberger W, Dodt HU (2002) Neocortical long-term potentiation and long-term depression: site of expression investigated by infrared-guided laser stimulation. J Neurosci 22:7558–7568
56. Dodt HU, Frick A, Kampe K et al (1998) NMDA and AMPA receptors on neocortical neurons are differentially distributed. Eur J Neurosci 10:3351–3357
57. Moyer JD, Brown TH (2007) Visually guided patch-clamp recordings in brain slices. In: Walz W (ed) Patch clamp analysis advanced techniques. Humana Press, Totowa, NJ, USA
58. Debanne D, Boudkkazi S, Campanac E et al (2008) Paired-recordings from synaptically coupled cortical and hippocampal neurons in acute and cultured brain slices. Nat Protoc 3:1559–1568
59. Davie JT, Kole MH, Letzkus JJ et al (2006) Dendritic patch-clamp recording. Nat Protoc 1:1235–1247

60. Qi G, Feldmeyer D (2010) Cell type-specific excitatory synaptic connections from layer 4 to layer 6A in rat barrel cortex. Acta Physiol (Oxf) 198:90–90
61. Helmstaedter M, Staiger JF, Sakmann B et al (2008) Efficient recruitment of layer 2/3 interneurons by layer 4 input in single columns of rat somatosensory cortex. J Neurosci 28:8273–8284
62. Feldmeyer D, Roth A, Sakmann B (2005) Monosynaptic connections between pairs of spiny stellate cells in layer 4 and pyramidal cells in layer 5A indicate that lemniscal and paralemniscal afferent pathways converge in the infragranular somatosensory cortex. J Neurosci 25:3423–3431
63. Feldmeyer D, Lübke J, Silver RA et al (2002) Synaptic connections between layer 4 spiny neurone-layer 2/3 pyramidal cell pairs in juvenile rat barrel cortex: physiology and anatomy of interlaminar signalling within a cortical column. J Physiol 538:803–822
64. Radnikow G, Lübke JR, Feldmeyer D (2010) Developmental changes in synaptic transmission between layer 4 spiny neurons in rat barrel cortex. Acta Physiol (Oxf) 198:179–179
65. Horikawa K, Armstrong WE (1988) A versatile means of intracellular labeling: injection of biocytin and its detection with avidin conjugates. J Neurosci Methods 25:1–11
66. Hsu SM, Raine L, Fanger H (1981) Use of avidin-biotin-peroxidase complex (ABC) in immunoperoxidase techniques: a comparison between ABC and unlabeled antibody (PAP) procedures. J Histochem Cytochem 29:577–580
67. Adams JC (1992) Biotin amplification of biotin and horseradish peroxidase signals in histochemical stains. J Histochem Cytochem 40:1457–1463
68. Adams JC (1981) Heavy metal intensification of DAB-based HRP reaction product. J Histochem Cytochem 29:775
69. Lübke J, Roth A, Feldmeyer D et al (2003) Morphometric analysis of the columnar innervation domain of neurons connecting layer 4 and layer 2/3 of juvenile rat barrel cortex. Cereb Cortex 13:1051–1063
70. Lübke J, Egger V, Sakmann B et al (2000) Columnar organization of dendrites and axons of single and synaptically coupled excitatory spiny neurons in layer 4 of the rat barrel cortex. J Neurosci 20:5300–5311
71. Marx M, Günter RH, Hucko W et al. (2011) An improved protocol for biocytin labeling and neuronal reconstruction. Nat Protoc (in press).
72. Osborn M, Weber K (1982) Immunofluorescence and immunocytochemical procedures with affinity purified antibodies: tubulin-containing structures. Methods Cell Biol 24:97–132
73. Osborn M, Born T, Koitsch HJ et al (1978) Stereo immunofluorescence microscopy: I. Three-dimensional arrangement of microfilaments, microtubules and tonofilaments. Cell 14:477–488
74. Rodriguez J, Deinhardt F (1960) Preparation of a semipermanent mounting medium for fluorescent antibody studies. Virology 12:316–317
75. Longin A, Souchier C, Ffrench M et al (1993) Comparison of anti-fading agents used in fluorescence microscopy: image analysis and laser confocal microscopy study. J Histochem Cytochem 41:1833–1840
76. Woolsey TA, Van der Loos H (1970) The structural organization of layer IV in the somatosensory region (SI) of mouse cerebral cortex. The description of a cortical field composed of discrete cytoarchitectonic units. Brain Res 17:205–242
77. Land PW, Simons DJ (1985) Cytochrome oxidase staining in the rat SmI barrel cortex. J Comp Neurol 238:225–235
78. Wong-Riley MT, Welt C (1980) Histochemical changes in cytochrome oxidase of cortical barrels after vibrissal removal in neonatal and adult mice. Proc Natl Acad Sci USA 77:2333–2337
79. Egger V, Nevian T, Bruno RM (2008) Subcolumnar dendritic and axonal organization of spiny stellate and star pyramid neurons within a barrel in rat somatosensory cortex. Cereb Cortex 18:876–889
80. Frick A, Feldmeyer D, Helmstaedter M et al (2008) Monosynaptic connections between pairs of L5A pyramidal neurons in columns of juvenile rat somatosensory cortex. Cereb Cortex 18:397–406
81. Feldmeyer D, Lübke J, Sakmann B (2006) Efficacy and connectivity of intracolumnar pairs of layer 2/3 pyramidal cells in the barrel cortex of juvenile rats. J Physiol 575:583–602

Part VI

Neuronal Network Analysis in Humans

Neuromethods (2012) 67: 435–457
DOI 10.1007/7657_2011_15

Published online: 27 November 2011

Combining Transcranial Magnetic Stimulation with Electroencephalography to Study Human Cortical Excitability and Effective Connectivity

Mario Rosanova, Silvia Casarotto, Andrea Pigorini, Paola Canali, Adenauer G. Casali, and Marcello Massimini

Abstract

Excitability and effective connectivity are key parameters of cortical circuits' functioning. Moreover, alterations of these parameters have been suggested to underlie neurologic and psychiatric conditions. Transcranial Magnetic Stimulation (TMS) combined with neuronavigation systems, electroencephalography (EEG), and reliable reconstruction of cortical generators allows to directly measure cortical excitability and effective connectivity, noninvasively and with the appropriate temporal resolution (milliseconds). Here, the main technical challenges that have to be dealt with to efficiently combine navigated TMS with EEG on human subjects and a step-by-step experimental procedure to perform a successful TMS/EEG measurement are presented. Finally, ad hoc analytical tools to extract indices of excitability and effective connectivity from TMS/EEG data are described together with examples of their application.

Key words: TMS/EEG, Neuronavigation, Cortical circuits, Perturbation, Noninvasive

1. Introduction

The human cerebral cortex is composed of several specialized regions that communicate by means of short-range and long-range bundles of axons. Corticocortical interactions occur on a subsecond timescale (hundreds of milliseconds) and allow the cerebral cortex, for instance, to rapidly integrate information across different sensory modalities and submodalities, to orient the focus of visual attention, to maintain objects' representations in the visual working memory, to retrieve memories and, more generally, to generate conscious experiences. Moreover, physiological and pathological alterations of the excitability and effective connectivity within cortical circuits have been suggested to underlie deep sleep, anesthesia and the most common neurological and psychiatric diseases. Thus, to study the neural correlates of higher order brain functions, neuroimaging techniques with good temporal and spatial resolution are required.

Electroencephalography (EEG) combined with source modeling provides the appropriate spatial and temporal resolution to study noninvasively the activity of the cerebral cortex. However, observing the ongoing EEG might not be informative enough to measure excitability and effective connectivity. In order to achieve these measurements, one must directly and noninvasively perturb different corticothalamic regions and record the immediate electrical responses locally and at distant sites. Currently, it is possible to combine EEG with Transcranial Magnetic Stimulation (TMS/EEG), a technique that can activate noninvasively selected subsets of cortical neurons. Hence, TMS/EEG allows recording the evoked electrical responses to a direct perturbation of the cortical tissue. In the following paragraphs we summarize the main advantages of TMS/EEG (Sect. 1.1), we briefly go through the history of TMS/EEG (Sect. 1.2), we review the principles underlying the technique and the available ways to perform TMS/EEG recordings combined with neuronavigation (Sects. 2.1–2.4). Finally, we describe in detail the steps necessary to perform a reliable and successful TMS/EEG measurement (Sects. 3.1–3.3), we describe up-to-date data processing procedures to analyze TMS-evoked EEG potentials (Sects. 3.4 and 3.5) and we show some applications (Sect. 3.6) and future technical evolutions of the technique (Sect. 3.7).

1.1. The Advantages of TMS/EEG

Before starting, it can be useful to highlight the advantages that TMS/EEG may offer when used to probe human cortical circuits compared to other, more established techniques:

1. Cortical activations evoked by TMS are intrinsically causal (1). Thus, unlike methods based on temporal correlations between neural, metabolic or hemodynamic signals, TMS/EEG allows a straightforward assessment of cortical effective connectivity.
2. TMS/EEG bypasses sensory pathways and subcortical structures to probe directly the thalamocortical system. Therefore, unlike peripherally evoked potentials and evoked motor activations, TMS/EEG does not depend on the integrity of sensory and motor systems and can access any patient (deafferentated or paralyzed). Moreover, with TMS one can stimulate virtually any cortical area (including associative cortices) employing several different parameters (intensity, angle, main current direction), thus probing different aspects of cortical reactivity.
3. TMS-evoked potentials can be recorded with a time resolution in the order of milliseconds, a time scale that is commensurate with neuronal responses to depolarizations and synaptic interactions.

4. TMS/EEG does not require the subject to be involved in a task. Consequently, neither the subject's willingness to participate nor his/her effort and performance can influence the observable TMS-evoked activations. Hence, this technique is well suited to assess the state and the level of interaction of corticothalamic regions, independently on behavior.
5. Equipments to perform TMS/EEG measurements can be made portable.

1.2. A Brief History of TMS/EEG

Earliest attempts to combine TMS with recordings of the brain's electrical activity can be traced back to 1989, when Cracco and colleagues (2) measured transcallosal responses by stimulating the human frontal cortices at both sides by means of a magnetic coil. A few years later, the same group recorded the response of the brain to cerebellar magnetic stimulation (3). These pioneering studies, however, were still strongly limited (only one coil orientation with respect to the EEG derivation was possible) by the large artifacts introduced by the coil discharge into the electrical traces. This problem was solved at the end of 1990s. By employing a specifically designed TMS-compatible EEG amplifier (see Sect. 2.2) containing a sample-and-hold circuit, Ilmoniemi and coworkers (4) were able to record the immediate EEG response to TMS applied over the motor and visual cortices. After reconstructing the cortical generators of the TMS-evoked EEG responses, Ilmoniemi and his colleagues showed that the cortical areas ipsilateral to the stimulation were responding within 3–7 ms, while contralateral areas were responding with longer latencies (24–28 ms). Since then, the same group studied several technical issues to improve the technique, such as the influence of muscle (5) or auditory (6) artifacts, as well as the changes of the cortical responses as a function of stimulation intensities and coil placement (7, 8).

Other research groups focused their attention on the contribution of cortical and subcortical structures to TMS/EEG responses or on the key factors responsible for their fundamental properties. For instance, Paus and coworkers studied the frequency content of the EEG oscillations triggered by TMS when applied over the primary hand area, revealing a prevalence of TMS-evoked rhythms in the beta band (15–30 Hz) (9). Interestingly, the same group suggested that thalamic nuclei were playing a key role in the generation of the TMS-evoked oscillations (10, 11). Bonato and colleagues further explored the impact of coil orientation and placement on the components of the TMS-evoked potentials (12), while the group of Massimini and Tononi studied how the site and the intensity of stimulation influences the frequency content of the TMS-evoked responses (13) (see also Fig. 6 and Sect. 3.7). Most importantly, the same group have recently observed, by means of TMS/EEG, significant changes in corticocortical connectivity when subjects lose consciousness during sleep and anesthesia (14–16) (see also Fig. 7 and Sect. 3.7). Additional experiments employed TMS/EEG to explore how the corticocortical dialogue changes when different cortical circuits are engaged by attention (17, 18), to what extent the cortical excitability depends on the subsecond state of the cortical circuits (19, 20) or what are the neural activations underlying the perception of TMS-induced visual flashes (phosphenes) (21). Finally, a few studies have been performed to study alterations of cortical circuits excitability and connectivity both in psychiatric (22) and neurological (23) disorders demonstrating the utility of this technology in the development of biomarkers for neuropsychiatric conditions.

2. Materials

2.1. Basic Principles of TMS

Transcranial magnetic stimulation is based on the physical principle of electromagnetic induction, discovered by Faraday in 1831. Faraday's experiments showed that currents (and voltages) were only induced by a changing, or "time varying" magnetic field, and not by a static field. If a pulse of current passing through a coil placed over a person's head has sufficient strength and short enough duration, rapidly changing magnetic pulses are generated that penetrate scalp and skull to reach the brain with negligible tissutal attenuation. These magnetic pulses induce secondary ionic currents in the brain (eddy currents), that penetrates the membranes of the neurons, resulting in an action potential or excitatory (or inhibitory) postsynaptic potential. The magnetic field falls off rapidly with distance from the coil (24), so it is usually assumed, unless the stimulus intensity is very high at the surface of the brain, that the stimulus activates neural elements in the cortex or in the subcortical white matter (limiting direct stimulation to the outer parts of the cerebral cortex under the skull).

The neuronal elements that are directly excited by the TMS are still unclear. However, experimental and modeling studies strongly suggest that axons, rather than cell bodies, are the most likely targets of the stimulus as they have the lowest threshold for activation to the brief electrical currents induced by TMS. In particular, most mathematical models of neuronal stimulation are further evolutions of the Hodgkin and Huxley model. In this class of models the transmembrane potential, V, is mathematically represented by the following equation:

$$\lambda^2 \frac{\mathrm{d}^2 V}{\mathrm{d}^2 x^2} - V = \tau \frac{\mathrm{d}V}{\mathrm{d}t},$$

where λ and τ represent respectively the axonal membrane space and the time constant.

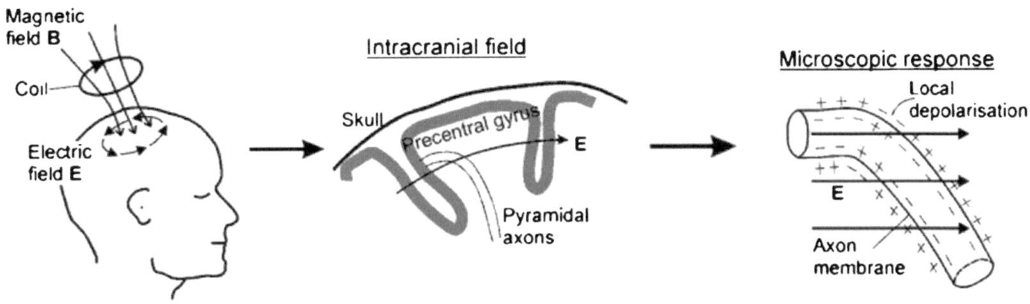

Fig. 1. Neural targets of TMS. From *left* to *right*: a schematic representation of the electric field (E) induced into the subject's head by the magnetic field (B) after a brief pulse of electrical current is passed through the TMS coil; the macroscopic (cortical gyri) and microscopic targets (bent axons) of the electric field E induced intracranially by the TMS coil (reproduced with permission from (26)).

The capacity of TMS to depolarize neurons depends on the "activating function", which causes a sufficient transmembrane current to flow and depolarize the membrane. In order to represent also this external current source applied to the axon, such as the one induced by the TMS, the previous equation can be modified by adding an "activating function" as follows:

$$\lambda^2 \frac{\mathrm{d}^2 V}{\mathrm{d}^2 x^2} - V = \tau \frac{\mathrm{d}V}{\mathrm{d}t} + \lambda^2 \frac{\mathrm{d}^2 A}{\mathrm{d}x\,\mathrm{d}t}$$

According to this mathematical model (24), stimulation will take place at the point where the spatial derivative of the induced electric field is maximal. A few modeling (25, 26) together with experimental studies conducted on peripheral axons in vitro (27) confirmed this model showing that axon fibers are more sensitive to electric fields induced by TMS where they bend, i.e., where the spatial derivative of the induced electric field is maximal (see Fig. 1).

From the site of cortical stimulation, neuronal excitation propagates along the available connection pathways, via intra and interhemispheric association fibers, to other cortical areas and deeper neural structures and, via projecting fibers, to subcortical structures and the spinal cord.

2.2. TMS/EEG

As described above, with TMS the cerebral cortex is stimulated directly by generating a brief but strong magnetic pulse (1 ms, 1–2 T) through a coil applied to the surface of the scalp. The rapid change in magnetic field strength induces a current flow in the tissue, which results in the activation of the underlying neuronal population. The synchronous volley of action potential thus initiated propagates along the available connection pathways and can produce significant electrical activations in target cortical regions. The main problem that one has to face when applying TMS simultaneously with EEG recordings is the powerful electric field that is induced by the discharge of the TMS coil in the electrical leads. Considering a typical pulse intensity of about 1 T and a rise time of 0.1 ms, the voltage induced in the electrodes underlying the stimulator can reach an amplitude of 10 V. This voltage, being several orders of magnitude larger than the signal produced by the brain (tens of microvolts) can cause large artifacts in the recordings and may put an ordinary EEG amplifier out of the operating range for a few seconds.

Today, it is possible to deal with the electric artifact induced by the TMS discharge in the EEG signal by employing specific hardware devices or offline artifact subtractions. Virtanen et al. (28) developed a 60-channel TMS-compatible EEG system that includes gain-control and sample-and-hold circuits to block the artifact induced by TMS in the leads. This system pins the acquired signal to a constant level for a couple of ms around the pulse and records TMS-induced EEG potentials that are completely free from artifacts. Similarly, Iramina and colleagues (29) recorded EEG responses to TMS by means of a TMS-compatible amplifier provided with a sample-and-attenuate stage that actively reduces the signal during the TMS pulse. An alternative way to deal with the TMS artifact has been implemented by Thut and his coworkers (30). They used a slew-limited amplifier that prevents the electronics from saturating during the TMS pulse resulting in a short-lasting artifact that decays within 30 ms. Finally, Bonato and colleagues (12) have recently used an MRI-compatible DC amplifier with a wide dynamic range to successfully record TMS-evoked potentials preceded by a short artifact lasting between 10 and 20 ms. With this method, recordings must be obtained without any filtering, as these might interact with the TMS artifact, producing ripples for up to a second.

Recently, Litvak and colleagues (31) have proposed an offline method to effectively reduce the TMS-induced artifact. This method may be in principle applied also to signals recorded by a wide dynamic range, 24-bit 64 channels EEG amplifier, yet no TMS-compatible. The method is based on a multiple step procedure as follows: (1) Single TMS/EEG trials were interpolated by means of a spline function (implemented in BESA package) within the 12 ms around the stimulus (−2 to 10 ms); (2) Poststimulus residual artifacts were removed by calculating the average TMS/EEG responses across all trials, computing their topographies based on principal component analysis (PCA) over a limited, manually defined time-window (also implemented in BESA package). The procedure has been used to compare, at a group level, local and spatially extended EEG responses to TMS recorded in schizophrenic patients and healthy controls, replicating results obtained by other labs by using a dedicated TMS/EEG equipment (32). Although the algorithm designed by Litvak and colleagues to remove artifacts from TMS/EEG recordings is not suitable for analyses at single subject level, at least in principle it can be employed by all those researchers and clinicians that do not own a TMS-compatible EEG amplifier, yet interested in studying connectivity and excitability in neuropsychiatric disorders. As a matter of fact, by now several works have clearly demonstrated that it is possible to record artifact-free TMS-evoked potentials that reliably reflect the state of excitability of underlying cortical circuits (5, 12, 14, 15, 33).

2.3. Navigated TMS/EEG

The precise and reproducible placement of the TMS coil is a challenging task. In order to measure local excitability and effective connectivity at different cortical sites and across different subjects or populations, it is worth to localize precisely the cortical structures to be targeted by the stimulator. Diverse methods can be applied to perform a precise and reproducible cortical targeting.

Up to a few years ago, only two "blind" methods were commonly employed to achieve reproducible targeting within and across individuals. The first method involves the estimation of the motor threshold (MT), i.e., the lower output of the TMS stimulator able to elicit a consistent and visible twitch of a peripheral muscle, typically of the thumb controlateral to the stimulated hemisphere. In order to stimulate other cortical areas than the motor cortex, the coil was moved referring to the location used to estimate the MT. The stimulation intensity was expressed in terms of percentage of MT. A second method is based on the 10–20 EEG electrode system and assumes that there is a consistent matching between EEG electrodes locations and underlying cortical regions across subjects. This method does not account for the interindividual variability of the skull shape and may lead to errors up to 20 mm in different directions (34). Nevertheless, this approach is fast, economical and may be viable when studying large cortical regions.

To target the same regions in different subjects with higher accuracy and on a finer anatomical scale, one must employ a neuroimage-guided, stereotactic neuronavigation system: this strategy requires the acquisition of either structural or functional MRIs or CT scans. In addition, single-subject functional neuroimaging or average functional atlases ("probabilistic approach") can be used to guide the brain navigator (35). Regardless of the neuroimaging data employed, the stereotactic navigation systems are able to locate the relative positions of the subject's head and the TMS coil by means of an optical, or magnetic, tracking system. The system may also take into account the individual's head shape, the coil position and scalp-to-cortex distance to calculate the electric field induced by TMS on the cortical surface. In this case, during the experiments, the TMS intensity can be adjusted according to the maximum electric field intensity (expressed in volts/meter) estimated on the cortical surface, rather than relying on individual motor threshold, or on the percentage of maximum stimulator output.

Although these estimations are indicative, they should always be interpreted cautiously, since they may include very large errors due to coregistration/navigation mismatches. In these cases, correcting maximal stimulator simply based on scalp-to-cortex distance (36) is the safest option. To further standardize stimulation parameters, the maximum electric field should be always kept on the convexity of the targeted gyrus with the induced current perpendicular to its main axis. With most systems, the coordinates of the stimulator are usually input to a virtual aiming device of the navigation software and can be used during the experiment to ensure stability of the position, angle, direction, and intensity of the stimulation. The neuronavigation systems allow also to digitize the electrode positions by means of a pen visible to the optical tracking system and to store this information in the navigation computer, at the end of each experimental session. This allows performing off-line, accurate source modeling of the TMS-evoked responses.

2.4. Equipment and Materials Employed for the Experiments Described in Sect. 3

- 60 channels TMS-compatible EEG amplifiers, electrodes and electrode cap: eximia EEG unit (Nexstim Ltd., Helsinki, Finland: http://www.nexstim.com)
- Navigated TMS: eximia Navigated Brain Stimulation (NBS) device (Nexstim Ltd, Helsinki, Finland)
- TMS coil: eximia TMS stimulator (Nexstim Ltd., Helsinki, Finland)
- Electroconductive gel for electrode–skin contacts: Electro-Gel (Electro-Cap International Inc., Eaton, Ohio, USA)
- Skin cleaner: LemonPrep (Mavidon Medical Products, Lake Worth, Florida, USA)
- Wooden cotton tips
- In-ear earplugs (Sony OEM In-The-Ear Headsets)
- Dermatological tape (3M)

3. Methods

Even when TMS-compatible EEG amplifiers and/or off-line procedures are effective in reducing or abolishing magnetic and electrical artifacts, TMS-evoked potentials may still be contaminated by spurious signals. Indeed, electrode motion and polarization artifacts, somatosensory and auditory evoked potentials, spontaneous and evoked muscle activations, as well as spontaneous or TMS-induced eye blinks can overlap with TMS-evoked potentials and mislead their interpretation. In this section, we describe in more detail the general experimental steps and the strategies that can be adopted to prevent and control the main confounding factors. Finally, we describe the procedures to preprocess TMS/EEG recordings and to calculate some indices of excitability and effective connectivity.

3.1. Preparing the Subject

The comfort of the subject is crucial to prevent him to assume compensatory positions by constantly contracting postural muscles of the neck or of the upper back. The electrical activity generated by these muscles is usually recorded by the EEG leads as electromyogram (EMG) and, due to its large amplitude (tens or hundreds of microvolts), can reduce the signal-to-noise ratio of TMS-evoked potentials. Thus, the subject should settle back preferably on an ergonomic chair or on a chair provided with armrest and an adjustable headrest. Moreover, the operator should encourage the subject to relax, to reduce the muscle contraction of the temporal and facial muscles that can introduce a large EMG in the temporal and frontal leads and, by lowering the threshold for the direct muscles activation, can result in a muscle artifact time-locked with the TMS.

Then, the subject will wear the EEG cap (Fig. 2) and the preparation of the electrode-to-scalp contacts will start. This step is crucial to further avoid TMS-induced artifacts. Indeed, the impedance of these contacts should be below 5 kΩ. To this aim, in correspondence of the electrode positions the hair should be displaced to carefully and delicately scrub the scalp by means of an abrasive prep applied on cotton tips. Once the scalp is clearly visible through the electrode's ring, a little amount of bubble-free electroconductive gel can be applied between the skin and the electrode pin.

Finally, the subject will wear the goggles that will be used by the navigation system to localize his head relatively to the coil. It is worth to ensure the stability of the goggles throughout the whole TMS/EEG session. Thus, before performing the coregistration between the fiducials selected on the individual MRI (usually nasion, left tragus, and right tragus) and the corresponding points on the subject's head, the operator can secure the goggles on the subject's head and nose by means of dermatological tape.

3.2. The Pilot TMS/EEG Session

Setting up coil position, coil orientation and stimulation intensity. Once the cortical target is selected and marked on the MRI and/or on the 3D rendering of the cortical surface (Fig. 3), the coil can be positioned accordingly and both its orientation and stimulation intensities can be set up. Ideally, the main direction of the current generated by the coil should be perpendicular to the targeted gyrus and the maximum of the induced electric field, as estimated by the navigation system, should be localized on the convexity of the gyrus. In this way, it is more likely to activate bent axons that give rise to bundles of white matter and are the structures more

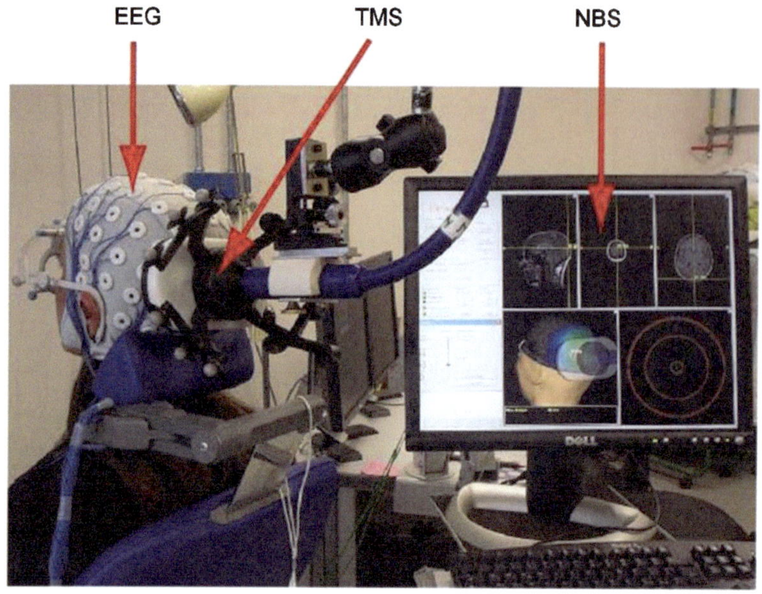

Fig. 2. TMS/EEG set-up. The subject is sitting on an ergonomic chair, while TMS is positioned over the occipital cortex. The *red arrows* indicate, from *left* to *right*, the three fundamental components that compose the TMS/EEG setup: a cap for high-density EEG recordings (60 channels) connected to a TMS-compatible amplifier; a focal figure-of-eight stimulating coil (TMS), held in place by a specifically designed mechanical arm; the display of the navigated brain stimulation system (NBS). The system shown here employs an infrared camera (not visible) to navigate and target TMS on a 3D reconstruction of the subject's MRI. NBS localizes in real time the electric field induced by the TMS on the cortical surface and estimates its intensity.

sensitive to electric fields induced by the TMS (see Sect. 2.1). Most of the commercial navigation systems are provided with a software that is able to roughly estimate and localize the maximal electric field induced on the cortical surface by TMS. Using this device, the output of the stimulator should be set to obtain a maximum electric field of about or beyond 120 V/m. As a matter of fact, this intensity of stimulation is known to be above motor threshold when applied over the motor cortex (37, 38) and to elicit well-characterized responses when applied over associative cortical regions (13, 39). It is generally advisable to maximize the impact of TMS on cortical axons to keep stimulation intensities low, thus reducing the chances of artifacts and of coil overheating. Before starting a recording on a new cortical target it is always worth launching a pilot TMS/EEG session during which only 20–30 TMS pulses will be delivered. This number of trials is usually sufficient to have a clear idea about the quality of the recording. Looking at the raw EEG during the TMS pulse can reveal fundamental problems with the recording and suggest optimal solutions. A large spike on the recording visible on several channels may point to a magnetic or a muscle artifact.

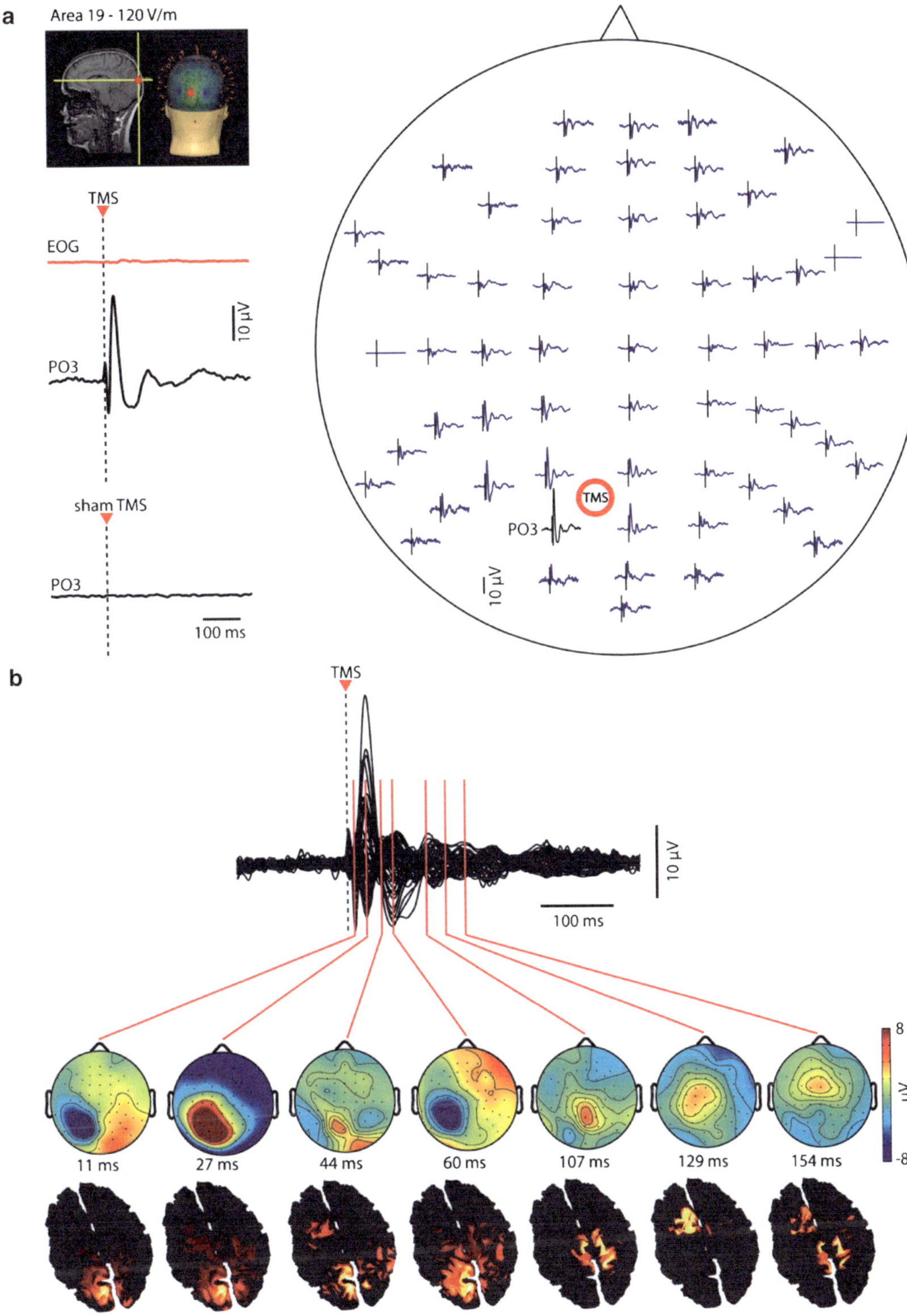

Fig. 3. EEG responses and cortical activations evoked by TMS of area 19. The *left* side of **a** shows the EOG trace (*red trace*) and the TMS evoked potential (*black trace*) recorded from the electrode under the stimulator (PO3) when Brodmann's area 19 is stimulated at 120 V/m (the *inset* shows the location of the maximal electric field induced by TMS on the cortical surface). On the *right* side of **a**, the evoked potentials recorded from all the 60 EEG channels are shown. In the *upper part* of **b** the butterfly plot of the 60 EEG channels is shown. In the *lower part* of **b** the voltage scalp maps and the corresponding cortical source reconstructions are represented at different time latencies from the TMS (reproduced with permission from (13)).

The former will be present even when the coil is placed far away from muscles (around the vertex) and may suggest to improve impedances (especially of the ground and reference electrodes), or to check the connections (trigger cables) that ensure TMS-compatibility. The latter will change dramatically by rotating/translating the coil and may suggest to reduce stimulation intensity or to stimulate over a more central scalp region. When possible, inspecting visually the average computed online at every channel across the performed trials can be useful to roughly estimate the signal-to-noise ratio of the TMS-evoked potential, the level of the background noise and the presence of large artifacts. In general, it is not worth starting an experiment until the raw/averaged EEG of most channels is clear from large spikes lasting for more than 10–20 ms.

Auditory evoked potentials. When a TMS pulse is delivered, the coil emits a loud "click" (tens of decibel). This noise, time-locked with the pulse, may elicit auditory evoked potentials (AEPs) that overlap with the TMS evoked responses. This confounding factor can be avoided by playing a noise-masking obtained by shuffling the time-varying spectral components of the "click" (see Fig. 4). This method proved to be effective in eliminating the AEP evoked by the coil's click in many studies (13, 14, 39, 40). The obliteration of AEPs through noise masking has been verified during sham experiments where the TMS coil is discharged while separated from the scalp by a 4-cm Plexiglas cube (see Figs. 3 and 4).

Before every TMS/EEG session, the operator should discharge a few TMS pulses while the noise-masking is being played and the coil kept in the chosen position. After that, the operator should ask the subject whether he/she is able to detect the "click". In the case of a positive response, the noisemasking should be played louder. Otherwise, the session may start, but the operator should check the presence of the AEP components that typically occur at 100 and 200 ms after the stimulus and appear larger at the vertex and temporal leads. The visual detection of AEP components will require a further increase of the noise-masking loudness.

Scalp muscle artifacts. Although a recent study, using intramuscular needle-electrode recordings, demonstrated that TMS delivered over the hand motor cortices at motor threshold intensity does not evoke EMG activity of temporal muscles (5), this artifact cannot be excluded a priori. Indeed, artifacts produced by the activation of temporal and frontal muscles can be elicited when TMS is delivered at high intensities over lateral aspects of the scalp. The operator should monitor the presence of this class of artifacts during the pilot session. Direct muscle activation typically appears as spikes, the so called M-waves, time-locked with the TMS trigger, with a voltage ranging from hundreds to thousands of

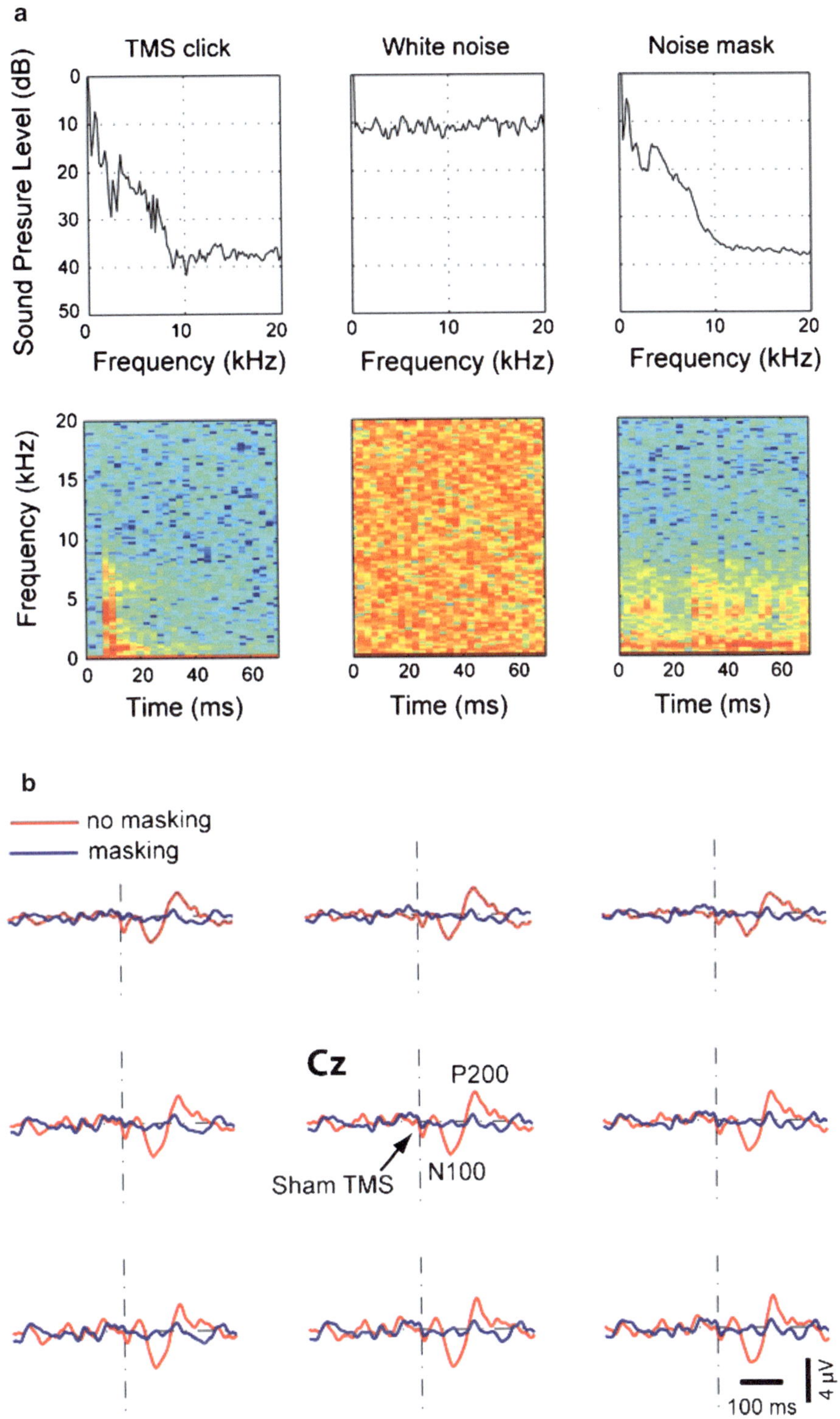

Fig. 4. Abolition of the auditory evoked potential (AEP) through noise masking evoked by the TMS "click". In **a** the spectral profile and the time-frequency characteristics of the TMS-associated "click" (*left*) are compared to white noise (*center*) and the synthetic waveform that we employ during the experiments (*right*). In **b** nine electrodes around the vertex are selected to show the evoked potentials obtained with (*blue traces*) and without (*red traces*) noise masking are compared while the coil, connected to the scalp by a plexiglass cube (to preserve bone conduction), was placed 5 cm above the vertex (Cz). Noise masking clearly abolishes the auditory N100-P200 component of the AEP (reproduced with permission from (14)).

microvolts (41). Scalp muscle artifacts are recordable mainly in the most lateral and frontal electrodes and lasting up to 50 ms and can be prevented by stimulating cortical areas underlying scalp regions close to the midline, by changing the coil orientation or by decreasing output level of the stimulator.

Eye movements and eye blinks. A few studies have demonstrated that single pulses of TMS applied over parietal or frontal oculomotor regions are unable to evoke eye movement (42). However, TMS can sometimes startle the subject and evoke eye blinks (43). The occurrence of systematic blinks or eye muscle reactions can be effectively prevented during the experiments by employing noise masking and by adjusting the coil location and the stimulation intensity. Indeed, it is worth to constantly check the Electroculogram (EOG), which is usually acquired concomitantly with EEG, and twitches of the subject's eyelids, which consistently occur time-locked with each pulse TMS.

3.3. The "Real" TMS/EEG Session: Number of Trials, Frequency of Stimulation and Online Monitoring of Coil Placement

After the pilot session has been performed and all the possible sources of artifacts have been controlled for, the real TMS/EEG session can take place. The number of trials that should be acquired to obtain a TMS-evoked potential with a high signal-to-noise ratio ranges from 150 up to 300, depending on the intensity of stimulation. Generally, the following empirical rule can be adopted: the highest the intensity, the lower the number of trials.

Throughout the measurement, it is worth to maintain constant the position of the coil relative to the head. TMS coils can be handheld by the operator or mounted on specifically designed coil holders. Indeed, in both cases the position of the subject's cortex relative to the coil could change either due to movements of the operator's hand or of the subject's head. This event can, in turn, transiently or steadily change some of the stimulation parameters, such as coil-to-cortex distance or coil orientation and hence change the strength and the impact of the TMS-induced electric field on the cortex.

The majority of the commercial navigation systems, such as the NBS, come complete with virtual aiming devices that allow compensating promptly for coil displacements from chosen position. Thus, if the aiming device signals a significant displacement, it is advisable to stop the stimulation for a few seconds, reposition the coil, and start again.

3.4. Off-Line Data Preprocessing

Off-line data preprocessing of TMS/EEG trials encompasses necessarily two steps that are similar to the ones usually applied to analyze sensory-evoked potentials:

1. Rejection of trials corrupted by spontaneous blinks, eye movements, or muscle artifacts, using a manual, semiautomatic, or automatic algorithm.

2. After trial rejection, EEG and EOG recordings can be averaged across trials. These average potentials can be further re-referenced against the average computed across all channels (average reference).

3.5. Measures of Cortical Excitability and Connectivity and Estimation of Sensitivity and Repeatability of TMS/EEG

In order to establish TMS/EEG as a neuroscientific method and to evaluate its potential as a diagnostic and monitoring technique normative data need to be collected. For this reason, the analysis, classification and interpretation of TMS-evoked potentials requires the implementation of a standardized, data-driven procedure to characterize the electrical response of the cerebral cortex to TMS. In a recent paper, Casali and colleagues (39), has developed an automatic TMS/EEG data analysis package able to output three synthetic indices: significant current density (SCD), phase-locking (PL), and significant current scattering (SCS) (Fig. 5). SCD sums up the amplitude of all significant currents induced by TMS, PL reflects the ability of TMS to reset the phase of ongoing cortical oscillations, while SCS measures the average distance of significantly activated sources from the site of stimulation. These indices are aimed at capturing different aspects of brain responsiveness, ranging from global cortical excitability towards global cortical connectivity. Casali et al. tested the three indices by analyzing the EEG responses to TMS of Brodmann's area 19 at increasing intensities in five healthy subjects. The spatial distribution and time course of SCD, PL, and SCS revealed a reproducible profile of excitability and connectivity, characterized by a local activation threshold around a TMS-induced electric field of 50 V/m and by a selective propagation of TMS-evoked activation from occipital to ipsilateral frontal areas that reached a maximum at 70–100 ms. In principle, SCD, PL, and SCS may be used to characterize the effects of TMS on any cortical area and to quantitatively evaluate cortical excitability and connectivity in physiological and pathological conditions. Similar indices that enable data reduction and straightforward comparisons of TMS-evoked activations across subjects and conditions should be developed. Moreover, to reliably evaluate the potential of TMS/EEG as a dependable tool to studying, diagnosing, and monitoring pathological conditions, some estimation of the technique sensitivity and repeatability is required. Indeed, TMS-evoked potentials would have limited scientific or clinical applications if they were found not to change when different cortical sites are stimulated or if they were found to vary randomly when stimulation parameters are maintained constant. Ideally, TMS-evoked potentials, recorded across separate sessions in a healthy brain, should always change significantly if stimulation parameters are varied (100% sensitivity) and should not change when stimulation parameters are kept constant (100% repeatability). Several studies have suggested that TMS-evoked potentials have a certain degree of sensitivity to changes in stimulation parameters, such as location, intensity, and direction of the induced

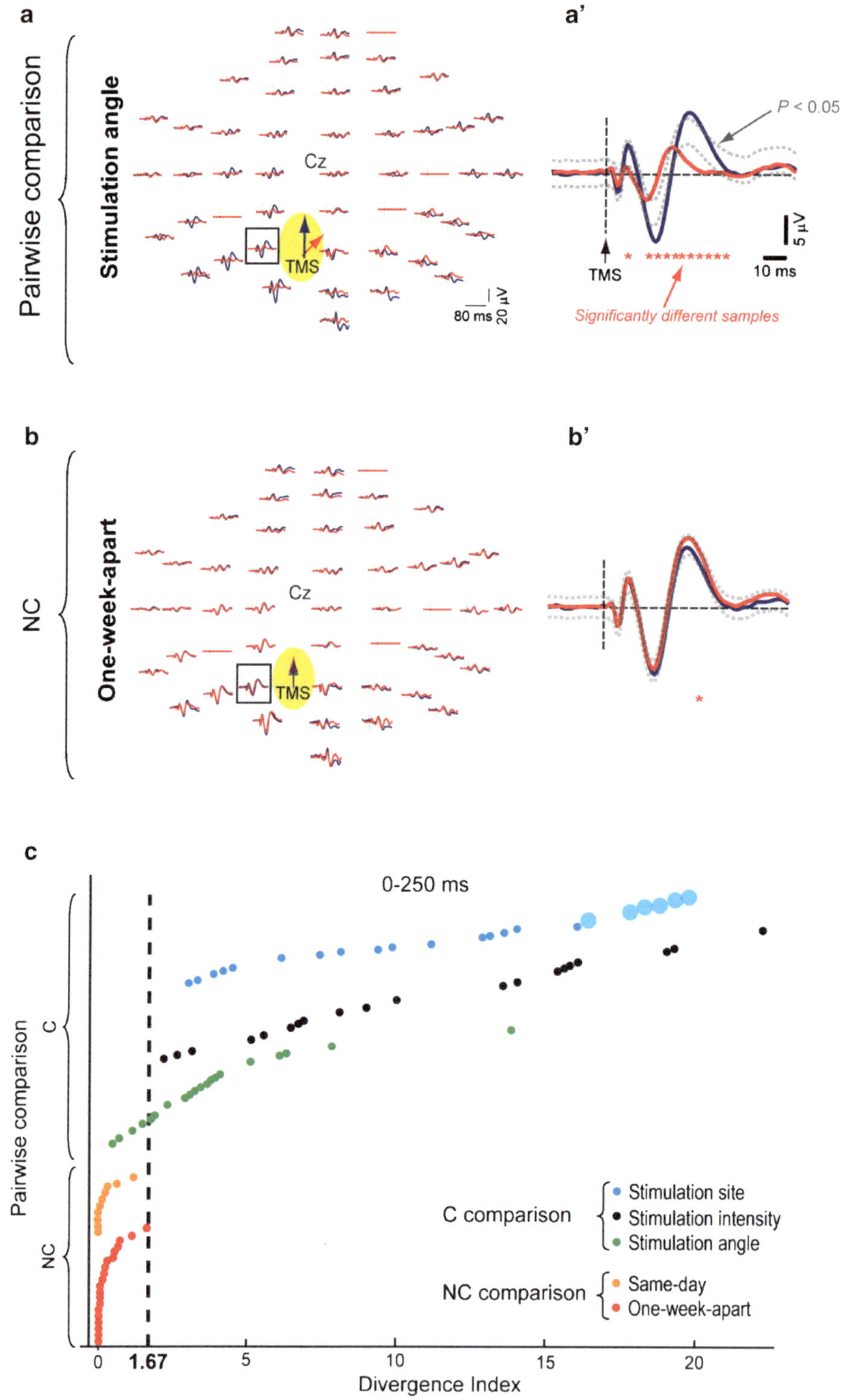

Fig. 5. Results of pairwise comparisons between EEG responses to TMS at the subject (**a**, **a**′, **b**, and **b**′) and population levels (**c**). EEG responses to stimulation of BA19 at 0° angle on day 1 (*blue traces*) are compared with brain responses recorded during different sessions (*red traces*), in which two parameters were varied one at a time, namely angle (45°) and

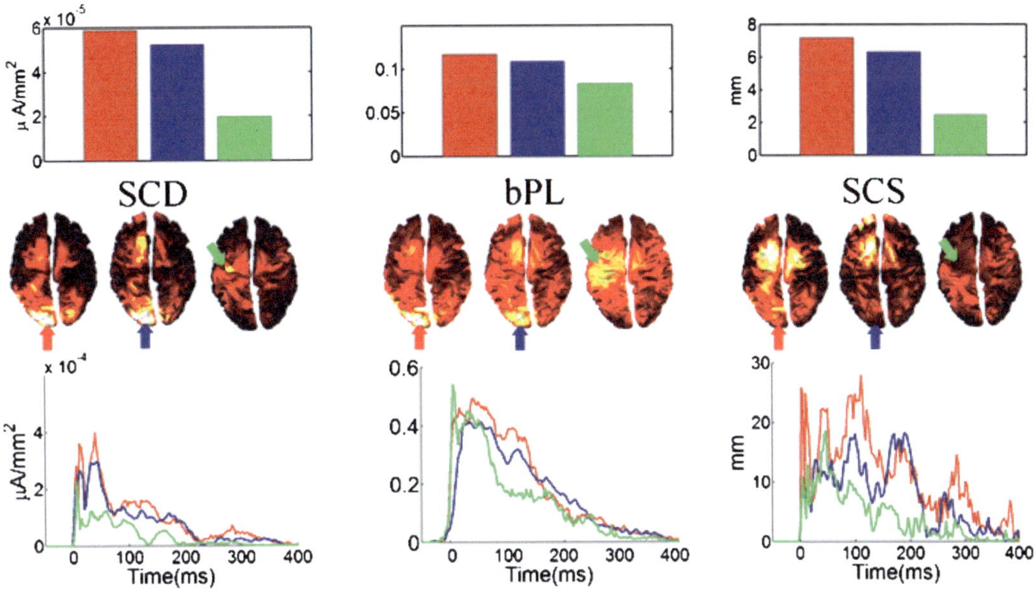

Fig. 6. Indices of cortical excitability and effective connectivity. Computation of synthetic indices of cortical responsiveness to TMS in a single subject at a stimulation intensity of 140 V/m delivered to the left occipital cortex in two different days (*red* and *blue colors*, respectively) and to the left hand motor area (*green color*). Source Current Density (SCD), Broadband Phase-Locking (bPL) and Source Current Scattering (SCS) values are cumulated either over the whole brain (time courses) or over the full poststimulus period (spatial maps) or over both the whole brain and the poststimulus period (*bars*) (reproduced with permission from (39)).

current with respect to the cortical surface. Moreover, a few works have demonstrated that TMS-evoked potentials can also detect changes in the state of cortical circuits, such as the ones induced by alcohol intake (5), by falling asleep (14) (see also the following section) and by the induction of cortical potentiation with repetitive TMS (33), while repeatability has been evaluated only in one work by Lioumis and his coworkers (44).

In a recent work, we performed a statistical joint evaluation of the sensitivity and repeatability of TMS/EEG measures (40). In ten volunteers, we performed 92 single-subject comparisons to evaluate the similarities/differences between pairs of TMS-evoked potentials recorded in the same/different stimulation conditions (Fig. 6). For each pairwise comparison, we used nonparametric

Fig. 5. (continued) day (day 8), resulting in one Change (C) comparison and one No Change (NC) comparison. For each comparison, superimposition of pairs of TMS-evoked potentials in all sensors is displayed in (**a**), while enlarged view of PO3 channel is shown in (**a′**), together with significance boundaries (*dotted gray traces*) and significantly different samples (*red stars*). Single DI values computed over the entire poststimulus period (250 ms) are shown in (**a**) with the following color-coding: DIs of the C comparisons for changes in the stimulation site, intensity and angle are represented by *cyan*, *black*, and *green dots*, respectively, while DIs of NC comparisons are depicted in *yellow* for same-day sessions and in *red* for 1-week-apart sessions (reproduced with permission from (40)).

statistics to calculate a Divergence Index (DI), i.e., the percentage of samples that differed significantly, considering all scalp locations and the entire poststimulus period. A receiver operating characteristic analysis showed that it was possible to find an optimal DI threshold of 1.67%, yielding 96.7% overall accuracy of TMS/hd-EEG in detecting whether a change in the perturbation parameters occurred or not. These results demonstrate that the EEG responses to TMS essentially reflect deterministic properties of the stimulated neuronal circuits as opposed to stereotypical responses or uncontrolled variability. To the extent that TMS-evoked potentials are sensitive to changes and repeatable over time, they may be employed to detect longitudinal changes in the state of cortical circuits.

3.6. The Estimation of Cortical Excitability and Effective Connectivity in Wakefulness and Sleep

Some recent studies employed TMS/EEG to measure cortical excitability and effective connectivity in healthy subjects during wakefulness, slow-wave sleep, REM sleep, and benzodiazepine-induced anesthesia.

In a first set of experiments (13) TMS was targeted over three different cortical sites (one occipital, one parietal and one frontal region) of six healthy subjects and the ensuing EEG oscillations were recorded (Fig. 7). As shown in Fig. 6 TMS-evoked potentials recorded at all 60 channels clearly differed in terms of amplitude, waveform and dominant frequency depending on the stimulated site. Indeed, TMS consistently evoked dominant alpha-band oscillations (8–12 Hz) in the occipital cortex,

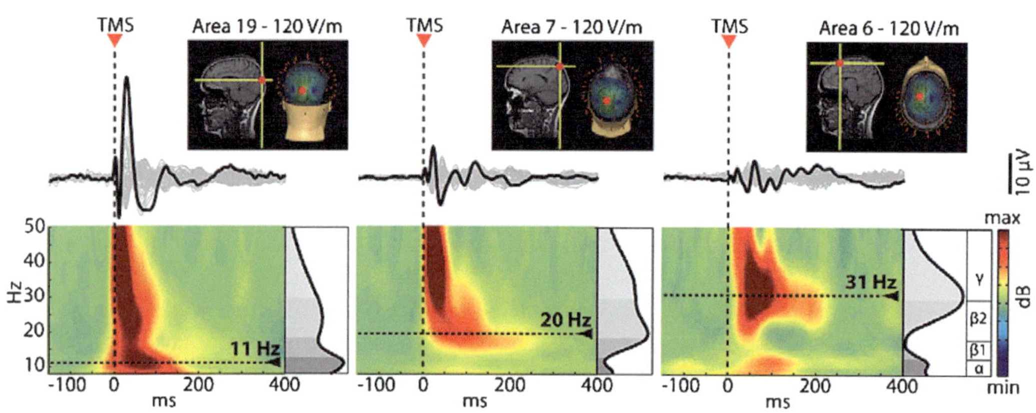

Fig. 7. EEG oscillations evoked by TMS are specific for the stimulated site. The *insets* illustrate the three cortical sites targeted by TMS (maximal TMS-induced electric field on the individual MRI) in one subject. The *traces* below represent butterfly plots, where the *black trace* highlights the electrode directly underlying the stimulator. The *bottom panels* show the Event Related Spectral Perturbation (ERSP) patterns calculated globally on the scalp (average of all electrodes). The *gray scale* graph plotted at the *right* of each ERSP depicts the power spectrum profile induced during the first 200 ms after TMS. The *dotted lines* highlight the frequency with maximum power. TMS elicited, from *left* to *right*, early gamma components immediately followed by prominent alpha-band oscillations after occipital stimulation, beta-band oscillations after parietal stimulation, and fast beta/gamma oscillations after perturbation of frontal cortex (reproduced with permission from (13)).

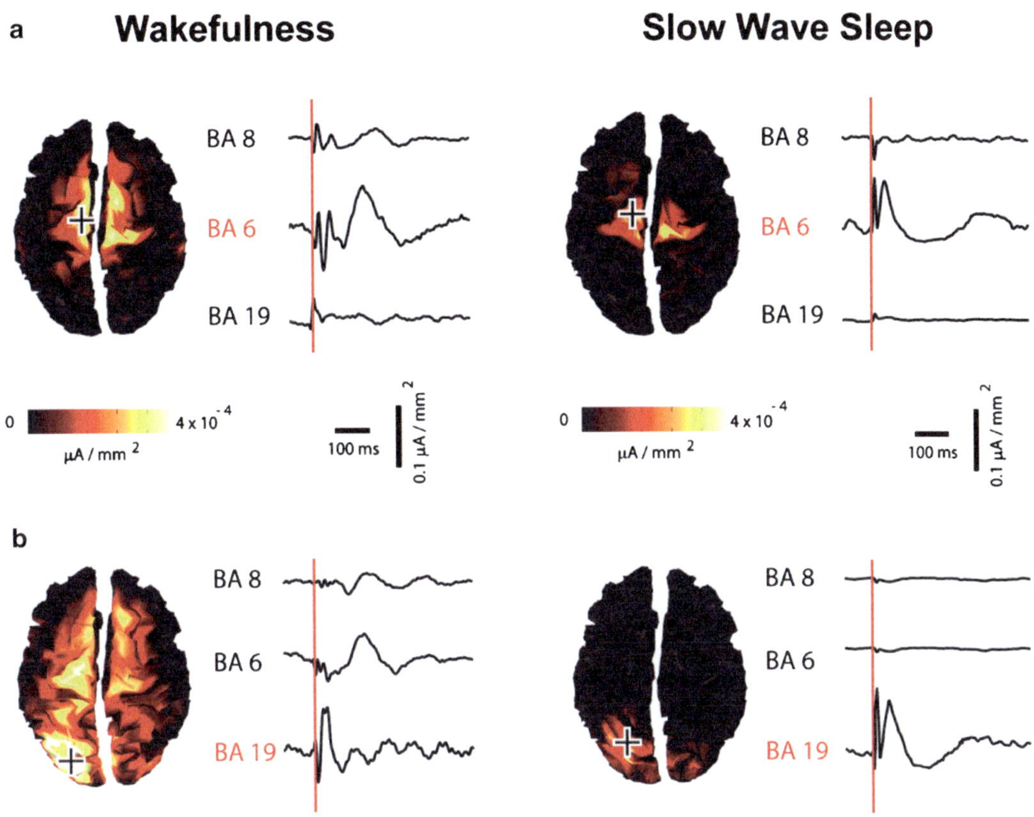

Fig. 8. Breakdown of cortical effective connectivity during slow-wave sleep. Premotor cortex (**a**) and visual cortex (**b**) are targeted by TMS during wakefulness (*left panels*) and during slow-wave sleep (*right panels*). After source modeling, nonparametric statistics is performed to detect the significant currents induced by TMS. For each condition, the significant currents recorded during the entire poststimulus interval are plotted on the cortical surface; on the right side of each cortical surface, the time series of the currents recorded from three selected areas (Brodmann areas (BA) 8, 6 and 19) are depicted (the time of TMS pulse is marked by a *red line*). With the transition from wakefulness to slow-wave sleep, distant cortical areas cease to be causally affected by the initial perturbation, indicating a breakdown in cortical effective connectivity. At the same time, cortical responses to TMS become stereotypical, indicating a loss of cortical differentiation (reproduced with permission from (16)).

beta-band oscillations (13–20 Hz) in the parietal cortex, and fast beta/gamma-band oscillations (21–50 Hz) in the frontal cortex.

In a second series of experiments, Massimini and colleagues (14) compared the EEG responses to TMS during wakefulness and deep sleep in the same subjects (Fig. 8). The study stems from some recent theoretical considerations suggesting that a breakdown of causal interactions within the cerebral cortex, thus of cortical effective connectivity, is the neural substrate of the loss of consciousness usually observed during deep sleep (45). This study showed that specific long-range response obtained during wakefulness broke down into a stereotypical and local response during sleep. A similar behavior of the human cerebral cortex has been recently showed when TMS-evoked potentials are compared in wakefulness and deep sedation induced by Midazolam (15). Interestingly, during REM sleep, TMS/EEG reveals a resurgence of complex spatiotemporal patterns of cortical activations (46).

These studies suggest that TMS/EEG may be used to study the neural correlates of loss of consciousness and its recovery during sleep, anesthesia and in brain-injured patients. Most important, in this context, they show that the EEG response to TMS, once all confounding factors are controlled for, reflects only the intrinsic properties of targeted cortical circuits rather than physical or biological artifacts.

3.7. Future Technical Developments

Currently, the main technical limitations of TMS/EEG are related to the unwanted activation of head and neck muscles by the TMS pulse. This activation may produce large and long-lasting electrical artifacts that obliterate the genuine neural response to TMS (see Sect. 3.2). Thus, it is very challenging to obtain clean EEG responses after stimulation of brain areas that lay under the peripheral aspects of the scalp. For instance, in most subjects it may be difficult, or impossible, to probe with TMS/EEG the excitability and the connectivity of a temporal language area or of the orbitofrontal cortex; similarly, a powerful muscle artifact may mask the response to TMS of the primary visual cortex. However, solutions, either complete or partial, to this problem can be envisaged. First, the development of smaller and more focal coil may help extending the cortical region that is currently accessible to TMS/EEG. Second, coil designs and shielding devices can be developed to reduce the electric field induced by TMS on the scalp (47). Third, different durations and shapes of the TMS pulse should be explored (48) to find the stimulation parameters that maximize the stimulation of cortical axons while minimizing the depolarization of peripheral fibers, taking into account their specific biophysical properties. Alternatively, sham coils should be developed that result in the same scalp activation, whether cortical stimulation is delivered or not (49). In this case, muscle artifact could be removed off-line by subtraction.

Further development of the technique may include improvement of the estimation of the electric field induced by TMS in the gray matter. In this case, the exact geometry and conductive properties of the tissues underlying the coil should be taken into account, possibly by building a finite element model (FEM). Such an improvement not only will result in a better estimation of the TMS dose in healthy subjects (50) but will also allow controlling the effectiveness of stimulation in patients with brain lesions, which may perturb and distort the current flow in the surrounding gray matter (51).

Acknowledgments

This work was supported by grants from European Grant Strep ICT-2007-224328 “Predict AD” (to M. M.), MIUR PRIN program 2008, and University of Milan PUR program 2009 (to M.R. and M.M.).

References

1. Paus T (2005) Inferring causality in brain images: a perturbation approach. Philos Trans R Soc Lond B Biol Sci 360:1109–1114
2. Cracco RQ, Amassian VE, Maccabee PJ, Cracco JB (1989) Comparison of human transcallosal responses evoked by magnetic coil and electrical stimulation. Electroencephalogr Clin Neurophysiol 74:417–424
3. Amassian VE, Cracco RQ, Maccabee PJ, Cracco JB (1992) Cerebello-frontal cortical projections in humans studied with the magnetic coil. Electroencephalogr Clin Neurophysiol 85:265–272
4. Ilmoniemi RJ, Virtanen J, Ruohonen J, Karhu J, Aronen HJ, Naatanen R, Katila T (1997) Neuronal responses to magnetic stimulation reveal cortical reactivity and connectivity. Neuroreport 8:3537–3540
5. Kahkonen S, Kesaniemi M, Nikouline VV, Karhu J, Ollikainen M, Holi M, Ilmoniemi RJ (2001) Ethanol modulates cortical activity: direct evidence with combined TMS and EEG. Neuroimage 14:322–328
6. Nikouline V, Ruohonen J, Ilmoniemi RJ (1999) The role of the coil click in TMS assessed with simultaneous EEG. Clin Neurophysiol 110:1325–1328
7. Kahkonen S, Komssi S, Wilenius J, Ilmoniemi RJ (2005) Prefrontal transcranial magnetic stimulation produces intensity-dependent EEG responses in humans. Neuroimage 24:955–960
8. Kahkonen S, Komssi S, Wilenius J, Ilmoniemi RJ (2005) Prefrontal TMS produces smaller EEG responses than motor-cortex TMS: implications for rTMS treatment in depression. Psychopharmacology (Berl) 181:16–20
9. Paus T, Sipila PK, Strafella AP (2001) Synchronization of neuronal activity in the human primary motor cortex by transcranial magnetic stimulation: an EEG study. J Neurophysiol 86:1983–1990
10. Van Der Werf YD, Paus T (2006) The neural response to transcranial magnetic stimulation of the human motor cortex. I. Intracortical and cortico-cortical contributions. Exp Brain Res 175:231–245
11. Van Der Werf YD, Sadikot AF, Strafella AP, Paus T (2006) The neural response to transcranial magnetic stimulation of the human motor cortex. II. Thalamocortical contributions. Exp Brain Res 175:246–255
12. Bonato C, Miniussi C, Rossini PM (2006) Transcranial magnetic stimulation and cortical evoked potentials: a TMS/EEG co-registration study. Clin Neurophysiol 117:1699–1707
13. Rosanova M, Casali A, Bellina V, Resta F, Mariotti M, Massimini M (2009) Natural frequencies of human corticothalamic circuits. J Neurosci 29:7679–7685
14. Massimini M, Ferrarelli F, Huber R, Esser SK, Singh H, Tononi G (2005) Breakdown of cortical effective connectivity during sleep. Science 309:2228–2232
15. Ferrarelli F, Massimini M, Sarasso S, Casali A, Riedner BA, Angelini G, Tononi G, Pearce RA (2010) Breakdown in cortical effective connectivity during midazolam-induced loss of consciousness. Proc Natl Acad Sci USA 107:2681–2686
16. Massimini M, Boly M, Casali A, Rosanova M, Tononi G (2009) A perturbational approach for evaluating the brain's capacity for consciousness. Prog Brain Res 177:201–214
17. Morishima Y, Akaishi R, Yamada Y, Okuda J, Toma K, Sakai K (2009) Task-specific signal transmission from prefrontal cortex in visual selective attention. Nat Neurosci 12:85–91
18. Akaishi R, Morishima Y, Rajeswaren VP, Aoki S, Sakai K (2010) Stimulation of the frontal eye field reveals persistent effective connectivity after controlled behavior. J Neurosci 30:4295–4305
19. Romei V, Brodbeck V, Michel C, Amedi A, Pascual-Leone A, Thut G (2008) Spontaneous fluctuations in posterior alpha-band EEG activity reflect variability in excitability of human visual areas. Cereb Cortex 18:2010–2018

20. Romei V, Gross J, Thut G (2010) On the role of prestimulus alpha rhythms over occipito-parietal areas in visual input regulation: correlation or causation? J Neurosci 30:8692–8697
21. Taylor PC, Walsh V, Eimer M (2010) The neural signature of phosphene perception. Hum Brain Mapp 31:1408–1417
22. Ferrarelli F, Massimini M, Peterson MJ, Riedner BA, Lazar M, Murphy MJ, Huber R, Rosanova M, Alexander AL, Kalin N, Tononi G (2008) Reduced evoked gamma oscillations in the frontal cortex in schizophrenia patients: a TMS/EEG study. Am J Psychiatry 165:996–1005
23. Julkunen P, Jauhiainen AM, Westeren-Punnonen S, Pirinen E, Soininen H, Kononen M, Paakkonen A, Maatta S, Karhu J (2008) Navigated TMS combined with EEG in mild cognitive impairment and Alzheimer's disease: a pilot study. J Neurosci Methods 172:270–276
24. Roth BJ, Saypol JM, Hallett M, Cohen LG (1991) A theoretical calculation of the electric field induced in the cortex during magnetic stimulation. Electroencephalogr Clin Neurophysiol 81:47–56
25. Abdeen MA, Stuchly MA (1994) Modeling of magnetic field stimulation of bent neurons. IEEE Trans Biomed Eng 41:1092–1095
26. Ruohonen J, Ilmoniemi RJ (1999) Modeling of the stimulating field generation in TMS. Electroencephalogr Clin Neurophysiol Suppl 51:30–40
27. Maccabee PJ, Amassian VE, Eberle LP, Cracco RQ (1993) Magnetic coil stimulation of straight and bent amphibian and mammalian peripheral nerve in vitro: locus of excitation. J Physiol 460:201–219
28. Virtanen J, Ruohonen J, Naatanen R, Ilmoniemi RJ (1999) Instrumentation for the measurement of electric brain responses to transcranial magnetic stimulation. Med Biol Eng Comput 37:322–326
29. Iramina K, Maeno T, Nonaka Y, Ueno S (2003) Measurement of evoked electroencephalography induced by transcranial magnetic stimulation. J Appl Phys 93:6718–6720
30. Thut G, Ives JR, Kampmann F, Pastor MA, Pascual-Leone A (2005) A new device and protocol for combining TMS and online recordings of EEG and evoked potentials. J Neurosci Methods 141:207–217
31. Litvak V, Komssi S, Scherg M, Hoechstetter K, Classen J, Zaaroor M, Pratt H, Kahkonen S (2007) Artifact correction and source analysis of early electroencephalographic responses evoked by transcranial magnetic stimulation over primary motor cortex. Neuroimage 37:56–70
32. Levit-Binnun N, Litvak V, Pratt H, Moses E, Zaroor M, Peled A (2010) Differences in TMS-evoked responses between schizophrenia patients and healthy controls can be observed without a dedicated EEG system. Clin Neurophysiol 121:332–339
33. Esser SK, Huber R, Massimini M, Peterson MJ, Ferrarelli F, Tononi G (2006) A direct demonstration of cortical LTP in humans: a combined TMS/EEG study. Brain Res Bull 69:86–94
34. Herwig U, Satrapi P, Schonfeldt-Lecuona C (2003) Using the international 10–20 EEG system for positioning of transcranial magnetic stimulation. Brain Topogr 16:95–99
35. Paus T, Jech R, Thompson CJ, Comeau R, Peters T, Evans AC (1997) Transcranial magnetic stimulation during positron emission tomography: a new method for studying connectivity of the human cerebral cortex. J Neurosci 17:3178–3184
36. Stokes MG, Chambers CD, Gould IC, Henderson TR, Janko NE, Allen NB, Mattingley JB (2005) Simple metric for scaling motor threshold based on scalp-cortex distance: application to studies using transcranial magnetic stimulation. J Neurophysiol 94:4520–4527
37. Epstein C, Schwartzberg D, Davey K, Sudderth D (1990) Localizing the site of magnetic brain stimulation in humans. Neurology 40:666–670
38. Thielscher A, Kammer T (2002) Linking physics with physiology in TMS: a sphere field model to determine the cortical stimulation site in TMS. Neuroimage 17:1117–1130
39. Casali AG, Casarotto S, Rosanova M, Mariotti M, Massimini M (2010) General indices to characterize the electrical response of the cerebral cortex to TMS. Neuroimage 49:1459–1468
40. Casarotto S, Romero Lauro LJ, Bellina V, Casali AG, Rosanova M, Pigorini A, Defendi S, Mariotti M, Massimini M (2010) EEG responses to TMS are sensitive to changes in the perturbation parameters and repeatable over time. PLoS One 5:e10281
41. Merletti R, Fiorito A, Lo Conte LR, Cisari C (1998) Repeatability of electrically evoked EMG signals in the human vastus medialis muscle. Muscle Nerve 21:184–193
42. Meyer BU, Diehl R, Steinmetz H, Britton TC, Benecke R (1991) Magnetic stimuli applied over motor and visual cortex: influence of coil position and field polarity on motor responses, phosphenes, and eye movements. Electroencephalogr Clin Neurophysiol Suppl 43:121–134

43. Corthout E, Uttl B, Juan CH, Hallett M, Cowey A (2000) Suppression of vision by transcranial magnetic stimulation: a third mechanism. Neuroreport 11:2345–2349
44. Lioumis P, Kicic D, Savolainen P, Makela JP, Kahkonen S (2009) Reproducibility of TMS-Evoked EEG responses. Hum Brain Mapp 30:1387–1396
45. Tononi G (2004) An information integration theory of consciousness. BMC Neurosci 5:42
46. Massimini M, Ferrarelli F, Murphy M, Huber R, Riedner B, Casarotto S, Tononi G (2010) Cortical reactivity and effective connectivity during REM sleep in humans. Cognit Neurosci 1:176–183
47. Davey KR, Riehl M (2006) Suppressing the surface field during transcranial magnetic stimulation. IEEE Trans Biomed Eng 53:190–194
48. Peterchev AV, Jalinous R, Lisanby SH (2008) A transcranial magnetic stimulator inducing near-rectangular pulses with controllable pulse width (cTMS). IEEE Trans Biomed Eng 55:257–266
49. Mennemeier M, Triggs W, Chelette K, Woods A, Kimbrell T, Dornhoffer J (2009) Sham transcranial magnetic stimulation using electrical stimulation of the scalp. Brain Stimul 2:168–173
50. Thielscher A, Opitz A, Windhoff M (2011) Impact of the gyral geometry on the electric field induced by transcranial magnetic stimulation. Neuroimage 54(1):234–243
51. Wagner T, Fregni F, Eden U, Ramos-Estebanez C, Grodzinsky A, Zahn M, Pascual-Leone A (2006) Transcranial magnetic stimulation and stroke: a computer-based human model study. Neuroimage 30:857–870

Neuromethods (2012) 67: 459–485
DOI 10.1007/7657_2011_16

Published online: 9 December 2011

Biophysically Principled Computational Neural Network Modeling of Magneto-/Electro-Encephalography Measured Human Brain Oscillations

Stephanie R. Jones

Abstract

Brain rhythms are the most prominent signal measured noninvasively in humans with magneto-/electro-encephalography (MEG/EEG). MEG/EEG measured rhythms have been shown to be functionally relevant and signature changes are used as markers of disease states. Despite the importance of understanding the underlying neural mechanisms creating these rhythms, relatively little is known about their in vivo origin in humans. There are obvious challenges in linking the extracranially measured signals directly to neural activity with invasive studies in humans, and although animal models are well suited for such studies, the connection to human brain function under cognitively relevant tasks is often lacking. Biophysically principled computational neural modeling provides an attractive means to bridge this critical gap. Here, we describe a method for creating a computational neural model capturing the laminar structure of cortical columns and how this model can be used to make predictions on the cellular and circuit level mechanisms of brain oscillations measured with MEG/EEG. Specifically, we describe how the model can be used to simulate current dipole activity, the common macroscopic signal inferred from MEG/EEG data. We detail the development and application of the model to study the spontaneous somatosensory mu-rhythm, containing mu-alpha (7–14 Hz) and mu-beta (15–29 Hz) components. We describe a novel prediction on the neural origin on the mu-rhythm that accurately reproduces many characteristic features of MEG data and accounts for changes in the rhythm with attention, detection, and healthy aging. While the details of the model are specific to the somatosensory system, the model design and application are based on general principles of cortical circuitry and MEG/EEG physics, and are thus amenable to the study of rhythms in other frequency bands and sensory systems.

Key words: Computational modeling, MEG imaging, Alpha rhythm, Beta rhythm, Current dipole

1. Introduction

In the past decade, a vast body of research has been devoted to studying the connection between rhythmic brain activity, cognition, and neural pathologies. Magneto- and electro-encephalography (MEG/EEG) provide noninvasive millisecond temporal resolution of human neural information, and increasingly improved spatial information from multimodal imaging techniques

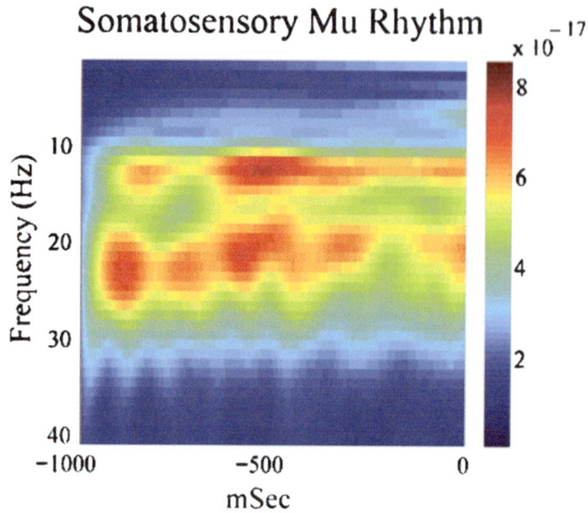

Fig. 1. Time–frequency representation of MEG measured somatosensory mu-rhythm, containing 10 Hz (mu-alpha) and 20 Hz (mu-beta) components, from an equivalent current dipole in the hand area of area 3b in SI (example subject, mean $N = 100$ trials). Am Physiol Soc, used with permission (48).

(combined MEG/EEG/MRI/fMRI). Oscillatory signals are the most prominent emergent behavior recorded with MEG/EEG (1, 2) (for example, see Fig. 1). However, despite the clear importance of these rhythms as a marker for changes in attention (3–5), perception (5–8), and disease states (9–13) our understanding of the underlying microscopic neural dynamics inducing them is limited, hindering the research and clinical use of these signals. Difficulty in filling this critical gap comes from the difficulty in obtaining cellular level neural information in humans. Animal studies are clearly useful for addressing these mechanisms (14, 15); however, information acquired in these preparations does not necessarily directly apply to humans, especially in tasks involving perception or other cognitive functions. A broadly recognized need in the field is for tools that allow us to transcend the simple observation of these rhythms, and understand more precisely how they reflect specific ongoing neural processes.

Computational neural network modeling is a powerful technique that can be used to make a direct connection between cellular-level neural dynamics and human brain rhythms, and their direct impact on cognitive process and disease. Further, predictions made by computational models can be used to focus and guide invasive experimental research. Neural network models of varying complexity have proven valuable to elucidating the mechanisms controlling brain oscillations. A paramount example is the role computational neural modeling has played in identifying that the time constant of GABAA inhibition as an essential factor in generating gamma frequency brain rhythms (40–80 Hz), and in synchronizing local gamma oscillations

(16–20). Computational modeling predictions on the interplay between excitation and inhibition on gamma rhythmicity have been verified experimentally in animal models in vitro (16) and recently in vivo, using advances in optogenetic recording techniques (21).

The computational complexity of neural network models can quickly become cumbersome, and it is often difficult to determine the level of detail needed to accurately study the data at hand. Minimally constructed neural models that retain salient features of the biology are often employed and have proven successful. The level of detail needs to be tailored to the question of interest and type of data the model reproduces (e.g., single unit spiking, local-field potential (LFP), MEG/EEG). Neural network models can be broadly categorized into three scales. (1) Models that lump parameters assumed to be proportional to some superposition of the model neurons' membrane potential that contribute to the mean field activity across the population. Such models include population firing rate models (22, 23) and "neural mass models" (24–27). Neural mass models are often used to study MEG/EEG data and have been applied, for example, to study sensory evoked response waveforms and to estimate cross-brain connectivity in time and frequency domains (28–30). However, these models do not take into account the details of spiking activity and their simplification makes assumptions about the level of synchrony in the network and can mask the importance of specific intrinsic ionic currents and electric field properties of laminated cortical circuit dynamics important when studying current dipole activity (see Sect. 1). (2) Network models whose individual cells are modeled with low-dimensions including, but not limited too, "theta" neurons (31, 32) firing rate neurons (33), and single compartment point neuron with various intrinsic ionic current simulated with Hodgkin-Huxley type kinetics to reproduce spiking activities (16–18, 34, 35). Models of this type have been particularly useful for studying rhythmic activity that has been tested in vitro but can also misrepresent important electric field properties of extended neurons and tissue. (3) Large-scale models that contain individual cells with multiple cell compartments representing detailed dendritic morphology across different layers and sometimes multiple brain areas (36–39). While the parameter space and computational requirements of these models are cumbersome, this level of detail may ultimately be necessary to study complex brain rhythms in vivo in humans. Specifically, to simulate cortical activity with units directly comparable to those estimated as current dipoles from MEG/EEG data [i.e., Ampere-meters (Am)], the network model should minimally include the morphology and physiology of large pyramidal neurons that span multiple cortical layers, as the postsynaptic intracellular current flow in these neurons is the primary generator of a current dipole signals (see Sect. 1). This is the category of model that is described here (Fig. 2).

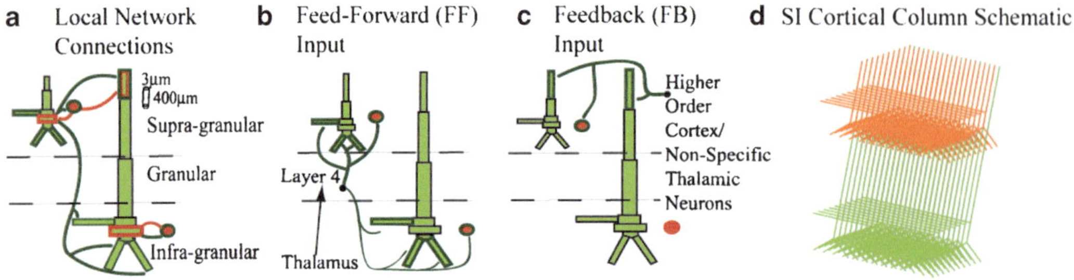

Fig. 2. Schematic of computational neural model of a cortical column in SI. (**a**) Individual cell morphology (PNs: *green*, INs: *red*) and local network connectivity (excitatory *green*, inhibitory *red*; targeted dendrites outlined, connections are schematic only as axons were not explicitly model). (**b**, **c**) Feed-forward (FF) and Feedback (FB) synaptic connection patterns, respectively. (**d**) Schematic of full column network containing 100 PNs and 30 INs in the supra- and infragranular layers. Am Physiol Soc, used with permission (48).

1.1. Modeling Macroscopically Measured Human Brain Oscillations Based on MEG/EEG Current Dipole Sources

MEG and EEG measure magnetic and electric field activity, respectively, outside of the head. The field activity is generated by cortical neurons that create electrical current sources in the brain. Since the neurons are at most a few millimeters long and the sensors are outside of the scalp at a distance of a few centimeters from the cortex, the electrical currents within an activated cortical region can be well approximated by a point or "primary" current sources, called the equivalent current dipole (Fig. 3). The main contributions to active "primary" currents are generated by spatially aligned long pyramidal neurons, which also drive ohmically conducted "return" currents through the conducting volume, i.e., the head. Calculating the activity of the "primary" current sources based on the recorded data and known features of the conduction volume is a task generally referred to as the MEG/EEG inverse problem.

There are varied techniques to solve the inverse problem using MEG, EEG, or combined MEG/EEG data (40–43). Solving the inverse problem based on MEG data is somewhat simplified because the volume currents are fully determined by the primary current sources and the conductivity geometry (e.g., shape of the skull, often modeled as a sphere). EEG inverse solutions must also consider conductivity in the cerebral spinal fluid and scalp. Thus, the "basic element" of MEG source patterns can be taken to be a current dipole (40). Further, MEG will only record activity from current sources oriented tangential to the sensors outside of the head, whereas EEG measures a combination of activity from both radial and tangential sources. Here, we focus on modeling oscillations of current dipole activity based on MEG data. However, because similar biophysical current sources underlie both MEG and the EEG scalp potentials, if the primary current sources are estimated with EEG or combined MEG/EEG our overall strategy remains applicable.

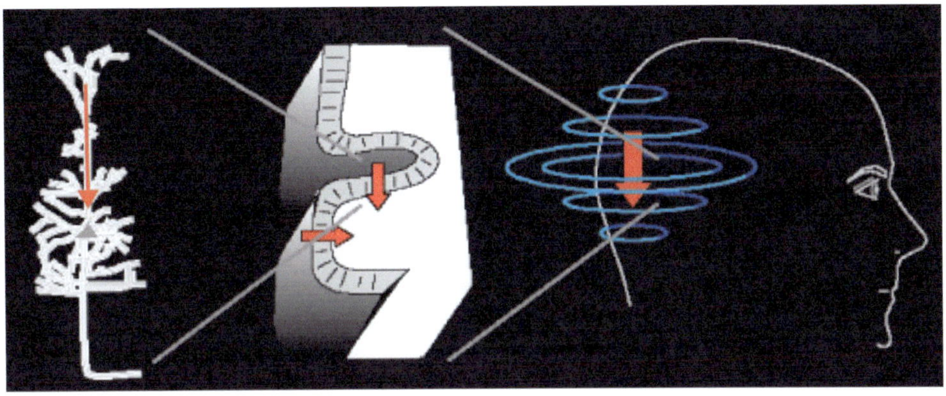

Fig. 3. Origin of MEG measured equivalent current dipole activity. The intracellular longitudinal current flow in large populations of synchronous pyramidal neurons, aligned in parallel and tangential to MEG sensors outside of the head, create a net primary current dipole (e.g., *red arrows* in sulcus in *middle panel*) large enough to be measured. Radially oriented dipoles (e.g., *red arrow* in gyrus in *middle panel*) will be visible with EEG but will not create a measurable magnetic field signal. Figure adapted with permission from original figure by Dr. Seppo Alphors.

Several lines of evidence suggest that post-synaptic intracellular longitudinal currents within the long apical dendrites of synchronized cortical pyramidal neurons are the main contributors to MEG primary current sources (14, 15, 40, 44–46). The length and parallel orientation of these dendrites create a net summed post-synaptic intracellular current flow across a large population that produces a perpendicular rotating magnetic field signal that is large enough to be recorded outside of the head with MEG/EEG (Fig. 3) (14, 40, 44, 45, 47). It has been estimated that areas as small as 10 mm^2 can produce currents measurable with MEG (14).

Based on this theory, current dipole activity, including rhythmic activity, can be simulated with biophysically principled models that include the morphology and physiology of cortical pyramidal neurons spanning multiple layers. The net current dipole is calculated from the postsynaptic intracellular currents in these neurons. This is a distinct advantage of models of current dipole activity compared to models of extracellular measures of neural activity, such as LFP or EEG sensor data. Accurately simulating such extracellular signals is an extremely complex and yet unsolved problem, and current methods typically make necessary assumptions and simplifications with success in reproducing signatures of the recorded signals (see Sect. 4).

Developing cortical models with laminar structure is a nontrivial task because of the complexity in neural circuitry and the large parameter space involved. Several groups have developed laminated cortical models and used them to investigate surface EEG or LFP oscillations (37–39). We have recently developed the first of its kind laminar cortical model of primary somatosensory cortex, based on a large literature of anatomy and physiology in rodents and mammals, designed specifically to reproduce current dipole activity

(Fig. 2) (36, 48, 49). This strategy enables us to make a direct comparison [same units of measure nano (n)-Am] between model activity and human MEG data. Our methods were derived from earlier models of current dipole activity based on hippocampal pyramidal neuron networks, and single pyramidal neuron models (44–46). These models accurately accounted for the magnetic field recorded outside of a hippocampal slice, and showed that active intrinsic cellular currents can be more important than synaptic currents in defining the shape and magnitude of the current dipoles; a finding that persisted in our analysis (36, 48).

1.2. Modeling Rhythmic Activity from a Current Dipole Source in Primary Somatosensory (SI) Neocortex

This chapter describes our methods to construct a laminar model of primary somatosensory cortex (SI) using the publicly available neural simulation software NEURON and how to use this model to calculate MEG measured rhythmic current dipole activity (36, 48, 49). In brief, the model contained pyramidal neurons (PNs) and inhibitory interneurons (INs) in the supragranular and infragranular layers (Fig. 2a). To maintain their morphology, the PNs were simulated with multiple dendritic compartments and the net current dipole activity was calculated from the intracellular current flow in a direction parallel to the apical dendrites across the PN populations. We note that the orientation of the PN in area 3b in SI, where our MEG signals were confirmed to be localized to, are such that the apical dendrites lie tangential to the MEG sensors (see (36, 48)). The INs were modeled with single compartments since their activity does not contribute directly to the current dipole signal (due to the radial orientation of their dendrites (40)) but is nonetheless important for the network dynamics (see Sect. 4 on more detailed INs). The granular layer, which does not contain long PNs, was not explicitly modeled but directly propagated output from simulated lemniscal thalamic-cortical activity. Exogenous inputs representing lemniscal thalamic "feedforward" (FF) excitatory synaptic drive to granular layers, and "feedback" (FB) excitatory drive from higher order cortex or nonspecific thalamic sources to supragranular layers were based on the laminar location of their postsynaptic effects (Fig. 2b, c) (50–52). The external driving inputs were simulated as spike trains with a predefined temporal distribution. The network was constructed into a grid of 100PN and 30IN in the supragranular and infragranular layers (Fig. 2d). The details of the model were based on a large body of literature of cortical somatosensory anatomy and physiology and most parameters were tuned and fixed based on reported data on connectivity and cell behavior, with only a few targeted parameters (namely, the timing and strength of exogenous excitatory drive to the network) varied to study the origin of specific features of MEG-measured neocortical rhythms.

We have employed this model to make detailed cellular and network level predictions on the neural origin of MEG-measured SI

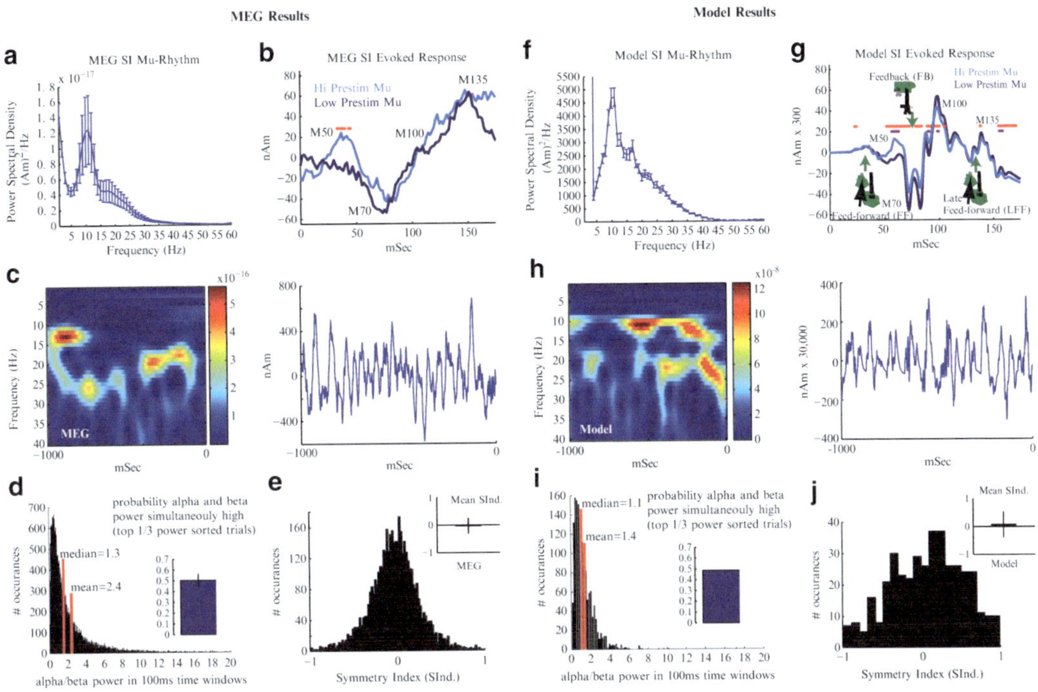

Fig. 4. Comparison of MEG and model results. (**a**) Power spectral density of spontaneous MEG SI current dipole mu-rhythm (mean and S.E. $N = 10$ subjects). (**b**) MEG SI current dipole-evoked response waveform from a brief perceptual-threshold level tap to the contralateral middle index finger-tip during states of high and low prestimulus mu power. During high mu there is a significant (*red asterisks*) increase in the magnitude of the M50 peak. (**c**) Example time-frequency representation (TFR; *left panel*) and corresponding waveform (*right panel*) over a 1-s epoch of the spontaneous mu-rhythm shows independent bouts of mu-alpha (7–14 Hz) and mu-beta (15–29 Hz) activity. (**d**) Histogram quantifying the ratio of mu-alpha to mu-beta power over short 100 ms time windows (grand average across $n = 10$ subjects). *Inset*: Probability that mu-alpha and mu-beta occur simultaneously in high power, which occurs 50% of the time on average. (**e**) Histogram showing the symmetry of the oscillation waveform around zero polarity. (**f–j**) Corresponding results in the model, where the mu-rhythm was created by two nearly simultaneous ~10 Hz inputs contacting the network with an FF and FB projection pattern (5 ms delay), see Fig. 5. In (**g**), *red asterisks* mark a significance difference at $p = 0.05$ and purple $p = 0.08$. There is remarkable consistency between the MEG data and model results for each quantified measure. Am Physiol Soc, adapted with permission (48).

spontaneous rhythmic activity, tactile-evoked responses and changes in these measures with perception, attention, and healthy aging (5, 36, 48, 49), see also Sect. 4. In each study, we have found remarkably close agreement between experimental and model results (e.g., see Fig. 4). Here, we detail the use of the model to study the cellular level neural origin of the commonly recorded somatosensory mu-rhythm.

1.3. Novel Predictions on the Origin of the Somatosensory Mu-Rhythm

The human somatosensory mu-rhythm is a two-component rhythm containing mu-alpha (7–14 Hz) and mu-beta (15–29 Hz) frequency bands (Figs. 1 and 4). In an effort to understand the origin of this rhythm, we have recently characterized the temporal overlap of mu-alpha and mu-beta in nonaveraged single trial data (48). We found that the mu-alpha and mu-beta components emerge separately on

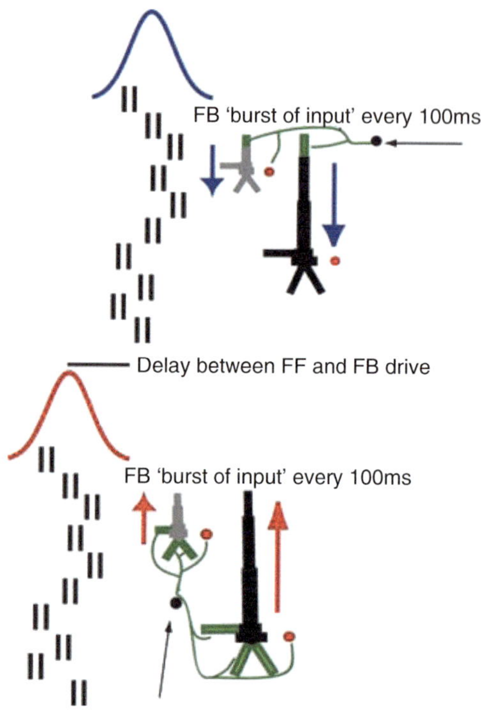

Fig. 5. Schematic of external drive to the network creating the somatosensory mu-rhythm. Every ~100 ms a "burst" of input provides excitatory drive to the SI network in an FF and FB connection pattern. Mu-beta emerges in the network in relevant proportions only when the mean delay between the two inputs in small (<5 ms). Am Physiol Soc, used with permission (48).

small time windows occurring simultaneously in high power in 100 ms epochs less than 50% of the time (Fig. 4c, d). This suggests that, while they share some common elements, they have separable mechanisms of generation.

In investigating the neural origin of the separable mu-alpha and mu-beta components, we first turned to previous theories of cortical alpha and beta activity. The long-standing view of the origin of cortical alpha activity is that it generated by ~10 Hz bursting activity in the thalamus that propagates to the cortex in a thalamocortical loop of activity (53). Other studies suggest that specific intrinsic currents in cortical PNs (h and T currents) are involved in 10-Hz rhythm generation or may help the circuit resonate at 10 Hz (18, 34, 54), and more recent work suggests that a specific class of interneuron producing slow inhibition are involved in alpha generation (20, 55) (see Sect. 4). Cortical beta activity has been thought to be dependent on a specific potassium current in PNs, namely an M current, in collaboration with local inhibition (17, 18, 56). However, this theory on beta emergence is based primarily on in vitro slice experiments combined with computational neural modeling and does not capture the complexity of in vivo human networks.

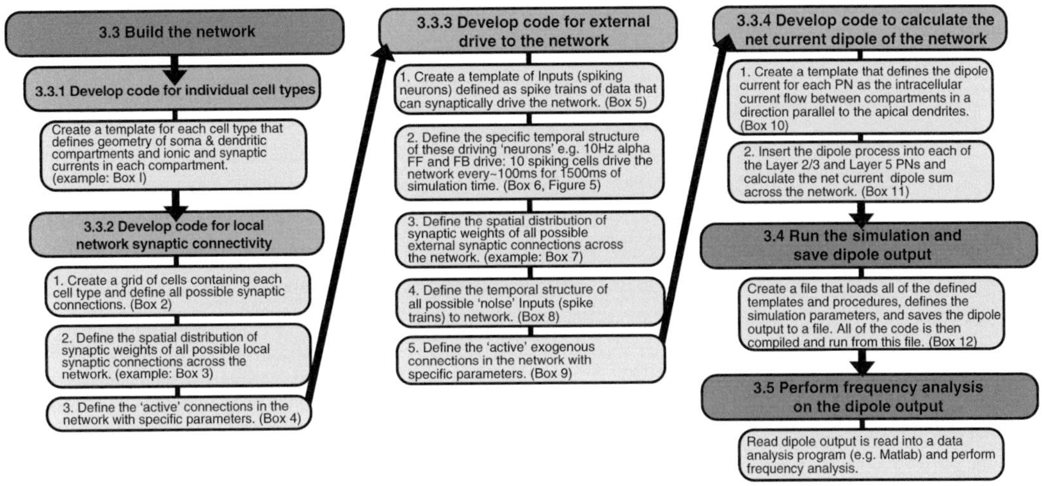

Fig. 6. Model blueprint outlining strategy for model development and analysis. Headings are numbered with chapter sections detailing each step.

Our model development and analysis of human MEG data has led to a new theory on the origin of cortical beta activity, and its relation to alpha activity in the generation of the somatosensory mu-rhythm. The model contained the essential elements based on prior theory to generate alpha and beta activity, including 10 Hz thalamocortical drive to granular layers, h-, T-, and M-currents in pyramidal neurons, and biophysically based local excitatory and inhibitory connections. However, these features alone were not enough to reproduce oscillatory current dipole patterns consistent with the MEG data. With 10 Hz lemniscal thalamocortical drive to the granular layers only, a predominantly 10 Hz rhythm emerged in the SI network (see Fig. 6b in Jones et al. (48)). Instead, we found that to reproduce the mu-beta component and to capture characteristic features of the two-component mu-rhythm, a second nearly simultaneous ~10 Hz drive to the supragranular layers was necessary (see Fig. 5 for schematic of exogenous drive). Based on the synaptic hierarchy of neural circuits, this input would represent the input that would arrive from higher order cortical areas or nonspecific thalamic sources that project directly to supragranular layers (52, 57). Importantly, the arrival times of the two 10 Hz inputs to the SI network was stochastic (input arrived with an expected wait time of 100 ms, with mean inter-spike-interval drawn from a Gaussian distribution) and mu-beta activity emerged in relevant proportions only when the mean difference in arrival time between the two inputs was small (<5 ms). The stochastic nature of the inputs across time created small bouts of alpha and beta activity as observed experimentally on single trials (compare Fig. 4c, left panel and Fig. 4d with Fig. 4h left panel and Fig. 4i). Further, the opposing dendritic location of the synaptic FF and FB connections drove alternating current flow up and down the pyramidal neuron dendrites creating a dipole waveform that oscillated symmetrically around zero, consistent with the experimental data (compare Fig. 4c, right panel and Fig. 4e with Fig. 4h, right panel and Fig. 4j). We note that model data was multiplied by a scaling factor representing the number of neurons contributing to the signal to simulate signals with the same order of magnitude as MEG data.

In addition to accurately reproducing many of the characteristic features of the mu-rhythm with two stochastic nearly simultaneous 10 Hz FF and FB drives, the model accurately reproduced differences in a tactile evoked response that were dependent on the power of the prestimulus rhythm. The model showed that mu-dependent changes in the ongoing level of depolarization of the PNs and INs induced the specific changes in peak responses observed in the MEG signal (compare Fig. 4b with Fig. 4g), see Jones et al. (36) for further detail on simulating tactile evoked responses. The model also made specific principled predictions on the origin changes in the mu-rhythm and tactile-evoked responses with healthy aging, and with functionally relevant cued-attention shifts. Specifically, the model predicted that the rhythmic FF and FB inputs become stronger with age (comparison of 22–32 years group with 33–45 years group), and that the FB inputs arrive more synchronously. These simulated neurophysiological changes were sufficient to account for MEG measured age-related differences in both prestimulus mu- and tactile-evoked responses (49). While the model has not been directly applied to simulate observed changes in MEG activity with cued-selective attention (e.g., cue to attend hand vs. attend foot), inferences made with the model suggested that observed focal decreases in mu-alpha in the somatosensory representation that attention was cued to with cued attention were mediated by changes in thalamic circuits providing FF alpha drive (5). Further, changes in the tactile-evoked response with cued attention could be accounted for by specific changes in the prestimulus mu-rhythm, as Jones et al. (48).

1.4. Model Validation and Future Directions

While there is remarkably close agreement between model and experimental results in several domains, our theory on the origin of the somatosensory mu-rhythm, and in particular human neocortical beta activity, has yet to be tested with intracranial recordings. Intracranial recordings in humans are possible in clinical settings, but present obvious difficult challenges. Since the general features of cortical physiology are preserved across species, testing in animal models is attractive. Further, recent advances in optogenetic techniques enable targeted stimulation of specific cell populations creating an ideal method for testing model predictions (21, 58). We have begun preliminary testing of our predictions on the origin of somatosensory mu-rhythms in rodent preparations using optogenetic stimulation of somatosensory projecting thalamic cortices with simultaneous SI recordings, and preliminary findings supports our hypothesis (59).

Future directions in the modeling of macroscopic human brain signals will include (1) expansion to a larger cortical networks (38), (2) incorporation of greater cellular-level details in thalamic areas (37), and the development of synaptic plasticity mechanisms (20, 35, 60) (see Sect. 4). As the size of the network grows, the simulation speeds increase and conversion to NEURON parallel computing processing capabilities will be beneficial (61). Our model development was based primarily on generalizable features of cortical physiology. As such, the model is amenable to study rhythmic current dipole activity in other frequency bands and other sensory systems with minimal adaptation. Ultimately, biophysically principled neural modeling provides a powerful tool to study the cellular level origin of rhythmic brain activity in healthy humans and maladaptive changes in these rhythms with disease.

2. Materials

2.1. Computer Hardware and Software

1. PC, Macintosh or Unix computer.
2. NEURON software shareware available for download at http://www.neuron.yale.edu/neuron.
3. Text editing software (e.g., Microsoft Word) and data analysis software (e.g., Matlab).

2.2. Model Blueprint

1. See Section 3.2 and Figure 6.

2.3. NEURON Code to Build the Cortical Network, See Section 3.3

1. See Section 3.3.

2.4. NEURON Code for Running the Simulation and Saving the Current Dipole Output

1. See Section 3.4

2.5. Methods for Frequency Analysis on Current Dipole Output

1. See Section 3.5

3. Methods

3.1. Computer Hardware and Software

The results described in this chapter were generated on a MacBook Pro running Mac OS X 10.5.8 with a 2.53 GHz Intel Core

2 processor running the neural simulation software NEURON, which is publicly available shareware. All code was developed in Microsoft Word and postprocessing frequency analysis was performed with Matlab software. The modeling strategy described is adaptable to other platforms running NEURON or to other programming languages (e.g., C programming) designed with differential equation solvers.

3.2. Model Blueprint

When developing a large and detailed computational neural model, it is useful to first develop a model blueprint that outlines the strategy for model development. Here, we describe our strategy to build a biophysically principled SI cortical column network model and to study rhythmic current dipole activity produced by the network. Figure 6 depicts the blueprint for our model strategy outlining the steps to build and analyze the network. We describe each of these steps in detail in the next subsections. The coding and analysis of the modeling described relies on a basic understanding of the NEURON simulation software environment (62), but the ideas are generalizable and can be adapted to other simulation techniques. We provide details for a subset of NEURON procedures within the complete code. The complete simulations require thousands of lines of code and are beyond the scope of this chapter, but our original model is freely available on the model DB Web site at http://senselab.med.yale.edu/modeldb/ShowMo del.asp?model=113732.

3.3. NEURON Code to Build the Cortical Network

3.3.1. Code for Individual Cell Types

Each cell type in the network was developed with a NEURON template. The template included details of individual cell geometry (i.e., compartmentalization), biophysics, and all postsynaptic processes (e.g., AMPA, NMDA, GABAA, GABAB) in each compartment. Arrays of cells of each neuron type were then made from the templates and synaptically connected using a separate procedure (see Sect. 3.2). Box 1 gives example code for the Layer II/III PN template (geometry shown in Fig. 2). The Layer V PN and inhibitory IN were similarly constructed. The parameters within the template were tuned so the individual cells produced spiking activity to injected current analogous to that observed experimentally based on literature and then fixed, see examples in Fig. 2b in Jones et al. (36).

3.3.2. Code for Network Connectivity

There are three main steps to constructing a network out of the individual neurons. Step 1—Box 2: Create a grid of cells and a procedure ("rconnect") that defines all possible synaptic connections from each cell types to each postsynaptic dendritic compartments, see Box 2. The synapses are attached to each dendritic compartment in the individual cell templates (Box 1) and the dynamics of the synapse types are in their own files (known as .mod files) whose dynamics can be inserted into any compartment and are compiled separately.

Box 1
NEURON template defining initialization, geometry, biophysics, and synaptic processes of Layer II/III pyramidal neurons. Similar templates define other cell types. Separate .mod files used are listed in Table 1

```
begintemplate Layer2_pyr

public init, topol, basic_shape, subsets, geom, biophys, synlist, x, y, z,
position, connect2target, ampa, nmda, gabaa, gabab, soma, dend, all,
somatic, dendritic, apical, basal, dend0, dend1,dend2, dend3, dend4, dend5,
dend6 ,pre, ampa, gabaa, gabab, nmda, gbar_ar, gbar_cat

objref synlist

proc init() {
 topol()
 subsets()
 geom()
 biophys()
 geom_nseg()
 synlist = new List()
 synapses()
 x = y = z = 0 }

create soma, dend[7]

proc topol() { local i
 connect dend(0), soma(1)
 for i = 1, 2 connect dend[i](0), dend(1)
 connect dend[3](0), dend[2](1)
 connect dend[4](0), soma(0)
  for i = 5, 6 connect dend[i](0), dend[4](1)
 basic_shape()}

proc basic_shape() {
 soma {pt3dclear() pt3dadd(-50, 765, 0, 1) pt3dadd(-50, 778, 0, 1)}
 dend {pt3dclear() pt3dadd(-50, 778, 0, 1) pt3dadd(-50, 813, 0, 1)}
 dend[1] {pt3dclear() pt3dadd(-50, 813, 0, 1) pt3dadd(-250, 813, 0, 1)}
 dend[2] {pt3dclear() pt3dadd(-50, 813, 0, 1) pt3dadd(-50, 993, 0, 1)}
 dend[3] {pt3dclear() pt3dadd(-50, 993, 0, 1) pt3dadd(-50, 1133, 0, 1)}
 dend[4] {pt3dclear() pt3dadd(-50, 765, 0, 1) pt3dadd(-50, 715, 0, 1)}
 dend[5] {pt3dclear() pt3dadd(-50, 715, 0, 1) pt3dadd(-156, 609, 0, 1)}
 dend[6] {pt3dclear() pt3dadd(-50, 715, 0, 1) pt3dadd(56, 609, 0, 1)}}

proc position() { local i
 soma for i = 0, n3d()-1 {
  pt3dchange(i, $1-x+x3d(i), $2-y+y3d(i), $3-z+z3d(i), diam3d(i))}
 x = $1  y = $2  z = $3}

proc connect2target() {  soma $o2 = new NetCon(&v(1), $o1)}
//double back slashes indicate a comment
//this procedure take is called with two variables
//$o1= target point process, $o2 =returned NetCon

objref ampa[8],nmda[8],gabaa[8],gabab[8]
proc synapses() {
 dend[0]{ ampa[0] = new AMPA() nmda[0] = new NMDA() gabaa[0] =
new GABAA() gabab[0] = new GABAB() }
 dend[1]{ ampa[1] = new AMPA() nmda[1] = new NMDA() gabaa[1] =
new GABAA() gabab[1] = new GABAB() }
 dend[2]{ ampa[2] = new AMPA() nmda[2] = new NMDA() gabaa[2] =
new GABAA() gabab[2] = new GABAB() }
 dend[3]{ ampa[3] = new AMPA() nmda[3] = new NMDA() gabaa[3] =
new GABAA() gabab[3] = new GABAB() }
 dend[4]{ ampa[4] = new AMPA() nmda[4] = new NMDA() gabaa[4] =
new GABAA() gabab[4] = new GABAB() }
 dend[5]{ ampa[5] = new AMPA() nmda[5] = new NMDA() gabaa[5] =
new GABAA() gabab[5] = new GABAB() }
 dend[6]{ ampa[6] = new AMPA() nmda[6] = new NMDA() gabaa[6] =
new GABAA() gabab[6] = new GABAB() }
 soma    { ampa[7] = new AMPA() nmda[7] = new NMDA() gabaa[7] =
new GABAA() gabab[7] = new GABAB() }}
```

```
  dend[6].L = 255
  soma.diam = 23.4
  dend.diam = 4.25
  dend[1].diam = 3.91
  dend[2].diam = 4.08
  dend[3].diam = 3.4
  dend[4].diam = 4.25
  dend[5].diam = 2.72
  dend[6].diam = 2.72}

proc geom_nseg() {
 soma area(.5)
 forsec all { if (L < 50) {nseg=1}
 else {nseg=int(L/50)} }}

proc biophys() {
forsec all {
  Ra = 200
  cm = 0.6195 }
  forsec somatic {
        insert hh
        gnabar_hh=0.18

        gkbar_hh=0.01
        gl_hh=0.0000426
        el_hh=-65
        insert km
        gbar_km=250
        insert cat
        gbar_cat=0.0
        insert ar
        gbar_ar=0.0}
forsec dendritic {
        insert hh
        gnabar_hh=0.15
        gkbar_hh=0.01
        gl_hh=0.0000426
        el_hh=-65
        insert km
        gbar_km=250
        insert cat
        gbar_cat=0.0
        insert ar
        gbar_ar=0.0}}

proc subsets() { local i
 objref all, somatic, dendritic, apical,
basal, dend0, dend1,  dend2, dend3,
dend4, dend5, dend6
 all = new SectionList()
  soma all.append()
  for i=0, 6 dend[i] all.append()
 somatic=new SectionList()
  soma somatic.append()
 dendritic=new SectionList()
  for i=0, 6 dend[i] dendritic.append()
 apical = new SectionList()
  for i=0, 3 dend[i] apical.append()
 basal = new SectionList()
  for i=4, 6 dend[i] basal.append()
 dend0 = new SectionList()
  dend dend0.append()
 dend1 = new SectionList()
  dend[1] dend1.append()
 dend2 = new SectionList()
  dend[2] dend2.append()
 dend3 = new SectionList()
```

(continued)

Box 1 (continued)

```
proc geom() {
  forsec all { }
   soma.L = 22.1
   dend.L = 59.5
   dend[1].L = 340
   dend[2].L = 306
   dend[3].L = 238
   dend[4].L = 85
   dend[5].L = 2

   dend[3] dend3.append()
  dend4 = new SectionList()
   dend[4] dend4.append()
  dend5 = new SectionList()
   dend[5] dend5.append()
  dend6 = new SectionList()
   dend[6] dend6.append()}

func is_art() { return 0 }

endtemplate Layer2_pyr
```

Box 2 NEURON code creating a 2-D grid of neurons and procedure to synaptically connect all possible cells and dendritic compartments

```
// Create a 10X10 2-D grid of PNs with interleaved INs
XD = 10 - 1
YD = 10 - 1
objref PL5[X_DIM][Y_DIM], PL2[X_DIM][Y_DIM],
IPL2[X_DIM][Y_DIM], IPL5[X_DIM][Y_DIM]

for i=0,XD{
for j=0,YD{
PL5[i][j] = new Layer5_pyr()
PL5[i][j].position(i*100,0,j*100)
PL2[i][j] = new Layer2_pyr()
PL2[i][j].position(i*100,535,j*100)} }

proc create_inhib(){local x, y
for y=0,YD{
for x=$1,XD{
IPL2[x][y] = new Inhib()
IPL5[x][y] = new Inhib()
x=x+$2 }}}

proc I_zig_zag(){local x, y, offset
// Creates a sparse matrix of inhibitory cells staggered
// so that the 1st element along the X-axis
// begins alternatly at position 1 or 0.
// Usage: I_zig_zag(space)
// space= number of blank X axis positions between cells
offset = 0
for y=0,YD{
for x=offset,XD{
IPL2[x][y] = new Inhib()
IPL5[x][y] = new Inhib()
x=x+$1 }
if (offset==0) {offset=1} else {offset=0}}}

I_zig_zag(2)

// Create synaptic connections in network
objref AMPAconnects, NMDAconnects, GABAAconnects,
GABABconnects, netcon

AMPAconnects = new List()
NMDAconnects = new List()
GABAAconnects = new List()
GABABconnects = new List()

proc rconnect(){
// usage: rconnect(source-cell, target-cell, target-section, rtype,
// weight, delay, threshold )
// rtype= 0=AMPA ||| 1=NMDA ||| 2=GABAa ||| 3= GABAb

if ( $4 == 0){ $o1.connect2target($o2.ampa[$3].s, netcon)
netcon.weight = $5 netcon.delay = $6 netcon.threshold = $7
 AMPAconnects.append(netcon) }

if ( $4 == 1){ $o1.connect2target($o2.nmda[$3].s, netcon)
netcon.weight = $5 netcon.delay = $6 netcon.threshold = $7
 NMDAconnects.append(netcon) }

if ( $4 == 2){ $o1.connect2target($o2.gabaa[$3].s, netcon)
netcon.weight = $5 netcon.delay = $6 netcon.threshold = $7
 GABAAconnects.append(netcon) }

if ( $4 == 3){ $o1.connect2target($o2.gabab[$3].s, netcon)
netcon.weight = $5 netcon.delay = $6 netcon.threshold = $7
 GABABconnects.append(netcon) }

if (( $4 > 3) || ($3 < 0)){ print "Receptor Type not supported" }}
```

Table 1
List of all .mod used in the model and employed in the example code given in Boxes, detailed code available at http://senselab.med.yale.edu/modeldb/ShowModel.asp?model=113732

ar.mod	kca.mod	pp_dipole.mod
ca.mod	kdr.mod	precall.mod
cad.mod	km.mod	seed.mod
cat.mod	na.mod	tchannel.mod
dipole.mod (see Box 10)		

For example, NEURON comes with a built in exponential decay synapse *expsyn.mod*, whose postsynaptic activation depends on the voltage in a defined compartment (e.g., soma) of a presynaptic cell (code not shown, see (62)). Please see Table 1 for a list of all .mod files used in our simulation and referenced in code in Boxes that are not automatically built into the NEURON simulation environment and are available for download at http://senselab.med.yale.edu/mod eldb/ShowModel.asp?model=113732. Step 2—Box 3: Define the spatial distribution of synaptic weights of all possible local synaptic connections across the network. Box 3 shows the procedure used for defining the spatial spread of all connections from Layer V PNs to Layer II/III PNs, which had a Gaussian maximal conductance profile. Similar profiles and procedures were used for all connection patterns. See Sect. 4 for a description of an alternate random connection pattern using a uniform distribution. Step 3—Box 4: Define the local connections that are turned on or "active." This was done by calling the procedures, as in Box 3, with specific parameter values in a separate NEURON file (known as a .hoc file) that is loaded when running the simulation (see Sect. 4). The procedures called for the local connection pattern used here (schematically illustrated in Fig. 2a) are listed in Box 4. The parameters used where tuned based on literature and to maintain low spontaneous firing rates with a baseline level of "noise" (see Sect. 3.3) and then fixed for all simulations.

3.3.3. Code for External Drive to Network

As described in Sect. 3, the MEG somatosensory mu-rhythm was reproduced in the model by driving the SI network with two stochastic nearly simultaneous ~10 Hz FF and FB inputs (shown schematically in Fig. 5) from exogenous brain areas (i.e., lemniscal thalamus and higher cortical areas or nonspecific thalamic sources, respectively). These external drives to local SI cortical network were not modeled with individual cell templates as in Box 1 but rather

Box 3
NEURON procedure defining Gaussian spatial distribution of synaptic weights between Layer V and Layer II/III PNs in the local SI network. Similar procedures define weights of other local connections

```
proc VtoII(){ local weight, delay, distX, distY, distance, preX, preY, postX, postY
// Usage: VtoII($1=Max_Weight, $2=Space constant, $3=post-synaptic compartment [0-7],
//$4=receptor type [0=ampa, 1=nmda], $5=minimum_delay, $6=delay_space_constant )

for preX=0,XD{
 for preY=0,YD{
  for postX=0,XD{
   for postY=0,YD{
    distX = abs(preX-postX)
    distY = abs(preY-postY)
    distance = sqrt(distX^2 + distY^2)
    weight = $1 * exp( -(distance^2) / ( $2^2) )
    delay = $5 * 1 / (exp( -(distance^2) / ( $6^2) ) )
    rconnect(PL5[preX][preY], PL2[postX][postY], $3, $4, weight, delay, 0) } } } } }
```

Box 4
NEURON code defining "active" local synaptic connections by calling procedures as in Box 3 with specific parameter values

```
//Each excitatory synaptic connection procedure
//called is created analogous to the example in Box 3

// excitatory to excitatory connections
IItoII(0.0005,3,1,0,1,3)
IItoII(0.0005,3,5,0,1,3)
IItoII(0.0005,3,6,0,1,3)
IItoII(0.0005,3,1,1,1,3)
IItoII(0.0005,3,5,1,1,3)
IItoII(0.0005,3,6,1,1,3)
IItoV(0.00025,3,1,0,3,3)
IItoV(0.00025,3,6,0,3,3)
IItoV(0.00025,3,7,0,3,3)
IItoV(0.00025,3,4,0,3,3)
VtoV(0.0005,3,6,0,1,3)
VtoV(0.0005,3,7,0,1,3)
VtoV(0.0005,3,1,0,1,3)
VtoV(0.0005,3,6,1,1,3)
VtoV(0.0005,3,7,1,1,3)
VtoV(0.0005,3,1,1,1,3)
```

```
//excitatory to inhibitory connections
IItoIL2(0.0005,3,1,3,0)
IItoIL5(0.00025,3,1,3,0)
VtoIL5(0.0005,3,1,3,0)

//The procedures for inhibitory synaptic connections
//to excitatory cells have the form
// XtoX(seg,rtype,maxW,WSC,minD,DSC)
IL2toII(7,2,0.05,50,1,50)
IL2toV(4,2,0.001,50,1,50)
IL5toV(8,2,0.025,70,1,70)
IL5toV(8,3,0.025,70,1,70)

//The procedures for inhibitory synaptic connections
//to inhibitory cells have the form
//XtoX(maxW,WSC,minD,DSC,rtype)
IL2toII(7,3,0.05,50,1,50) // "
IL2toIL2(0.02,20,1,20,2) //
IL5toIL5(0.02,20,1,20,2)
```

with a predefined spike pattern of output that directly activated SI synapses at defined postsynaptic dendritic locations (see Fig. 2b, c and Fig. 5). There are four steps to incorporating this external drive. Step 1—Box 5: Create a spike train for each of the external driving "neurons." In NEURON, predefined spiking "cells" (i.e.,

Box 5
NEURON template defining pre-synaptic external drives to local SI network as spike trains using the NEURON Netstim function and connects it as a Netcon to postsynaptic neurons using the procedure connect2target

```
// To create an artificial NetSTtim cell the following code is used
create acell_home_
access acell_home_

L=1
diam=1

//Below is the template for each external FF and FB drive
begintemplate FeedX

public pp, connect2target, x, y, z, is_art
external acell_home_
objref pp

proc init() {
  acell_home_ pp = new NetStim(.5)
    pp.start = 50
    pp.number = 0
    pp.interval = 10
    pp.noise = 0}

func is_art() { return 1 }

proc connect2target() { $o2 = new NetCon(pp, $o1) }

endtemplate FeedX
```

spike trains) can be coupled into a network using the built in NETSTIM function that can be defined as a template. For example, Box 5 shows a template that defines a NETSTIM function *FeedX*, which creates a poison process spike train with default parameter values, to be called when the network is driven externally. Step 2—Box 6: Define the specific external drives. For the simulation described here, we drove the network with an ~10 Hz (alpha frequency) "bursty" spike patterns that contacted the network in an FF and FB connection pattern (Fig. 5). This pattern was created by defining 150 presynaptic NETSTIM neuron objects referred to as *AlphaFF* and *AlphaFB*, as in Box 6. Ten each of these *AlphaFF/FB* inputs drove the network every ~100 ms for a total simulation time of 1,500 ms (10 × 15 = 150 *AlphaFF/FB* objects). The specific parameters for the *AlphaFF/FB* poison processes were defined in a procedure called *make_MuBursts* (Box 6), where each of the 150 NETSTIMS produced 2 spikes with an ISI of 10 ms (Fig. 5). Stochastic processes in the drive were set by the mean timing of each of the ten drives (every ~100 ms), which was

Box 6

NEURON code defining the temporal structure of the external drives to SI as ~10 Hz alpha frequency FF and FB input (see schematic in Fig. 5). Ten "bursting" cells (2-spike bursts with 10 ms ISI) drive the SI network every ~100 ms in an FF and FB connection pattern for 1,500 ms. The mean of the presynaptic spike times is chosen from a Gaussian distribution with variance 400 ms

```
objref AlphaFF[150], AlphaFB[150], GaussRand
GaussRand=new Random()

proc make_MuBursts(){local i
// Feed-Forward Alpha Frequency Drive
for i=0,149{
AlphaFF[i] = new FeedX() //see Box 5
AlphaFF[i].pp.noise=0
AlphaFF[i].pp.number=2
AlphaFF[i].pp.interval = 10
AlphaFF[i].pp.start = 0}

for i=0,9{AlphaFF[i].pp.number = 0}
// inputs start at 150ms to allow network to settle to steady state
for i=10,19{AlphaFF[i].pp.start=GaussRand.normal(150, 400)}
for i=20,29{AlphaFF[i].pp.start= GaussRand.normal(250, 400)}
for i=30,39{AlphaFF[i].pp.start= GaussRand.normal(350, 400)}
for i=40,49{AlphaFF[i].pp.start= GaussRand.normal(450, 400)}
for i=50,59{AlphaFF[i].pp.start= GaussRand.normal(550 , 400)}
for i=60,69{AlphaFF[i].pp.start= GaussRand.normal(650, 400)}
for i=70,79{AlphaFF[i].pp.start= GaussRand.normal(750, 400)}
for i=80,89{AlphaFF[i].pp.start= GaussRand.normal(850, 400)}
for i=90,99{AlphaFF[i].pp.start= GaussRand.normal(950, 400)}
for i=100,109{AlphaFF[i].pp.start= GaussRand.normal(1050,
400)}
for i=110,119{AlphaFF[i].pp.start= GaussRand.normal(1150,
400)}
for i=120,129{AlphaFF[i].pp.start= GaussRand.normal(1250,
400)}
for i=130,139{AlphaFF[i].pp.start= GaussRand.normal(1350,
400)}
for i=140,149{AlphaFF[i].pp.start= GaussRand.normal(1450,
400)}

// FeedBack Alpha Frequency Drive
for i=0,149{
AlphaFF[i] = new FeedX()
AlphaFF[i].pp.noise=0
AlphaFF[i].pp.number=2
AlphaFF[i].pp.interval = 10
AlphaFF[i].pp.start = 0}

for i=0,9{AlphaFB[i].pp.number = 0}
// inputs start at 150ms to allow network to settle to steady state
for i=10,19{AlphaFB[i].pp.start=GaussRand.normal(150, 400)}
for i=20,29{AlphaFB[i].pp.start= GaussRand.normal(250, 400)}
for i=30,39{AlphaFB[i].pp.start= GaussRand.normal(350, 400)}
for i=40,49{AlphaFB[i].pp.start= GaussRand.normal(450, 400)}
for i=50,59{AlphaFB[i].pp.start= GaussRand.normal(550 , 400)}
for i=60,69{AlphaFB[i].pp.start= GaussRand.normal(650, 400)}
for i=70,79{AlphaFB[i].pp.start= GaussRand.normal(750, 400)}
for i=80,89{AlphaFB[i].pp.start= GaussRand.normal(850, 400)}
for i=90,99{AlphaFB[i].pp.start= GaussRand.normal(950, 400)}
for i=100,109{AlphaFB[i].pp.start= GaussRand.normal(1050,
400)}
for i=110,119{AlphaFB[i].pp.start= GaussRand.normal(1150,
400)}
for i=120,129{AlphaFB[i].pp.start= GaussRand.normal(1250,
400)}
for i=130,139{AlphaFB[i].pp.start= GaussRand.normal(1350,
400)}
for i=140,149{AlphaFB[i].pp.start= GaussRand.normal(1450,
400)}}
```

Box 7

NEURON procedure defining Gaussian spatial distribution of synaptic weights for external FF and FB inputs to Layer V PNs. Similar procedures define weights of external drives to other SI neurons

```
proc FeedtoV(){ local weight,delay, distX, distY, distance, preX, preY, postX, postY
// Usage: FeedtoV( L.5 Pyram_compartment [0-8], receptor type
// [ 2=GABAa, 3-GABAb ], max_weight, weight_space_constant [>1],
//  min-delay [ms.] , delay-space-constant [>1], FeedX object(eg. FF) )

preX=4 //coordinate for all feeds is grid middle
preY=4
 for postX=0,XD{
  for postY=0,YD{
   distX = abs(preX-postX)
   distY = abs(preY-postY)
   distance = sqrt(distX^2 + distY^2)
   weight = $3 * exp( -(distance^2) / ( $4^2) )
   delay = $5 * 1 / (exp( -(distance^2) / ( $6^2) ) )
   rconnect($o7, PL5[postX][postY], $1, $2, weight, delay, 0) } } }
```

Box 8
NEURON template defining noise process in any compartment as a randomly active current clamp with mean magnitude chosen from a uniform random process. The procedure UnoiseV inserts the noise process into all compartments of the Layer V PNs. Similar procedures define noise process in other cell types

```
begintemplate UNOISE

public setmean
objref r, pp

proc init() {
pp=new IClamp(0.5)
pp.dur=5000
r=new Random()
r.MCellRan4()}

proc setmean() {
r.uniform($1,$2)
r.play(&pp.amp)}

endtemplate UNOISE

objref UnoisyV[X_DIM * Y_DIM * 10]
proc UnoiseV(){
// Inserts uniform noise into all compartments of the Pyramidal V cells
j=0
for postX=0,XD{
 for postY=0,YD{ forsec PL5[postX][postY].all
 {UnoisyV[j]=newUNOISE() UnoisyV[j].setmean($1,$2) j=j+1 }}}}
```

Gaussian normal randomized on every cycle with variance set at 400 ms. Step 3—Box 7: Define the spatial distribution of synaptic weights from all possible FF and FB inputs to the network. Box 7 shows an example procedure for inputs to the Layer V pyramidal neurons, external drive to the other cells were defined with a similar procedure. Step 4—Box 8: Define the temporal structure of "noise" input into the network. This was done by creating a uniform noise, *UNOISE*, template that defines the noise as a stochastic current clamp in the middle of each compartment of each cell that was turned on for the duration of the simulation and whose strength was chosen from a uniform random distribution at every point in time. The max and min of this distribution were set during the call to a procedure, *UnoiseX*, defined for each cell type. Box 8 shows an example noise procedure for the Layer V PNs, *UnoiseV*, the procedures for the other cell types were analogous. Step 5—Box 9: Define the external driving inputs that are turned on or "active." This was done by calling each procedure with specific parameters in a separate .hoc file that was loaded when

Box 9
NEURON code defining "active" external FF and FB synaptic connections to SI by calling procedures as in Boxes 7 and 9 with specific parameter values

```
//Each procedure called is created analogous to the examples in
//Box 7 and 8

//10 Hz FF and FB Drive
for i=1, 150{
 weight=0.00004 //pSiemens
 fbdelay=5/ /FB delay after FF input (ms)

//Feedforward connections to Layer II
FeedtoII(1,0, weight,100,0,100,AlphaFF[i])
FeedtoII(5,0, weight,100,0,100,AlphaFF[i])
FeedtoII(6,0, weight, 100,0,100,AlphaFF[i])
FeedtoIL2(0, weight *2, 100,0,100,AlphaFF[i])

//Feedforward connections to Layer V
FeedtoV(1,0, weight, 100,1,100,AlphaFF[i])
FeedtoV(6,0, weight, 100,1,100,AlphaFF[i])
FeedtoV(7,0, weight, 100,1,100,AlphaFF[i])
FeedtoIL5(0, weight *2,100,1,100,AlphaFF[i])

//Feedback connections to Layer II and V
FeedtoII(3,0, weight,100,fbdelay,100,AlphaFB[i])
FeedtoIL2(0, weight, 100, fbdelay,100,AlphaFB[i])
FeedtoV(4,0, weight, 100, fbdelay,100,AlphaFB[i]) }

//Noise Drive
UnoiseV(-0.3,0.3)
UnoiseII(-0.3,0.3)
UnoiseIPL5(-0.3,0.3)
UnoiseIPL2(-0.3,0.3)
```

running the simulation (see Sect. 4). The specific exogenous drive used for the results in Fig. 4a, h, i, and j are listed in Box 9. We note that parameter choices for the external drive described here were tuned to create dipole data in close agreement with the observed MEG results. For changes in results with variation in parameters, see Fig. 8 in Jones et al. (48).

3.3.4. Code for Calculating the Net Current Dipole of the Network

The net current dipole produced by the network is calculated from the intracellular current flow across the Layer V and Layer II/III PN populations in the direction parallel to the apical dendrites (see Sect. 1). This is done in two steps. Step 1—Box 10: Create a template for calculating the dipole current of each cell, *DipoleForCell*. Box 10 shows the template used in our simulations that created the initialization of the dipole for each cell compartment in procedure *init*, set up the dipole across each compartment in a procedure

Box 10
NEURON template and mod file defining the net current dipole produced by a cell as the intracellular current in a direction parallel with the apical dendrite

```
begintemplate DipoleForCell

public bd, seclist, rotmat
objref bd,  seclist, point

proc init() {
        point = new Vector(3)
        seclist = new SectionList()
        seclist.wholetree //each compartment in the cell
        bd=new List()
        forsec seclist {
                insert dipole
                bd.append(new Dipole(1))}
        setup_dipole()
        fill_ztan() }

proc setup_dipole() { local i, j, x, xold  localobj d
        j=0
        forsec seclist {
                d= bd.object(j)
                j += 1
                d.ri=ri(1)
                setpointer d.pv,v(0.999)
                i += 1
                xold=0
                for(x, 0) {
                        ri_dipole(x)=ri(x)
                        setpointer pv_dipole(x),v(xold)
                        setpointer Qsum_dipole(x), d.Qsum
                        xold=x}}}

proc fill_ztan() {local i, j, x  localobj xvec, z3dvec, a3dvec, d
        j=0
        forsec seclist {
                xvec = new Vector(nseg+2)
                z3dvec = new Vector(n3d())
                a3dvec = new Vector(n3d())
                i = 0
                for(x) { xvec.x[i] = x*L
                        i += 1 }
                for i=0, n3d()-1 {
                        point.x[0] = x3d(i)
                        point.x[1] = y3d(i)
                        point.x[2] = z3d(i))
                        z3dvec.x[i] = point.x[1]
                        a3dvec.x[i] = arc3d(i) }
                z3dvec.interpolate(xvec, a3dvec)
                z3dvec.deriv(1,1)
                i = 0
                for (x, 0) { ztan_dipole(x) = z3dvec.x[i]  i += 1 }
                d= bd.object(j)
                j += 1
        d.ztan = z3dvec.x[i]} }
endtemplate DipoleForCell
```

```
//dipole.mod
NEURON {
SUFFIX dipole
RANGE ri, ia, Q, ztan
POINTER pv
POINTER Qsum

UNITS {
(nA) = (nanoamp)
(mV) =(millivolt)
(Mohm) = (megaohm)
(um) = (micrometer)
(Am)= (amp meter)
(fAm) = (femto amp meter)}

ASSIGNED {
ia (nA)
ri (Mohm)
pv (mV)
v (mV)
ztan (um)
Q  (fAm)
Qsum (fAm) }

AFTER SOLVE {
        ia=(pv-v)/ri
        Q=ia*ztan
        Qsum = Qsum + Q}

AFTER INITIAL {
        ia=(pv-v)/ri
        Q=ia*ztan
        Qsum = Qsum + Q}
```

Box 11
NEURON code inserting the dipole process into each of the Layer V and Layer II/III pyramidal neurons and calculating the net dipole sum across the separate populations

```
objref dipoleL5, dipoleL2
dipoleL5 = new List()
dipoleL2 = new List()

for i=0,XD{
for j=0,YD{
 PL5[i][j].soma dipoleL5.append(new DipoleForCell() )
 PL2[i][j].soma dipoleL2.append(new DipoleForCell() )} }

dL5=0
func L5_dipole(){ local dp
dp=0
for i=0,XD{
for j=0,YD{
forsec PL5[i][j].all {dp += Qsum_dipole} } }
dL5 = dp
return dp}
```

```
dL2=0
func L2_dipole(){ local dp
dp=0
for i=0,XD{
for j=0,YD{
forsec PL2[i][j].all {
dp += Qsum_dipole} } }
dL2=dp
return dp}

func dset() {local s
    s=0
    forsec $o1 {s+=Qsum_dipole}
    return s}
```

setup_dipole, and calculated the contribution from the direction parallel with the apical dendrites in procedure *fill_ztan*. The *init* procedure inserted the dipole process into each compartment. The calculation of the dipole current as the intracellular longitudinal currents between dendritic compartments was defined in separate NEURON file, *dipole.mod*, following convention (e.g., all ionic current process are defined in separate .mod files, see Table 1) and shown in Box 10. Step 2—Box 11: Insert the dipole process into each of the Layer II/III and Layer V PNs and calculate the net summed current dipole across the two populations. This is the dipole output of the program from which rhythmic activity is investigated.

3.4. NEURON Code for Running the Simulations and Saving the Dipole Output

Once all of the templates and procedures are defined, a main .hoc file, *batch.hoc*, is created to load them and run the simulation from the command line in “batch” mode, see Box 12. We note that the templates and procedures used call several processes defined in separate .mod files, many of them default in NEURON, which are compiled before running the network simulation with the command line *mknrndl* (see (62) for further details). All nonbuilt in .mod files used are listed in Table 1. The main .hoc file also defines the simulation parameters (i.e., length of simulation, number of times to run simulation, time step (dt) for difference equations solver), and saves the dipole data to a separate file, referred to as *batch-data.dat* in Box 12. The command to compile

Box 12
NEURON .hoc file that loads the previous code described in Boxes 1–11, defines the parameters to run the simulation, and saves the net dipole sum output. The simulation is run with the terminal line command nrniv batch.hoc

```
//file batch.hoc
//each of the templates and procedures defined in Boxes 1-11
//need to be loaded in here (code omitted)

tstop=1500 //Total time for one run (ms)
dt=0.2 // time increment for differential equation solver
NRUN = 25 // Number of times to run the simulation

proc runonce(){
//
init()
while (t<tstop){
        fadvance()
        fprint( "%f %f \n", \t, \L2_dipole() + L5_dipole() ) }
} //end runonce()

proc runit(){ local i
wopen("batch-data.dat") //dipole output saved in this file
for i=0, NRUN-1 { runonce()
print(i)}
wopen()}

runit()
```

and run this file in a terminal window is *nrniv batch.hoc*. It is also possible to compile and run the simulation in a NEURON GUI that enables real-time visualization of the model output; however, those methods are not described here.

3.5. Methods for Frequency Analysis on Dipole Output

The net current dipole output of the simulation (i.e., net summed current dipole across the Layer V and Layer II/III PNs) saved in *batch-data.dat* (Box 12) is then read into a data analysis program to calculate the frequency spectrum of the signal. For the results in Fig. 4, this was done in Matlab where a time frequency representation of the data was calculated using a wavelet transform, see Jones et al. (48) for specific details of frequency analysis. NEURON also has default methods for frequency analysis included in its software package (62); however, these methods were not employed for our study.

4. Notes

We note that we have used a derivate of the laminar model described to study LFP signals measured in rodents. In particular, the model was used study the neural mechanisms of contrasting gamma frequency and low frequency (near 8 Hz) resonances in the somatosensory cortex during selective optogenetic drive of fast-spiking interneurons and regular spiking pyramidal neurons, respectively (20, 21). In this work, the Layer II/III network in the current model was extended to include more detailed inhibitory neurons simulated with multiple dendritic compartments and representing two distinct populations, fast-spiking, and low-threshold spiking. The model also contained uniformly random synaptic connections across the network based on known probabilities of connections between cell types and synaptic plasticity mechanisms were included. LFP signals were simplified as the mean membrane potential in the mid-apical dendrites across the pyramidal neuron population. Results suggested that the two populations of inhibitory neurons may be necessary to capture the experimentally observed dual resonances with selective drive in a single network model (20).

Acknowledgments

I thank my long-term close collaborator Dr. Christopher Moore whose insight was crucial to shaping the concepts for this work. Dr. Moore's expertise in somatosensory neural dynamics was essential to model development, MEG experimental design, and interpretation of simulated and experimental data. I thank Drs. Matti Hamalainen and Steve Stufflebeam for expertise in MEG methods and analysis and theoretical developments to compare MEG and model results, Seppo Ahlfors for Figure 4. Mike Sikora for consultant work on model implementation and analysis in NEURON. Dorea Vierling-Claasen for thoughtful reading of manuscript and further model extension. Dominique Pritchett for implementation of MEG experimental paradigms and Michael Halassa for preliminary testing of model predictions with optogenetic techniques in rodents. This work was supported by National Institute of Mental Health Grant K25-MH-072941, and the Athinoula A. Martinos Center for Biomedical Imaging, Mass. General Hospital.

References

1. Berger H (1969) On the electroencephalogram of man. Electroencephalogr Clin Neurophysiol 28(Suppl):37
2. Hari R, Salmelin R (1997) Human cortical oscillations: a neuromagnetic view through the skull. Trends Neurosci 20:44–49
3. Worden MS, Foxe JJ, Wang N et al (2000) Anticipatory biasing of visuospatial attention indexed by retinotopically specific alpha-band electroencephalography increases over occipital cortex. J Neurosci 20:RC63
4. Bauer M, Oostenveld R, Peeters M et al (2006) Tactile spatial attention enhances gamma-band activity in somatosensory cortex and reduces low-frequency activity in parieto-occipital areas. J Neurosci 26:490–501
5. Jones SR, Kerr CE, Wan Q et al (2010) Cued spatial attention drives functionally-relevant modulation of the mu rhythm in primary somatosensory cortex. J Neurosci 30 (41):13760–13765
6. Linkenkaer-Hansen K, Nikulin VV, Palva S et al (2004) Prestimulus oscillations enhance psychophysical performance in humans. J Neurosci 24:10186–10190
7. Palva S, Linkenkaer-Hansen K, Naatanen R et al (2005) Early neural correlates of conscious somatosensory perception. J Neurosci 25:5248–5258
8. Zhang Y, Ding M (2009) Detection of a Weak Somatosensory Stimulus: Role of the Prestimulus Mu Rhythm and Its Top-Down Modulation. J Cogn Neurosci 22(2):307–322
9. Brown P (2006) Bad oscillations in Parkinson's disease. J Neural Transm 70(Suppl):27–30
10. Tecchio F, Zappasodi F, Tombini M et al (2006) Brain plasticity in recovery from stroke: an MEG assessment. Neuroimage 32 (3):1326–1334
11. Li Y, Fleming IN, Colpan ME et al (2008) Neuronal desynchronization as a trigger for seizure generation. IEEE Trans Neural Syst Rehabil Eng 16:62–73
12. Dockstader C, Gaetz W, Cheyne D et al (2008) MEG event-related desynchronization and synchronization deficits during basic somatosensory processing in individuals with ADHD. Behav Brain Funct 4:8
13. Siekmeier PJ, Stufflebeam SM (2010) Patterns of spontaneous magnetoencephalographic activity in patients with schizophrenia. J Clin Neurophysiol 27:179–190
14. Okada YC, Wu J, Kyuhou S (1997) Genesis of MEG signals in a mammalian CNS structure. Electroencephalogr Clin Neurophysiol 103:474–485
15. Ikeda H, Wang Y, Okada YC (2005) Origins of the somatic N20 and high-frequency oscillations evoked by trigeminal stimulation in the piglets. Clin Neurophysiol 116:827–841
16. Whittington MA, Traub RD, Kopell N et al (2000) Inhibition-based rhythms: experimental and mathematical observations on network dynamics. Int J Psychophysiol 38:315–336
17. Kopell N, Ermentrout GB, Whittington MA et al (2000) Gamma rhythms and beta rhythms have different synchronization properties. Proc Natl Acad Sci USA 97:1867–1872
18. Pinto DJ, Jones SR, Kaper TJ et al (2003) Analysis of state-dependent transitions in frequency and long-distance coordination in a model oscillatory cortical circuit. J Comput Neurosci 15:283–298
19. Borgers C, Epstein S, Kopell NJ (2008) Gamma oscillations mediate stimulus competition and attentional selection in a cortical network model. Proc Natl Acad Sci USA 105:18023–18028
20. Vierling-Claassen D, Cardin J, Moore CI et al (2010) Computational modeling of distinct neocortical oscillations driven by cell-type selective optogenetic drive: separable resonant circuits controlled by low-threshold spiking and fast-spiking interneurons. Front Hum Neurosci 4:198
21. Cardin JA, Carlen M, Meletis K et al (2009) Driving fast-spiking cells induces gamma rhythm and controls sensory responses. Nature 459:663–667
22. Wilson HR, Cowan JD (1972) Excitatory and inhibitory interactions in localized populations of model neurons. Biophys J 12:1–24
23. Pinto DJ, Brumberg JC, Simons DJ et al (1996) A quantitative population model of whisker barrels: re-examining the Wilson–Cowan equations. J Comput Neurosci 3:247–264
24. David O, Friston KJ (2003) A neural mass model for MEG/EEG: coupling and neuronal dynamics. Neuroimage 20:1743–1755
25. David O, Kiebel SJ, Harrison LM et al (2006) Dynamic causal modeling of evoked responses in EEG and MEG. Neuroimage 30: 1255–1272

26. Marreiros AC, Kiebel SJ, Friston KJ (2010) A dynamic causal model study of neuronal population dynamics. Neuroimage 51:91–101
27. Friston KJ, Dolan RJ (2010) Computational and dynamic models in neuroimaging. Neuroimage 52:752–765
28. David O, Harrison L, Friston KJ (2005) Modelling event-related responses in the brain. Neuroimage 25:756–770
29. David O, Kilner JM, Friston KJ (2006) Mechanisms of evoked and induced responses in MEG/EEG. Neuroimage 31:1580–1591
30. Moran RJ, Kiebel SJ, Stephan KE et al (2007) A neural mass model of spectral responses in electrophysiology. Neuroimage 37:706–720
31. Borgers C, Kopell N (2003) Synchronization in networks of excitatory and inhibitory neurons with sparse, random connectivity. Neural Comput 15:509–538
32. Vierling-Claassen D, Siekmeier P, Stufflebeam S et al (2008) Modeling GABA alterations in schizophrenia: a link between impaired inhibition and altered gamma and beta range auditory entrainment. J Neurophysiol 99:2656–2671
33. Vladimirski BB, Tabak J, O'Donovan MJ et al (2008) Episodic activity in a heterogeneous excitatory network, from spiking neurons to mean field. J Comput Neurosci 25:39–63
34. Jones SR, Pinto DJ, Kaper TJ et al (2000) Alpha-frequency rhythms desynchronize over long cortical distances: a modeling study. J Comput Neurosci 9:271–291
35. Garabedian CE, Jones SR, Merzenich MM et al (2003) Band-pass response properties of rat SI neurons. J Neurophysiol 90:1379–1391
36. Jones SR, Pritchett DL, Stufflebeam SM et al (2007) Neural correlates of tactile detection: a combined MEG and biophysically based computational modeling study. J Neurosci 27:10751–10764
37. Traub RD, Contreras D, Cunningham MO et al (2005) Single-column thalamocortical network model exhibiting gamma oscillations, sleep spindles, and epileptogenic bursts. J Neurophysiol 93:2194–2232
38. Markram H (2006) The blue brain project. Nat Rev Neurosci 7:153–160
39. Izhikevich EM, Edelman GM (2008) Large-scale model of mammalian thalamocortical systems. Proc Natl Acad Sci USA 105:3593–3598
40. Hamalainen M, Hari R, Ilmoniemi RJ et al (1993) Magnetoencephalography: theory, instrumentation, and applications to noninvasive studies of the working human brain. Rev Mod Phys 65:413–497
41. Sharon D, Hamalainen MS, Tootell RB et al (2007) The advantage of combining MEG and EEG: comparison to fMRI in focally stimulated visual cortex. Neuroimage 36:1225–1235
42. Ou W, Hamalainen MS, Golland P (2009) A distributed spatio-temporal EEG/MEG inverse solver. Neuroimage 44:932–946
43. Liu H, Tanaka N, Stufflebeam S et al (2010) Functional mapping with simultaneous MEG and EEG. J Vis Exp 40(pii):1668
44. Murakami S, Zhang T, Hirose A et al (2002) Physiological origins of evoked magnetic fields and extracellular field potentials produced by guinea-pig CA3 hippocampal slices. J Physiol 544:237–251
45. Murakami S, Hirose A, Okada YC (2003) Contribution of ionic currents to magnetoencephalography (MEG) and electroencephalography (EEG) signals generated by guinea-pig CA3 slices. J Physiol 553:975–985
46. Murakami S, Okada Y (2006) Contributions of principal neocortical neurons to magnetoencephalography and electroencephalography signals. J Physiol 575:925–936
47. Lu ZL, Williamson SJ (1991) Spatial extent of coherent sensory-evoked cortical activity. Exp Brain Res 84:411–416
48. Jones SR, Pritchett DL, Sikora MA et al (2009) Quantitative analysis and biophysically realistic neural modeling of the MEG mu rhythm: rhythmogenesis and modulation of sensory-evoked responses. J Neurophysiol 102:3554–3572
49. Ziegler DA, Pritchett DL, Hosseini-Varnamkhasti P et al (2010) Transformations in oscillatory activity and evoked responses in primary somatosensory cortex in middle age: a combined computational neural modeling and MEG study. Neuroimage 52:897–912
50. Rockland KS, Pandya DN (1979) Laminar origins and terminations of cortical connections of the occipital lobe in the rhesus monkey. Brain Res 179:3–20
51. Felleman DJ, Van Essen DC (1991) Distributed hierarchical processing in the primate cerebral cortex. Cereb Cortex 1:1–47
52. Jones EG (2001) The thalamic matrix and thalamocortical synchrony. Trends Neurosci 24:595–601
53. Hughes SW, Crunelli V (2005) Thalamic mechanisms of EEG alpha rhythms and their pathological implications. Neuroscientist 11:357–372

54. Silva LR, Amitai Y, Connors BW (1991) Intrinsic oscillations of neocortex generated by layer 5 pyramidal neurons. Science 251:432–435
55. Fanselow EE, Richardson KA, Connors BW (2008) Selective, state-dependent activation of somatostatin-expressing inhibitory interneurons in mouse neocortex. J Neurophysiol 100:2640–2652
56. Roopun AK, Middleton SJ, Cunningham MO et al (2006) A beta2-frequency (20–30 Hz) oscillation in nonsynaptic networks of somatosensory cortex. Proc Natl Acad Sci USA 103:15646–15650
57. Jones EG (1998) Viewpoint: the core and matrix of thalamic organization. Neuroscience 85:331–345
58. Cardin J, Carlén M, Meletis K et al (2010) Targeted optogenetic stimulation and recording of neurons in vivo using cell type-specific expression of channelrhodopsin-2. Nat Protoc 5:247–254
59. Jones SR, Halassa M, Pritchett DL et al (2010) A neocortical beta origin hypothesis: computational modeling, electrophysiological, and optogentic studies. Program No. 683. 2010 Neuroscience Meeting Planner. Society for Neuroscience, San Diego, CA
60. Varela JA, Sen K, Gibson J et al (1997) A quantitative description of short-term plasticity at excitatory synapses in layer 2/3 of rat primary visual cortex. J Neurosci 17:7926–7940
61. Hines ML, Carnevale NT (2008) Translating network models to parallel hardware in NEURON. J Neurosci Methods 169:425–455
62. Carnevale NT, Hines ML (2006) The NEURON book. Cambridge University Press, Cambridge

Neuromethods (2012) 67: E1
DOI 10.1007/978-1-61779-633-3_17

Erratum to: Voltage-Sensitive Dye Imaging of Cortical Function In Vivo

Eugene F. Civillico

Erratum to: Neuromethods
DOI 10.1007/7657_2011_2

Revised affiliation of Prof. Eugene F. Civillico was not updated in the Chapter and Contributors section during production process.

The correct affiliation is as follows:

Eugene F. Civillico • Laboratory of Biophysics, Division of Physics, U.S. Food and Drug Administration, Center for Devices and Radiological Health, Office of Science and Engineering Laboratories, Silver Spring, MD, USA

The online version of the original chapter can be found under DOI 10.1007/7657_2011_2

Neuromethods (2012) 67: 487–490
DOI 10.1007/7657_2011

INDEX

A

AAV. *See* Adeno-associated virus (AAV)
Acousto-optic deflector (AOD) 6–8
Action potentials (APs) 6, 12–15, 23, 34, 35, 38, 85, 95, 112, 116, 152, 153, 192, 195, 199, 205, 207, 219, 220, 239, 246–248, 259, 315, 317, 324, 342, 352, 371, 417, 418
Adeno-associated virus (AAV) 30, 327–329, 341, 343, 344, 348, 349
AEP. *See* Auditory evoked potentials (AEP)
Alpha rhythm 460, 465–468
α-Amino-3-hydroxy-5-methyl-4-isoxazole-propionate (AMPA) 67, 131, 158, 176, 179, 470
Anesthetics 25, 29, 46
AOD. *See* Acousto-optic deflector (AOD)
APs. *See* Action potentials (APs)
Archaerhodopsin 308, 319, 322
Astrocytes 12, 21, 22, 27, 29, 32–34, 252
Auditory evoked potentials (AEP) 442, 446, 447
Axonal domain 425–427
Axon terminals 196, 379, 383, 385

B

Bacteriorhodopsins (BR) 308, 318–319, 322–326
Bayesian statistical methods 261
Beta rhythm 437, 460, 465–468
Biocytin 199, 226, 229, 231, 407, 408, 412, 413, 418–425
Biomechanics 284
BR. *See* Bacteriorhodopsins (BR)

C

CA1 ... 32, 108, 128–132, 134–136, 138–141, 192, 193, 195, 196, 200
CA3 .... 33, 129–132, 134–141, 176
Caged compound 61, 115, 179
Calbindin 377
Calcium (Ca^{2+})
  imaging 13, 23, 67, 68, 75
  indicator 5, 11–13, 15, 22, 23, 26, 28, 38, 45, 51, 70, 75, 87
Calcium Green-1 .... 27, 31
C. elegans 327
Cell morphology 418, 462
Cell type specificity 31, 318, 340–342, 349–350
Cerebellum 128, 135, 210, 211, 225, 227, 246, 396
Channelrhodopsin 8, 68, 115, 116, 196, 201, 306, 308, 311–315, 318, 319, 323, 326, 340, 350, 407
Chronic recording 29, 352, 360, 370
Computational modeling 461
Computerised reconstructions 413
Computer vision 297
Connectivity 3, 4, 104, 108, 109, 116–118, 196, 339, 377, 379, 381, 382, 405, 407–410, 414, 415, 417–420, 425–427, 435–454, 461, 462, 464, 470–473
Cortex 10, 39, 47, 85–96, 108, 109, 116, 128, 131, 132, 149, 178, 192, 193, 206, 209, 210, 213, 215, 224, 227, 228, 230, 237–241, 245, 247, 252, 260, 317, 328, 339, 341, 343, 349, 351, 352, 358–360, 370, 376–380, 382, 383, 386, 401, 402, 408, 422, 425, 435, 436, 438, 439, 441, 442, 444, 448, 449, 451–454, 462–466, 482
Cortical
  circuits 149–168, 209, 339–354, 435, 436, 438, 441, 451, 452, 454
  networks 95, 109, 149, 405–428, 469
Craniotomy 25, 29, 39, 48, 50–58, 91, 92, 215, 216, 222–226, 228–230, 232, 233, 245, 331, 346, 348
Cre-loxP 340, 341
Current dipole 460–465, 467, 469, 470, 478–481
Current source-density (CSD) 196, 205–216
Cytosol 22, 23, 28, 37, 87, 105, 309, 317

D

Dendritic domain 425, 427
Diffractive optical element (DOE) 9, 72–74, 78–79
Dissection 25, 135, 156, 339–354, 375–387, 392, 393, 395, 410, 414
DOE. *See* Diffractive optical element (DOE)
Drosophila 12, 327
Dynamic clamp 149–168

E

EEG. *See* Electroencephalography (EEG)
Electrical stimulation 95 96, 111–113, 116, 117, 131, 196, 201, 357–358
Electroencephalography (EEG) 91, 128, 206, 237, 238, 240–243, 246, 248, 249, 350, 369, 435–454, 459–482

Electrophysiology 47, 67, 105, 117, 150, 160, 166, 191, 192, 198, 201, 220, 226, 238, 239, 285, 300, 314, 342, 345, 350–351, 357–371, 415, 416
Electroporation 11, 13, 15, 29, 30, 115, 199, 328, 340, 386
Endoplasmic reticulum (ER) 309, 317, 319, 320, 350
Entorhinal cortex 108, 131, 132
Extracellular recordings 127–145, 191–193, 196, 197, 201, 220, 238, 241–244, 345, 352

F

Fast perfusion 173–186
fEPSPs. *See* Field excitatory postsynaptic potentials (fEPSPs)
Fiber optic 11, 197, 360–371
Field excitatory postsynaptic potentials (fEPSPs) 33, 127, 130, 134, 136–143, 145
Fixation 46, 49, 223, 284, 392–394, 397, 400, 419, 420
Flavoprotein autofluorescence (FA) 116–118
Fluo-4 ....31, 75–77
Fluorescence
  lifetime imaging 24
  resonance energy transfer 31
Functional magnetic resonance imaging (fMRI) 197, 206, 460
Fura-2 ....22, 23

G

GCaMP 12, 13, 22, 23, 24, 30, 31, 51
Genetically encoded indicator 11
Glutamate (GABA) 67, 68, 70, 76, 77, 105, 106, 111, 113–118, 130, 131, 133, 141, 153, 158, 174, 176, 177, 179, 207, 377, 382, 407
Glycoprotein 328, 383–386

H

Habituation 223, 225, 227–228, 232
Halorhodopsin 8, 197, 201, 306, 308, 311, 312, 316–323, 326, 340, 348, 350
Head restrained 11, 45–59, 199, 219–234, 238
Hippocampus 9, 10, 32, 108, 128–131, 134, 138–141, 145, 176, 192–194, 196, 200, 210, 224, 240, 260, 266, 328, 349, 358

I

Imaging 4, 5, 8–15, 21–39, 45–59, 61–63, 67, 68, 74–77, 83–96, 115–117, 197, 206, 224, 233, 259, 285, 288, 289, 298, 306, 308, 379, 407, 416, 417, 436, 459–460
Immunohistochemistry 251
Implantation 50, 223–228, 232, 233, 244, 245, 359, 370
Indo-1 31
Injection 6, 13, 27, 35, 39, 111–114, 117, 150, 152, 154, 157, 167, 221, 307, 324, 343, 346–351, 370, 379–381, 383, 385, 406
Innervation domain 426–428
Interneurons 27, 30, 88, 115, 130, 139, 149, 173, 207, 224, 241, 251, 266, 269, 339–343, 407, 464, 482
Intracellular recording 67, 84, 87, 167, 191, 192, 199–201, 210, 220–224, 229, 238, 241, 345, 349, 350, 352, 382, 412
Iontophoresis 113, 114
Isoflurane 27, 56, 212, 215, 225–228, 244, 344, 346, 347, 414
Isotropic fractionator 391–402

J

Juxtacellular recording 198–199

K

Kainate receptors (KAR) 131, 176, 177
Ketamine 25, 27, 50, 53, 54, 90, 105, 106, 232, 244

L

Laser
  photolysis 178, 179
  photostimulation 114, 116–118
  pulse 4, 6, 8, 9, 31, 79
  scanning 3–8, 12, 26, 52–53, 58, 61, 62, 70–72, 74, 75, 77, 114, 318, 330
Lentiviral vectors 30, 341
LFP. *See* Local field potential (LFP)
Ligand-gated ion channels 173–186
Light-sensitive
  cation channel 340
  chloride pump 306, 311, 318, 319, 323, 326, 340
  proton pump 308, 311, 312, 318, 322–326, 340
Local field potential (LFP) 84, 85, 87, 88, 91, 94, 110, 192, 195–197, 206–209, 211–215, 223, 238, 239, 240, 242–244, 248, 273, 331, 332, 343, 348, 350, 352, 353, 359, 371, 461, 463, 481, 482
Long-term depression (LTD) 32, 33, 106, 127–130, 132–135, 137–145
Long-term potentiation (LTP) 32, 140, 196

M

Matlab 164, 166, 249, 251, 289, 297, 300, 467, 469, 470, 481
MEG imaging 459
Micromanipulator 25, 26, 29, 30, 133, 182, 232, 296, 415, 416

Mossy fibers 132, 139, 141, 176
Multiphoton imaging 46–47, 58
Multiunit activity 195, 214, 237–253

N

Neocortex 85, 109, 195, 196, 199, 200, 210, 211, 214, 224, 228, 237–253, 376, 408, 464–465
NeuN. *See* Neuronal nuclear antigen (NeuN)
Neural
  activity 3, 8, 15, 45–59, 112–118, 216, 268, 283, 305–308, 314, 323, 332, 342, 343, 352, 463
  circuits 3–15, 305, 306, 308, 339, 340, 376, 386, 463, 467
  ensemble 45, 260, 266, 267, 270
  ensemble recording 261, 266
Neuronal
  reconstructions 420, 423–425
  tracers 379–381, 383
Neuronal nuclear antigen (NeuN) 392–394, 397–399
Neuronavigation 436, 441, 442
Neurotransmitter 3, 4, 6, 8, 15, 111, 153, 173–176, 178, 179, 181, 185, 219
NMDA-Rs 132, 133
Noninvasive 116, 117, 128, 436, 459
Non-linear bleaching 24
Nuclei 9, 177, 178, 210, 357, 360, 392–402, 437

O

Olfactory bulb 21–39, 49, 51, 227
Opsins 15, 196, 306–313, 315, 317, 326–330
Optical recording 45–47, 93, 95, 96
Optogenetics 68, 111, 115–116, 178, 196, 197, 201, 285, 306, 309–312, 339–354, 357–361, 371, 375, 461, 468, 482
Optogenetic stimulation 111, 115–116, 201, 345, 348–350, 468
Oregon Green BAPTA-1 22, 23, 75
Oscillations
  gamma 195, 196, 340, 342, 352, 360
  slow waves 110
  theta (θ) 195, 196, 267

P

Paired recordings 308, 406, 407, 409, 413, 415–417
Parvalbumin 115, 343, 377
Patch-clamp 26, 30, 144, 199, 219–234
Permeability 309, 313, 319–320, 322–323
Perturbation 12, 330, 436, 452, 453
Photorelease 114, 179
Photostimulation 8, 63, 67, 72, 76, 113–118, 376, 382
Piezoelectric devices 184
Poisson processes 262
Post-synaptic receptors 173–176, 179, 406
Prefrontal cortex 128

R

Rabies virus 328, 376, 381–387
Resting membrane potential 24, 103, 111, 150, 207
Retrograde transport 381

S

Signal-to-noise 11–13, 15, 34, 35, 37, 38, 68, 76, 83, 95, 96, 117, 246, 443, 448
Silicon probe 193, 195, 209, 214, 220, 352, 360
Sleep
  homeostasis 195, 237–253
  REM 128, 130, 232, 238, 240, 248, 452, 454
Slice
  maintenance 109–110
  preparation 67, 76, 85, 103–108, 110, 112, 117, 118, 133, 135–136, 144, 405–428
Slow-wave activity (SWA) 238, 242, 243
Somatosensory cortex 108, 109, 213, 224, 377, 382, 422, 463, 464, 482
Spatial light modulator 8, 61–79
Spike
  sorting 239, 247–252, 266
  train decoding 263, 265
Spine 3, 4, 8, 67, 76, 115, 423, 424
Stereology 391, 392
Stereotrodes 193, 194, 345, 352, 353, 360, 369, 370
Stratum radiatum 32, 33, 129, 136, 138, 143, 195
Structured light 62, 65, 67–69, 75–77
Structure-function correlations 407
SWA. *See* Slow-wave activity (SWA)
Synaptic
  contacts 406–408, 424, 426–428
  plasticity 111, 127–145, 156, 469, 482
  transmission 32, 33, 46, 105, 106, 111, 127, 137, 157, 175, 176, 326, 406, 409, 411, 414, 428

T

Thalamus 9, 104, 108–110, 116, 118, 210, 241, 358–360, 466, 473
Theta pipette 174, 175, 180–185
TMS. *See* Transcranial magnetic stimulation (TMS)
Transcranial magnetic stimulation (TMS) 435–454
Two-photon microscopy 4–6, 8–11, 14

U

Uncaging 4, 8, 63, 67, 68, 75–77, 114–116, 178, 179

V

Vibrissa....224, 284–287
Videography.... 283–301
Viral vectors....30, 31, 115, 340–342, 347, 349
Voltage indicator.... 11, 13–15
Voltage-sensitive dye imaging (VSDI)....83–96

W

Wakefulness 222, 232, 237, 240, 242, 243, 452–454
Whiskers....15, 86, 93, 95, 109, 209, 210, 224, 230, 231, 283–301
Whole-cell patch-clamp....199, 219–234
Wireless recording....198

X

Xylazine....25, 27, 50, 53, 54, 90, 232, 244

Y

YC2.12....24
Yellow chameleon 3.6 (YC3.6)....24

MIX
Papier aus verantwortungsvollen Quellen
Paper from responsible sources
FSC® C105338

If you have any concerns about our products,
you can contact us on
ProductSafety@springernature.com

In case Publisher is established outside the EU,
the EU authorized representative is:
Springer Nature Customer Service Center GmbH
Europaplatz 3, 69115 Heidelberg, Germany

Printed by Libri Plureos GmbH
in Hamburg, Germany